Texts in Applied Mathematics

Volume 83

The mathematization of all sciences, the fading of traditional scientific boundaries, the impact of computer technology, the growing importance of computer modelling and the necessity of scientific planning all create the need both in education and research for books that are introductory to and abreast of these developments. The aim of this series is to provide such textbooks in applied mathematics for the student scientist. Books should be well illustrated and have clear exposition and sound pedagogy. Large number of examples and exercises at varying levels are recommended. TAM publishes textbooks suitable for advanced undergraduate and beginning graduate courses, and complements the Applied Mathematical Sciences (AMS) series, which focuses on advanced textbooks and research-level monographs.

Yuri Kuznetsov · Odo Diekmann ·
Wolf-Jürgen Beyn

Dynamical Systems Essentials

An Application Oriented Introduction to Ideas, Concepts, Examples, Methods, and Results

Yuri Kuznetsov
Department of Mathematics
Utrecht University
Utrecht, The Netherlands

Department of Applied Mathematics
University of Twente
Enschede, The Netherlands

Odo Diekmann
Department of Mathematics
Utrecht University
Utrecht, The Netherlands

Wolf-Jürgen Beyn
Department of Mathematics
Bielefeld University
Bielefeld, Germany

ISSN 0939-2475 ISSN 2196-9949 (electronic)
Texts in Applied Mathematics
ISBN 978-3-032-04082-4 ISBN 978-3-032-04083-1 (eBook)
https://doi.org/10.1007/978-3-032-04083-1

Mathematics Subject Classification: 34Cxx, 37Cxx, 37Dxx, 34Dxx, 37Gxx, 37Nxx

This Springer imprint is published by the registered company Springer Nature Switzerland AG
The registered company address is: Gewerbestrasse 11, 6330 Cham, Switzerland

To Lioudmila
Ketje (and in loving memory of Marleen and Liesbeth)
Ellen

Preface

What is a Dynamical System?

Science aims at uncovering the cause-effect relationship between mechanisms and phenomena. As the world is large, it makes sense to isolate a part of it. Like, for instance, a pendulum or a continuously stirred tank reactor with certain dissolved chemicals. The underlying idea is, of course, that the understanding of an isolated small part may serve as a building block for later more encompassing analyses. Such an isolated part of the world, and its abstraction by way of a model, we call a *system*. The human perception of the world is ordered by time, so a frequent objective is to describe how a system changes in the course of time. The *state* of the system at a particular moment in time should summarize all information about the history that is relevant for a prediction of the future. For the pendulum, the combination of the displacement and the velocity qualifies as such and for the reactor the concentrations of the various chemicals, supplemented, perhaps, by the temperature of the fluid. A *dynamical system* is a, possibly rather implicitly specified, rule for constructing future states from a given initial one.

What is This Book About?

Experimental investigation and idealized mathematical description are complementary approaches for studying specific systems. They reinforce each other through the challenge of their contrast. For both one needs methods and tools. This book presents a basic course on the qualitative theory of dynamical systems, which means that it deals with mathematical ideas, concepts, methods, and results concerning a qualitative (and, to a lesser extent, also quantitative) description of how the state of a system evolves deterministically in the course of time. For example, the description may be that in the long run a periodic pattern develops. Such a possible mode of behaviour may change qualitatively when the system is slightly changed, for instance if the value of some control parameter is altered. Such events are called *bifurcations*.

To a large extent, the theory deals with the classification of possibilities and the formulation of criteria to decide, for a specific system, which of these possibilities applies. For example, as will be explained at length (including the terminology) in the following sections, a steady state may be an attractor, a saddle point, or a repellor, and, apart from special situations, the position in the complex plane of the spectrum of the linearized system can be used to identify the case at hand. The special situations correspond to bifurcations. For example, a steady state can lose its stability, giving rise to a periodic motion of small amplitude for nearby parameter values (this is the famous Andronov-Hopf bifurcation).

Thus the theory yields both a catalogue of pictures (ranging from the literal, i.e., planar drawings, to abstract mental images) and a set of guidelines and criteria for finding pictures describing properties of a given system. This last aspect is very relevant for applications. Unfortunately it can be rather difficult. Largely by way of examples, we shall provide some training, thus justifying the adjective "application oriented". The book is organized in seven chapters.

Chapter 1. "Introduction". Here we introduce basic terminology. We define dynamical systems generated by autonomous ordinary differential equations (ODEs) and maps, introduce orbits, phase portraits, stationary points, periodic orbits (cycles), and other invariant sets, and define their stability, as well as various equivalence relationships between dynamical systems. Many examples of dynamical systems, including symbolic dynamics, are given.

Chapter 2. "Linear Maps and ODEs". This chapter treats the simplest dynamical systems, namely those generated by linear autonomous ODEs and linear maps. We describe the dynamics in the eigenspaces and establish the topological classification of hyperbolic linear ODEs and maps. Whenever possible, we employ invariant techniques that avoid explicit coordinates. Several practical methods to compute projections to the eigenspaces, including those based on partial fraction decomposition, are illustrated in detail.

Chapter 3. "Local Behaviour of Nonlinear Systems". In this chapter we give complete proofs of the Linearized Stability Principle for fixed points and equilibria of smooth nonlinear maps and ODEs, i.e., we derive sufficient conditions for stability and instability of such points based on properties of the corresponding linearized system. Then we introduce Poincaré maps near periodic orbits in ODEs and characterize their stability in terms of the linearization of the Poincaré maps and the variational equation about the corresponding periodic solutions. In particular, the complete but simple discusson of the relationships between the eigenvalues of the linearization of the Poincaré map and the Floquet multipliers of the corresponding periodic solutions is given. Finally, we formulate and prove in detail the classical

Grobman-Hartman Theorems for maps and ODEs, i.e., we show that any smooth nonlinear system is locally topologically equivalent to its linearization at a steady state, provided that the linearized system is hyperbolic. All necessary tools, including the Contraction Mapping Principle and the Lipschitz Inverse Function Theorem, are also developed in this chapter. Notice that the Stable Invariant Manifold Theorem is not proven here but its variant follows from the results established in Chap. 6.

Chapter 4. "Planar ODEs". This chapter presents the classical qualitative theory of planar smooth autonomous ODEs. Here we prove in detail the famous Poincaré-Bendixson Theorem that classifies the asymptotic behaviour of bounded orbits in planar ODEs. We derive the Bendixson-Dulac conditions ensuring existence or nonexistence of periodic orbits in planar systems, and we introduce Pontryagin's method to locate periodic orbits in planar ODEs, when the system is close to a special (Hamiltonian) system. We analyse several prey-predator ecological models and, in particular, give the complete proof of the uniqueness of the periodic orbit in the classical Rosenzweig-MacArthur system with logistic prey and Holling Type II predator functional response. Finally, we briefly discuss essential properties of multidimensional Hamiltonian systems.

Chapter 5. "Local Bifurcations in Minimal Dimensions". This chapter is devoted to the simplest bifurcations, i.e., those happening in small neighborhoods of steady states or cycles, in dynamical systems depending on one parameter. We study all generic bifurcations of equilibria (fixed points) and cycles, i.e., fold, Andronov-Hopf, period-doubling, and Neimark-Sacker bifurcations in ODEs and maps. In this chapter, these bifurcations are analyzed in a state space of the minimal possible dimension using a normal form approach in which a generic system is transformed near the bifurcation to the simplest form. Where applicable, we give complete proofs of local topological equivalence of generic systems to the normal forms. In particular, simple proofs of topological stability of the truncated normal forms for fold and Hopf bifurcations of equilibria in ODEs, as well as fold and period-doubling bifurcations of fixed points of iterated maps, are given. In the Neimark-Sacker case, existence and uniqueness of a closed invariant curve in a full neighbourhood of the critical fixed point are established.

Chapter 6. "Centre Manifold Reduction". Here we turn our attention to state spaces of arbitrary finite dimension and prove that essential events near a local bifurcation occur on a low-dimensional invariant curve or surface in the state space, the so called centre manifold. We establish the existence and smoothness of centre-unstable, centre-stable, and centre manifolds and show how to compute the normal forms on the centre manifold in practice, by combining the computation of this manifold with the transformation of the restricted system to the normal form. The study of limit cycle bifurcation in ODEs is reduced to that of fixed points of Poincaré maps.

Chapter 7. "Symbolic Dynamics and Global Bifurcations". This final chapter is an introduction to complicated behaviour of dynamical systems, also known as "chaos". We discuss in detail similarities between the dynamics generated by nonlinear maps and ODEs and shifts of infinite sequences of several symbols. We begin with one-dimensional maps, for which we prove the Li-Yorke part of the Sharkovskii Theorem, saying that "period three implies all periods". The classical bifurcation scenario leading to the strange attractor in the Lorenz system is also discussed. The Smale horseshoe map is then introduced and studied in detail, revealing the role of cone conditions. Then we prove that complicated dynamics generated by the horseshoe map is present near two global objects—an orbit homoclinic to a saddle-focus equilibrium and a transverse homoclinic orbit to a hyperbolic limit cycle. In the former case, Shilnikov's homoclinic chaos can emerge, while in the latter case, the corresponding return map exhibits the Poincaré homoclinic tangle. We also consider the general problem of bifurcation of a periodic orbit from a homoclinic orbit to a hyperbolic saddle in ODEs, for which Exponential Dichotomies are introduced and extensively used. Moreover, details on the symbolic description of the dynamics near transverse homoclinic orbits in maps are also provided, using the techniques of orbit shadowing and Exponential Dichotomies for maps.

This book intends to provide the minimal knowledge required to describe, classify, analyze, and understand the dynamical behaviour of concrete systems arising in an applied context. It gives a conceptual framework and introduces tools. It prepares the reader to confront more advanced books on dynamical systems. We do not treat systematically global bifurcations of dynamical systems and avoid considering dimensional and metric properties of chaotic attractors, as well as their reconstruction from time-series. The reader will find here neither the general theory of hyperbolic invariant sets nor ergodic theory. Only occasionally we do consider infinite-dimensional dynamical systems. We restrict ourselfs to bifurcations happening in generic one-parameter dynamical systems, i.e. treat only codimension one phenomena. No numerical methods for bifurcation analysis are presented here. All these and other topics should be studied systematically using other textbooks for which the present one could serve as a serious introduction. There are some topics, e.g. study of planar polynomial systems at infinity or the Lyapunov-Schmidt reduction technique for branching of stationary solutions, which are not treated in full generality but illustrated with examples.

We provide a number of carefully selected exercises in each chapter. Many of them were elaborated by our students during computer sessions accompanying the course. In these sessions, as well as while making their homework, our students used several standard software packages, including MAPLE, CONTENT, and MATCONT.

Each chapter has a section called References. It should be stressed that we give neither a historical overview nor acknowledgements of priority there. Instead, we exclusively refer to English-language textbooks, monographs, and reviews listed in the Bibliography. Of course, our selection is determined by personal preferences. Often, these books provide alternative treatments of some topics and contain material for further reading. Usually, we refer to the latest editions of the textbooks. References to models used in examples are given as footnotes.

How This Book was Written and for Whom?

The book combines Russian, Dutch, and German mathematical traditions in applied mathematics. This text originated from the Lecture Notes *Dynamische Systeme. Vorlesungsskriptum WS* for a course given in 1991/ 1992 at the Mathematics Faculty of Bielefeld University (Germany) and from the online Lecture Notes for a course on nonlinear differential equations and dynamical systems of 1999 at the Department of Mathematics of Utrecht University (The Netherlands). We expect from our audience only a basic knowledge of analysis, linear algebra, and differential equations. We introduce more advanced mathematical tools when they are really needed, with necessary details and rigour. Thus, the level of this book can be characterized as "advanced elementary".

What we have written is definitely not a "first course" on differential equations. We assume, indeed, that fundamental material on ODEs such as elementary methods to solve scalar differential equations, existence and uniqueness theorems, properties of n-dimensional linear ODE systems with constant coefficients and methods to solve these equations explicitly, as well as basic computer techniques to approximate solutions numerically and visualise them, is familiar to the readers. These topics are covered in the standard Bachelor courses on ODEs, and we do not pretend to substitute for such a course.

The presented text constitutes a "second" course on differential equations, where the notion of a dynamical system becomes central and where qualitative properties of such systems generated by iterated maps and ODEs are studied as functions of their parameters. Depending on the adopted curriculum, such a course is given either at the end of Bachelor studies ("upper-level undergraduate") or in the very beginning of Master programs ("low-level graduate"). The audience of such a course consists usually of two groups: (1) mathematics students who will continue their studies either in pure or applied mathematics; (2) other students (e.g., Master students from physics, biology, etc.) for whom this could be the last mathematical course on dynamical systems and who want to apply dynamical systems theory in their fields. The selected material should, therefore, suit both groups and provide both with a solid mathematical basis for more advanced courses on dynamical systems and with concepts and tools to understand the modern applied research literature and to analyze dynamical systems from applications.

Our book is designed to be used in a one-semester course. That is why we have included only really necessary material. However, we tried to avoid a dangerous trend in recent textbooks of not giving complete proofs. Thus, we do give self-contained proofs of major fundamental results (like Poincaré-Bendixson theorem, Lyapunov linearized stability/instability theorem, Grobman-Hartman theorem, existence of local (centre-)stable invariant manifolds, all topological normal forms for codimension-one bifurcations of maps and ODEs, existence and uniqueness of a closed invariant curve near the Neimark-Sacker bifurcation, essential part of Li-Yorke theorem, existence of symbolic dynamics near certain homoclinic bifurcations, etc.). We believe that these proofs should be known to an average graduate student who studies dynamical systems. Of course, the proofs should be as simple as possible and one should be able to separate them from the rest of the text, so that a student could be able to skip them, if necessary. Hopefully, the proofs that we present satisfy these requirements. More technical parts are often set in a smaller font and can be skipped at the first reading. In fact, we try to find a balance between "elementary geometrical" and "more abstract" proofs. Our teaching experience suggests that "more abstract" techniques originated in Functional Analysis (such as spectral projectors, equivalent norms, Fredholm alternative, contractions in Banach spaces—all of which are carefully developed in the text) actually simplify the proofs and make them more understandable. Numerous figures should also help to understand and memorize the presented results. Since we cover many classical topics, the text could not be entirely new. Its novelty mainly lies in the selection of the material and its uniform, compact, but detailed and self-contained presentation combining Geometry and Analysis.

This book is a result of a collaborative interaction with our colleagues and friends. Many useful remarks on various chapters have been made by Heinz Hanßmann and Ferdinand Verhulst (Utrecht University) and by Ale Jan Homburg (University of Amsterdam). We are also thankful to all readers, who reported misprints, errors, and discrepancies in early versions of this text published on the Web. We do realize that a text is never perfect and so we certainly welcome additional remarks and suggestions. Yet we do hope that the present text is solid enough to be studied with pleasure by those who want to become acquainted with the fascinating world of dynamical systems.

Utrecht, The Netherlands — Yuri Kuznetsov
Utrecht, The Netherlands — Odo Diekmann
Bielefeld, Germany — Wolf-Jürgen Beyn
May 2025

Competing Interests The authors have no competing interests to declare that are relevant to the content of this manuscript.

Typographical Conventions and Notations

We use $\square$ to indicate the end of a **Proof** (or its absence) of a **Theorem** or a **Lemma** or a **Proposition** or a **Corollary**, while $\Diamond$ and $\Diamond$ are placed at the end of **Examples** and **Remarks**, respectively.

$x \in X$	x is element of the set X
$x \notin X$	x is not an element of the set X
$X \subset Y$	X is a subset of Y
$X \cup Y$	Union of the sets X and Y
$X \cap Y$	Intersection of the sets X and Y
$X \backslash Y$	Set of all elements in X not belonging to Y
$X \times Y$	Set of all pairs (x, y) where $x \in X$ and $y \in Y$
A^{T}	Transpose of the matrix A
$\det(A)$	Determinant of the matrix A
$\mathrm{Tr}(A)$	Trace of the matrix A (the sum of the diagonal elements of A)
$\mathbb{R}^n$	Linear space of one-column real matrices with n elements (n-vectors)
$\mathbb{C}^n$	Complex-linear space of complex n-vectors
$\mathbb{R}^{n \times m}$	Linear space of real matrices with n rows of m elements
$\mathbb{C}^{n \times m}$	Complex-linear space of complex matrices with n rows of m elements
$\bar{x}$	Complex-conjugate of complex number (vector, matrix) x
$\mathrm{Re}\, x$	Real part of complex number (vector, matrix) x
$\mathrm{Im}\, x$	Imaginary part of complex number (vector, matrix) x
$\langle x, y \rangle$	Scalar (inner) product of two vectors: $\bar{x}^{\mathrm{T}} y$
$\lVert x \rVert$	Standard (Euclidean) norm of vector x: $\sqrt{\langle x, x \rangle}$
$X \oplus Y$	Direct sum of the linear subspaces X and Y
$A\vert_T$	Restriction of the map A to an invariant subspace T
$\sigma(A)$	Set of all eigenvalues of A (spectrum)
$r(A)$	Spectral radius of A
S^{n-1}	Unit sphere in $\mathbb{R}^n$: $\{x \in \mathbb{R}^n : \lVert x \rVert = 1\}$
$\mathrm{Int}\, S$	Open set bounded by a compact (hyper)surface S
$\mathrm{Ext}\, S$	Complement to $\mathrm{Int}\, S$

∂X	Boundary of a compact set $X \subset \mathbb{R}^n$ or $\mathbb{C}^n$
$f \circ g$	Composition of maps g and $f: (f \circ g)(x) = f(g(x))$
fU	Image $f(U)$ of the set $U \subset X$ under the map $f: X \to Y$
$f^{-1}V$	Preimage $f^{-1}(V)$ of the set $V \subset Y$ for the map $f: X \to Y$
$D^k f$	k-th full differential of a smooth map $f: \mathbb{R}^n \to \mathbb{R}^m$

Notions related to dynamical systems are introduced in **Definitions**, while some general mathematical terms and key words are set in *italic*.

We use the following notation for n-vectors: $(z_1, z_2, \ldots, z_n) := (z_1\, z_2\, \ldots\, z_n)^{\mathrm{T}}$

Contents

Chapter 1
Introduction

Abstract To make a conceptual start, we describe the representation of time, the notion of state space, and the idea that a dynamical system consists of a collection of maps sending the current state to the state at a later time. We explain what is meant by generator and how that relates, in continuous time, to differential equations. We discuss orbits, phase portraits, and equivalence of dynamical systems. We present various examples, including symbolic dynamics.

1.1 Fundamental Definitions

1.1.1 Time, State Space, and Evolution

We use real numbers (denoted by symbols like $t, t_0, s, \tau, \ldots$) to label the moments in time. In other words, *time* is represented by the set $\mathbb{R}$, equipped with the operation of addition, the natural order relation "$\geq$", and the absolute value $|\cdot|$ to measure the distance between the time moments. Relative to time t, the *future* is the set (t, ∞) and the *past* is the set $(-\infty, t)$. We use the notation $\mathbb{R}_+ = [0, \infty)$ and $\mathbb{R}_- = (-\infty, 0]$.

So far we think of time as *continuous*, as flowing. The stroboscopic view of time arises when one monitors (e.g. by opening ones eyes) rhythmically, say every time the clock strikes the hour or the calendar shows a specific day of the year. We represent this so-called *discrete* time by the set of integers $\mathbb{Z}$, with, relative to time zero, the future corresponding to positive and the past to negative integers. We denote the set of nonnegative integers by $\mathbb{Z}_+$ (or $\mathbb{N}$) and the set of nonpositive integers by $\mathbb{Z}_-$.

We shall use the symbol $\mathbb{T}$ to denote a set of real numbers that is closed under addition, i.e. $t, s \in \mathbb{T}$ implies $t + s \in \mathbb{T}$. In this book, $\mathbb{T} = \mathbb{R}, \mathbb{R}_+, \mathbb{Z}$, or $\mathbb{Z}_+$.

Yu. Kuznetsov et al., *Dynamical Systems Essentials*, Texts in Applied Mathematics 83, https://doi.org/10.1007/978-3-032-04083-1_1

The set of all conceivable states of a system is called the *state* or *phase space* and we shall usually denote it by the symbol X. Its elements will be denoted by symbols like $x, y, \xi, \eta, \ldots$. The key idea about the notion of "state" is that it should summarize, in a condensed way, the information about the past that is relevant for predicting the future. In most of this text we consider systems whose states can be described by specifying n real numbers and then X is (a nice subset of) $\mathbb{R}^n$ for some positive $n \in \mathbb{N}$. In this case, $x \in X$ is a *vector* (one-column matrix):

$$x = \begin{pmatrix} x_1 \\ x_2 \\ \vdots \\ x_n \end{pmatrix} =: (x_1, x_2, \ldots, x_n),$$

with $x_i \in \mathbb{R}$ for $i = 1, 2, \ldots, n$. Sometimes it is more convenient to describe a state by *complex numbers* (in particular, replace $\mathbb{R}^n$ by $\mathbb{C}^n$), even if our ultimate interest is in real quantities only. Complex state spaces occur naturally in quantum physics.

It could also happen that a finite number of real numbers is insufficient to characterize the state of a system, i.e. X must have *infinite dimension.* Occasionally we shall use state spaces that contain infinite sequences of numbers or more general functions.

Naturally, the theory becomes richer if X is equipped with some additional structure. All the spaces we will work with are *metric spaces*, i.e. a *distance* (or *metric*) $d(x, y)$ between points $x, y \in X$ is defined such that it has all the required properties.[1] This allows to define the distance between a point $x \in X$ and a subset $S \subset X$ as $\text{dist}(x, S) = \inf_{y \in S} d(x, y)$, as well as the ε-neighbourhood of S as the set of all points $x \in X$ with $\text{dist}(x, S) < \varepsilon$. It is also convenient if X is *complete*[2] with respect to d.

If X is a (subset of) a complex linear space with a *norm* $\|\cdot\|$ satisfying the standard requirements,[3] the distance can be defined as $d(x, y) = \|x - y\|$. If the linear space X is complete in this norm, it is called a *Banach space.* Finally, in a linear space X endowed with a *scalar product*[4] $\langle \cdot, \cdot \rangle$, a norm can be defined via $\|x\| = \sqrt{\langle x, x \rangle}$. As an example, take $X = \mathbb{C}^n$. This is a linear space with the standard scalar product

---

[1] For any $x, y, z \in X$ should hold: (i) $d(x, y) \geq 0$ and $d(x, y) = 0$ if and only if $x = y$; (ii) $d(x, y) = d(y, x)$; (iii) $d(x, y) \leq d(x, z) + d(z, y)$.

[2] X is called complete if every Cauchy sequence in X has a limit that belongs to X. Recall that $\{x_k\}$ is a *Cauchy sequence* if for every $\varepsilon > 0$ there exists integer N such that $d(x_n, x_m) < \varepsilon$ for all $n, m \geq N$.

[3] For any $x, y \in X$ and $\lambda \in \mathbb{C}$ should hold: (i) $\|x\| \geq 0$ and $\|x\| = 0$ if and only if $x = 0$; (ii) $\|\lambda x\| = |\lambda| \, \|x\|$; (iii) $\|x + y\| \leq \|x\| + \|y\|$.

[4] In general, a scalar product has the following properties. For any $x, y, z \in X$ and $\lambda \in \mathbb{C}$ holds: (i) $\langle x, y + \lambda z \rangle = \langle x, y \rangle + \lambda \langle x, z \rangle$; (ii) $\langle x, y \rangle = \overline{\langle y, x \rangle}$; (iii) $\langle x, x \rangle > 0$ for $x \neq 0$.

$$\langle x, y \rangle = \bar{x}^{\mathrm{T}} y = \sum_{k=1}^{n} \bar{x}_k y_k,$$

where $\bar{x}^{\mathrm{T}} = (\bar{x}_1 \, \bar{x}_2 \, \cdots \, \bar{x}_n)$ is the one-row matrix with the elements which are complex conjugate to those of x and where the standard matrix multiplication is used. Then,

$$d(x, y) = \sqrt{\langle x - y, x - y \rangle}$$

is the standard distance in $\mathbb{C}^n$. For the real subspace $\mathbb{R}^n$ of $\mathbb{C}^n$ this leads to the familiar Euclidean distance.

Now we have to describe the *evolution* of the system, i.e. the change of its state in time. A natural way to do this is to consider a map $x : \mathbb{T} \to X$,

$$t \mapsto x(t),$$

such that $x(t)$ can be interpreted as the state of the system at time t. Here our understanding is that the system can only be in one state at any moment of time, the map x is single-valued, i.e. a function.

But not all functions will do, there are constraints. We will consider only *deterministic* systems: By assumption the specification of a state $x_0 = x(t_0)$ at time t_0 uniquely determines the state $x(t)$ at any time $t > t_0$, so that

$$x(t) = \psi(t, t_0, x_0)$$

for $t \geq t_0$, with some function ψ that describes the internal properties of the system and the input from the environment, the external world in which it evolves. The word "input" should not be taken too literally, as it may refer to such things as cooling or harvesting.

To make things even simpler, we concentrate on *autonomous* systems, meaning that both the system properties and the input should not vary in time (in other words, we describe them by constant *parameters*). The consequence is that it does not matter at what time t_0 we specify the initial state x_0, only time differences matter and the system is time-translation invariant:

$$x(t) = \varphi(t - t_0, x_0).$$

In that case, we can take without loss of generality $t_0 = 0$ as the moment to specify the initial state $x_0 = x(0)$ of the system.

Often, it is convenient to single out the role of time in the notation and to write

$$x(t) = \varphi^t(x_0),$$

i.e. for given sets $\mathbb{T}$ and X, we consider a collection $\Phi = \{\varphi^t\}_{t \in \mathbb{T}}$ of maps from X to X, parametrized by the elements of $\mathbb{T}$:

$$\varphi^t : X \to X.$$

Let us immediately point out that $\varphi^t(x)$ need not be defined for all $(x,t) \in X \times \mathbb{T}$.

Systems for which the map φ^t is defined for $t \geq 0$ as well as $t < 0$ are called *invertible systems.* For such systems, the initial state x_0 fixes not only future but also past states. If one can predict future states given an initial state, but past states are not reconstructible, the system will be called *non-invertible.* For such systems, φ^t has meaning (in the sense of being defined and single-valued) only when $t \geq 0$, so that we should choose $\mathbb{T} = \mathbb{Z}_+$ or $\mathbb{R}_+$. Moreover, even if $\mathbb{T} = \mathbb{R}$, $\varphi^t(x)$ might exist only for $t \in (-a(x), b(x)) \subset \mathbb{R}$, for some $a(x), b(x) > 0$. Such systems are called *locally defined.* As we shall see later, this is not unusual for dynamical systems associated to nonlinear differential equations. Finally, depending on the smoothness of φ^t for fixed t, one distinguishes *continuous* and *smooth* dynamical systems. When time is continuous we usually have joint continuity in t and x.

Of course, $\varphi^t(x)$ may very well be defined for all $x \in X$ and all $t \in \mathbb{T}$. An important class of such globally defined systems are those associated to linear ordinary differential equations or invertible linear maps in $\mathbb{R}^n$.

Let us return to the properties of the maps $\Phi = \{\varphi^t\}_{t\in\mathbb{T}}$ defining the evolution of a deterministic system. There are two general properties that these maps should possess, namely:

$$\varphi^0(x) = x; \tag{1.1}$$

$$\varphi^{t+s}(x) = \varphi^t(\varphi^s(x)), \tag{1.2}$$

for all $x \in X$ and $t, s \in \mathbb{T}$ such that all quantities at both sides of the above relations are defined.[5] Property (1.1) expresses that x is the prescribed state at time zero, while the so-called *group property* (1.2) expresses that the state after time $t + s$ when starting at x, is identical to the state at time t when starting at $\varphi^s(x)$ (see Fig. 1.1). In case $\mathbb{T} = \mathbb{Z}_+, \mathbb{R}_+$ Eq. (1.2) is usually called the *semigroup property.* Note that (1.2) expresses the combined effect of uniqueness and translation invariance.

Definition 1.1 *A* ***dynamical system*** *is a triple* $\{\mathbb{T}, X, \varphi^t\}$, *where* $\mathbb{T}$ *is a time set,* X *is a state space, and* $\{\varphi^t\}_{t\in\mathbb{T}}$ *is a family of evolution operators satisfying* (1.1) *and* (1.2).

As it is often clear from the context what $\mathbb{T}$ and X are, we shall also call $\{\varphi^t\}$ a dynamical system. The map φ^t is called the *evolution operator* or the *t-shift map.* When time is continuous, the collection of evolution operators is

[5] Using the symbol "$\circ$" for map composition and denoting the identity map by I, i.e. $I(x) = x$ for all $x \in X$, one can rewrite these properties as $\varphi^0 = I$ and $\varphi^{t+s} = \varphi^t \circ \varphi^s$ where defined, respectively.

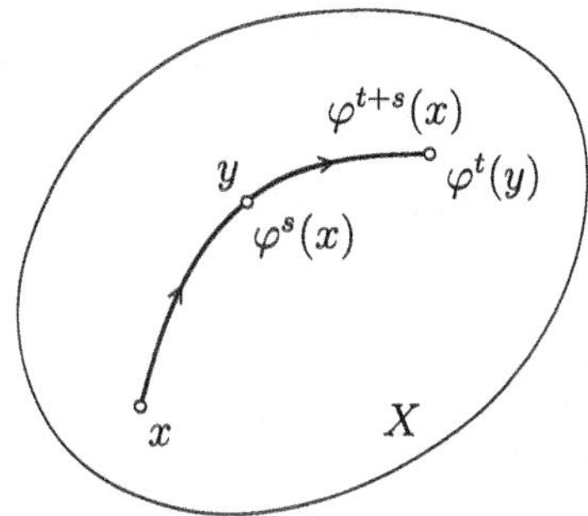

Fig. 1.1 Group property: If $y = \varphi^s(x)$, then $\varphi^t(y) = \varphi^{t+s}(x)$

called a *flow* (or *semiflow* if $\mathbb{T} = \mathbb{R}_+$). Thus, in the continuous time case the words "dynamical system" and "flow" are synonymous.

Example 1.2 (Translation dynamics) There are many examples of dynamical systems with infinite-dimensional state spaces, among which we only mention *translation on BC*, i.e.

$$\{\mathbb{R}, BC(\mathbb{R}, \mathbb{R}), \theta^t\},$$

where $BC(\mathbb{R}, \mathbb{R})$ is the Banach space of bounded continuous scalar functions on $\mathbb{R}$ equipped with the supremum norm

$$\|f\|_\infty = \sup_{-\infty < x < +\infty} |f(x)|$$

and

$$[\theta^t(f)](x) = f(x+t),$$

for $x, t \in \mathbb{R}$. This is a continuous-time, invertible dynamical system. ◊

1.1.2 The Concept of a Generator

Consider a dynamical system with $\mathbb{T} = \mathbb{Z}_+$. Define $f : X \to X$ by $f(x) = \varphi^1(x)$. We call f the *generator* of the dynamical system $\{\mathbb{Z}_+, X, \varphi^k\}$ to express that the collection $\Phi = \{\varphi^k\}_{k \in \mathbb{Z}_+}$ can be constructed from f by the operation of map composition. Indeed, (1.2) implies that

$$\varphi^2 = \varphi^1 \circ \varphi^1 = f \circ f$$

and, by induction for $k \in \mathbb{Z}_+$,

$$\varphi^k = \underbrace{f \circ f \circ \cdots \circ f}_{k \text{ times}}. \tag{1.3}$$

The key point about this observation is that we can, vice versa, for any given map $f : X \to X$ define a dynamical system with $\mathbb{T} = \mathbb{Z}_+$ by means of these formulas. So, a discrete-time dynamical system corresponds to the iteration of a map and the super-index in φ^t can be rightfully interpreted as a power with respect to composition!

Example 1.3 (Linear maps in $\mathbb{R}^n$) Let A be a $n \times n$ matrix with real elements a_{ij}, $i, j = 1, 2, \ldots, n$; we also write $A \in \mathbb{R}^{n \times n}$. For $x \in \mathbb{R}^n$ the vector $Ax \in \mathbb{R}^n$ is defined by

$$(Ax)_i = \sum_{j=1}^{n} a_{ij} x_j, \quad i = 1, 2, \ldots, n.$$

Thus $x \mapsto Ax$ is a linear map from $\mathbb{R}^n$ into $\mathbb{R}^n$. This map generates the discrete-time linear dynamical system $\{\mathbb{Z}, \mathbb{R}^n, A^k\}$. Here A^k denotes the map obtained by applying A k-times in a row. Of course, the corresponding matrix is the kth power of the matrix A. $\diamondsuit$

Consider now an invertible dynamical system with $\mathbb{T} = \mathbb{Z}$ and define $f = \varphi^1$ as before. Then the map f has to be invertible. Indeed, by the fundamental properties (1.1) and (1.2) we have

$$\varphi^{-1} \circ f = \varphi^{-1} \circ \varphi^1 = \varphi^0 = I$$

and

$$f \circ \varphi^{-1} = \varphi^1 \circ \varphi^{-1} = \varphi^0 = I,$$

which shows that f^{-1} exists and is given by φ^{-1}. Now it is clear that any invertible map $f : X \to X$ defines an invertible dynamical system $\{\mathbb{Z}, X, \varphi^k\}$, where φ^k is specified for integer $k < 0$ by the formula

$$\varphi^k = \underbrace{f^{-1} \circ f^{-1} \circ \cdot \circ f^{-1}}_{|k| \text{ times}}. \tag{1.4}$$

Example 1.4 (Symbolic dynamics) Important examples of discrete-time dynamical systems are provided by *symbolic dynamics.*

Take the set Ω_m of doubly-infinite sequences of symbols chosen from a collection of m symbols, say the integer numbers $\{0, 1, 2, \ldots, m-1\}$. An element $\omega \in \Omega_m$ is a sequence (or string) with a distinguished position zero

$$\omega = \ldots \omega_{-2}\, \omega_{-1}\, \omega_0\, \omega_1\, \omega_2\, \ldots,$$

where ω_i is an integer number satisfying $0 \le \omega_i \le m - 1$. The space Ω_m is a compact and complete metric space with respect to a distance between $\omega, \theta \in \Omega_m$ defined by the formula:

$$d(\omega, \theta) = \sum_{j \in \mathbb{Z}} \frac{|\omega_j - \theta_j|}{2^{|j|}}.$$

According to this formula, two sequences are considered to be close if they have a long block of coinciding elements centered at position zero.

The *shift map* $\sigma : \Omega_m \to \Omega_m$ is defined by the formula $\sigma(\omega) = \theta$, where

$$\theta_j = \omega_{j+1}, \; j \in \mathbb{Z}. \tag{1.5}$$

The triple $\{\mathbb{Z}, \Omega_m, \sigma^k\}$ is a discrete-time, invertible dynamical system called the *shift dynamics.*

We can also take any closed subset $\Omega \subset \Omega_m$ that is invariant under the shift σ, that is $\sigma(\Omega) = \Omega$. Such a subset Ω is called a *subshift.* Then, clearly $\{\mathbb{Z}, \Omega, \sigma^k\}$ is again a dynamical system, called a *subshift dynamics.* An interesting subshift Ω_A can be defined by a *transition matrix*, i.e. a matrix $A = (a_{ij})_{i,j=1,\ldots,m}$ with elements $a_{ij} \in \{0, 1\}$,

$$\Omega_A = \{\omega \in \Omega_m : a_{\omega_j, \omega_{j+1}} = 1 \text{ for all } j \in \mathbb{Z}\}. \tag{1.6}$$

The subshift dynamics $\{\mathbb{Z}, \Omega_A, \sigma^k\}$ is called a *topological Markov chain* associated with A. As an example consider the $m \times m$ matrix

$$A = \begin{pmatrix} 1 & 1 & 0 & \ldots & 0 \\ 0 & 0 & 1 & \ddots & \vdots \\ \vdots & & \ddots & \ddots & 0 \\ 0 & & & \ddots & 1 \\ 1 & 0 & \ldots & \ldots & 0 \end{pmatrix}. \tag{1.7}$$

Then any sequence $\omega \in \Omega_A$ can contain arbitrarily long blocks of 0's. Such a block either extends indefinitely or terminates with a sequence $\ldots 012 \ldots (m-1)0 \ldots$, after which another block of 0's can follow, etc.

One may view the transition matrix A as the *adjacency matrix* of a directed graph and then note that any sequence $\omega \in \Omega_A$ corresponds to an infinite travel through this graph. Figure 1.2 shows a graph with the adjacency matrix

$$A = \begin{pmatrix} 1\,1\,0\,0\,0 \\ 0\,0\,1\,0\,0 \\ 0\,0\,0\,1\,0 \\ 0\,0\,0\,0\,1 \\ 1\,0\,0\,0\,0 \end{pmatrix}. \tag{1.8}$$

Here $m = 5$. $\Diamond$

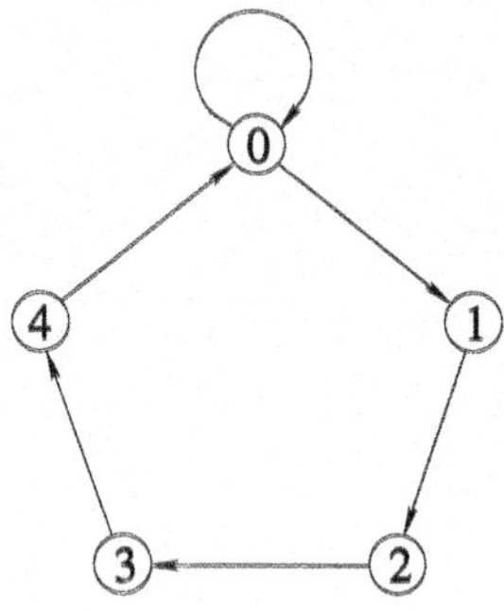

Fig. 1.2 The directed graph with the adjacency matrix (1.8)

In continuous time, the notion of generator is far more subtle. Assume that $T = \mathbb{R}_+$ or $\mathbb{R}$ and that X is a (subset of) a linear normed space. Also, let us supplement (1.1) and (1.2) by the continuity condition

$$\lim_{t \downarrow 0} \varphi^t(x) = x, \tag{1.9}$$

which excludes jumps. The question still is: What is the minimal amount of information needed to specify a dynamical system? And to get some feeling for the answer we again begin by working in the other direction, i.e. from a given/known dynamical system towards a condensed representation of the essential information.

In the discrete time case we considered the smallest possible time step (equal to 1), but now there is no such step, as time is continuous. The identity (1.9) states that we obtain only trivial (i.e. system unspecific) information when letting the time interval shrink to zero. The key idea is now to compute the *rate* of change of the state and to define (if possible) a function $f : X \to X$ by

$$f(x) = \lim_{t \downarrow 0} \frac{1}{t}(\varphi^t(x) - x). \tag{1.10}$$

Next we ask the question: Can we recover $\Phi = \{\varphi^t\}_{t \in T}$ from f? If the answer is positive, we call f the (infinitesimal) *generator* of the system. And, most importantly, we apply the "reconstruction procedure" to functions f, for which it is not yet known that they generate a dynamical system, in order to construct continuous-time dynamical systems from minimal and, from a modelling point of view, natural information. It is here that differential equations enter the scene.

A function $f : \mathbb{R}^n \to \mathbb{R}^n$ is called a *vector field*. The following theorem – which we formulate without proof – will allow us to conclude that smooth vector fields define continuous-time, invertible, smooth dynamical systems on $\mathbb{R}^n$.

Theorem 1.5 *Consider a system of autonomous ordinary differential equations*

$$\dot{x} = f(x), \quad x \in \mathbb{R}^n,$$

where $f : \mathbb{R}^n \to \mathbb{R}^n$ *is smooth in an open region* $U \subset \mathbb{R}^n$. *Then*

(a) *there is an open domain*

$$\Omega = \{(t, x_0) : x_0 \in U,\ t \in J(x_0) := (-\delta_1(x_0), \delta_2(x_0)) \text{ with } \delta_{1,2}(x_0) > 0\}$$

and a unique function $\psi : \Omega \to U, x = \psi(t, x_0)$ *that is smooth with respect to* (t, x_0), *and satisfies, for each* $x_0 \in U$, *the following conditions:*

(i) $\psi(0, x_0) = x_0$;
(ii) $\frac{\partial}{\partial t}\psi(t, x_0) = f(\psi(t, x_0))$ *for* $t \in J(x_0)$;

(b) *for each* $y_0 \in U$, *there exists an open neighborhood* $U_0 \subset U$ *and* $\delta_0 > 0$ *such that*

$$\varphi^t(x_0) := \psi(t, x_0)$$

is defined for all $x_0 \in U_0$ *and* $|t| \leq \delta_0$. *Moreover, the second-order mixed derivatives* $\frac{\partial}{\partial t} D_{x_0} \varphi^t(x_0)$ *and* $D_{x_0} \frac{\partial}{\partial t} \varphi^t(x_0)$ *do exist and coincide for* $x_0 \in U_0$ *and* $|t| \leq \delta_0$.

(c) *the map* φ^t *is as smooth as the vector field* f, *i.e. if* $f \in C^k(\mathbb{R}^n, \mathbb{R}^n)$, *then* $\varphi^t \in C^k(U_0, \mathbb{R}^n)$ *for* $|t| \leq \delta_0$. □

Thus, the triple $\{\mathbb{R}, \mathbb{R}^n, \varphi^t\}$ is a continuous-time, invertible, smooth dynamical system with infinitesimal generator f. In general, this system is only locally defined in time.

As far as continuous time dynamical systems are concerned, we shall in most of this book restrict our attention to those defined on (a subset of) $\mathbb{R}^n$ and generated by a system of ordinary differential equations (*ODEs*). There is also a modelling related motivation to base the description of a dynamical system on differential equations. Many physical, chemical, biological, etc. laws give expressions for the *rate* of change in time of quantities that describe the state. Moreover, contributions of different mechanisms to the rate of change can be simply added, whereas such contributions may interact in subtle ways when one considers the changes over a finite time interval.

1.2 Orbits and Phase Portraits

Definition 1.6 *The* ***time-series*** *starting at* $x_0 \in X$ *is*

$$\Theta(x_0) = \{(t, x) : x = \varphi^t(x_0) \text{ for } t \in \mathbb{T} \text{ such that } \varphi^t(x_0) \text{ is defined}\}.$$

The ***orbit*** *starting at* $x_0 \in X$ *is*

$$\Gamma(x_0) = \{x : x = \varphi^t(x_0) \text{ for } t \in \mathbb{T} \text{ such that } \varphi^t(x_0) \text{ is defined}\}.$$

Note that $\Theta(x_0) \subset \mathbb{T} \times X$, while $\Gamma(x_0) \subset X$. Also note that the orbit starting at $y_0 \in \Gamma(x_0)$ coincides with $\Gamma(x_0)$ (see Exercise 1.5). Accordingly we can speak about orbits without specifying a particular starting point. The orbits are oriented by the increase (advance) of time. Finally note that the orbits of continuous time dynamical systems are *curves*, provided $\varphi^t(x)$ depends continuously on t.

Example 1.7 (Scalar maps) Let $f : \mathbb{R} \to \mathbb{R}$, $x \mapsto x' = f(x)$ be a smooth map. There is a useful technique to visualize orbits of the corresponding discrete-time dynamical system: Staircase/cobweb diagrams, sometimes called *Lemeray's diagrams* (Fig. 1.3). Consider the graph of f in the (x, x')-plane together with the line $x' = x$. Take a point (a, a) on this line and then produce the sequence of points

$$(a, a),\ (a, f(a)),\ (f(a), f(a)),\ (f(a), f^2(a)),\ (f^2(a), f^2(a)), \ldots$$

by "jumping" vertically and horizontally between the graph of f and the line $x' = x$. Now notice that the sequence obtained by taking the first coordinate of all odd points composes the forward part

$$a, f(a), f^2(a), \ldots,$$

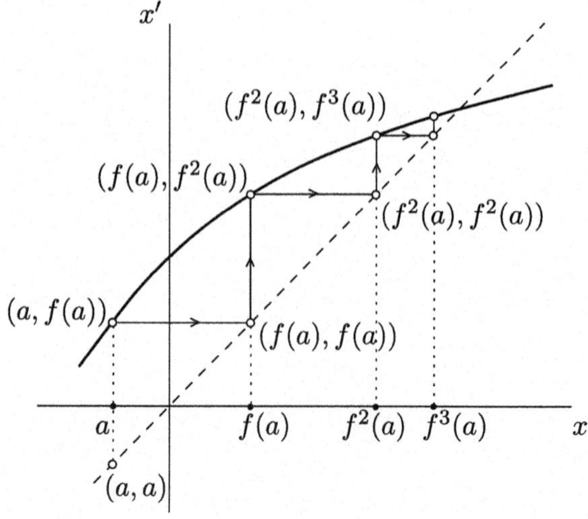

Fig. 1.3 Staircase diagram

of the orbit starting at $a \in \mathbb{R}$. If f is invertible, the backward part can be constructed in a similar manner by jumping between the line $x' = x$ and the graph of f, that is, in the opposite order. Notice that constant orbits

$$\dots, a, a, a, a, \dots$$

correspond to those points, where the graph of f intersects the line $x' = x$: $f(a) = a$.

◊

Example 1.8 (Scalar autonomous ordinary differential equations) Another example, where one could plot and analyse orbits easily, is a continuous-time dynamical system generated by a scalar differential equation $\dot{x} = f(x)$ with a smooth $f : \mathbb{R} \to \mathbb{R}$. Looking at the graph of the function $y = f(x)$ in the (x, y)-plane (Fig. 1.4), one concludes immediately that any orbit is either a root of the equation $f(x) = 0$ or an open segment of the x-axis, bounded by such roots or extending to infinity. The orientation of a nontrivial orbit is determined by the sign of $f(x)$ in the corresponding interval. ◊

The previous example serves to motivate the following definitions.

Definition 1.9 *The **phase portrait** of a dynamical system is the collection of its oriented orbits.*

Whenever the system is invertible, the phase portrait induces a partitioning of the state space (see Exercise 1.5). When we draw the phase portrait of a continuous-time planar dynamical system, we only sketch a few orbits qualitatively. However, we will do this in such a way that one can infer from the orbits drawn how the others should look. In particular, we will try to depict all orbits with exceptional properties, which bound "cells" composed of orbits having similar properties. Moreover, we draw at least one representative orbit in each cell.

Definition 1.10 *A point $x^0 \in X$ is called a **fixed point** (equilibrium) if $\varphi^t(x^0) = x^0$ for all $t \in \mathbb{T}$.*

We use the term "fixed point" primarily in the discrete-time case, while the term "equilibrium" is usually reserved for the continuous-time case. Both fixed points and equilibria are also called *steady states*. Fixed points of a discrete-time dynamical system generated by a map $x \mapsto f(x)$ satisfy the equation $x = f(x)$, while equilibria of a continuous-time system generated by an ODE $\dot{x} = f(x)$ satisfy $f(x) = 0$. In both cases one may think of $x, f(x) \in \mathbb{R}^n$. The shift dynamics with m symbols (see Example 1.4) has only m fixed points (find them!), while the equilibria of the translation dynamics from Example 1.2 are all constant functions $f(x) = f_0$.

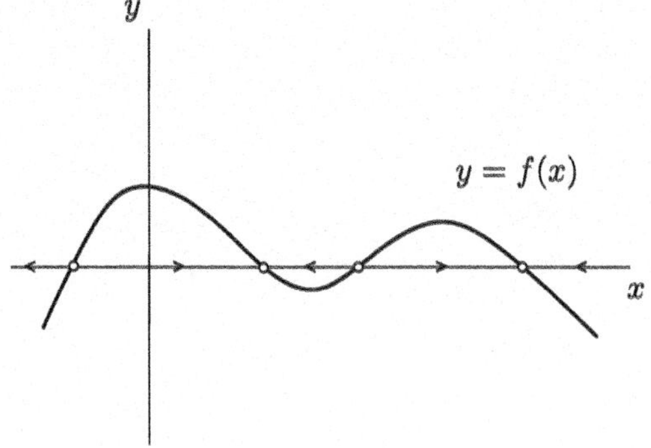

Fig. 1.4 A scalar differential equation

Definition 1.11 *A* ***periodic orbit*** *(cycle) of period $p > 0$ is a nonequilibrium orbit L_0, such that for any point $x_0 \in L_0$ and for all $t \in \mathbb{T}$ holds*

$$\varphi^{t+p}(x_0) = \varphi^t(x_0).$$

Suppose p_0 is the minimal value of p for which the identity in the definition holds. In a continuous-time dynamical system generated by a smooth system of ODEs, each periodic orbit is a smooth closed curve that is traversed exactly once in a time interval of length p_0 (Fig. 1.5a). For such systems, an isolated periodic orbit is often called a *limit cycle.* In a discrete-time system generated by a map f, periodic orbits are finite sets of points of the form

$$\{x_0, f(x_0), f^2(x_0), \ldots, f^{p_0-1}(x_0)\}$$

with $f^{p_0}(x_0) = x_0$ for an integer period p_0 (Fig. 1.5b). The shift dynamic has an infinite number of periodic orbits. Indeed, any periodic sequence is a starting point of a periodic orbit, since it is mapped exactly onto itself by shifting it over the period.

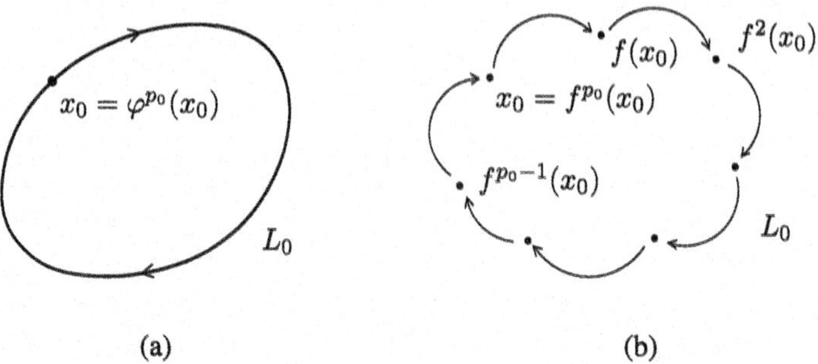

Fig. 1.5 Periodic orbits

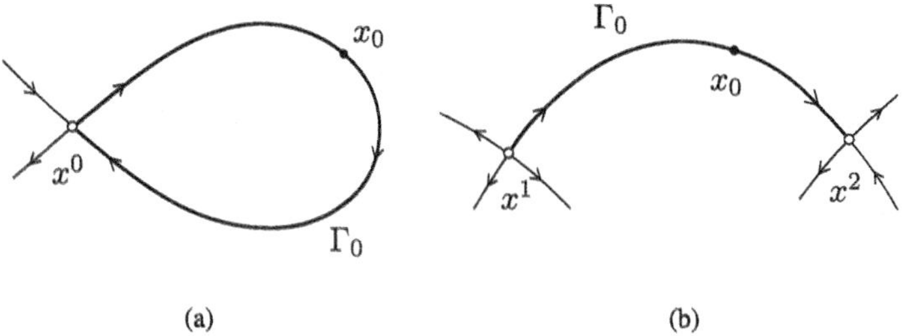

Fig. 1.6 Homoclinic (**a**) and heteroclinic (**b**) orbits to equilibria

There are other interesting types of special orbits. Let d be a metric in the state space X.

Definition 1.12 *An orbit Γ_0 is called* **homoclinic to a fixed point** *(equilibrium) x^0 if for any point $x_0 \in \Gamma_0$: $d(\varphi^t(x_0), x^0) \to 0$, for $t \to \pm\infty$.*

An orbit Γ_0 is called **heteroclinic to fixed points** *(equilibria) x^1 and x^2 if for any point $x_0 \in \Gamma_0$: $d(\varphi^t(x_0), x^1) \to 0$, for $t \to -\infty$, and $d(\varphi^t(x_0), x^2) \to 0$, for $t \to +\infty$.*

Figure 1.6 illustrates these notions for continuous time systems in the plane. Similarly one can define orbits homoclinic to a periodic orbit L_0, as well as heteroclinic orbits that tend to different periodic orbits $L^{1,2}$ as $t \to \pm\infty$ (or to an equilibrium as $t \to -\infty$ and to a periodic orbit as $t \to +\infty$, etc.). Together homoclinic and heteroclinic orbits are called *connecting orbits.*

Definition 1.13 *An* **invariant set** *of a dynamical system is a set $S \subset X$ entirely composed of orbits which are defined for all time, i.e. $x \in S$ if and only if $\varphi^t(x) \in S$ for all $t \in \mathbb{T}$.*

Definition 1.14 *An invariant set S is called* **(Lyapunov) stable** *if for any neighbourhood $U \supset S$ there exists a neighbourhood $V \supset S$ such that $\varphi^t(x) \in U$ for all $x \in V$ and $t > 0$.*

An invariant set which is not stable is called *unstable.*

Definition 1.15 *An invariant set S is called* **asymptotically stable** *if it is stable and there exists a neighbourhood $U \supset S$ such that* $\operatorname{dist}(\varphi^t(x), S) \to 0$ *for all $x \in U$, as $t \to +\infty$. If we can take $U = X$, we call S* **globally asymptotically stable**.

The fixed point of a contraction $f : X \to X$ in a complete metric space X is globally asymptotically stable (see Theorem 3.20 in Chap. 3).

A compact[6] asymptotically stable invariant set is often called an *attractor.* If a bounded invariant set becomes asymptotically stable after time reversal $t \mapsto -t$, it is called a *repellor.* An attractor (in general, invariant set) is called *strange* if it has a Cantor (fractal) structure, cf. Chap. 7.

[6] In finite-dimensional spaces, compact sets are those which are bounded and closed.

1.3 Equivalence of Dynamical Systems

Definition 1.16 *A dynamical system* $\{\mathbb{T}, X, \varphi^t\}$ *is* ***equivalent*** *to a dynamical system* $\{\mathbb{T}, Y, \psi^t\}$ *if there is an invertible surjective map* $h : X \to Y$ *which maps orbits of the first system onto orbits of the second system, preserving the direction of time.*

By "preserving the direction of time" we mean that for any $x \in X$ there is a strictly increasing function $t \mapsto \alpha_x(t)$ such that $h(\varphi^t(x)) = \psi^{\alpha_x(t)}(h(x))$. Recall that a map $h : X \to Y$ is *surjective* if it maps X onto Y. If h and h^{-1} are continuous maps, i.e. h is a *homeomorphism*, the equivalence is called *topological equivalence.* If h and h^{-1} are continuously-differentiable maps, i.e. h is a *diffeomorphism*, the equivalence is called *smooth equivalence.*

Example 1.17 (Equivalence of planar systems) Two continuous-time dynamical systems in $\mathbb{R}^2$ with linear evolution operators, given by the matrices

$$\varphi^t = e^{-t}\begin{pmatrix} 1 & 0 \\ 0 & 1 \end{pmatrix}$$

and

$$\psi^t = e^{-t}\begin{pmatrix} \cos t & -\sin t \\ \sin t & \cos t \end{pmatrix},$$

are topologically equivalent (each has a globally asymptotically stable equilibrium at the origin, see Fig. 1.7). A proof is indicated in Exercises 1.13 and 1.14. ◊

If h preserves not only the direction of time, but also the time parametrization along orbits, the systems are called (topologically, smooth) *conjugate.* In this case:

$$\psi^t = h \circ \varphi^t \circ h^{-1}, \tag{1.11}$$

where defined, see Fig. 1.8. The two systems from Example 1.17 are not only topologically equivalent but also topologically conjugate.

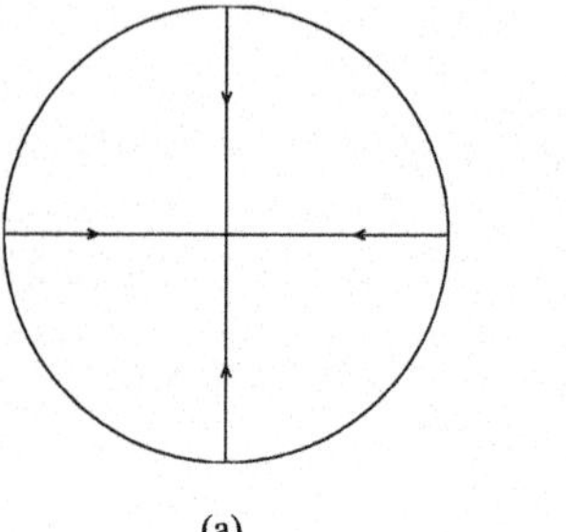

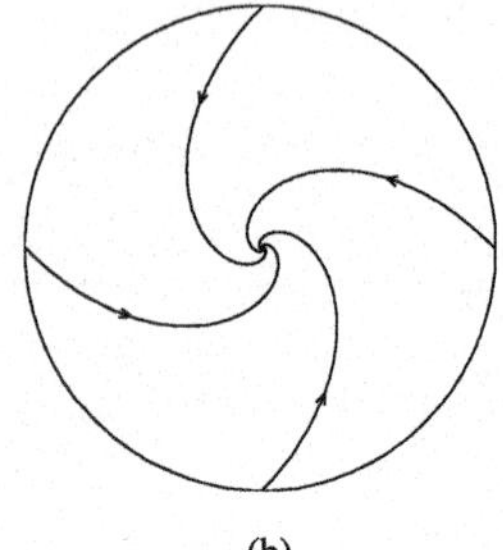

Fig. 1.7 Node (**a**) and focus (**b**) are topologically equivalent

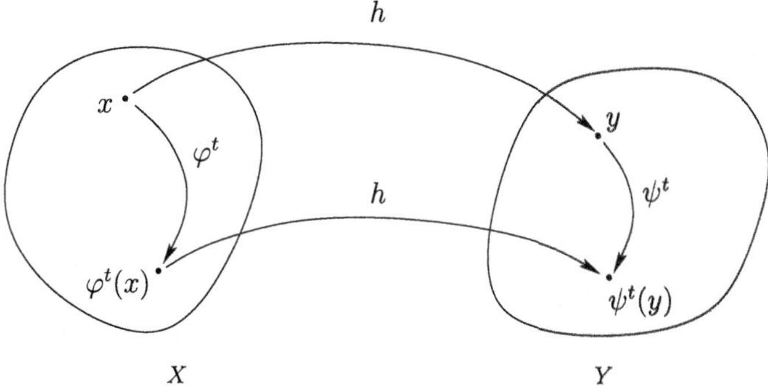

Fig. 1.8 Conjugate systems: $h(\varphi^t(x)) = \psi^t(h(x))$

Topologically (smoothly) equivalent discrete-time systems are always conjugate. Moreover, if $f = \varphi^1$ and $g = \psi^1$, then the topological conjugacy of the corresponding systems follows from

$$g = h \circ f \circ h^{-1}$$

by induction. When both f and g are invertible, their inverses are conjugate via the same map h:

$$g^{-1} = h \circ f^{-1} \circ h^{-1}.$$

Smooth conjugacy of the flows generated by two smooth autonomous systems of ODEs can be expressed in terms of the vector fields. Introduce the generators

$$\left.\frac{d}{dt}\varphi^t(x)\right|_{t=0} = f(x), \quad \left.\frac{d}{dt}\psi^t(y)\right|_{t=0} = g(y),$$

then differentiate $\psi^t \circ h = h \circ \varphi^t$ with respect to t at $t = 0$ to obtain

$$g(h(x)) = Dh(x)f(x).$$

Since h is a diffeomorphism, the Jacobian matrix $Dh(x)$ with elements

$$[Dh(x)]_{ij} = \frac{\partial h_i(x)}{\partial x_j}, \quad i, j \in \{1, 2, \ldots, n\},$$

is invertible and we find the relation

$$f(x) = [Dh(x)]^{-1} g(h(x)). \tag{1.12}$$

Two smooth autonomous systems $\dot{x} = f(x)$ and $\dot{y} = g(y)$ for which the relation (1.12) holds for some diffeomorphism h are called *diffeomorphic*. Changing variables by substituting $y = h(x)$ in $\dot{y} = g(y)$ and using (1.12) we immediately obtain $\dot{x} = f(x)$. Above we have shown that smooth conjugacies lead to diffeomorphic ODE's. Conversely, it is obvious that the flows generated by two smoothly diffeomorphic ODE's are smoothly conjugate.

Two smooth autonomous systems $\dot{x} = f(x)$ and $\dot{x} = g(x)$ for which

$$f(x) = \mu(x)g(x),$$

where $\mu = \mu(x) > 0$ is a smooth positive function, are called in this book *orbitally equivalent*. In this case, h in Definition 1.16 can be defined by $h(x) = x$ and the systems have identical orbits with different time-parametrization.

If a system $\dot{x} = f(x)$ is diffeomorphic to a system $\dot{y} = \nu(y)g(y)$, where $\nu = \nu(y) > 0$ is a smooth function, then the systems $\dot{x} = f(x)$ and $\dot{y} = g(y)$ are called *smoothly orbitally equivalent*. The vector field g is obtained from f by a smooth change of variables combined with a multiplication by a smooth positive function $\mu = 1/\nu$.

If h is defined in some neighbourhood $U \subset X$ of an invariant set S of the first system, while h^{-1} is defined in the neighbourhood $V = h(U) \subset Y$ of the invariant set $h(S)$ of the second system, the equivalence is called *local equivalence* near the invariant sets. Notice that stability is a topologically invariant notion, i.e. the corresponding invariant sets (S and $h(S)$) of two topologically equivalent systems are both stable or both unstable.

The (overly ambitious) aim of the qualitative theory of dynamical systems is to provide a catalogue of equivalence classes, as well as rules to determine the equivalence class from the generator. The pragmatic and more realistic version concentrates on the neighbourhood of special orbits (like those introduced in the previous section). In Chap. 3 we shall see that "linearization" is a powerful method when trying to determine the phase portrait near an equilibrium or a fixed point. Clearly then, we should carry out the classification program for *linear* systems, and this is exactly what the next chapter is all about.

1.4 References

There are many texts on dynamical systems, where all fundamental notions introduced in this chapter are carefully discussed, e.g. Irwin (1980), Amann (1990), Hale and Koçak (1991), Alligood et al. (1997), Perko (2001), Hasselblatt and Katok (2003), see also Anosov et al. (1988, Chap. 1).

1.5 Exercises

Exercise 1.1 (Nonautonomous systems) In the non-autonomous case we have a map

$$\psi : \mathbb{T} \times \mathbb{T} \times X \to X$$

such that $\psi(t, t_0, x_0)$ is the state of the system at time t, given that its state at time t_0 is x_0.

(i) Formulate the analogue of the properties (1.1) and (1.2). (*Hint*: First formulate them in words and only then re-express the statements in mathematical symbols.)

(ii) Show that ψ leads to an autonomous dynamical system on the extended phase space $Y = \mathbb{T} \times X$ via

$$\Phi(s, y) = (t + s, \psi(t + s, y)), \quad \text{where} \quad y = (t, x).$$

(iii) In the smooth continuous time case how would you define the infinitesimal generator of ψ^t and what will be its relation to the generator of the extended flow Φ^t?

Exercise 1.2 (Continuous-time system with discontinuous orbits) For $x \in \mathbb{R}, t \geq 0$ define

$$\varphi^t(x) = \begin{cases} x + t + 1 & \text{if} \quad x < 0 \leq x + t, \\ x + t & \text{otherwise.} \end{cases}$$

Show that this defines a dynamical system $\{\mathbb{R}_+, \mathbb{R}, \varphi^t\}$ for which orbits are right continuous (i.e. $\lim_{t \downarrow 0} \varphi^t(x) = x$ for all $x \in \mathbb{R}$) but not continuous. Can this system be extended to an invertible one on $X = \mathbb{R}$?

Exercise 1.3 (Orbits of maps) Consider the *Ricker map*[7]

$$x \mapsto x' = \alpha x e^{-x}.$$

(i) Set $\alpha = 6$. Pick an initial point, say $x_0 = 0.2$, and construct graphically the orbit of the map starting at this point (see Fig. 1.9a). Do you see why one also speaks of cobweb (rather than staircase) diagrams? Calculate the fixed point.

(ii) Construct an orbit starting at the same initial point x_0, but for $\alpha = 9$ (see Fig. 1.9b). Find (approximately) the coordinates of the two points on the period-2 cycle.

(iii) Explain first in graphical terms the difficulty of going backwards in time and then give an analytic reformulation.

Exercise 1.4 (Orbits of ODEs) Using one of the standard ODE solvers, construct numerically an orbit of the *Rössler system*[8]:

$$\begin{cases} \dot{x}_1 = -x_2 - x_3, \\ \dot{x}_2 = x_1 + A x_2, \\ \dot{x}_3 = B x_1 - C x_3 + x_1 x_3, \end{cases}$$

with $A = 0.36, B = 0.4, C = 4.5$, starting at $x_0 = (0.74, -0.41, 0.06)$ on the time interval $0 \leq t \leq 200$ (see Fig. 1.10). Think about the accuracy of the approximation of the solution! What kind of invariant set do you expect in this system?

---

[7] Ricker, W. (1954). Stock and recruitment. *Journal of the Fisheries Research Board of Canada, 211*, 559–663.

[8] Rössler, O. E. (1979). Continuous chaos–four prototype equations. In: Bifurcation Theory and Applications in Scientific Disciplines (Vol. 316, pp. 376–392). New York: Annals of the New York Academy of Sciences.

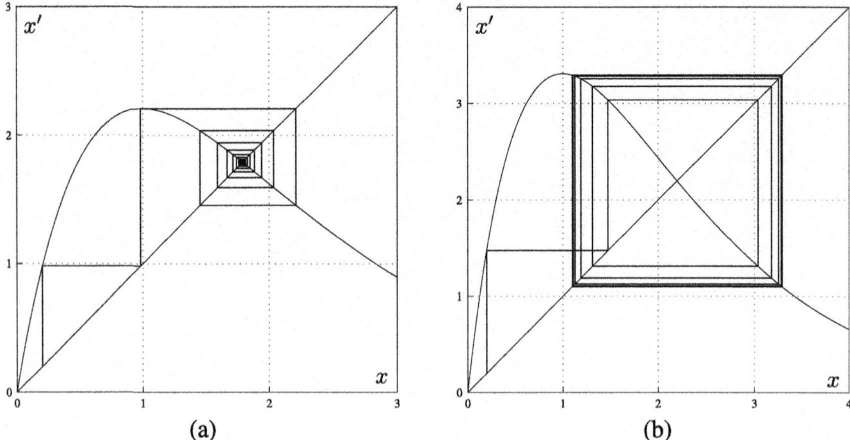

Fig. 1.9 Orbits of Ricker map: (**a**) $\alpha = 6$; (**b**) $\alpha = 9$

Fig. 1.10 An orbit of the Rössler system

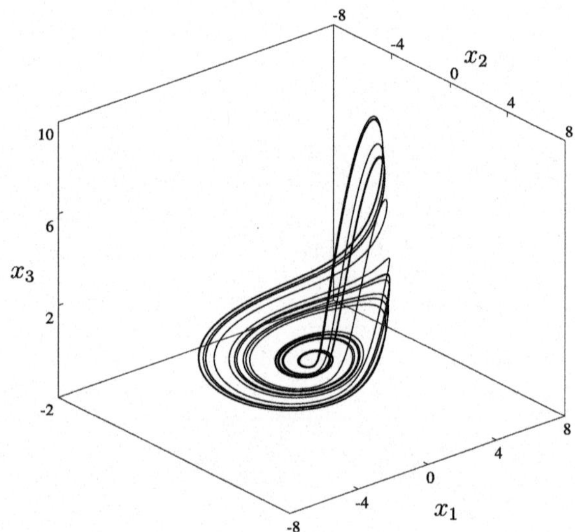

Exercise 1.5 (Orbits and time-series) Consider an invertible dynamical system. Write $y \sim x$ when the point y belongs to the orbit starting at the point x.

(i) Show that "$\sim$" is an equivalence relation, i.e. prove that (1) $x \sim x$ (reflexivity); (2) $y \sim x$ implies $x \sim y$ (symmetry); (3) $x \sim y$ and $y \sim z$ implies $x \sim z$ (transitivity).

Accordingly, we can without any ambiguity speak about "orbits" without specifying a particular starting point. Every point belongs to exactly one orbit. A mathematical way of expressing this fact is to say that orbits *partition* the state space X.

(ii) Assume that x_0 and x_1 belong to the same orbit. Show that the time-series starting at, respectively, x_0 and x_1 are translates of each other.

Exercise 1.6 (Higher-order recursions) Fibonacci's law of replication reads

$$x_{t+1} = x_t + x_{t-1}, \quad t \in \mathbb{N},$$

with two initial values x_0, x_1. Write this as a dynamical system in $\mathbb{R}^2$. What would you do if x_{t+1} depends on x_t and k states $x_{t-1}, \dots, x_{t-k}$ from the past?

Exercise 1.7 (Generator of translation dynamics) Compute the infinitesimal generator of the translation dynamics from Example 1.2. This generator is only defined on a subspace of $BC(\mathbb{R}, \mathbb{R})$ (which one?).

Exercise 1.8 (Periodic and connecting orbits of shift dynamics) (i) Compute the number $N(p)$ of period-p cycles in the symbolic dynamics $\{\mathbb{Z}, \Omega_2, \sigma^k\}$.

(ii) For the symbolic dynamics $\{\mathbb{Z}, \Omega_2, \sigma^k\}$, describe all orbits homoclinic to the zero sequence and all heteroclinic orbits connecting the zero sequence to the sequence of ones.

Exercise 1.9 (Different metrics on spaces of symbols) Show that the metric in Example 1.4 is equivalent to any of the metrics

$$d_q(\omega, \theta) = \sum_{k \in \mathbb{Z}} \frac{|\omega_k - \theta_k|}{q^{|k|}},$$

where $q > 1$ is a constant. (*Hint:* Two metrics $d^{(1)}$ and $d^{(2)}$ on a space X are called equivalent if for any $\omega \in X$ and $\varepsilon > 0$ there is a $\delta > 0$ such that $d^{(1)}(\omega, \theta) \le \delta$ implies $d^{(2)}(\omega, \theta) \le \varepsilon$ and vice versa).

Exercise 1.10 (Topological equivalence of scalar ODEs) Consider two *scalar* autonomous ODEs defined by smooth functions f_1 and f_2, respectively. Give a graphical criterion for the topological equivalence of the dynamical systems generated by f_1 and f_2.

Exercise 1.11 (Topological equivalence of monotone scalar maps) Consider two *continuous monotone* scalar functions f_1 and f_2. Give a graphical criterion for the topological equivalence of the discrete-time dynamical systems generated by iterating f_1 and f_2. *Hint*: Use the method of fundamental domains, see Sect. 2.3.2 in Chap. 2.

Exercise 1.12 (Smooth equivalence of linear ODEs) If two linear systems $\dot{x} = Ax$ and $\dot{y} = By$ are diffeomorphic through a linear change $y = Sx$ of coordinates, what is the relation between the eigenvalues (or more generally the Jordan normal forms) of A and B?

Exercise 1.13 (Topological equivalence of planar linear ODEs) Elaborate Example 1.17 from Sect. 1.3.

(i) Show that two continuous-time dynamical systems in $\mathbb{R}^2$ with linear evolution operators, given by the matrices

$$\varphi^t = e^{-t} \begin{pmatrix} 1 & 0 \\ 0 & 1 \end{pmatrix} \quad \text{and} \quad \psi^t = e^{-t} R(t), \quad R(t) = \begin{pmatrix} \cos t & -\sin t \\ \sin t & \cos t \end{pmatrix},$$

are topologically conjugate in $\mathbb{R}^2$. Compute a conjugating map h and its inverse h^{-1} explicitly. *Hint:* Assume that $h(x)$ is the identity $h(x) = x$ on the circle $\|x\| = 1$ and determine it elsewhere from the relation

$$h(e^{-t}x) = e^{-t}R(t)h(x).$$

(ii) Discuss the differentiability at the origin of the map h thus constructed.

Exercise 1.14 (Topological equivalence of planar linear ODEs revisited) Reconsider Exercise 1.13 by working with generators and show topological conjugacy in the unit disc $U = \{x \in \mathbb{R}^2 : x_1^2 + x_2^2 \leq 1\}$.

(i) By computing the infinitesimal generators, show that the above evolution operators are associated with the following linear systems

$$\begin{cases} \dot{x}_1 = -x_1, \\ \dot{x}_2 = -x_2, \end{cases} \tag{1.13}$$

and

$$\begin{cases} \dot{x}_1 = -x_1 - x_2, \\ \dot{x}_2 = x_1 - x_2. \end{cases} \tag{1.14}$$

(ii) Write Eqs. (1.13) and (1.14) in polar coordinates (ρ, θ),

$$\begin{cases} x_1 = \rho \cos \theta, \\ x_2 = \rho \sin \theta. \end{cases}$$

Verify that the solution of (1.13) in these coordinates is given by

$$\rho(t) = \rho_0 e^{-t},$$
$$\theta(t) = \theta_0,$$

while that of (1.14) is

$$\rho(t) = \rho_0 e^{-t},$$
$$\theta(t) = \theta_0 + t.$$

Here (ρ_0, θ_0) corresponds to an initial point.

(iii) For any point $x \in U$ with polar coordinates (ρ_0, θ_0), $\rho_0 \neq 0$, construct a point $y \in U$ with polar coordinates (ρ_1, θ_1) as follows (see Fig. 1.11). Consider the time τ required to move, along an orbit of system (1.13), from the point on the boundary with polar coordinates $(1, \theta_0)$ to the point x. Then consider an orbit of system (1.14) starting at the boundary point with polar coordinates $(1, \theta_0)$, and let $y = (\rho_1, \theta_1)$ be the point at which this orbit arrives after $\tau(\rho_0)$ units of time. Compute explicitly in polar coordinates the map $x \mapsto y = h(x)$ thus defined.

(iv) Define $h(0, 0) = (0, 0)$ and show that the resulting map $h : U \to U$ is a homeomorphism that maps orbits of (1.13) in U onto orbits of (1.14) in U, meaning that these two systems are topologically conjugate.

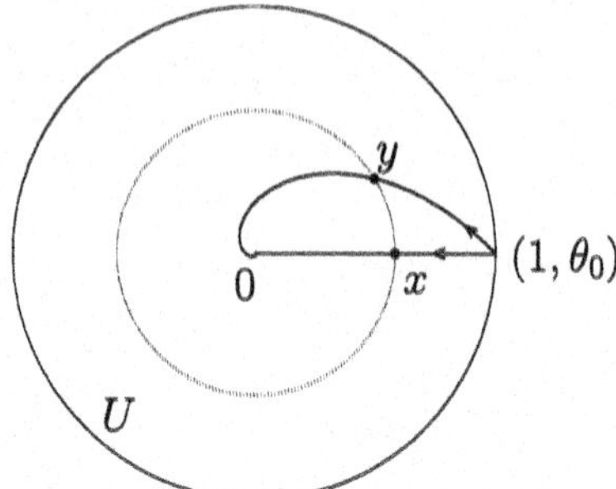

Fig. 1.11 Construction of the homeomorphism in Exercise 1.14: $x \mapsto y = h(x)$

(v) Discuss the differentiability at the origin of map h. *Hint:* Use the original coordinates (x_1, x_2).

(vi) Show how a similar procedure can be used to construct a conjugating homeomorphism h outside U in $\mathbb{R}^2$.

Chapter 2
Linear Maps and ODEs

Abstract The aim of this chapter is to study *linear* dynamical systems and in particular

- to obtain insight into the qualitative features of their phase portrait and how these relate to eigenvalues of the defining matrices;
- to derive estimates that settle the stability issue for the zero steady state;
- to discuss the more subtle issue of topological equivalence (conjugacy) of these systems.

2.1 Linear Maps in $\mathbb{R}^n$

Consider a linear map

$$x \mapsto Ax, \quad x \in \mathbb{R}^n, \tag{2.1}$$

where $A \in \mathbb{R}^{n \times n}$ (i.e. $A = (a_{ij})_{i,j=1,\dots,n}$ is an $n \times n$ real matrix) and Ax is the product of the matrix A and the vector $x = (x_1, x_2, \dots, x_n)$:

$$(Ax)_i = \sum_{j=1}^{n} a_{ij} x_j, \quad i = 1, \dots, n.$$

The map (2.1) will be denoted by A as well. Note however, that if another basis in $\mathbb{R}^n$ is selected, the same map will be represented by a different matrix. We extend (2.1) to complex vectors by the rule

$$A(u + iv) = Au + iAv, \quad u + iv \in \mathbb{C}^n, \quad u, v \in \mathbb{R}^n.$$

Now consider a discrete-time dynamical system $\{\mathbb{N}, \mathbb{R}^n, A^k\}$ or $\{\mathbb{Z}, \mathbb{R}^n, A^k\}$ defined by iteration of the map A (and—if defined—its inverse).

Yu. Kuznetsov et al., *Dynamical Systems Essentials*, Texts in Applied Mathematics 83, https://doi.org/10.1007/978-3-032-04083-1_2

2.1.1 Dynamics in Eigenspaces

Recall that an *eigenvalue* of A is a number $\lambda \in \mathbb{C}$ satisfying

$$Av = \lambda v,$$

for a nonzero vector $v \in \mathbb{C}^n$, which is called a corresponding *eigenvector*. The eigenvalues of a real matrix A can be complex but then occur in complex-conjugate pairs. The set of all eigenvalues of A is denoted by $\sigma(A)$ and is called the *spectrum* of A. It should be known to the reader that $\lambda \in \mathbb{C}$ is an eigenvalue of A if and only if it is a root of the *characteristic polynomial*

$$h(\lambda) = \det(\lambda I_n - A), \tag{2.2}$$

where I_n is the unit $n \times n$ matrix. The polynomial h and hence the *characteristic equation*

$$h(\lambda) = 0 \tag{2.3}$$

are independent of the choice of the basis, so one can speak about the eigenvalues of a linear map A. Note that $h(A) = 0$ (*Hamilton–Cayley Theorem*).

First consider algebraically simple eigenvalues of A, i.e. simple roots of the characteristic polynomial.

(*Simple real eigenvalue*) Suppose that λ is a simple real eigenvalue of A with the eigenvector $v \in \mathbb{R}^n$, i.e. $Av = \lambda v$. The line

$$X_\lambda = \{x \in \mathbb{R}^n : x = sv, \ s \in \mathbb{R}\}$$

is a linear subspace of $\mathbb{R}^n$ called the *eigenspace* of A corresponding to the eigenvalue λ. The eigenspace X_λ is invariant under A, since $x \in X_\lambda$ implies $Ax \in X_\lambda$. If $x = sv$, then

$$Ax = A(sv) = sAv = \lambda sv.$$

This means that on the line X_λ the dynamics is simply the multiplication by the number λ in every time step: if $x = s_0 v$ then $A^k x = s_k v$ where $s_k = \lambda s_{k-1}$, so that

$$s_k = \lambda^k s_0.$$

If $|\lambda| < 1$, $|s_k| \to 0$ as $k \to +\infty$, while $|s_k| \to +\infty$ as $k \to -\infty$ if $s_0 \neq 0$ and $\lambda \neq 0$. In contrast, if $|\lambda| > 1$ and $s_0 \neq 0$, then $|s_k| \to +\infty$ as $k \to +\infty$, while $|s_k| \to 0$ as $k \to -\infty$. Notice that if $\lambda > 0$, then the time-series is monotone, while if $\lambda < 0$, the time-series exhibits oscillation. Also notice that for $\lambda = 0$ the decay of s_k to zero is instantaneous. Finally, note that we did not use the assumption that λ is simple. The assumption is nevertheless made, because for simple eigenvalues the formulated results provide the whole story, while for multiple eigenvalues they capture only part of it, see below.

(*Simple complex-conjugate pair of eigenvalues*) Suppose $\lambda, \bar{\lambda}$ is a pair of simple complex eigenvalues with eigenvectors $v, \bar{v} \in \mathbb{C}^n$, i.e.

$$Av = \lambda v, \quad A\bar{v} = \bar{\lambda}\bar{v}.$$

(Again, the observations below are correct even if λ and $\bar{\lambda}$ are not simple.) Since $\lambda \neq \bar{\lambda}$ the vectors v and $\bar{v}$ are complex-linearly-independent (see Exercise 2.32) and the plane in $\mathbb{R}^n$

$$X_{\lambda,\bar{\lambda}} = \{x \in \mathbb{R}^n : x = zv + \bar{z}\bar{v}, \quad z \in \mathbb{C}\}$$

is a linear subspace of $\mathbb{R}^n$ called the *eigenspace* of A corresponding to the pair of eigenvalues $\lambda, \bar{\lambda}$. It is invariant under A. Indeed, if $x = zv + \bar{z}\bar{v}$ then

$$Ax = A(zv + \bar{z}\bar{v}) = zAv + \bar{z}A\bar{v} = \lambda zv + \bar{\lambda}\bar{z}\bar{v}.$$

This means that A acts as

$$z \mapsto z' = \lambda z,$$

i.e. as a multiplication by $\lambda \in \mathbb{C}$. If $\lambda = \rho e^{i\psi}$ and $z = re^{i\varphi}$, then

$$z' = (\rho r)e^{i(\psi+\varphi)}.$$

The motion in $X_{\lambda,\bar{\lambda}}$ is therefore a superposition of the rotation through the angle $\psi = \arg\lambda$ and the radial expansion ($\rho > 1$) or contraction ($\rho < 1$) with factor $\rho = |\lambda|$.

Alternatively (but equivalently) we can work with real coordinates (a, b) and write

$$X_{\lambda,\bar{\lambda}} = \{x \in \mathbb{R}^n : x = a \text{ Re } v - b \text{ Im } v, \ a, b \in \mathbb{R}\}.$$

The relationship between the coordinates is given by

$$z = \frac{1}{2}(a + ib).$$

Writing $Av = \lambda v$ as

$$A(\text{Re } v + i \text{ Im } v) = (\text{Re } \lambda + i \text{ Im } \lambda)(\text{Re } v + i \text{ Im } v)$$

we obtain

$$A(\text{Re } v) = \text{Re } \lambda \text{ Re } v - \text{Im } \lambda \text{ Im } v,$$
$$A(\text{Im } v) = \text{Im } \lambda \text{ Re } v + \text{Re } \lambda \text{ Im } v.$$

Therefore, in the coordinates (a, b) the map A restricted to $X_{\lambda,\bar{\lambda}}$ is represented by the matrix

$$\begin{pmatrix} \text{Re } \lambda & -\text{Im } \lambda \\ \text{Im } \lambda & \text{Re } \lambda \end{pmatrix} = \rho \begin{pmatrix} \cos\psi & -\sin\psi \\ \sin\psi & \cos\psi \end{pmatrix}.$$

Whenever the characteristic polynomial (2.2) has n distinct (and therefore necessarily simple) roots λ_i, $i = 1, \ldots, n$, we can combine our understanding of the dynamics in each of the invariant subspaces X_{λ_i} (for real λ_i) or $X_{\lambda_i, \bar{\lambda}_i}$ (for complex λ_i) to obtain an overall description of the dynamics. The key point is that $\mathbb{R}^n$ can be decomposed into these subspaces.

Definition 2.1 *Let $X_1, \ldots, X_r$ be linear subspaces of a linear space X. We say that X is the* **direct sum** *of $X_1, \ldots, X_r$, and write*

$$X = X_1 \oplus X_2 \oplus \cdots \oplus X_r,$$

whenever every $x \in X$ can be written in a unique way as

$$x = x_1 + x_2 + \cdots + x_r$$

with $x_i \in X_i$ for $i = 1, 2, \ldots, r$.

Elements in one of the eigenspaces X_{λ_i} or $X_{\lambda_i, \bar{\lambda}_i}$ are linearly independent of elements in another eigenspace. This implies that the eigenspaces corresponding to different eigenvalues intersect only in zero and their sum is a direct sum with the dimension equal to the sum of the dimensions. When the characteristic equation has n distinct roots, the sum of the dimensions equals n, the dimension of the space itself. So an arbitrary $x \in \mathbb{R}^n$ can be uniquely written as a linear combination of elements in X_{λ_i} and $X_{\lambda_i, \bar{\lambda}_i}$. The action of A on x can then be simply reconstructed from that on the invariant subspaces by superposition. Since a generic matrix A has only simple eigenvalues, the above consideration completely describes the dynamics generated by such a map. If there are no eigenvalues with $|\lambda| = 1$, we get exponential contraction/expansion on the linear eigenspaces, combined with rotation in the case of nonreal eigenvalues.

What happens when the characteristic equation (2.3) has multiple roots?

(*Real multiple eigenvalue*) First assume that the multiple (i.e. not simple) eigenvalue λ is *real*. The preceding analysis still applies when there are n linearly-independent eigenvectors. But this is an exceptional case. Generically, there are $m \geq 1$ *Jordan chains* of linearly-independent vectors $v^{(j,k)} \in \mathbb{R}^n$ such that

$$\begin{cases} Av^{(j,1)} &= \lambda v^{(j,1)}, \\ Av^{(j,2)} &= \lambda v^{(j,2)} + v^{(j,1)}, \\ \quad \cdots & \\ Av^{(j,n_j)} &= \lambda v^{(j,n_j)} + v^{(j,n_j-1)}, \end{cases} \tag{2.4}$$

for $j = 1, 2, \ldots, m$. The integer n_j is called the *length* of the Jordan chain. The number of Jordan chains with length l can be computed by the formula

$$N(l,\lambda) = R_{l+1} - 2R_l + R_{l-1} = R_{l+1} - R_l - (R_l - R_{l-1}),$$

where $R_0 = n$ and $R_k = \text{rank}\,(A - \lambda I_n)^k$ for $k \geq 1$. The vectors $v^{(j,1)}$ are the eigenvectors of A corresponding to the eigenvalue λ. The vectors $v^{(j,2)}, v^{(j,3)}, \ldots, v^{(j,n_j)}$ are called *generalized eigenvectors* corresponding to the eigenvalue λ. It follows from (2.4) that these vectors are null-vectors of the matrix $(A - \lambda I_n)^{n_j}$.

We now have an n_j-dimensional subspace of $\mathbb{R}^n$ defined by

$$X_\lambda^{(j)} = \{x \in \mathbb{R}^n : x = \sum_{k=1}^{n_j} c_k^{(j)} v^{(j,k)}, \quad c_k^{(j)} \in \mathbb{R}\} \tag{2.5}$$

that is invariant under (2.1). A coordinate vector $c = (c_1, c_2, \ldots, c_{n_j}) \in \mathbb{R}^{n_j}$ is mapped to

$$c' = Jc,$$

where the $n_j \times n_j$ matrix J is given by the *Jordan block*

$$J = \begin{pmatrix} \lambda & 1 & & 0 \\ & \lambda & 1 & \\ & & & \\ & & \lambda & 1 \\ 0 & & & \lambda \end{pmatrix}. \tag{2.6}$$

Therefore, to understand how orbits of (2.1) behave in $X_\lambda^{(j)}$, we must consider J^k.

The low-dimensional examples (see Exercise 2.38 for the general case)

$$\begin{pmatrix} \lambda & 1 \\ 0 & \lambda \end{pmatrix}^k = \begin{pmatrix} \lambda^k & k\lambda^{k-1} \\ 0 & \lambda^k \end{pmatrix}$$

and

$$\begin{pmatrix} \lambda & 1 & 0 \\ 0 & \lambda & 1 \\ 0 & 0 & \lambda \end{pmatrix}^k = \begin{pmatrix} \lambda^k & k\lambda^{k-1} & \frac{k(k-1)}{2}\lambda^{k-2} \\ 0 & \lambda^k & k\lambda^{k-1} \\ 0 & 0 & \lambda^k \end{pmatrix}$$

clearly indicate that all elements of J^k tend to zero as $k \to +\infty$ when $|\lambda| < 1$, whereas elements of J^k on and above the main diagonal diverge as $k \to +\infty$ when $|\lambda| > 1$. Moreover, in the critical case $|\lambda| = 1$ only higher multiplicity can and, for suitable initial vectors, will lead to divergent orbits.

The direct sum of all linear subspaces $X_\lambda^{(j)}$,

$$X_\lambda = X_\lambda^{(1)} \oplus X_\lambda^{(2)} \oplus \cdots \oplus X_\lambda^{(m)},$$

is an invariant linear subspace of $\mathbb{R}^n$ called the *generalized eigenspace* of A corresponding to the eigenvalue λ. Its dimension is $n_1 + \cdots + n_m$. Note that X_λ is the null-space of $(A - \lambda I_n)^l$ for $l = \max_{1\leq j\leq m} n_j$.

(*Complex multiple eigenvalue*) A similar construction applies when λ is a multiple complex eigenvalue. In this case, $v^{(j,k)}$, $j = 1, 2, \ldots, m$, $k = 1, 2, \ldots, n_j$, are complex vectors and the Jordan chains corresponding to the eigenvalue $\bar{\lambda}$ can be composed of complex-conjugate vectors $\bar{v}^{(j,k)}$. Then

$$X_{\lambda,\bar{\lambda}}^{(j)} = \{x \in \mathbb{R}^n : x = \sum_{k=1}^{n_j} \left(c_k^{(j)} v^{(j,k)} + \bar{c}_k^{(j)} \bar{v}^{(j,k)}\right), \quad c_k^{(j)} \in \mathbb{C}\} \tag{2.7}$$

is a real $2n_j$-dimensional invariant subspace of $\mathbb{R}^n$ parameterized by $c \in \mathbb{C}^{n_j}$. Also in this case, $\|J^k c\| \to 0$ as $k \to +\infty$ when $|\lambda| < 1$, and $\|J^k c\| \to \infty$ as $k \to +\infty$ when $|\lambda| > 1$ and $c \neq 0$. The direct sum of all linear subspaces $X_{\lambda,\bar{\lambda}}^{(j)}$,

$$X_{\lambda,\bar{\lambda}} = X_{\lambda,\bar{\lambda}}^{(1)} \oplus X_{\lambda,\bar{\lambda}}^{(2)} \oplus \cdots \oplus X_{\lambda,\bar{\lambda}}^{(m)},$$

is an invariant linear subspace of $\mathbb{R}^n$ called the *generalized eigenspace* of A corresponding to the eigenvalues $\lambda, \bar{\lambda}$. Its dimension is $2(n_1 + \cdots + n_m)$.

It is also possible to define

$$X_\lambda^{(j)} = \{x \in \mathbb{C}^n : x = \sum_{k=1}^{n_j} c_k^{(j)} v^{(j,k)}, \quad c_k^{(j)} \in \mathbb{C}\}$$

and consider the generalized eigenspace corresponding to λ

$$X_\lambda = X_\lambda^{(1)} \oplus X_\lambda^{(2)} \oplus \cdots \oplus X_\lambda^{(m)}$$

as a complex-linear subspace of $\mathbb{C}^n$. Then $X_{\lambda,\bar{\lambda}} = (X_\lambda \oplus X_{\bar{\lambda}}) \cap \mathbb{R}^n$.

Since $\mathbb{R}^n$ can be decomposed into the direct sum of all generalized eigenspaces X_λ for real eigenvalues λ and $X_{\lambda,\bar{\lambda}}$ for complex-conjugated eigenvalue pairs $\lambda, \bar{\lambda}$, we may summarize the main results of this section as follows.

Theorem 2.2 *If all eigenvalues of A are located strictly inside the unit circle in the complex plane, then for any $x \in \mathbb{R}^n$ we have that*

$$\|A^k x\| \to 0$$

as $k \to +\infty$.

If there is an eigenvalue outside the unit circle, or at least one generalized eigenvector corresponding to some eigenvalue on the unit circle, then there exists $x \in \mathbb{R}^n$ such that

$$\|A^k x\| \to \infty$$

as $k \to +\infty$. □

In the next section we shall prove a somewhat stronger result using more general techniques. Moreover, each statement of the above theorem has a suitable converse, see Exercise 2.42.

Example 2.3 (Dynamics generated by planar linear maps) When $n = 2$, the map (2.1) takes the form

$$\begin{pmatrix} x_1 \\ x_2 \end{pmatrix} \mapsto \begin{pmatrix} a_{11} & a_{12} \\ a_{21} & a_{22} \end{pmatrix} \begin{pmatrix} x_1 \\ x_2 \end{pmatrix}$$

and defines a linear dynamical system $\{\mathbb{N}, \mathbb{R}^2, A^k\}$, where

$$A = \begin{pmatrix} a_{11} & a_{12} \\ a_{21} & a_{22} \end{pmatrix} \tag{2.8}$$

Let $\tau = \text{Tr}(A) = a_{11} + a_{22}$, $\Delta = \det(A) = a_{11}a_{22} - a_{21}a_{12}$. Then the characteristic equation (2.3) can be written as

$$\lambda^2 - \tau\lambda + \Delta = 0$$

and has two (possibly complex) solutions

$$\lambda_{1,2} = \frac{\tau}{2} \pm \sqrt{\frac{\tau^2}{4} - \Delta},$$

which coalesce to a double solution if $\tau^2 - 4\Delta = 0$.

It can easily be shown that both eigenvalues λ_j satisfy $|\lambda| < 1$ if and only if the point (τ, Δ) falls inside the triangle

$$\{(\tau, \Delta) : \Delta < 1,\ \Delta > -\tau - 1,\ \Delta > \tau - 1\}$$

depicted in Fig. 2.1. In this case, in accordance with Theorem 2.2, $\|A^k x\| \to 0$ for all $x \in \mathbb{R}^2$.

The reader is also invited to verify that A has

- an eigenvalue $\lambda_1 = 1$ along the boundary where $\Delta = \tau - 1$;
- an eigenvalue $\lambda_1 = -1$ along the boundary where $\Delta = -\tau - 1$;
- eigenvalues $\lambda_{1,2} = e^{\pm i\theta}$ with $\cos\theta = \frac{1}{2}\tau$ along the boundary where $\Delta = 1$.

◊

2.1.2 *Growth Estimates and the Spectral Radius*

For any norm in $\mathbb{R}^n$ the associated *operator norm* for linear maps is defined as

$$\|A\| = \sup_{x \neq 0} \frac{\|Ax\|}{\|x\|} = \sup_{\|x\|=1} \|Ax\|. \tag{2.9}$$

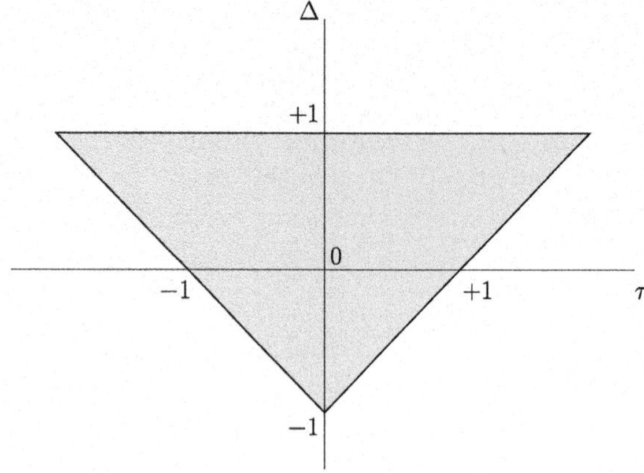

Fig. 2.1 Stability triangle for planar linear maps

The number $\|A\|$ describes a rate of expansion/contraction of the map A that depends on the choice of the norm in $\mathbb{R}^n$. A norm independent measure is given by the spectral radius.

Definition 2.4 *The* ***spectral radius*** *of a linear map A is defined by*

$$r(A) = \sup_{\lambda \in \sigma(A)} |\lambda|.$$

Theorem 2.2 implies that $\|A^k\| \to 0$ as $k \to \infty$ if $r(A) < 1$, while $\|A^k\| \to \infty$ as $k \to \infty$ if $r(A) > 1$. A more precise relation between $\{\|A^k\|\}_{k\in\mathbb{N}}$ and $r(A)$ is specified by the following lemma, which once again shows that the eigenvalues of A yield information about the growth- or decay-rate of the time-series obtained by iterating A.

Lemma 2.5 (Gelfand's formula)

$$r(A) = \lim_{k\to\infty} \|A^k\|^{1/k} = \inf_{k\geq 1} \|A^k\|^{1/k}. \tag{2.10}$$

Proof Define

$$r = \inf_{k\geq 1} \|A^k\|^{1/k}.$$

Then for all $\varepsilon > 0$ there exists $m = m(\varepsilon)$ such that $\|A^m\|^{1/m} \leq r + \varepsilon$. For arbitrary k, put $k = pm + q$ with $0 \leq q \leq m - 1$. Then

$$\|A^k\|^{1/k} \leq \|A^m\|^{p/k}\|A\|^{q/k} \leq (r+\varepsilon)^{pm/k}\|A\|^{q/k}.$$

As $k \to \infty$, necessarily $q/k \to 0$ and $pm/k \to 1$, hence

$$\limsup_{k\to\infty} \|A^k\|^{1/k} \leq r + \varepsilon.$$

By letting $\varepsilon \downarrow 0$ we deduce that

$$\limsup_{k\to\infty} \|A^k\|^{1/k} \le r = \inf_{k\ge 1} \|A^k\|^{1/k},$$

from which we conclude that the limit of $\|A^k\|^{1/k}$ for $k \to \infty$ exists and that this limit equals the infimum.

For $|\lambda| > r$ the series

$$\lambda^{-1} \sum_{k=0}^{\infty} \lambda^{-k} A^k \tag{2.11}$$

converges absolutely. If we multiply (2.11), either from the left or from the right, by $(\lambda I - A)$, we obtain

$$\sum_{k=0}^{\infty} \lambda^{-k} A^k - \sum_{k=0}^{\infty} \lambda^{-(k+1)} A^{k+1} = \lambda^0 A^0 = I.$$

Thus the series (2.11) defines the inverse $(\lambda I - A)^{-1}$ and, accordingly, $\lambda \notin \sigma(A)$. It follows that $r(A) \le r$.

It remains to exclude the possibility that $r(A) < r$. The matrix-valued function

$$f(z) = (I - zA)^{-1}$$

is analytic in $\{z \in \mathbb{C} : |z| < 1/r(A)\}$. Since $f^{(k)}(0) = k!A^k$, *Cauchy's integral formula* yields the identity

$$A^k = \frac{1}{2\pi i} \int_\gamma \frac{f(z)}{z^{k+1}}\, dz,$$

where γ is a circle centered at the origin with radius $(r(A) + \varepsilon)^{-1}$ for arbitrary $\varepsilon > 0$. It follows that

$$\|A^k\| \le (r(A) + \varepsilon)^k \sup_{z\in\gamma} \|f(z)\|$$

and hence

$$r = \lim_{k\to\infty} \|A^k\|^{1/k} \le r(A) + \varepsilon.$$

By letting $\varepsilon \downarrow 0$, we obtain that $r \le r(A)$. □

Remark The matrix-valued function $\lambda \mapsto (\lambda I - A)^{-1}$ is called the *resolvent* of A. The series representation (2.11) is the Taylor series at $\lambda = \infty$. By working with $z = \lambda^{-1}$ we avoid having to deal with the point at infinity. ◊

Theorem 2.6 *For every $\rho > r(A)$ there exists $M \ge 1$ such that*

$$\|A^k\| \le M\rho^k$$

for all $k \ge 1$.

Proof If $\rho > r(A)$ then Lemma 2.5 implies $\|A^k\| \le \rho^k$ for *large* k and consequently

$$M = \sup_{k\ge 0} \rho^{-k} \|A^k\| < \infty.$$

Thus *for all* $k \ge 0$ the inequality

$$\|A^k\| \le M\rho^k$$

holds. For $k = 0$ this gives $M \ge 1$, since $A^0 = I$ and $\|I\| = 1$. □

Using (2.9), we find that

$$\|A^k x\| \le M\rho^k \|x\|, \quad x \in \mathbb{R}^n, k \in \mathbb{N}. \tag{2.12}$$

This estimate establishes the global asymptotic stability of the origin when $r(A) < 1$. Indeed, in this case we can choose ρ such that $r(A) < \rho < 1$. Then for any point $x \in \mathbb{R}^n$ with $\|x\| \le \varepsilon/M$ the estimate $\|A^k x\| \le \rho^k \varepsilon$ holds, which implies the (global) asymptotic stability of $x = 0$. In Exercise 2.45 the reader is asked to explore the geometric manifestation of the fact that, possibly, $M\rho > 1$.

We now show that one can introduce an equivalent norm,[1] tailor-made for the linear map A, such that M reduces to one (the advantage being that the estimate for $k = 1$ carries all information, in the sense that the estimate for arbitrary $k > 1$ is obtained by iteration of the one for $k = 1$).

Theorem 2.7 *Let $\rho > r(A)$. There exists an equivalent norm $\|\cdot\|_1$ on $\mathbb{R}^n$ such that $\|A\|_1 \le \rho$.*

Proof Define $\|\cdot\|_1$ for $x \in \mathbb{R}^n$ by the formula:

$$\|x\|_1 = \sum_{k=0}^{\infty} \rho^{-k} \|A^k x\|.$$

Formula (2.10) implies that this series converges. Indeed, for k sufficiently large and some $q < 1$,

$$\rho^{-1} \|A^k\|^{1/k} \le q$$

and hence

$$\rho^{-k} \|A^k x\| \le \rho^{-k} \|A^k\| \, \|x\| \le \|x\| q^k.$$

Clearly, $\|x\|_1 \ge 0$ for all $x \in \mathbb{R}^n$ and $\|x\|_1 = 0$ if and only if $x = 0$. Likewise the properties $\|\alpha x\|_1 = |\alpha| \|x\|_1$ and $\|x + y\|_1 \le \|x\|_1 + \|y\|_1$ for $x, y \in \mathbb{R}^n$ and $\alpha \in \mathbb{R}$ follow straightaway from the corresponding properties of $\|\cdot\|$. So $\|\cdot\|_1$ is a norm on $\mathbb{R}^n$. Moreover, we have

$$\|x\| \le \|x\|_1 \le \left(\sum_{k=0}^{\infty} \left(\rho^{-1} \|A^k\|^{1/k} \right)^k \right) \|x\|,$$

i.e. $\|\cdot\|_1$ is *equivalent* to $\|\cdot\|$ on $\mathbb{R}^n$. Now, for $x \in \mathbb{R}^n$,

---

[1] Recall that two norms $\|\cdot\|$ and $\|\cdot\|_1$ on a linear space E are called *equivalent*, if there are constants $C_2 \ge C_1 > 0$, such that for all $x \in E$: $C_1\|x\| \le \|x\|_1 \le C_2\|x\|$.

$$\|Ax\|_1 = \sum_{k=0}^{\infty} \rho^{-k}\|A^{k+1}x\| = \rho(\|x\|_1 - \|x\|)$$

so that $\|Ax\|_1 \leq \rho\|x\|_1$ for $x \in \mathbb{R}^n$. □

Note that by induction the estimate $\|A\|_1 \leq \rho$ implies that

$$\|A^k x\|_1 \leq \rho^k \|x\|_1, \quad x \in \mathbb{R}^n,$$

for all $k \geq 0$. An alternative proof of Theorem 2.7 is indicated in Exercise 2.41.

Definition 2.8 *The map A is called a (strict)* ***linear contraction*** *if for some ρ with $0 < \rho < 1$ we have*

$$\|Ax\|_1 \leq \rho\|x\|_1, \quad x \in \mathbb{R}^n.$$

Since a linear contraction obviously has the origin as its asymptotically stable fixed point, we have proved the following criterion, which we have already mentioned before.

Theorem 2.9 *If all eigenvalues of a matrix A are located strictly inside the unit circle, then A defines a linear contraction with respect to an appropriate norm and the origin is globally asymptotically stable.* □

As with Theorem 2.2 a certain converse of the statements above holds (see Exercise 2.42).

Since $\sigma(A^{-1}) = \{\lambda^{-1} : \lambda \in \sigma(A)\}$ whenever $0 \notin \sigma(A)$, we can obtain similar bounds for A^{-1} in terms of $\inf_{\lambda\in\sigma(A)} |\lambda|$.

Theorem 2.10 *Assume that $|\lambda| > \tilde{\rho} > 0$ for all $\lambda \in \sigma(A)$. Then there exists $M \geq 1$ such that*

$$\|A^{-k}\| \leq M\tilde{\rho}^{-k}, \quad k \geq 0.$$

Moreover, there exists an equivalent norm $\|\cdot\|_1$ on $\mathbb{R}^n$ such that

$$\|A^{-1}\|_1 \leq \tilde{\rho}^{-1}$$

and hence $\|A^{-k}\|_1 \leq \tilde{\rho}^{-k}$ for all $k \geq 0$. □

2.1.3 Hyperbolic Linear Maps and Spectral Projectors

In this section we focus on the situation when

$$\sigma(A) = \sigma_s(A) \cup \sigma_u(A),$$

where $\sigma_s(A)$ is located strictly inside the unit circle but does not contain zero, while $\sigma_u(A)$ is located strictly outside the unit circle.

Definition 2.11 *A linear map $x \mapsto Ax$ in $\mathbb{R}^n$ is called* ***hyperbolic*** *if A is invertible and has no eigenvalue λ with $|\lambda| = 1$.*

The direct sum of all generalized eigenspaces of A corresponding to $\sigma_s(A)$ forms an invariant subspace T^s of A, called the *stable eigenspace*, while the direct sum of all generalized eigenspaces of A corresponding to σ_u forms the *unstable eigenspace* T^u of A. Moreover, we have $T^s \oplus T^u = \mathbb{R}^n$, meaning that any $x \in \mathbb{R}^n$ can be uniquely decomposed as

$$x = x_s + x_u$$

with $x_s \in T^s$ and $x_u \in T^u$. Clearly, $\sigma(A|_{T^s}) = \sigma_s(A)$ and $\sigma(A|_{T^u}) = \sigma_u(A)$.

There are explicit methods to find x_s (and hence x_u) for a given x, using the so called *spectral projectors.* We begin by assuming that $\sigma_s(A)$ consists of just one simple eigenvalue $\lambda \in \mathbb{R}$ with $|\lambda| < 1$. Let $v \in \mathbb{R}^n$ be the corresponding eigenvector, so that

$$Av = \lambda v.$$

Let A^{T} denote as usual the transposed matrix corresponding to A. It also has a simple eigenvalue λ with some eigenvector $w \in \mathbb{R}^n$,

$$A^{\mathrm{T}} w = \lambda w.$$

Since the eigenvalue λ is simple, the *Fredholm decomposition* for linear maps (see Lemma 6.22 in Chap. 6) tells us that $\langle w, v\rangle \neq 0$ and therefore w can be selected to satisfy the normalization condition

$$\langle w, v\rangle = 1.$$

Now the map

$$x \mapsto Px = \langle w, x\rangle v$$

assigns to each $x \in \mathbb{R}^n$ a vector $x_s = Px \in T^s \subset \mathbb{R}^n$. The map P is linear and has obviously the property $P^2 = P$. We say that P defines a *projection* of $\mathbb{R}^n$ onto T^s, and call the map P *projector.* Note that

$$T^s = \{x \in \mathbb{R}^n : Px = x\} = \{sv : s \in \mathbb{R}\}.$$

The linear map $Q = I_n - P$ or, in more detail,

$$x \mapsto Qx = x - \langle w, x\rangle v$$

assigns to each $x \in \mathbb{R}^n$ a vector $x_u = Qx \in T^u \subset \mathbb{R}^n$, i.e. defines a projection of $\mathbb{R}^n$ onto T^u, and we have

$$T^u = \{x \in \mathbb{R}^n : Px = 0\} = \{x \in \mathbb{R}^n : \langle w, x\rangle = 0\} = w^{\perp}.$$

Clearly, $Q^2 = Q$ and $PQ = QP = 0$.

If σ_s is composed of n_s eigenvalues $\lambda_1, \lambda_2, \ldots, \lambda_{n_s} \in \mathbb{R}$, all of which are simple and satisfy $|\lambda_i| < 1$, the corresponding projectors can be defined by

$$P = \sum_{i=1}^{n_s} P_i, \quad Q = I_n - P,$$

where

$$P_i x = \langle w^{(i)}, x \rangle v^{(i)}, \tag{2.13}$$

defines the projection of $\mathbb{R}^n$ onto the one-dimensional invariant eigenspace X_{λ_i} corresponding to the eigenvalue λ_i. Here $Av^{(i)} = \lambda_i v^{(i)}$ and $A^{\mathrm{T}} w^{(i)} = \lambda_i w^{(i)}$ for $i = 1, 2, \ldots, n_s$, and

$$\langle w^{(i)}, v^{(j)} \rangle = \begin{cases} 1, & i = j, \\ 0, & i \neq j. \end{cases}$$

The projectors defined by (2.13) have the following properties:

$$P_i^2 = P_i, \quad P_i P_j = 0 \text{ for } j \neq i. \tag{2.14}$$

Notice that the eigenvalues (and the corresponding eigenvectors) can be complex, provided that $\langle \cdot, \cdot \rangle$ is interpreted as the standard scalar product in $\mathbb{C}^n$ and $w^{(i)}$ satisfy $A^{\mathrm{T}} w^{(i)} = \bar{\lambda}_i w^{(i)}$.

There is an alternative method to construct a projector P_i onto the generalized eigenspace corresponding to an eigenvalue $\lambda_i \in \mathbb{C}$, which is applicable for both simple and multiple eigenvalues and does not require computing any eigenvector or generalized eigenvector. In this method, the projection matrix P_i is constructed as a *polynomial function* of the matrix A. We describe this method briefly below. Actually, we explain how to compute P_i corresponding to any eigenvalue λ_i of A.

Suppose that the characteristic polynomial $h(\lambda) = \det(\lambda I_n - A)$ has d distinct roots: $\lambda_1, \lambda_2, \ldots, \lambda_d \in \mathbb{C}$. Then, it can be written as

$$h(\lambda) = \prod_{i=1}^{d} (\lambda - \lambda_i)^{m_i},$$

where m_i are the algebraic multiplicities of the eigenvalues and hence $m_1 + \cdots + m_d = n$. This implies that the following *partial fraction decomposition* holds:

$$\frac{1}{h(\lambda)} = \sum_{i=1}^{d} \frac{g_i(\lambda)}{(\lambda - \lambda_i)^{m_i}}, \tag{2.15}$$

where $g_i(\lambda)$ are polynomials of degree $m_i - 1$ which can be found, for example, by the method of unknown coefficients.

Multiplying both sides of (2.15) by $h(\lambda)$ leads to the equation

$$1 = \sum_{i=1}^{d} p_i(\lambda), \tag{2.16}$$

where

$$p_i(\lambda) = \frac{h(\lambda)g_i(\lambda)}{(\lambda - \lambda_i)^{m_i}} = g_i(\lambda)\prod_{j\neq i}(\lambda - \lambda_j)^{m_j}. \tag{2.17}$$

Each p_i is a polynomial of degree less than n in λ. Moreover, all the polynomials $p_i^2 - p_i$ and $p_i p_j$ with $j \neq i$ are *divisible* by the characteristic polynomial h. Indeed,

$$p_i(\lambda)p_j(\lambda) = \frac{g_i(\lambda)g_j(\lambda)h^2(\lambda)}{(\lambda - \lambda_i)^{m_i}(\lambda - \lambda_j)^{m_j}} = h(\lambda)\left(g_i(\lambda)g_j(\lambda)\prod_{k\neq i, k\neq j}(\lambda - \lambda_k)^{m_k}\right).$$

Moreover,

$$p_i^2(\lambda) - p_i(\lambda) = p_i(\lambda)(p_i(\lambda) - 1) = -p_i(\lambda)\sum_{j\neq i} p_j(\lambda) = -\sum_{j\neq i} p_i(\lambda)p_j(\lambda),$$

with each term in the sum divisible by h due to the previous formula.

Define now

$$P_i = p_i(A), \quad i = 1, 2, \ldots, d. \tag{2.18}$$

Since $h(A) = 0$ by the Hamilton–Cayley Theorem, we see that the linear maps P_i thus constructed satisfy (2.14), i.e. are projectors. Because each P_i is a polynomial function of A, it defines a matrix that commutes with A. This implies that each range $P_i(\mathbb{C}^n)$ is an invariant subspace of A. The formula (2.17) implies that $P_i v = p_i(A)v = 0$ if $(A - \lambda_j I_n)^{m_j} v = 0$ with $j \neq i$, in particular, $P_i v = 0$ if $Av = \lambda_j v$ where $j \neq i$. Therefore, since $\sum_{i=1}^{d} P_i = I_n$, P_i must project $\mathbb{C}^n$ onto the generalized eigenspace of A corresponding to the eigenvalue λ_i:

$$X_{\lambda_i} = P_i(\mathbb{C}^n).$$

If there are d_s distinct eigenvalues $\lambda_1, \lambda_2, \ldots, \lambda_{d_s}$ satisfying $|\lambda_i| < 1$, then the projectors onto T^s and T^u are given by

$$P = \sum_{i=1}^{d_s} P_i \quad \text{and} \quad Q = I_n - P,$$

respectively.

Example 2.12 (Practical computation of spectral projectors) Consider

$$A = \begin{pmatrix} -1 & -3 \\ \frac{3}{2} & \frac{7}{2} \end{pmatrix}.$$

Then

$$\lambda I_2 - A = \begin{pmatrix} \lambda + 1 & 3 \\ -\frac{3}{2} & \lambda - \frac{7}{2} \end{pmatrix}$$

so that the characteristic polynomial is

$$h(\lambda) = \det(\lambda I_2 - A) = \lambda^2 - \frac{5}{2}\lambda + 1 = (\lambda - \frac{1}{2})(\lambda - 2)$$

and the eigenvalues of A are $\lambda_1 = \frac{1}{2}$ and $\lambda_2 = 2$. Hence A is hyperbolic.

The vector

$$v = \begin{pmatrix} 2 \\ -1 \end{pmatrix}$$

is an eigenvector corresponding to the eigenvalue $\lambda_1 = \frac{1}{2}$, while the vector

$$u = \begin{pmatrix} -1 \\ 1 \end{pmatrix}$$

is an eigenvector corresponding to the eigenvalue $\lambda_2 = 2$. The transposed matrix

$$A^{\mathrm{T}} = \begin{pmatrix} -1 & \frac{3}{2} \\ -3 & \frac{7}{2} \end{pmatrix}$$

has

$$w = \begin{pmatrix} 1 \\ 1 \end{pmatrix}$$

as an eigenvector corresponding to its eigenvalue $\lambda_1 = \frac{1}{2}$. Notice that

$$\langle w, v \rangle = 1, \quad \langle w, u \rangle = 0.$$

For any vector

$$x = \begin{pmatrix} x_1 \\ x_2 \end{pmatrix} \in \mathbb{R}^2,$$

we have $\langle w, x \rangle = x_1 + x_2$ and, therefore,

$$Px = \langle w, x \rangle v = \begin{pmatrix} 2x_1 + 2x_2 \\ -x_1 - x_2 \end{pmatrix}.$$

This gives

$$P = \begin{pmatrix} 2 & 2 \\ -1 & -1 \end{pmatrix} \tag{2.19}$$

which is indeed a projection matrix:

$$P^2 = \begin{pmatrix} 2 & 2 \\ -1 & -1 \end{pmatrix} \begin{pmatrix} 2 & 2 \\ -1 & -1 \end{pmatrix} = \begin{pmatrix} 2 & 2 \\ -1 & -1 \end{pmatrix} = P.$$

The map $x \mapsto Px$ projects the space $\mathbb{R}^2$ onto the eigenline T^s spanned by the eigenvector v. The matrix

$$Q = I_2 - P = \begin{pmatrix} -1 & -2 \\ 1 & 2 \end{pmatrix}$$

defines a projection $x \mapsto Qx$ onto the eigenline T^u spanned by the eigenvector u. Clearly, $PQ = QP = 0$.

To find the spectral projectors P and Q using the partial fraction decomposition (2.15), write

$$\frac{1}{h(\lambda)} = \frac{1}{(\lambda - \frac{1}{2})(\lambda - 2)} = \frac{-\frac{2}{3}}{\lambda - \frac{1}{2}} + \frac{\frac{2}{3}}{\lambda - 2}.$$

This gives $1 = p_1(\lambda) + p_2(\lambda)$, where

$$p_1(\lambda) = -\frac{2}{3}(\lambda - 2), \quad p_2(\lambda) = \frac{2}{3}(\lambda - \frac{1}{2}).$$

Therefore

$$P = p_1(A) = -\frac{2}{3}(A - 2\, I_2) = -\frac{2}{3}\begin{pmatrix} -3 & -3 \\ \frac{3}{2} & \frac{3}{2} \end{pmatrix} = \begin{pmatrix} 2 & 2 \\ -1 & -1 \end{pmatrix},$$

which (re-assuringly) is the same projector (2.19) as before. For the projector Q, we have

$$Q = p_2(A) = \frac{2}{3}(A - \frac{1}{2}\, I_2) = \frac{2}{3}\begin{pmatrix} -\frac{3}{2} & -3 \\ \frac{3}{2} & 3 \end{pmatrix} = \begin{pmatrix} -1 & -2 \\ 1 & 2 \end{pmatrix}.$$

Note that $P + Q = I_2$, so that the computation of Q as $p_2(A)$ is redundant: It can be found by merely evaluating $I_2 - P$. $\diamondsuit$

Remark There is yet another alternative formula for the projector P_i, namely:

$$P_i = \frac{1}{2\pi i}\int_{\gamma_i} (\lambda I_n - A)^{-1}\, d\lambda. \tag{2.20}$$

Here the integral is interpreted (for each matrix element) as the contour integral in the complex plane along a circle γ_i around λ_i oriented counter-clockwise and excluding all eigenvalues $\lambda_j \neq \lambda_i$. It can be computed by the method of *residues*, leading to the same results as the above techniques. Notice that the formula (2.20) is valid in case of multiple eigenvalues as well. Moreover, by a suitable choice of contour γ it works for any spectral set, i.e. any subset of $\sigma(A)$. For example, $\gamma = S^1$, where S^1 is the unit circle, gives the projector onto T^s of a hyperbolic map A. The formula (2.20) can also be generalized to an operator A in a Banach space and has many application in the analysis of infinite-dimensional dynamical systems. $\diamondsuit$

Theorem 2.13 *For any hyperbolic map A, there is an equivalent norm $\|\cdot\|_1$ in which $A|_{T^s}$ and $A^{-1}|_{T^u}$ become linear contractions, i.e. for some ρ with $0<\rho<1$ we have*

$$\|Ax\|_1 \le \rho\|x\|_1, \quad x \in T^s, \tag{2.21}$$

and

$$\|A^{-1}x\|_1 \le \rho\|x\|_1, \quad x \in T^u. \tag{2.22}$$

Proof Take some ρ such that $0 \le r(A|_{T^s}) < \rho < 1$. According to Theorem 2.7, there exists an equivalent norm $\|\cdot\|_1$ on T^s such that (2.21) holds.

Next note that $A^{-1}|_{T^u}$ is defined and that $\sigma(A^{-1}|_{T^u})$ is contained strictly inside the unit circle. Therefore, it follows from Theorem 2.10 that there exists an equivalent norm $\|\cdot\|_1$ on T^u such that

$$\|A^{-1}x\|_1 \le \tilde{\rho}\|x\|_1, \quad x \in T^u,$$

for some $\tilde{\rho}<1$. To obtain the estimates (2.21) and (2.22) with the same ρ, we can simply take $\rho := \max\{\rho, \tilde{\rho}\}$.

Finally, extend the norm to all of $\mathbb{R}^n$ by setting

$$\|x\|_1 = \|x_s\|_1 + \|x_u\|_1 \quad \text{for} \quad x = x_s + x_u \in T^s \oplus T^u. \qquad \square$$

Remarks

(1) Note that, by induction,

$$\|A^k x\|_1 \le \rho^k \|x\|_1, \quad x \in T^s,$$

and

$$\|A^{-k} x\|_1 \le \rho^k \|x\|_1, \quad x \in T^u.$$

(2) Note that we did not use in the proof the invertibility of A, i.e. the absence of eigenvalue 0. All that matters for the estimates is that there are no eigenvalues on the unit circle. Indeed, we only used $A^{-1}|_{T^u}$ and we can replace this by $(A|_{T^u})^{-1}$ since $0 \notin \sigma(A|_{T^u})$. $\diamondsuit$

Due to Theorem 2.13 we can give an alternative definition of a hyperbolic linear map, which does not use eigenvalues.

Definition 2.14 *An invertible linear continuous map $A: X \to X$ on a Banach space X is called* ***hyperbolic*** *if $X = T^s \oplus T^u$, where both T^s and T^u are invariant under A, and with respect to an equivalent norm $\|\cdot\|_1$ we have*

$$\|A|_{T^s}\|_1 \le \rho, \quad \|A^{-1}|_{T^u}\|_1 \le \rho,$$

for some $0<\rho<1$.

Remark Not all authors include "invertibility" in the definition of "hyperbolicity". We do this to facilitate the formulation of a topological equivalence result ahead (see Theorem 2.30). For linear maps, the sign of an eigenvalue has no impact on growth or decay of the image length, but it does have an impact on orientation. In particular, the border between orientation preserving and orientation reversing linear contractions corresponds to matrices with a zero eigenvalue. ◇

2.2 Linear Autonomous Systems of ODEs

Consider an autonomous system of real linear ODEs

$$\dot{x} = Ax, \quad x \in \mathbb{R}^n. \tag{2.23}$$

The associated flow is given by $\varphi^t(x) = e^{tA}x$ where

$$e^{tA} := \sum_{k=0}^{\infty} \frac{t^k}{k!} A^k.$$

(Note in particular the group property $e^{(t+s)A} = e^{tA}e^{sA}$.) Now consider the linear continuous-time dynamical system $\{\mathbb{R}, \mathbb{R}^n, \varphi^t\}$.

For the spectra we have

$$\sigma(e^{tA}) = e^{t\sigma(A)} = \{\mu \in \mathbb{C} : \mu = e^{t\lambda}, \ \lambda \in \sigma(A)\},$$

for any $t \in \mathbb{R}$, so in particular zero is never an eigenvalue of e^{tA} (indeed, e^{tA} has inverse e^{-tA}).

Note that $z \mapsto e^{tz}$ maps the imaginary axis onto the unit circle and the left half plane inside the unit circle. Accordingly, the condition $|\lambda| < 1$ translates into the condition Re $\lambda < 0$.

2.2.1 Dynamics in the Eigenspaces

First consider the dynamics in one- and two-dimensional invariant eigenspaces associated to, respectively, a simple real eigenvalue and a simple complex pair of eigenvalues.

(*Simple real eigenvalue*) Let λ be a simple real eigenvalue of A with eigenvector $v \in \mathbb{R}^n$, *i.e.* $Av = \lambda v$. The line

$$X_\lambda = \{x \in \mathbb{R}^n : x = sv, \ s \in \mathbb{R}\}$$

is invariant with respect to the flow, since $x \in X_\lambda$ implies $\dot{x} = Ax \in X_\lambda$. The function $x(t) = s(t)v \in X_\lambda$ satisfies Eq. (2.23) if and only if

$$\dot{s} = \lambda s, \tag{2.24}$$

since

$$\dot{s}v = \dot{x} = Ax = Asv = sAv = \lambda sv.$$

Thus we find that

$$s(t) = s_0 e^{\lambda t}$$

and that $s(t) \to 0$ as $t \to +\infty$ if $\lambda < 0$, while $s(t) \to 0$ as $t \to -\infty$ if $\lambda > 0$.

(*Simple complex-conjugate pair of eigenvalues*) Let $\lambda, \bar{\lambda}$ be simple nonreal eigenvalues with the eigenvectors $v, \bar{v} \in \mathbb{C}^n$, *i.e.*

$$Av = \lambda v, \quad A\bar{v} = \bar{\lambda}\bar{v}.$$

The plane

$$X_{\lambda,\bar{\lambda}} = \{x \in \mathbb{R}^n : x = zv + \bar{z}\bar{v}, \quad z \in \mathbb{C}\} \subset \mathbb{R}^n$$

is invariant with respect to the flow generated by (2.23). Moreover, $x(t) = z(t)v + \bar{z}(t)\bar{v}$ is a solution if and only if

$$\dot{z} = \lambda z.$$

This implies

$$z(t) = z_0 e^{\lambda t}.$$

If $\lambda = \alpha + i\omega$ and $z_0 = r_0 e^{i\varphi_0}$, then

$$z(t) = (r_0 e^{\alpha t}) e^{i(\varphi_0 + \omega t)}.$$

If $\alpha \neq 0$, the motion in $X_{\lambda,\bar{\lambda}}$ is a superposition of rotation (with angular speed ω) and radial divergence (if $\alpha > 0$) or convergence (if $\alpha < 0$).

If $\alpha = \text{Re}\,\lambda = 0$ then the invariant subspace $X_{\lambda,\bar{\lambda}}$ is filled with periodic orbits parametrized by $r_0 > 0$. The corresponding real periodic solutions are given by

$$x(t) = r_0 \left(e^{i(\varphi_0 + \omega t)} v + e^{-i(\varphi_0 + \omega t)} \bar{v} \right),$$

where the phase φ_0 determines the position of $x(0)$ on the closed orbit. Of course, the period of the orbit is $p = \frac{2\pi}{\omega}$.

Since a generic $n \times n$ matrix A has n different simple eigenvalues and therefore n linearly independent eigenvectors, the above consideration completely describes the dynamics of a generic linear system (2.23). If there are no eigenvalues with $\text{Re}\,\lambda = 0$, we get exponential contraction/expansion on the linear eigenspaces, combined with rotations in the case of nonreal eigenvalues.

There is a possibility, however, that the matrix A has multiple eigenvalues.

(*Real multiple eigenvalue*) Assume first that the eigenvalue $\lambda \in \mathbb{R}$. If there is a basis of eigenvectors in the corresponding invariant subspace, then the solutions behave as in the simple real case. Otherwise, there are Jordan chains (2.4) of vectors $v^{(j,k)} \in \mathbb{R}^n$, $j = 1, 2, \ldots, m$, $k = 1, 2, \ldots, n_j$. The n_j-dimensional subspace $X_\lambda^{(j)}$ defined by (2.5) is invariant and $c = (c_1, c_2, \ldots, c_{n_j}) \in \mathbb{R}^{n_j}$ satisfies

$$\dot{c} = Jc,$$

where J is again given by (2.6). Therefore, the i-th solution component is the sum of scalar multiples of $t^k e^{\lambda t}$ with $k \leq n_j - i$ (see Exercise 2.38 for an explicit expression for e^{tJ}).

(*Complex multiple eigenvalue*) A similar construction applies when $\lambda \in \mathbb{C}$ is a multiple eigenvalue. In this case, $v^{(j,k)}$ are complex vectors and the Jordan chains corresponding to the eigenvalue $\bar{\lambda}$ can be composed of complex-conjugate vectors $\bar{v}^{(j,k)}$. Then $X_{\lambda,\bar{\lambda}}$ defined by (2.7) is a real $2n_j$-dimensional invariant subspace parameterized by $c \in \mathbb{C}^{n_j}$. All components of Re $c(t)$ and Im $c(t)$ are the sums of scalar multiples of $t^k e^{\alpha t} \sin(\omega t), t^k e^{\alpha t} \cos(\omega t)$, where $\lambda = \alpha + i\omega$, $k < n_j$.

Thus, the following theorem holds.

Theorem 2.15 *If all eigenvalues* $\lambda_1, \lambda_2, \ldots, \lambda_n$ *of the matrix* A *satisfy* Re $\lambda_i < 0$, *then for any* $x \in \mathbb{R}^n$

$$\|e^{tA}x\| \to 0,$$

as $t \to \infty$.

If at least one eigenvalue of A *satisfies* Re $\lambda > 0$ *or there is at least one generalized eigenvector with corresponding eigenvalue* Re $\lambda = 0$, *then there exists* $x \in \mathbb{R}^n$ *such that*

$$\|e^{tA}x\| \to \infty,$$

as $t \to \infty$. □

Remark Notice that when $\|x(t)\| \to 0$ as $t \to \infty$, the decay need not to be monotone, see Exercise 2.46. Moreover, as in the discrete time case the statements in this theorem can be reversed, see Exercise 2.43. ◇

Example 2.16 (Phase portraits of planar linear autonomous ODEs) When $n = 2$, Eq. (2.23) takes the form

$$\begin{pmatrix} \dot{x}_1 \\ \dot{x}_2 \end{pmatrix} = \begin{pmatrix} a_{11} & a_{12} \\ a_{21} & a_{22} \end{pmatrix} \begin{pmatrix} x_1 \\ x_2 \end{pmatrix}$$

or

$$\begin{cases} \dot{x}_1 = a_{11}x_1 + a_{12}x_2, \\ \dot{x}_2 = a_{21}x_1 + a_{22}x_2. \end{cases} \tag{2.25}$$

Let

$$\tau = \mathrm{Tr}(A) = a_{11} + a_{22}, \;\; \Delta = \det(A) = a_{11}a_{22} - a_{21}a_{12}.$$

Then the characteristic equation (2.3) can be written as

$$\lambda^2 - \tau\lambda + \Delta = 0$$

and has two (possibly complex) solutions

$$\lambda_{1,2} = \frac{\tau}{2} \pm \sqrt{\frac{\tau^2}{4} - \Delta},$$

which coalesce to a double solution if $\tau^2 - 4\Delta = 0$. It follows right away that both eigenvalues λ_j satisfy $\mathrm{Re}\,\lambda < 0$ if and only if

$$\tau < 0 \quad \text{and} \quad \Delta > 0.$$

Under this condition, Theorem 2.15 implies that $\|e^{tA}x\| \to 0$ as $t \to \infty$ for any $x \in \mathbb{R}^2$.

Traditionally, three generic cases (*saddles*, *nodes*, and *foci*) are distinguished, characterized by inequalities in terms of τ and Δ. While saddles are always unstable, nodes and foci can be stable or unstable (see Fig. 2.2, where the stable area is grey). Let us study the mentioned cases in detail.

(*Saddle*: $\Delta < 0$) The matrix A has one positive and one negative eigenvalue: $\lambda_2 < 0 < \lambda_1$. Let $v^{(1)}$ and $v^{(2)}$ be the corresponding eigenvectors:

$$Av^{(j)} = \lambda_j v^{(j)}, \;\; j = 1, 2.$$

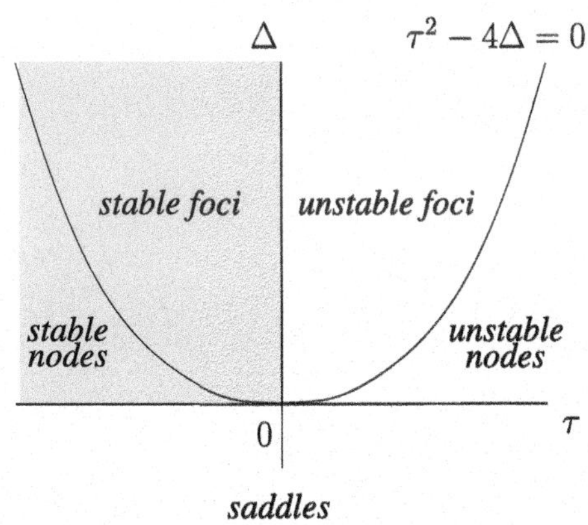

Fig. 2.2 Generic equilibria in planar linear ODEs

Let us use these eigenvectors as new basis vectors in $\mathbb{R}^2$, i.e. write any $x \in \mathbb{R}^2$ in the form

$$x = s_1 v^{(1)} + s_2 v^{(2)},$$

with some $s_{1,2} \in \mathbb{R}$. Notice that $s_{1,2}$ can be considered as new coordinates in $\mathbb{R}^2$. In these new coordinates, the system (2.25) takes a particularly simple form, namely

$$\begin{cases} \dot{s}_1 = \lambda_1 s_1, \\ \dot{s}_2 = \lambda_2 s_2. \end{cases} \tag{2.26}$$

So each of the equations can be solved independently. This gives

$$s_1(t) = s_1(0)e^{\lambda_1 t}, \quad s_2(t) = s_2(0)e^{\lambda_2 t},$$

yielding a phase portrait in the (s_1, s_2)-plane presented in Fig. 2.3a. The phase portrait in the (x_1, x_2)-plane (see Fig. 2.3b) appears then by applying a linear transformation to the phase portrait in the (s_1, s_2)-plane. The equilibrium at the origin is a *saddle*.

(*Node*: $\tau^2 > 4\Delta > 0$) The matrix A has either two positive eigenvalues $\lambda_1 > \lambda_2 > 0$ (when $\tau > 0$) or two negative eigenvalues $\lambda_2 < \lambda_1 < 0$ (when $\tau < 0$). The analysis here is very similar to that in the saddle case. Using the eigenvectors $v^{(j)}$ as new basis vectors in $\mathbb{R}^2$, we obtain once again system (2.26) and solve it as above. When $\tau < 0$, this leads to a phase portrait shown in Fig. 2.4a, since both $s_1(t)$ and $s_2(t)$ tend to zero exponentially fast as $t \to \infty$. Notice, however, that due to the difference in the convergence rates, a generic orbit of (2.26) tends to the origin horizontally. Indeed, in this case, along a solution we have $s_2(t) = C[s_1(t)]^\nu$, where C is determined by the initial conditions and $\nu = \frac{\lambda_2}{\lambda_1} > 1$. The phase portrait in the (x_1, x_2)-plane (see Fig. 2.4b) is obtained by a linear transformation. The equilibrium in the origin is a *stable node*. The case $\tau > 0$ gives an *unstable node*. The corresponding phase portrait can be obtained from that of the stable node by reversing time.

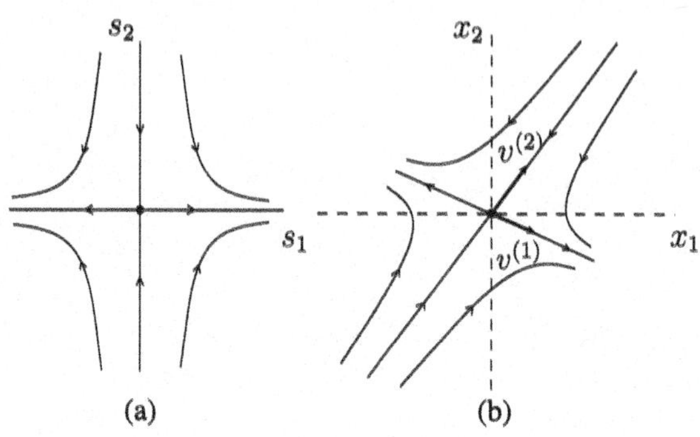

Fig. 2.3 Canonical (**a**) and original (**b**) saddles

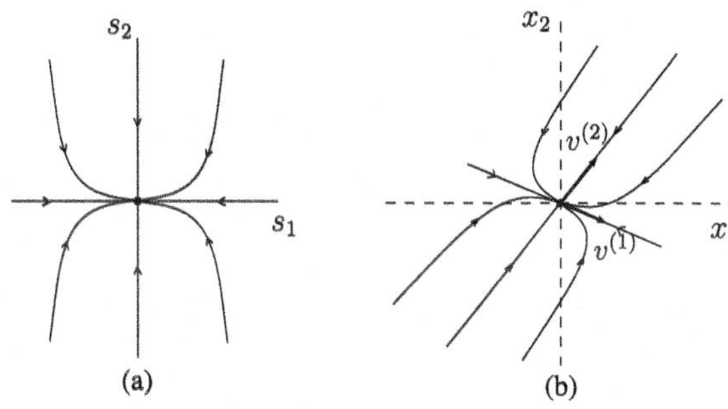

Fig. 2.4 Canonical (**a**) and original (**b**) stable nodes

(*Focus*: $0 < \tau^2 < 4\Delta$) The matrix A has now two complex-conjugate eigenvalues $\lambda_{1,2} = \mu \pm i\omega$ with

$$\mu = \frac{\tau}{2}, \quad \omega = \frac{1}{2}\sqrt{4\Delta - \tau^2}.$$

The corresponding eigenvectors $v^{(1)}$ and $v^{(2)}$ are also complex but can be selected to be complex-conjugate:

$$v^{(1)} = \bar{v}^{(2)} = w \in \mathbb{C}^2.$$

The vector w can be used to introduce a new coordinate $z \in \mathbb{C}$ which specifies a point $x \in \mathbb{R}^2$ according to the formula

$$x = zw + \bar{z}\bar{w}.$$

As we have already seen, this coordinate satisfies

$$\dot{z} = \lambda z$$

with $\lambda = \mu + i\omega$, so that

$$z(t) = r_0 e^{\mu t} e^{i(\varphi_0 + \omega t)},$$

where r_0 and φ_0 are specified by the initial point at $t = 0$. For $s_1 = \text{Re } z$ and $s_2 = \text{Im } z$, we get oscillatory behaviour superimposed with growth ($\mu > 0$) or decay ($\mu < 0$):

$$\begin{cases} s_1(t) = r_0 e^{\mu t} \cos(\varphi_0 + \omega t), \\ s_2(t) = r_0 e^{\mu t} \sin(\varphi_0 + \omega t). \end{cases} \tag{2.27}$$

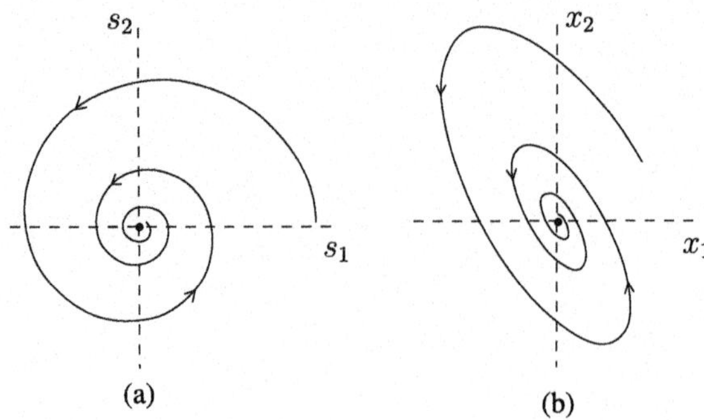

Fig. 2.5 Canonical (**a**) and original (**b**) stable foci

When $\tau < 0$, we have $\mu < 0$ and the phase portrait in the (s_1, s_2)-plane is as shown in Fig. 2.5a. All orbits spiral to the origin. A linear transformation produces the phase portrait in the (x_1, x_2)-plane (see Fig. 2.5b). The equilibrium at the origin is a *stable focus*. When $\tau > 0$, one obtains an *unstable focus*. As we have already noted (see Example 1.17 and Exercises 1.13 and 1.14 in Chap. 1), a node and a focus of the same stability type are topologically equivalent. We return to this issue in Sect. 2.3.1 ahead.

We leave it to the reader to carry out the analysis of all critical cases (defined by at least one equality relation involving τ and/or Δ). Here we only mention the so-called *center*, which occurs when $\tau = 0$ but $\Delta > 0$ (find this set in Fig. 2.2). In this case both eigenvalues of A are purely imaginary: $\lambda_{1,2} = \pm i\omega$ with $\omega = \sqrt{\Delta} > 0$. From (2.27) it follows that all solutions of (2.25) are p_0-periodic, where

$$p_0 = \frac{2\pi}{\omega}$$

and, therefore, all nonequilibrium orbits of (2.25) are closed (see Fig. 2.6). ◊

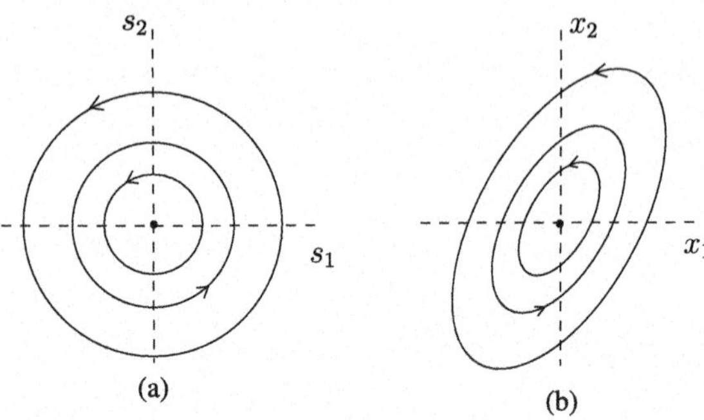

Fig. 2.6 Canonical (**a**) and original (**b**) centers

2.2.2 Growth Estimates and the Spectral Bound

Motivated by the observations so far, we can provide the following definition.

Definition 2.17 *The* **spectral bound** *of a linear map A is defined by*

$$s(A) = \max_{\lambda \in \sigma(A)} \mathrm{Re}(\lambda).$$

Theorem 2.15 implies that $\|e^{tA}\| \to 0$ as $t \to \infty$ if $s(A) < 0$, while $\|e^{tA}\| \to \infty$ as $t \to \infty$ if $s(A) > 0$. The following theorem provides more information on the growth/decay of $\|e^{tA}\|$ for large $t \geq 0$.

Theorem 2.18 *For every $\mu > s(A)$ there exists $M \geq 1$ such that*

$$\|e^{tA}\| \leq Me^{\mu t}$$

for $t \geq 0$.

Proof Define $B = -\mu I + A$. Since $\sigma(e^B) = e^{-\mu + \sigma(A)}$ and $|e^\lambda| = e^{\mathrm{Re}(\lambda)}$, we know that all eigenvalues of e^B are inside the unit circle. Let ρ be such that $r(e^B) < \rho < 1$. By Theorem 2.7 we know that

$$\|x\|_1 = \sum_{k=0}^{\infty} \rho^{-k} \|e^{kB} x\|$$

defines an equivalent norm such that $\|e^B\|_1 \leq \rho$.

We claim that, for some $\tilde{M} \geq 1$, the inequality

$$\|e^{tB}\|_1 \leq \rho^t \tilde{M}$$

holds for all $t \geq 0$. Indeed, write $t = m - \nu$ with $m \in \mathbb{N}$ and $\nu \in (0, 1]$ and define

$$\tilde{M} = \sup_{0 < s \leq 1} \|e^{-sB}\|_1.$$

Then $1 \leq \tilde{M} < \infty$ and, moreover,

$$\|e^{tB}\|_1 = \|e^{(m-\nu)B}\|_1 \leq \|e^{mB}\|_1 \|e^{-\nu B}\|_1 \leq \rho^m \tilde{M} \leq \rho^{m-\nu} \tilde{M} = \rho^t \tilde{M}.$$

It follows that

$$\|e^{tA}\|_1 \leq e^{(\mu + \ln \rho)t} \tilde{M} \leq e^{\mu t} \tilde{M}$$

(here we use that $\ln \rho < 0$). Finally, the equivalence of $\|\cdot\|_1$ and $\|\cdot\|$ yields the estimate for $\|e^{tA}\|$ with an adjusted constant M. □

Like in the discrete time case, we can eliminate the constant M by modifying the norm on $\mathbb{R}^n$ such that it is tailor-made for A.

Theorem 2.19 *Let $\mu > s(A)$. There exists an equivalent norm $\|\cdot\|_2$ on $\mathbb{R}^n$ such that*

$$\|e^{tA}\|_2 \leq e^{\mu t}$$

for all $t \geq 0$.

Proof As in the proof of Theorem 2.18, let

$$\|x\|_1 = \sum_{k=0}^{\infty} \rho^{-k} \|e^{kB} x\|$$

where $B = -\mu I + A$. Now define

$$\|x\|_2 = \int_0^{\infty} \|e^{\tau B} x\|_1 d\tau.$$

The integral converges and

$$\begin{aligned} \|x\|_2 &= \int_0^{\infty} \|e^{\tau B} x\|_1 d\tau \leq \int_0^{\infty} \|e^{\tau B}\|_1 \|x\|_1 d\tau \leq \tilde{M} \left(\int_0^{\infty} \rho^{\tau} d\tau \right) \|x\|_1 \\ &= -\frac{\tilde{M}}{\ln(\rho)} \|x\|_1, \end{aligned}$$

so that

$$\|x\|_2 \leq -\frac{\tilde{M}}{\ln(\rho)} \|x\|_1,$$

where $\tilde{M}$ is defined in the proof of Theorem 2.18. Moreover, since $x = e^{-tB} e^{tB} x$, we have

$$\|x\|_1 \leq \|e^{-tB}\|_1 \|e^{tB} x\|_1 \leq e^{t\|B\|_1} \|e^{tB} x\|_1$$

and hence

$$\|e^{tB} x\|_1 \geq e^{-t\|B\|_1} \|x\|_1.$$

It follows that

$$\|x\|_2 \geq \left(\int_0^{\infty} e^{-\tau \|B\|_1} d\tau \right) \|x\|_1 = \frac{1}{\|B\|_1} \|x\|_1.$$

Combining the two estimates above, we obtain

$$\frac{1}{\|B\|_1} \|x\|_1 \leq \|x\|_2 \leq -\frac{\tilde{M}}{\ln(\rho)} \|x\|_1,$$

meaning that $\|\cdot\|_2$ is equivalent to $\|\cdot\|_1$.

Finally, we compute

$$\begin{aligned}\|e^{tA}x\|_2 &= \int_0^\infty \|e^{\tau B}e^{tA}x\|_1 d\tau = \int_0^\infty \|e^{-\mu\tau}e^{(t+\tau)A}x\|_1 d\tau \\ &= \int_t^\infty \|e^{-\mu(\theta-t)}e^{\theta A}x\|_1 d\theta = e^{\mu t}\int_t^\infty \|e^{-\mu\theta}e^{\theta A}x\|_1 d\theta \\ &\le e^{\mu t}\int_0^\infty \|e^{-\mu\theta}e^{\theta A}x\|_1 d\theta \le e^{\mu t}\int_0^\infty \|e^{\theta B}x\|_1 d\theta = e^{\mu t}\|x\|_2,\end{aligned}$$

from which it follows that $\|e^{tA}\|_2 \le e^{\mu t}$ for all $t \ge 0$. □

Exercise 2.44 indicates an alternative proof of Theorem 2.19.

Definition 2.20 *A norm $\|\cdot\|_2$ in $\mathbb{R}^n$ is called a* ***Lyapunov norm*** *for a linear system $\dot{x} = Ax$ if*

$$\|e^{tA}x\|_2 \le e^{-\alpha t}\|x\|_2, \quad t \ge 0,$$

for some $\alpha > 0$.

Theorem 2.21 *If all eigenvalues of a matrix A have negative real part, then there is an equivalent Lyapunov norm $\|\cdot\|_2$ on $\mathbb{R}^n$ for the linear system $\dot{x} = Ax$ and the origin is a globally asymptotically stable equilibrium of this system.*

Proof When all eigenvalues of A have negative real part, we have by definition $s(A) < 0$. Thus, there exists $\mu < 0$ such that $s(A) < \mu < 0$. Theorem 2.19 implies that $\|\cdot\|_2$ is a Lyapunov norm with $\alpha = -\mu > 0$. The global asymptotical stability of $x = 0$ in this case is obvious. □

As in the discrete time case one can show that stability can be fully characterized in terms of eigenvalue conditions, see Exercise 2.43 for details. By time reversal, we obtain the following theorem.

Theorem 2.22 *Assume that* Re $\lambda > \tilde{\mu}$ *for all $\lambda \in \sigma(A)$. Then there exists $M \ge 1$ such that*

$$\|e^{-tA}\| \le Me^{-\tilde{\mu}t}, \quad t \ge 0,$$

and there exists an equivalent norm $\|\cdot\|_2$ on $\mathbb{R}^n$ such that

$$\|e^{-tA}\|_2 \le e^{-\tilde{\mu}t}, \quad t \ge 0.$$ □

2.2.3 Hyperbolic Linear ODEs

Definition 2.23 *A linear ODE $\dot{x} = Ax$ is called* ***hyperbolic*** *if A has no eigenvalue λ with* Re $\lambda = 0$.

The stable and unstable invariant subspaces $T^{s,u}$ of a linear hyperbolic ODE are defined like those of a linear hyperbolic map. Namely, we can write as before

$$\sigma(A) = \sigma_s(A) \cup \sigma_u(A),$$

where now

$$\sigma_s(A) = \{\lambda \in \sigma(A) : \text{Re}\ \lambda < 0\} \ \text{ and } \ \sigma_u(A) = \{\lambda \in \sigma(A) : \text{Re}\ \lambda > 0\}.$$

Then T^s and T^u will be spanned by all eigenvectors and generalized eigenvectors of A corresponding to $\sigma_s(A)$ and $\sigma_u(A)$, respectively. Spectral projectors can also be computed as before, with γ_i in the formula (2.20) replaced by any contour encircling $\sigma_s(A)$ and excluding $\sigma_u(A)$.

The dynamics on T^s is described by Theorem 2.21. There is no need to study the dynamics on T^u separately, since the results can be obtained from those about T^s by reversal of time, i.e. the substitution $t \mapsto -t$. Thus, we arrive at the following theorem (we leave it to the reader to provide the details of the proof).

Theorem 2.24 *Let $\dot{x} = Ax$ be a hyperbolic system and let $\mu_s, \mu_u > 0$ satisfy*

$$\mu_s < -\max_{\lambda \in \sigma_s(A)} \text{Re}\ \lambda, \quad \mu_u < \min_{\lambda \in \sigma_u(A)} \text{Re}\ \lambda.$$

Then there is a constant $M \geq 0$ such that for all $t \geq 0$

$$\|e^{tA}x\| \leq Me^{-\mu_s t}\|x\|, \quad x \in T^s,$$
$$\|e^{-tA}x\| \leq Me^{-\mu_u t}\|x\|, \quad x \in T^u.$$

Moreover, there is an equivalent norm $\|\cdot\|_2$ on $\mathbb{R}^n$ such that for all $t \geq 0$

$$\|e^{tA}x\|_2 \leq e^{-\mu_s t}\|x\|_2, \quad x \in T^s,$$
$$\|e^{-tA}x\|_2 \leq e^{-\mu_u t}\|x\|_2, \quad x \in T^u. \qquad \square$$

2.3 Topological Classification of Linear Systems

When are two linear systems topologically equivalent? For a hyperbolic system, the answer can be given in terms of the eigenvalues of the defining matrix.

2.3.1 *Topological Equivalence of Hyperbolic Linear ODEs*

We consider first the problem of topological classification of linear ODEs. We advise the reader to make Exercises 1.13 and 1.14 before proceeding.

Theorem 2.25 *The linear system*

$$\dot{x} = Ax, \quad x \in \mathbb{R}^n, \tag{2.28}$$

where all eigenvalues of A have negative real part, is topologically conjugate to the system

$$\dot{y} = -y, \quad y \in \mathbb{R}^n. \tag{2.29}$$

Proof Let $\|\cdot\|_2$ be a Lyapunov norm: $\|e^{tA}x\|_2 \le e^{-\alpha t}\|x\|_2, \quad \alpha > 0$. Any nontrivial orbit of (2.28) crosses

$$\Sigma^{n-1} = \{x \in \mathbb{R}^n : \|x\|_2 = 1\}$$

exactly once. To see this, note that the function $r : t \mapsto \|e^{tA}x\|_2$ is strictly monotone decreasing, since we have for $\tau > 0$

$$r(t+\tau) = \|e^{\tau A}e^{tA}x\|_2 \le e^{-\alpha\tau}r(t).$$

Further, $r(t) \to 0$ as $t \to \infty$ and $r(t) \to \infty$ as $t \to -\infty$ holds due to

$$e^{-\alpha t}\|x\|_2 = e^{-\alpha t}\|e^{-tA}e^{tA}x\|_2 \le \|e^{tA}x\|_2 = r(t), \quad t \le 0.$$

By the Intermediate Value Theorem we obtain for any $x \neq 0$ a unique time $\tau(x)$ (positive or negative) such that $r(-\tau(x)) = 1$, i.e.

$$\tilde{x} = e^{-\tau(x)A}x \in \Sigma^{n-1}. \tag{2.30}$$

Denote by $H : \Sigma^{n-1} \to S^{n-1}$ the map that assigns to the point of intersection of the ray $\{sv : s > 0\}$ with Σ^{n-1} the point of intersection of the same ray with S^{n-1} (see Fig. 2.7), i.e.

$$H(x) = \frac{x}{\|x\|}$$

where the standard (Euclidean) norm $\|\cdot\|$ in $\mathbb{R}^n$ is used.

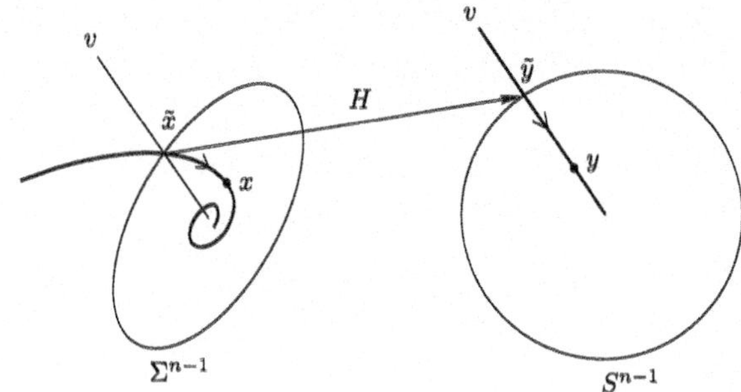

Fig. 2.7 Construction of the homeomorphism $x \mapsto y = h(x)$

Then we map $\tilde{x}$ from (2.30) to $\tilde{y} = H(\tilde{x})$ and define the map $h : \mathbb{R}^n \to \mathbb{R}^n$ as follows

$$y = h(x) = e^{-\tau(x)}\tilde{y} = e^{-\tau(x)}H(e^{-\tau(x)A}x).$$

Let $h(0) = 0$. The map $h : \mathbb{R}^n \to \mathbb{R}^n$ thus constructed has the following properties:

(i) h is a homeomorphism on $\mathbb{R}^n$ (a detailed check of this is left to the reader);

(ii) h maps orbits of (2.23) onto orbits of (2.29), preserving time parametrization (use the fact that $\tau(e^{sA}x) = \tau(x) + s$). □

Suppose that

$$\varphi_1^t : \mathbb{R}^{n_1} \to \mathbb{R}^{n_1}, \quad \psi_1^t : \mathbb{R}^{n_1} \to \mathbb{R}^{n_1},$$

are two topologically conjugate (global) flows, i.e. $\varphi_1^t = h_1^{-1} \circ \psi_1^t \circ h_1$ for all $t \in \mathbb{R}$ with some homeomorphism $h_1 : \mathbb{R}^{n_1} \to \mathbb{R}^{n_1}$. Suppose also that

$$\varphi_2^t : \mathbb{R}^{n_2} \to \mathbb{R}^{n_2}, \quad \psi_2 : \mathbb{R}^{n_2} \to \mathbb{R}^{n_2},$$

are two other topologically conjugate flows, i.e. $\varphi_2 = h_2^{-1} \circ \psi_2 \circ h_2$ for all $t \in \mathbb{R}$ with some homeomorphism $h_2 : \mathbb{R}^{n_2} \to \mathbb{R}^{n_2}$. Then obviously the direct product flows $\varphi^t, \psi^t : \mathbb{R}^{n_1} \times \mathbb{R}^{n_2} \to \mathbb{R}^{n_1} \times \mathbb{R}^{n_2}$ defined for each t coordinate-wise,

$$\varphi^t \begin{pmatrix} x_1 \\ x_2 \end{pmatrix} = \begin{pmatrix} \varphi_1^t(x_1) \\ \varphi_2^t(x_2) \end{pmatrix}, \quad \psi^t \begin{pmatrix} x_1 \\ x_2 \end{pmatrix} = \begin{pmatrix} \psi_1^t(x_1) \\ \psi_2^t(x_2) \end{pmatrix},$$

are topologically conjugate. Indeed,

$$\varphi^t = h^{-1} \circ \psi^t \circ h, \quad t \in \mathbb{R},$$

where $h : \mathbb{R}^{n_1} \times \mathbb{R}^{n_2} \to \mathbb{R}^{n_1} \times \mathbb{R}^{n_2}$ is a homeomorphism defined by

$$h : \begin{pmatrix} x_1 \\ x_2 \end{pmatrix} \mapsto \begin{pmatrix} h_1(x_1) \\ h_2(x_2) \end{pmatrix}.$$

The topology in the product space can be induced, for example, by the norm

$$\|x\| = \|x_1\|_1 + \|x_2\|_2,$$

where $\|\cdot\|_{1,2}$ are norms in $\mathbb{R}^{n_{1,2}}$, respectively.

We combine this procedure with Theorem 2.25 to deal with two hyperbolic linear systems $\dot{x} = Ax$ and $\dot{y} = By$, where A and B have the same number of eigenvalues with negative real part. First transform $\dot{x} = Ax$ by a similarity transformation (which is a smooth conjugacy) into a block-diagonal system

$$\begin{cases} \dot{x}_1 = A_1 x_1, & x_1 \in \mathbb{R}^{n_s}, \\ \dot{x}_2 = A_2 x_2, & x_2 \in \mathbb{R}^{n_u}, \end{cases}$$

where A_1 and $-A_2$ have eigenvalues with negative real part. Then apply Theorem 2.25 to $\dot{x}_1 = A_1 x_1$ and to the time reversed system $\dot{x}_2 = -A_2 x_2$ to establish topological conjugacies of $\dot{x}_1 = A_1 x_1$ to $\dot{x}_1 = -x_1$ and of $\dot{x}_2 = A_2 x_2$ to $\dot{x}_2 = x_2$. Note that the system

$$\begin{cases} \dot{x}_1 = -x_1, & x_1 \in \mathbb{R}^{n_s}, \\ \dot{x}_2 = x_2, & x_2 \in \mathbb{R}^{n_u}, \end{cases}$$

is called the *standard saddle*. Thus, $\dot{x} = Ax$ is topologically conjugate to the standard saddle. In the next step, combine the conjugacies as above and repeat the whole process for the system $\dot{y} = By$. Composing the conjugacies leads to a topological conjugacy of the original systems and thus to the following result.

Theorem 2.26 *Two hyperbolic linear systems in* $\mathbb{R}^n$, $\dot{x} = Ax$ *and* $\dot{y} = By$, *are topologically conjugate if the matrices* A *and* B *have the same number of eigenvalues with negative real part.* □

Remark Actually, the inverse result is also valid, i.e. one can write "if and only if" in Theorem 2.26. However, this fact is only of theoretical interest, since it is rarely known a priori that two linear ODE systems are topologically conjugate. ◇

2.3.2 Topological Equivalence of Hyperbolic Linear Maps

When are two hyperbolic linear maps topologically equivalent? We begin with an example that illustrates that we can use essentially the same idea as used in the continuous time case, but, because now we deal with "jumps" rather than with a continuous orbit, we need to replace the boundary of a ball by a set with non-empty interior.

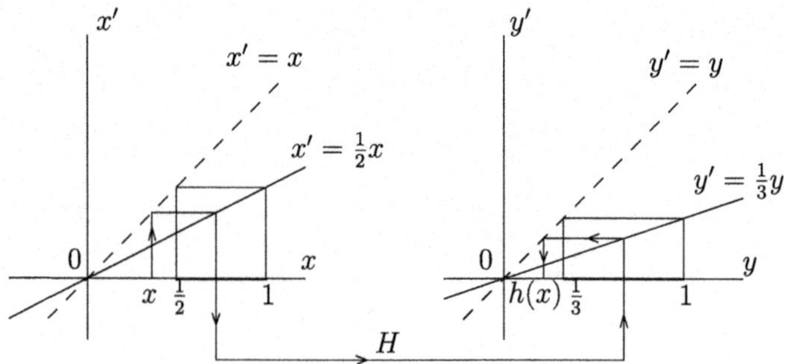

Fig. 2.8 The maps $f : x \mapsto \frac{1}{2}x$ and $g : y \mapsto \frac{1}{3}y$ are topologically conjugate

Example 2.27 (Topologically equivalent scalar linear maps) Consider two real linear scalar maps:

$$f : x \mapsto \frac{1}{2}x \quad \text{and} \quad g : y \mapsto \frac{1}{3}y.$$

These maps are not smoothly conjugate, since they have different eigenvalues, i.e. $\frac{1}{2}$ and $\frac{1}{3}$ (see Exercise 1.12). However, they are topologically conjugate. Indeed, the following construction (see Fig. 2.8) gives a conjugating homeomorphism $h : \mathbb{R} \to \mathbb{R}$. The underlying idea is to use not only the desired conjugacy $h \circ f = g \circ h$ but also the conjugacy $h \circ f^k = g^k \circ h$ of the iterates.

Take any, for example affine, homeomorphism

$$H : \left[\frac{1}{2}, 1\right] \to \left[\frac{1}{3}, 1\right],$$

satisfying $H(\frac{1}{2}) = \frac{1}{3}$, $H(1) = 1$. The intervals are called the *fundamental domains.*

For $x \in (0, \frac{1}{2})$ take

$$h(x) = g^{k(x)}(H(f^{-k(x)}(x))),$$

where $k(x)$ is the minimal[2] positive integer k, such that

$$f^{-k}(x) \in \left[\frac{1}{2}, 1\right].$$

For $x > 1$ take

$$h(x) = g^{-l(x)}(H(f^{l(x)}(x))),$$

[2] The adjective "minimal" is added to be specific but it is, in fact, superfluous. Ambiguity can only arise when $x = 2^{-n}$ but then both choices $k(x) = n - 1$ and $k(x) = n$ yield $h(x) = 3^{-n}$ so, after all, the ambiguity is harmless.

where $l(x)$ is the minimal positive integer l, such that

$$f^l(x) \in \left[\frac{1}{2}, 1\right].$$

Make the analogous construction for $x < 0$. Set $h(0) = 0$. The map h constructed in this manner has the following properties:

(i) $h : \mathbb{R} \to \mathbb{R}$ is a homeomorphism; if H is affine, h is piecewise-affine.

(ii) h sends orbits of f to orbits of g, i.e. $f = h^{-1} \circ g \circ h$.

This means that $f \sim g$.

Now consider another map g defined by

$$g : y \mapsto -\frac{1}{3}y.$$

This map is not topologically equivalent to f. Indeed, suppose that there exists a continuous map $h : \mathbb{R} \to \mathbb{R}$ sending orbits of f to orbits of g. In particular, it should map the orbit of f

$$\dots, 1, \frac{1}{2}, \frac{1}{4}, \dots$$

to the orbit of g

$$\dots, a, -\frac{a}{3}, \frac{a}{9}, \dots,$$

where $a \neq 0$. Therefore, we must have

$$h(1) = a, \; h\left(\frac{1}{2}\right) = -\frac{a}{3}, \; h\left(\frac{1}{4}\right) = \frac{a}{9},$$

showing that h is not monotone. Therefore, h is not invertible. ◊

Lemma 2.28 *Let*

$$A : x \mapsto Ax, \quad x \in \mathbb{R}^n, \tag{2.31}$$

be an invertible linear map which is a contraction with respect to a norm $\|\cdot\|$ *in* $\mathbb{R}^n$. *Then any linear map* B,

$$B : y \mapsto By, \quad y \in \mathbb{R}^n, \tag{2.32}$$

with $\|B - A\|$ *sufficiently small, is topologically conjugate to* A.

Proof We have

$$\|Ax\| \leq \rho\|x\|, \quad x \in \mathbb{R}^n,$$

for some $\rho < 1$. Take the unit sphere

$$S^{n-1} = \{x \in \mathbb{R}^n : \|x\| = 1\}$$

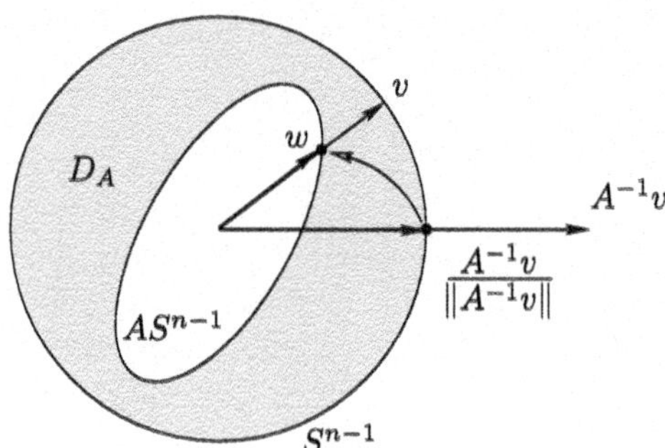

Fig. 2.9 Fundamental domain for the contraction A

and consider its image $AS^{n-1} \subset \text{Int}S^{n-1}$. Define a *fundamental domain* for A:

$$D_A = S^{n-1} \cup (\text{Int } S^{n-1} \setminus \text{Int } AS^{n-1}).$$

This is a closed annulus between S^{n-1} and AS^{n-1} (see Fig. 2.9).

If $\|B - A\| \leq \varepsilon$ with sufficiently small ε, we obtain by the triangle inequality

$$\|B\| = \|B - A + A\| \leq \|B - A\| + \|A\| \leq \varepsilon + \rho =: \rho_1 < 1$$

and therefore

$$\|By\| \leq \rho_1 \|y\|, \quad y \in \mathbb{R}^n.$$

Define the fundamental domain for B by

$$D_B = S^{n-1} \cup (\text{Int } S^{n-1} \setminus \text{Int } BS^{n-1}).$$

Suppose that we can construct a homeomorphism

$$H : D_A \to D_B,$$

which maps the corresponding boundaries of the fundamental domains D_A and D_B into each other and which "agrees" with the maps A and B on them, in the sense that

$$H|_{S^{n-1}} = \text{id}. \tag{2.33}$$

and

$$H|_{AS^{n-1}} = BA^{-1} \tag{2.34}$$

(see Fig. 2.10). Then we can define a map

$$h : \mathbb{R}^n \to \mathbb{R}^n$$

by setting first $h|_{D_A} = H$ and then defining for $x \in \text{Int } AS^{n-1}, x \neq 0$,

$$h(x) = B^{k(x)} H(A^{-k(x)} x),$$

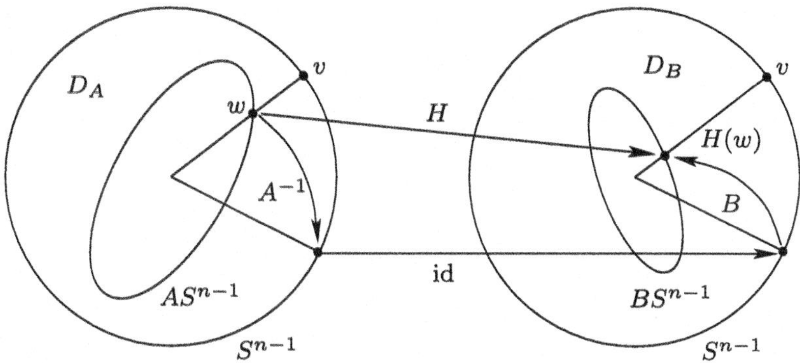

Fig. 2.10 Constraints on the map $H : H|_{S^{n-1}} = \text{id},\ H|_{AS^{n-1}} = BA^{-1}$

where $k(x)$ is the minimal integer $k > 0$, such that $A^{-k}x \in D_A$. Similarly define $h(x)$ for $x \in \text{Ext}\ S^{n-1}$, and finally set $h(0) = 0$. A careful reasoning using (2.33) and (2.34) shows that the resulting map $h : \mathbb{R}^n \to \mathbb{R}^n$ is a homeomorphism with the property

$$Ax = h^{-1}(Bh(x)), \quad x \in \mathbb{R}^n,$$

meaning that B is topologically conjugate to A. Therefore, the lemma will be proved if we construct a homeomorphism $H : D_A \to D_B$, satisfying (2.33) and (2.34). The rest of the proof is devoted to this construction.

First, we notice that any rayzz

$$\{x \in \mathbb{R}^n : x = sv,\ s \in (0, \infty)\}$$

for some given $v \in \mathbb{R}^n$ with $\|v\| = 1$, intersects AS^{n-1} at a unique point

$$w = A\left(\frac{A^{-1}v}{\|A^{-1}v\|}\right) = \frac{v}{\|A^{-1}v\|}$$

(see Fig. 2.10). This allows us to introduce new coordinates (v, t) in D_A. Namely, represent $x \in D_A$ by a point $\Phi(x) = (v, t) \in S^{n-1} \times [0, 1]$ as follows. For each $x \in D_A$, take

$$v = \frac{x}{\|x\|} \in S^{n-1},$$

compute

$$a = \frac{1}{\|A^{-1}v\|} = \frac{\|x\|}{\|A^{-1}x\|} \in \mathbb{R},$$

so that $w = av$, and then set

$$t = \frac{a - \|x\|}{a - 1} \in [0, 1].$$

(In Exercise 2.49 below you are asked to compute the inverse of the map $x \mapsto \Phi(x) = (v,t)$.) Make a similar construction for D_B, representing each point $y \in D_B$ by $\Psi(y) = (\eta, \tau) \in S^{n-1} \times [0,1]$.

The boundary constraints (2.33) and (2.34) motivate us to define two homeomorphisms

$$\delta_{1,0} : S^{n-1} \to S^{n-1},$$

by $\delta_1 = \mathrm{id}$ and

$$\delta_0(v) = \frac{BA^{-1}v}{\|BA^{-1}v\|}.$$

From the fact that B is close to A and, thus, BA^{-1} is close to the unit matrix, we infer that δ_0 is close to the identity map. The map

$$\delta^t(v) = \frac{(1-t)\delta_0(v) + t\delta_1(v)}{\|(1-t)\delta_0(v) + t\delta_1(v)\|},$$

is a homeomorphism on S^{n-1} for all $t \in [0,1]$ (Please, take some time to prove this!). Therefore, the map $\Delta : S^{n-1} \times [0,1] \to S^{n-1} \times [0,1]$ defined by the formula

$$\Delta(v,t) = (\delta^t(v), t)$$

is a homeomorphism. Finally, the homeomorphism Δ induces, via the above constructed maps, a homeomorphism

$$H : D_A \to D_B, \quad H = \Psi^{-1} \circ \Delta \circ \Phi,$$

with the required properties (2.33) and (2.34). Indeed, the property (2.33) is obvious, while the property (2.34) can be verified as follows. Take $v \in S^{n-1}$ and consider the point

$$w = \frac{v}{\|A^{-1}v\|} \in AS^{n-1}$$

that has coordinates $(v, 0) \in S^{n-1} \times [0,1]$ in D_A. Note that any point in AS^{n-1} can be uniquely represented in this way. Then

$$BA^{-1}w = BA^{-1}\left(\frac{v}{\|A^{-1}v\|}\right) = \frac{BA^{-1}v}{\|A^{-1}v\|} \in BS^{n-1}.$$

On the other hand,

$$\delta_0(v) = \frac{BA^{-1}v}{\|BA^{-1}v\|} \in S^{n-1},$$

which implies

$$\eta = \delta^0(v) = \delta_0(v) = \frac{BA^{-1}v}{\|BA^{-1}v\|} \in S^{n-1}.$$

The point with coordinates $(\eta, 0) \in S^{n-1} \times [0,1]$ corresponds to the point

$$H(w) = \frac{\eta}{\|B^{-1}\eta\|} = \frac{BA^{-1}v}{\|A^{-1}v\|} \in BS^{n-1},$$

from which we conclude that $H(w) = BA^{-1}w$ for $w \in AS^{n-1}$. □

Theorem 2.29 *Two linear invertible contractions in $\mathbb{R}^n$ that either both preserve or both reverse the orientation of $\mathbb{R}^n$ are topologically conjugate.*

Proof Let $x \mapsto Ax$ and $x \mapsto Bx$ be two such maps. By the assumption, the determinants of the corresponding matrices A and B are both positive or negative. Therefore, there exists a continuous family of matrices $M(t), t \in [0,1]$, such that $M(0) = A, M(1) = B$, and

$$\det M(t) \neq 0, \quad r(M(t)) < 1$$

for all $t \in [0,1]$ (see Exercise 2.20). By Lemma 2.28, any two linear maps $M(t')$ and $M(t'')$ are topologically conjugate if $|t'' - t'|$ is sufficiently small. Since $[0,1]$ can be covered with finitely many such intervals, A is topologically conjugate to B. □

Similar to the ODE case, suppose that

$$f_1 : \mathbb{R}^{n_1} \to \mathbb{R}^{n_1}, \quad g_1 : \mathbb{R}^{n_1} \to \mathbb{R}^{n_1},$$

are two topologically equivalent maps, i.e. $f_1 = h_1^{-1} \circ g_1 \circ h_1$ for some homeomorphism $h_1 : \mathbb{R}^{n_1} \to \mathbb{R}^{n_1}$. Suppose also that

$$f_2 : \mathbb{R}^{n_2} \to \mathbb{R}^{n_2}, \quad g_2 : \mathbb{R}^{n_2} \to \mathbb{R}^{n_2},$$

are two other topologically equivalent maps, i.e. $f_2 = h_2^{-1} \circ g_2 \circ h_2$ for some homeomorphism $h_2 : \mathbb{R}^{n_2} \to \mathbb{R}^{n_2}$. Then obviously the direct product maps $f, g : \mathbb{R}^{n_1} \times \mathbb{R}^{n_2} \to \mathbb{R}^{n_1} \times \mathbb{R}^{n_2}$ defined coordinate-wise,

$$f : \begin{pmatrix} x_1 \\ x_2 \end{pmatrix} \mapsto \begin{pmatrix} f_1(x_1) \\ f_2(x_2) \end{pmatrix}, \quad g : \begin{pmatrix} x_1 \\ x_2 \end{pmatrix} \mapsto \begin{pmatrix} g_1(x_1) \\ g_2(x_2) \end{pmatrix},$$

are topologically equivalent. Indeed,

$$f = h^{-1} \circ g \circ h,$$

where $h : \mathbb{R}^{n_1} \times \mathbb{R}^{n_2} \to \mathbb{R}^{n_1} \times \mathbb{R}^{n_2}$ is a homeomorphism defined by

$$h : \begin{pmatrix} x_1 \\ x_2 \end{pmatrix} \mapsto \begin{pmatrix} h_1(x_1) \\ h_2(x_2) \end{pmatrix}.$$

Let $n_s(M)$ and $n_u(M)$ denote the number of eigenvalues of the matrix M inside and outside the unit circle, respectively (taking multiplicities into account). Further, let $p_s(M)$ and $p_u(M)$ denote the products of the eigenvalues inside and outside the unit circle, respectively. Recall that topological conjugacy of two (in our case linear) invertible maps is equivalent to the conjugacy of their inverses (with the same conjugating homeomorphism). Applying Theorem 2.29 to the restriction of a hyperbolic linear map to its stable eigenspace, and to the inverse of the restriction to the unstable eigenspace, we obtain the following theorem.

Theorem 2.30 *Two linear hyperbolic maps in* $\mathbb{R}^n$, *say* $x \mapsto Ax$ *and* $y \mapsto By$, *are topologically conjugate if the following properties hold simultaneously:*

$$n_s(A) = n_s(B), \; p_s(A)p_s(B) > 0, \; p_u(A)p_u(B) > 0.$$

□

Remark As in the continuous time case, the inverse for Theorem 2.30 is also valid but again of minor interest. ◊

2.4 References

There are many good books, where linear autonomous ODEs are treated. The standard references here are Arnol'd (1973), Hirsch and Smale (1974). The analysis of linear maps can be done using similar techniques, see, for example Kato (1980, Chap. I). The construction of spectral projectors using the partial fraction decomposition is inspired by unpublished notes of Joop Kolk (Utrecht University).

The fundamental domains are used in Katok and Hasselblatt (1995) to prove the local topological equivalence of a map to its linearization, assuming that the linear part is hyperbolic.

2.5 Exercises

Exercise 2.31 (Exceptional cases of real simple eigenvalue) Describe the dynamics generated by the linear map A on the line X_λ when
(i) $\lambda = 1$;
(ii) $\lambda = -1$;
(iii) $\lambda = 0$.

Exercise 2.32 (Linear algebra refresher) (a) Let λ be a nonreal eigenvalue of A with eigenvector v. Show that:
(i) v and $\bar{v}$ are linearly independent in $\mathbb{C}^n$;
(ii) Re v and Im v are linearly independent in $\mathbb{R}^n$.
(b) Let λ and μ be eigenvalues of A with eigenvectors v and w, respectively. Show that $\lambda \neq \mu$ implies that these vectors are linearly independent.

(c) Let $Av^{(i)} = \lambda_i v^{(i)}$ with $v^{(i)} \neq 0$ and $\lambda_i \neq \lambda_j$ for $i \neq j$, where $i, j = 1, 2, \ldots, n$. Prove that the vectors $v^{(i)}$ are linearly independent.

Exercise 2.33 (Spectral projection) Consider

$$A = \begin{pmatrix} \frac{11}{2} & -3 \\ 6 & -\frac{7}{2} \end{pmatrix}$$

as a generator of a discrete-time dynamical system in $\mathbb{R}^2$. Compute the projectors corresponding to the decomposition $\mathbb{R}^2 = T^s \oplus T^u$.

Exercise 2.34 (Stability conditions for linear maps in $\mathbb{R}^2$ and $\mathbb{R}^3$)

(a) Prove all statements formulated in Example 2.3.
(b) Consider a three-dimensional linear map

$$x \mapsto Ax, \quad x \in \mathbb{R}^3$$

and write the characteristic equation $\det(\lambda I_3 - A) = 0$ in the form

$$\lambda^3 + a_0\lambda^2 + a_1\lambda + a_2 = 0.$$

Show that all its roots lie strictly inside the unit circle if and only if the following inequalities hold simultaneously:

$$1 + a_0 + a_1 + a_2 > 0, \quad 1 - a_0 + a_1 - a_2 > 0, \quad |a_2| < 1, \quad \text{and} \quad 1 - a_2^2 > a_1 - a_0 a_2.$$

Also verify that $\lambda_1 = 1$ is a root if $1 + a_0 + a_1 + a_2 = 0$, that $\lambda_1 = -1$ is a root if $1 - a_0 + a_1 - a_2 = 0$, and that there are two roots $\lambda_{1,2}$ with $\lambda_1\lambda_2 = 1$ if

$$1 - a_2^2 = a_1 - a_0 a_2.$$

Show that in the latter case $\lambda_{1,2} = e^{\pm i\theta}$ with $0 < \theta < \pi$ where $\cos\theta = \frac{1}{2}(a_2 - a_0)$, provided that $|a_2 - a_0| < 2$.

Exercise 2.35 (Operator norm of matrices) Let $\|x\|$ be the standard Euclidian norm of $x \in \mathbb{R}^n$. Consider a matrix $A \in \mathbb{R}^{n \times n}$ and define its *operator norm* $\|A\|$ by the formula (2.9) with Ax understood as the matrix-vector product.

(a) Prove that $\|A\| = \sqrt{\mu_{\max}}$, where $\mu_{\max}$ is the largest eigenvalue of the symmetric matrix $A^{\mathrm{T}}A$. (*Hint*: $\|Ax\|^2 = \langle Ax, Ax\rangle = \langle A^{\mathrm{T}}Ax, x\rangle$, where $\langle x, y\rangle = x^{\mathrm{T}}y$, the standard scalar product of $x, y \in \mathbb{R}^n$. Recall that a symmetric matrix has an orthonormal basis consisting of eigenvectors.)

(b) Using (a), show that for a (2×2)-matrix

$$A = \begin{pmatrix} a & b \\ c & d \end{pmatrix}, \quad a, b, c, d \in \mathbb{R},$$

the operator norm satisfies

$$\|A\|^2 = \frac{1}{2}\left[a^2 + b^2 + c^2 + d^2 + \sqrt{[(a-d)^2 + (b+c)^2][(a+d)^2 + (b-c)^2]}\right].$$

(c) Apply the formula derived in (b) to prove that

$$\|A^{-1}\| = \frac{1}{|\det A|}\|A\|$$

for any nonsingular 2×2 real matrix A. Is this result valid for $n > 2$?

Exercise 2.36 (Fibonacci numbers) Define a sequence of integers by the recurrence relation

$$a_k = a_{k-1} + a_{k-2}, \quad k \geq 2,$$

with initial condition $a_0 = a_1 = 1$. Find an explicit expression for a_k.

Hint: Write

$$\begin{pmatrix} a_k \\ a_{k+1} \end{pmatrix} = A \begin{pmatrix} a_{k-1} \\ a_k \end{pmatrix},$$

where

$$A = \begin{pmatrix} 0 & 1 \\ 1 & 1 \end{pmatrix},$$

and consider a discrete-time dynamical system generated by the linear map $x \mapsto Ax$ in $\mathbb{R}^2$ (cf. Exercise 1.6 in Chap. 1). Find eigenvalues λ and μ and corresponding eigenvectors v and w of A and compute the associated projection operator(s).

Answer:

$$a_k = \frac{1}{2^{k+1}\sqrt{5}}\left[(1+\sqrt{5})^{k+1} - (1-\sqrt{5})^{k+1}\right].$$

Exercise 2.37 (Quantum oscillations) In Quantum Mechanics, a system with two observable states is characterized by a vector

$$\psi = \begin{pmatrix} a_1 \\ a_2 \end{pmatrix} \in \mathbb{C}^2,$$

where a_k are complex numbers called *amplitudes*, satisfying the condition

$$|a_1|^2 + |a_2|^2 = 1.$$

The probability of finding the system in the kth state is equal to $p_k = |a_k|^2, k = 1, 2$.

The behaviour of the system between observations is governed by the *Heisenberg equation*:

$$i\hbar\frac{d\psi}{dt} = H\psi,$$

where the real symmetric matrix

$$H = \begin{pmatrix} E_0 & -A \\ -A & E_0 \end{pmatrix}, \quad E_0, A > 0,$$

is the Hamiltonian matrix of the system and $\hbar$ is Planck's constant divided by 2π.

(i) Write the Heisenberg equation as a system of two linear complex differential equations for the amplitudes.

(ii) Integrate the obtained system and show that the probability p_k of finding the system in the kth state oscillates periodically in time.

(iii) How does $p_1 + p_2$ behave?

Exercise 2.38 (Jordan blocks) Write the Jordan block from (2.6) as $J = \lambda I_{n_j} + N$ with the matrix

$$N = \begin{pmatrix} 0 & 1 & & & 0 \\ & 0 & 1 & & \\ & & & & \\ & & & 0 & 1 \\ 0 & & & & 0 \end{pmatrix} \in \mathbb{R}^{n_j \times n_j}.$$

Compute N^k for $k = 0, 1, 2, \dots, n_j - 1$, and verify that $N^{n_j - 1} = 0$ (i.e. N is nilpotent). Using the fact that I and N commute, calculate explicitly

$$\exp(tJ) = e^{\lambda t} \exp(tN) = e^{\lambda t} \sum_{k=0}^{\infty} \frac{t^k}{k!} N^k.$$

Exercise 2.39 (Jordan decomposition) It is known[3] that each $n \times n$ complex matrix A can be uniquely written as

$$A = S + N,$$

where S is diagonal in some basis (*semisimple*) and $N^m = 0$ for some $m < n$ (*nilpotent*), and such that

$$SN = NS$$

(Jordan decomposition).

Let P_i be the spectral projectors onto the generalized eigenspaces corresponding to the distinct eigenvalues $\lambda_1, \lambda_2, \cdots, \lambda_d$ of A.

(i) Prove that $S = \sum_{i=1}^{d} \lambda_i P_i$ is semisimple.

(ii) Define $N = A - S$ and prove that it is nilpotent.

(iii) Using the definition of $\exp(tJ)$ from Exercise 2.38, prove that

$$\exp(tS) = \sum_{i=1}^{d} e^{\lambda_i t} P_i, \quad \text{and} \quad \exp(tN) = \sum_{j=1}^{m-1} \frac{t^j}{j!} N^j .$$

(iv) Show that

$$\exp(tA) = \sum_{i=1}^{d} e^{\lambda_i t} \left(\sum_{j=1}^{m-1} \frac{t^j}{j!} N^j \right) P_i.$$

(v) Combine the formula from step (iv) with the partial fraction decomposition method to compute P_i and obtain an efficient algorithm for finding flows of linear autonomous systems.

Exercise 2.40 (Gelfand's formula revisited) (i) Use the definition of the eigenvectors of a matrix A to prove that $r(A) \leq \|A^k\|^{1/k}$ for all $k \geq 1$.

(ii) Assume that $r(A) > 0$ and introduce matrix $\hat{A} = (r(A) + \varepsilon)^{-1} A$, where $\varepsilon > 0$ is sufficiently small. Prove that $\|\hat{A}^k\| \to 0$ as $k \to \infty$. (*Hint*: $r(\hat{A}) < 1$.)

(iii) Use step (ii) to show that there exists $N \geq 1$ such that $\|A^k\|^{1/k} < r(A) + \varepsilon$ for all $k \geq N$.

(iv) Combine the results from (i) and (iii) to prove that $\lim_{k \to \infty} \|A^k\|^{1/k} = r(A)$.

Exercise 2.41 (Another proof of Theorem 2.7) (i) Consider the maximum norm $\|x\|_\infty = \max_{i=1}^{n} |x_i|$ in $\mathbb{R}^n$ resp. $\mathbb{C}^n$. Show that the associated operator norm $\|A\|_\infty$ of a real resp. complex matrix A is given by

---

[3] See, for example, Sect. 6.2 in Hirsch, M. W., & Smale, S. (1974). *Differential Equations, Dynamical Systems, and Linear Algebra*. New York-London: Academic.

$$\|A\|_\infty = \max_{i=1,\dots,n} \sum_{j=1}^{n} |a_{ij}|.$$

This matrix norm is called the *row sum norm*.

(ii) Let $J = \lambda I + N$ be an $n_j \times n_j$ Jordan block as in Exercise 2.38. For $\varepsilon > 0$ let $D_\varepsilon = \operatorname{diag}(1, \varepsilon, \dots, \varepsilon^{n_j - 1})$. Show that

$$\|D_\varepsilon^{-1} J D_\varepsilon\|_\infty = |\lambda| + \varepsilon.$$

(iii) Consider the Jordan normal form of a real resp. complex $n \times n$-matrix A

$$S^{-1}AS = J, \quad J = \begin{pmatrix} J_1 & \cdots & 0 \\ \vdots & \ddots & \vdots \\ 0 & \cdots & J_d \end{pmatrix}, \quad J_j = \lambda_j I_{n_j} + N_j$$

Find an $n \times n$ diagonal matrix $D = D(\varepsilon)$ such that $\|D^{-1}JD\|_\infty = r(A) + \varepsilon$ and show that the operator norm associated to $\|x\|_\varepsilon = \|D^{-1}S^{-1}x\|_\infty$ satisfies $\|A\|_\varepsilon \le r(A) + \varepsilon$.

Exercise 2.42 (Stability and eigenvalues for maps) (i) Extend Theorem 2.9 by showing that asymptotic stability of the origin implies that all eigenvalues of A lie strictly inside the unit circle.

(ii) Prove that for a map $x \mapsto Ax$ the following are equivalent

(a) The origin is stable.

(b) Each eigenvalue either lies strictly inside the unit circle or lies on the unit circle but has no generalized eigenvector.

(c) There exists a norm $\|\cdot\|_1$ in $\mathbb{R}^n$ such that $\|A\|_1 \le 1$.

Hint: (a) $\Rightarrow$ (b) follows from Theorem 2.2 and for (b)$\Rightarrow$ (c) prove the boundedness of $A^k (k \ge 0)$ and define $\|x\|_1 = \sup_{k \ge 0} \|A^k x\|$.

Exercise 2.43 (Stability and eigenvalues for flows) Make an analog of Exercise 2.42 for ODE's $\dot{x} = Ax$.

(i) Asymptotic stability of the origin implies that all eigenvalues of A have negative real part.

(ii) Lyapunov stability holds if and only if all eigenvalues either have negative real part or have zero real part but no generalized eigenvectors. Another equivalent condition says that there exists an equivalent norm $\|\cdot\|_1$ in $\mathbb{R}^n$ such that $\|e^{tA}\|_1 \le 1$ for all $t \ge 0$.

Hint: For the construction of a proper norm use $\|x\|_1 = \sup_{t \ge 0} \|e^{tA}x\|$.

Exercise 2.44 (Another proof of Theorem 2.19) Use Theorem 2.18 to show that

$$\|x\|_3 = \sup_{t \ge 0} e^{-\mu t} \|e^{tA}x\|$$

defines a norm which is equivalent to $\|\cdot\|$ and which satisfies for $t \ge 0$

$$\|e^{tA}\|_3 \le e^{\mu t}.$$

Exercise 2.45 (Asymptotic stability and contraction) Consider two maps in $\mathbb{R}^2$ generated by the matrices

$$A = \begin{pmatrix} \frac{1}{2} & 0 \\ 0 & \frac{1}{4} \end{pmatrix} \quad \text{and} \quad B = \begin{pmatrix} \frac{1}{2} & -\frac{1}{2} \\ 1 & \frac{1}{4} \end{pmatrix}.$$

(i) Show that the origin is an asymptotically stable fixed points for both maps.

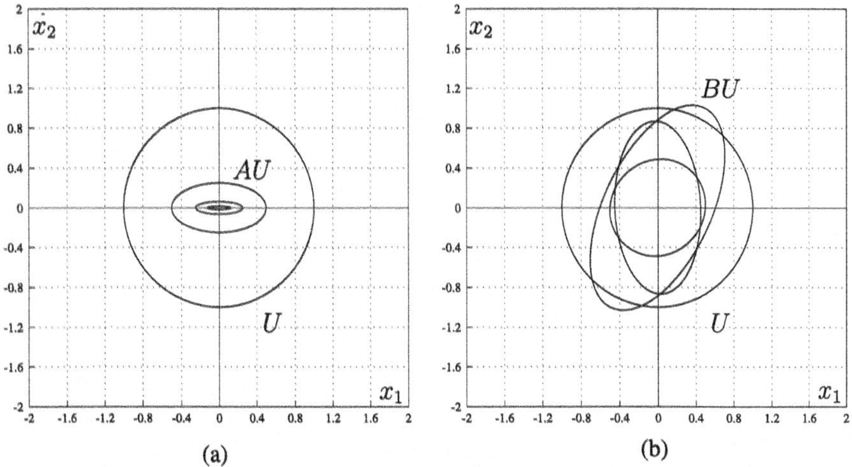

Fig. 2.11 Asymptotically stable fixed points: (**a**) a linear contraction; (**b**) not a contraction

(ii) Which of these maps is a linear contraction with respect to the standard norm $\|x\| = \sqrt{x_1^2 + x_2^2}$ in $\mathbb{R}^2$?

Hint: Consider the images AU and BU of the unit disk

$$U = \{x \in \mathbb{R}^2 : x_1^2 + x_2^2 \leq 1\}$$

under the maps A and B, respectively. Show that AU is located strictly inside U, while BU does not have this property (see Fig. 2.11).

(iii) Construct several first elements of the sequences of the images

$$U, AU, A^2U, A^3U, \ldots$$

and

$$U, BU, B^2U, B^3U, \ldots$$

and convince yourself that both sequences contract towards the origin.

Exercise 2.46 (Asymptotic stability and monotonicity) Consider a linear ODE

$$\dot{x} = Ax, \quad x \in \mathbb{R}^2, \tag{2.35}$$

defined by the matrix

$$A = \begin{pmatrix} -\frac{2}{5} & 2 \\ -\frac{1}{2} & -\frac{2}{5} \end{pmatrix}.$$

(i) Prove that $x = 0$ is an asymptotically stable equilibrium of (2.35). What kind of equilibrium is it?

(ii) Prove that the orbit starting at $x_0 = (1, 0)$ crosses the unit circle $\|x\|^2 = x_1^2 + x_2^2 = 1$ several times (see Fig. 2.12a).

(iii) Define $\|x\|_2^2 = x_1^2 + 4x_2^2$. Prove that $\|\cdot\|_2$ is a norm in $\mathbb{R}^2$ that is equivalent to $\|\cdot\|$.

(iv) Prove that any "circle" $\|x\|_2 = R_0 > 0$ is crossed by any orbit of the system only once (see Fig. 2.12b).

Hint: Show that

$$\frac{d}{dt}\|x(t)\|_2^2 = -\frac{4}{5}\|x(t)\|_2^2$$

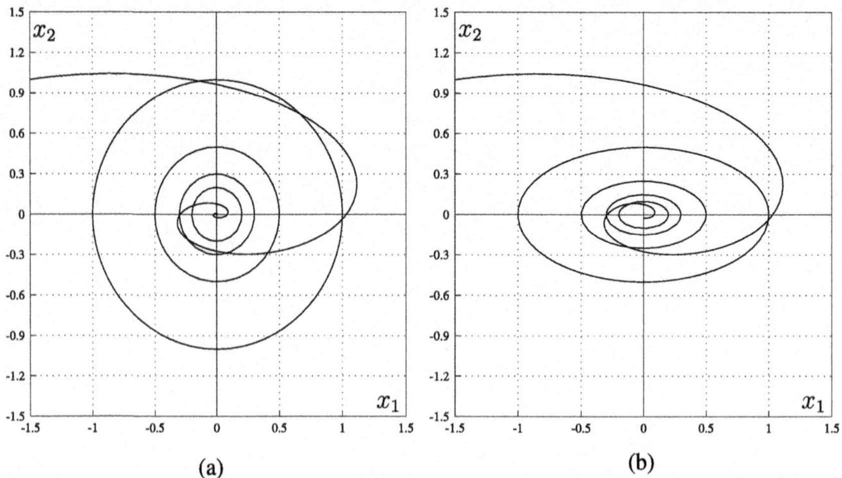

Fig. 2.12 Nonmonotone (**a**) and monotone (**b**) convergence

along any nonequilibrium solution of (2.35). Find $R(t) = \|x(t)\|_2$ by integrating this differential equation and demonstrate that it converges to zero monotonously. Is $\|\cdot\|_2$ a Lyapunov norm for (2.35)?

Exercise 2.47 (Orientation and phase portraits) In the two-dimensional focus case derive a criterion for the matrix coefficients that characterizes whether orbits are rotating clockwise or counterclockwise. Compare with the phase portraits from the previous exercise.

Exercise 2.48 (Explicit conjugating homeomorphism) Verify that the map $h : \mathbb{R} \to \mathbb{R}$ defined by the formula

$$h(x) = \begin{cases} x^\nu, \; x \geq 0, \\ -|x|^\nu, \; x < 0, \end{cases}$$

defines a homeomorphism conjugating the maps f and g from Example 2.27 for a suitable ν (which one ?).

Exercise 2.49 (Inverse map in Lemma 2.28**)** Compute the inverse of the map $x \mapsto (v, t)$ defined in the proof of Lemma 2.28.

Answer:

$$(v, t) \mapsto \left(t \left(1 - \frac{1}{\|A^{-1}v\|} \right) + \frac{1}{\|A^{-1}v\|} \right) v.$$

Exercise 2.50 (Connectedness of contractions) Show that the set of contracting and positively oriented matrices $M = \{A : \det(A) > 0, r(A) < 1\}$ is connected, i.e. for any two $A, B \in M$ there is a continuous path $M(t) \in M, 0 \leq t \leq 1$ such that $M(0) = A$ and $M(1) = B$.

Hint: It is sufficient to connect any A to the special matrix $B = \frac{1}{2}I \in M$. If A has no negative eigenvalues, then take the line $tB + (1-t)A$. If A has negative eigenvalues, they must occur in pairs. Isolate the invariant subspaces belonging to them and move the eigenvalues to the positive real axis by a suitable rotation.

Chapter 3
Local Behaviour of Nonlinear Systems

Abstract In Chap. 2 we explained how, for *linear* dynamical systems, the qualitative as well as the quantitative behaviour can be described in terms of eigenvalues and (generalized) eigenvectors. Thus, there is a systematic and feasible method for deducing the behaviour of orbits from the matrix that generates the dynamical system. For *nonlinear* dynamical systems, no such method exists in general. However, to characterize *local* dynamics near simple orbits (in particular, steady states and cycles), one can, as will be shown in detail in this chapter, rely generically on the information provided by a linear system that gives the best possible local approximation of the nonlinear system. So this chapter is all about *linearization* and conclusions that can be drawn concerning nonlinear systems from an analysis of linear systems obtained via linearization.

3.1 Principle of Linearized Stability

The first problem we address is: How to ascertain the stability character of a steady state of a map or ODE? The answer will be: Compute the Jacobian matrix at the steady state and then determine its eigenvalues. Whenever the test for stability or instability for the linear system defined by this matrix gives a clear answer, this answer also applies to the nonlinear system. Only in critical cases, i.e. at the verge of instability (when the linear system is not hyperbolic), one cannot draw a definite conclusion from just the linearization.

3.1.1 Linearized Stability of Fixed Points of Maps

Example 3.1 (Stable fixed point) We begin by considering a simple nonlinear map

$$g : x \mapsto \tfrac{1}{2}x + bx^2, \quad x \in \mathbb{R},$$

Yu. Kuznetsov et al., *Dynamical Systems Essentials*, Texts in Applied Mathematics 83, https://doi.org/10.1007/978-3-032-04083-1_3

where $b \neq 0$. Its linear part obviously has a stable fixed point $x = 0$, which is also a fixed point of the nonlinear map. For all sufficiently small $x > 0$,

$$0 < \tfrac{1}{2}x + bx^2 < x,$$

while for all sufficiently small $x < 0$,

$$x < \tfrac{1}{2}x + bx^2 < 0.$$

Thus, any sufficiently small interval containing the origin is mapped strictly into itself by g. Therefore $x = 0$ is stable. Moreover, since orbits starting in this interval are monotone and bounded, they converge. Since the limit is necessarily a fixed point, it must be zero, proving that $x = 0$ is (locally) asymptotically stable. ◊

The following theorem shows in general that adding higher-order terms does not affect the stability of a hyperbolic fixed point.

Theorem 3.2 (Principle of Linearized Stability for Maps) *Consider a C^1-map*

$$x \mapsto g(x), \quad x \in \mathbb{R}^n,$$

with $g(0) = 0$. Let $A = Dg(0)$, the Jacobian matrix of g evaluated at $x = 0$.
(i) *If $r(A) < 1$ then the fixed point $x = 0$ is asymptotically stable.*
(ii) *If $r(A) > 1$ then the fixed point $x = 0$ is unstable.*

Proof
(i) Take any ρ satisfying $r(A) < \rho < 1$. By Lemma 2.13 from Chap. 2, there is a norm $\|\cdot\|_1$, which is equivalent to $\|\cdot\|$ and for which

$$\|Ax\|_1 \leq \rho\|x\|_1, \quad x \in \mathbb{R}^n.$$

Since g is a C^1-map, for any small $\varepsilon > 0$, there is $\delta > 0$, such that

$$\|g(x) - Ax\|_1 \leq \varepsilon\|x\|_1,$$

when $\|x\|_1 \leq \delta$. Then, for all such x,

$$\|g(x)\|_1 = \|Ax + g(x) - Ax\|_1 \leq \|Ax\|_1 + \|g(x) - Ax\|_1 \leq (\rho + \varepsilon)\|x\|_1.$$

Since $\rho < 1$ and $\varepsilon > 0$ is arbitrarily small, we can achieve that $\rho_1 = \rho + \varepsilon < 1$, which implies that g maps the ball

$$\overline{B}_\delta = \{x \in \mathbb{R}^n : \|x\|_1 \leq \delta\}$$

into itself for all sufficiently small $\delta > 0$, so the fixed point $x = 0$ is stable.
By induction:

$$\|g^k(x)\|_1 \le \rho_1^k \|x\|_1,$$

showing that $g^k(x) \to 0$ as $k \to +\infty$ for any x with $\|x\|_1 \le \delta$. Therefore, the fixed point $x = 0$ is asymptotically stable. The convergence rate ρ_1 is arbitrarily close to the spectral radius $r(A)$.

In the original (or any other equivalent) norm:

$$\|g^k(x)\| \le M\rho_1^k \|x\|,$$

for some constant $M \ge 1$.

(ii) The instability of the fixed point $x = 0$ in the presence of unstable eigenvalues is not surprising. However, below we give independent instability proofs in three cases:

(1) all eigenvalues λ of A satisfy $|\lambda| > 1$;

(2) A has both eigenvalues λ with $|\lambda| < 1$ and eigenvalues λ with $|\lambda| > 1$, but no eigenvalues on the unit circle;

(3) A has both eigenvalues λ with $|\lambda| \le 1$ and eigenvalues λ with $|\lambda| > 1$.

Of course, it would be sufficient to consider case (3) only, but we will treat all of them separately for didactic reasons.

(1) Assume that all eigenvalues of A are outside the unit circle. We shall show that $x = 0$ is a repellor. The Inverse Function Theorem implies that g^{-1} is defined on a neighbourhood of the origin and that $[g^{-1}]_x(0) = A^{-1}$. Since all eigenvalues of A^{-1} are *inside* the unit circle, we have the estimate

$$\|g^{-k}(x)\|_1 \le \rho_1^k \|x\|_1$$

for some $0 < \rho_1 < 1$ and all x satisfying $\|x\|_1 \le \delta$ (see Part (i) above). Now suppose that y is such that

$$\|g^k(y)\|_1 \le \delta$$

for all $k \ge k_0 > 0$. Then, by taking $x = g^k(y)$ we find that

$$\|y\|_1 \le \rho_1^k \|g^k(y)\|_1 \le \rho_1^k \delta,$$

for $k \ge k_0$. As the right-hand side converges to zero as $k \to \infty$, we must have $y = 0$. So, any orbit starting at $y \ne 0$ must leave the δ-ball (with respect to the $\|\cdot\|_1$-norm), no matter how small $\|y\|_1$ is. Hence, $x = 0$ is a repellor.

(2) When there are eigenvalues of A inside as well as outside, but not on, the unit circle, the instability of the fixed point $x = 0$ follows from the existence of a *local forward-invariant cone* around T^u. We proceed by constructing such a cone.

For some norm in $\mathbb{R}^n$, we have the estimates

$$\|Ax_s\|_1 \le \rho \|x_s\|_1, \quad x_s \in T^s,$$

and

$$\|Ax_u\|_1 \ge \frac{1}{\rho} \|x_u\|_1, \quad x_u \in T^u,$$

with some $0 < \rho < 1$. Notice that such estimates hold in the hyperbolic case by definition, but are valid also when A has a zero eigenvalue.

We may assume that

$$\|x\|_1 = \|x_s\|_1 + \|x_u\|_1,$$

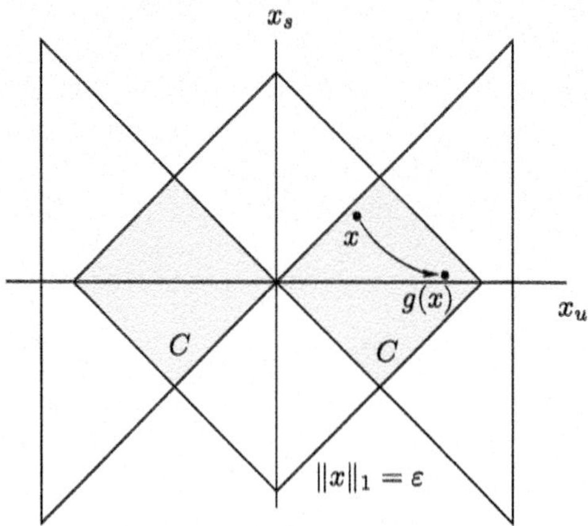

Fig. 3.1 The conditionally invariant cone segment C

where

$$x = x_s + x_u, \quad x_s = P_s x \in T^s, x_u = P_u x \in T^u,$$

and $P_{s,u}$ are the spectral projectors introduced in Sect. 2.1.2.

Since $\|g(x) - Ax\|_1 = o(\|x\|_1)$, we can choose $\varepsilon > 0$ such that for $\|x\|_1 \leq \varepsilon$ we have

$$\|g(x) - Ax\|_1 \leq \frac{1}{4}\left(\frac{1}{\rho} - \rho\right)\|x\|_1.$$

Define the *cone segment*:

$$C = \{x \in \mathbb{R}^n : \|x_s\|_1 \leq \|x_u\|_1, \ \|x\|_1 \leq \varepsilon\}$$

(see Fig. 3.1). We show that it is *conditionally invariant*. This means that if $x \in C$ and $\|g(x)\|_1 \leq \varepsilon$ then $g(x) \in C$. In other words, an orbit cannot leave C by violating the condition $\|x_s\|_1 \leq \|x_u\|_1$ without violating the condition $\|x\|_1 \leq \varepsilon$.

(a) Note first that for $x \in C$ we have $\|x\|_1 \leq 2\|x_u\|_1$.

(b) Then

$$\begin{aligned}\|P_u g(x)\|_1 &\geq \|P_u(Ax)\|_1 - \|P_u(g(x) - Ax)\|_1 \geq \|Ax_u\|_1 - \|g(x) - Ax\|_1 \\ &\geq \frac{1}{\rho}\|x_u\|_1 - \frac{1}{4}\left(\frac{1}{\rho} - \rho\right)\|x\|_1 \geq \frac{1}{2}\left(\frac{1}{\rho} + \rho\right)\|x_u\|_1.\end{aligned}$$

(c) Similarly, taking into account that $0 < \rho < 1$ and $x \in C$, we get

$$\begin{aligned}\|P_s g(x)\|_1 &\leq \|P_s(Ax)\|_1 + \|P_s(g(x) - Ax)\|_1 \leq \|Ax_s\|_1 + \|g(x) - Ax\|_1 \\ &\leq \rho\|x_s\|_1 + \frac{1}{4}\left(\frac{1}{\rho} - \rho\right)\|x\|_1 \leq \frac{1}{2}\left(\frac{1}{\rho} + \rho\right)\|x_u\|_1.\end{aligned}$$

Together (b) and (c) imply that

$$\|P_u g(x)\|_1 \geq \|P_s g(x)\|_1,$$

meaning that $g(x) \in C$ if $\|g(x)\|_1 \leq \varepsilon$.

Now suppose $x = 0$ is stable. Then there exists $\eta > 0$ such that

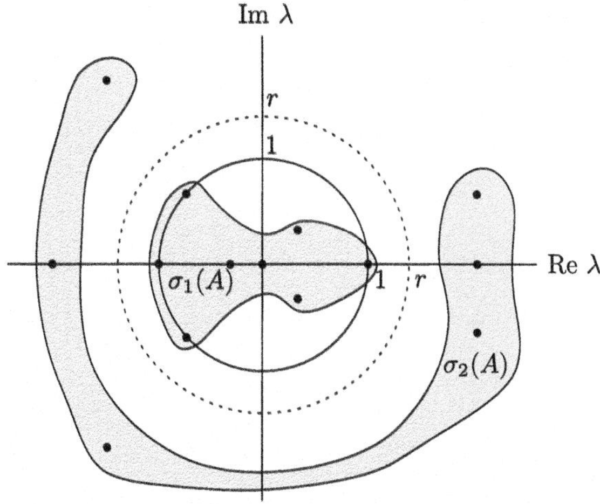

Fig. 3.2 Spectral decomposition: $\sigma(A) = \sigma_1(A) \cup \sigma_2(A)$

$$\|g^k(x)\|_1 \le \varepsilon$$

for all $k > 0$, provided that $\|x\|_1 \le \eta$. However, if $x \in C$ and $\|x\|_1 \le \eta$, then by (b) and (c) above

$$\|g^k(x)\|_1 \ge \|P_u(g^k(x))\|_1 \ge \left[\frac{1}{2}\left(\frac{1}{\rho} + \rho\right)\right]^k \|x_u\|_1 \ge \frac{1}{2}\left[\frac{1}{2}\left(\frac{1}{\rho} + \rho\right)\right]^k \|x\|_1.$$

If $x \neq 0$, the right-hand side tends to infinity for $k \to \infty$, since

$$\frac{1}{\rho} + \rho > 2 \text{ for } 0 < \rho < 1.$$

This is a clear contradiction to $\|g^k(x)\|_1 \le \varepsilon$ for all $k > 0$. So our assumption that $x = 0$ is stable must be false.

(3) In this general case, there are eigenvalues of A on the unit circle and (possibly) inside the unit circle, as well as outside the unit circle. Then there exists $r > 1$ such that

$$\sigma(A) = \sigma_1(A) \cup \sigma_2(A),$$

where

$$\sigma_1(A) = \{\lambda \in \sigma(A) : |\lambda| < r\}, \ \sigma_2(A) = \{\lambda \in \sigma(A) : |\lambda| > r\},$$

and $\sigma_2(A)$ is nonempty (see Fig. 3.2). The linear map

$$x \longmapsto Bx = \frac{1}{r}Ax, \quad x \in \mathbb{R}^n,$$

has a number of eigenvalues inside the unit circle, while all other eigenvalues are located outside the unit circle. Let T_1 and T_2 be the corresponding eigenspaces of B. Then

$$\mathbb{R}^n = T_1 \oplus T_2$$

and any $x \in \mathbb{R}^n$ can be decomposed as

$$x = x_1 + x_2, \quad x_1 = P_1 x \in T_1, x_2 = P_2 x \in T_2.$$

Since scaling with a positive factor does not change the eigenspaces of a matrix, T_1 consists of linear combinations of all (generalized) eigenvectors of A corresponding to $\sigma_1(A)$, while T_2 consists of linear combinations of all (generalized) eigenvectors of A corresponding to $\sigma_2(A)$.

Furthermore, there is a norm $\|\cdot\|_1$ in $\mathbb{R}^n$, such that

$$\|Bx_1\|_1 \leq \rho\|x_1\|_1, \quad x_1 \in T_1,$$

and

$$\|Bx_2\|_1 \geq \frac{1}{\rho}\|x_2\|_1, \quad x_2 \in T_2,$$

for some ρ with $0 < \rho < 1$. For the original map A this implies

$$\|Ax_1\|_1 \leq \rho r\|x_1\|_1, \quad x_1 \in T_1,$$

and

$$\|Ax_2\|_1 \geq \frac{r}{\rho}\|x_2\|_1, \quad x_2 \in T_2.$$

The rest of the analysis is similar to case (2) above. As before, we may assume that

$$\|x\|_1 = \|x_1\|_1 + \|x_2\|_1$$

and introduce the *cone segment* C defined by

$$C = \{x \in \mathbb{R}^n : \|x_1\|_1 \leq \|x_2\|_1, \ \|x\|_1 \leq \varepsilon\},$$

where $\varepsilon > 0$ is chosen such that

$$\|g(x) - Ax\|_1 \leq \frac{r}{4}\left(\frac{1}{\rho} - \rho\right)\|x\|_1.$$

(a) We now have $\|x\|_1 \leq 2\|x_2\|_1$ for $x \in C$.
(b) Then

$$\begin{aligned}\|P_2 g(x)\|_1 &\geq \|P_2(Ax)\|_1 - \|P_2(g(x) - Ax)\|_1 \geq \|Ax_2\|_1 - \|g(x) - Ax\|_1 \\ &\geq \frac{r}{\rho}\|x_2\|_1 - \frac{r}{4}\left(\frac{1}{\rho} - \rho\right)\|x\|_1 \geq \frac{r}{2}\left(\frac{1}{\rho} + \rho\right)\|x_2\|_1.\end{aligned}$$

(c) Similarly, taking into account that $0 < \rho < 1$ and $x \in C$,

$$\begin{aligned}\|P_1 g(x)\|_1 &\leq \|P_1(Ax)\|_1 + \|P_1(g(x) - Ax)\|_1 \leq \|Ax_1\|_1 + \|g(x) - Ax\|_1 \\ &\leq \rho r\|x_1\|_1 + \frac{r}{4}\left(\frac{1}{\rho} - \rho\right)\|x\|_1 \leq \frac{r}{2}\left(\frac{1}{\rho} + \rho\right)\|x_2\|_1.\end{aligned}$$

Together (b) and (c) imply

$$\|P_2 g(x)\|_1 \geq \|P_1 g(x)\|_1.$$

This means that $g(x) \in C$ if $\|g(x)\|_1 \leq \varepsilon$.

As in case (2), suppose $x = 0$ is stable. Then there exists $\eta > 0$ such that

$$\|g^k(x)\|_1 \leq \varepsilon$$

for all $k > 0$, provided that $\|x\|_1 \leq \eta$. However, if $x \in C$ and $\|x\|_1 \leq \eta$, then by (b) and (c) above

$$\|g^k(x)\|_1 \geq \|P_2(g^k(x))\|_1 \geq \left[\frac{r}{2}\left(\frac{1}{\rho}+\rho\right)\right]^k \|x_2\|_1 \geq \frac{1}{2}\left[\frac{r}{2}\left(\frac{1}{\rho}+\rho\right)\right]^k \|x\|_1.$$

If $x \neq 0$, the right-hand side tends to infinity for $k \to \infty$, which contradicts with the stability assumption. Thus, $x = 0$ is unstable. □

Remarks

(1) If $r(A) = 1$, one cannot decide about the stability character of $x = 0$ on the basis of information about A only: Depending on the nonlinear terms, the fixed point can be either stable or unstable (see Chap. 5).

(2) There are efficient algebraic criteria to determine whether $r(A) < 1$ for a given matrix. These so-called *Jury's criteria* are formulated in terms of the coefficients of the characteristic polynomial (see Exercise 2.34 for $n = 2$ and 3). There are also conditions directly in terms of the elements of A, based on the *bialternate matrix product.* ◇

3.1.2 Linearized Stability of ODEs

Consider a system

$$\dot{x} = F(x), \quad x \in \mathbb{R}^n,$$

where F is a C^1-function and $F(0) = 0$. This system can be rewritten in the form

$$\dot{x} = Ax + f(x), \quad x \in \mathbb{R}^n, \tag{3.1}$$

where $A = DF(0)$ and f is a C^1-function satisfying $f(0) = 0$ and $\|f(x)\| = o(\|x\|)$ as $x \to 0$. The following auxiliary results will be used:

(1) The variation of constants formula for a solution $x(t)$ to (3.1):

$$x(t) = e^{tA}x_0 + \int_0^t e^{(t-\tau)A} f(x(\tau))d\tau\,. \tag{3.2}$$

(Notice that this is an equation, not an explicit formula as the name suggests, and it is valid only for those t at which the solution exists.)

(2) The following Lemma:

Lemma 3.3 (Gronwall's Inequality) *Let $T > 0$ and let u, α, and β be continuous real-valued functions defined on $[0, T]$. Assume that for $t \in [0, T]$ it holds that $\beta(t) \geq 0$ and*

$$u(t) \leq \alpha(t) + \int_0^t \beta(\tau)u(\tau)\, d\tau, \quad t \in [0, T]. \tag{3.3}$$

Then

$$u(t) \leq \alpha(t) + \int_0^t e^{\int_\tau^t \beta(\eta)d\eta}\alpha(\tau)\beta(\tau)\, d\tau, \quad t \in [0, T]. \tag{3.4}$$

Moreover, if α is non-decreasing on $[0,T]$ then

$$u(t) \le \alpha(t) e^{\int_0^t \beta(\tau)d\tau}, \quad t \in [0,T]. \tag{3.5}$$

Proof Denote

$$v(t) := \int_0^t \beta(\tau)u(\tau)d\tau$$

and introduce $\varphi(t) := u(t) - v(t)$, so that (3.3) implies $\varphi(t) \le \alpha(t)$ for $t \in [0,T]$. The function v is continuously differentiable and its derivative can be written as

$$\dot{v}(t) = \beta(t)u(t) = \beta(t)v(t) + \beta(t)\varphi(t) .$$

Thus, v is the solution of the scalar inhomogeneous linear differential equation

$$\dot{v} = \beta(t)v + \beta(t)\varphi(t)$$

with the initial condition $v(0) = 0$. Therefore,

$$v(t) = \int_0^t e^{\int_\tau^t \beta(\eta)d\eta} \beta(\tau)\varphi(\tau)d\tau .$$

Since $\varphi(t) \le \alpha(t)$ for $t \in [0,T]$, we get

$$u(t) = \varphi(t) + v(t) = \varphi(t) + \int_0^t e^{\int_\tau^t \beta(\eta)d\eta} \beta(\tau)\varphi(\tau)\, d\tau \le \alpha(t) + \int_0^t e^{\int_\tau^t \beta(\eta)d\eta} \alpha(\tau)\beta(\tau)\, d\tau$$

for $t \in [0,T]$. This proves (3.4).

When α in non-decreasing on $[0,T]$, i.e. $\alpha(\tau) \le \alpha(t)$ for $0 \le \tau \le t \le T$, the inequality (3.4) implies

$$\begin{aligned} u(t) &\le \alpha(t) + \alpha(t) \int_0^t e^{\int_\tau^t \beta(\eta)d\eta} \beta(\tau)\, d\tau = \alpha(t) + \alpha(t) \int_0^t \frac{d}{d\tau}\left(-e^{\int_\tau^t \beta(\eta)d\eta}\right) d\tau \\ &= \alpha(t) + \alpha(t)\left(-e^{\int_\tau^t \beta(\eta)d\eta}\right)\Big|_{\tau=0}^{\tau=t} = \alpha(t) e^{\int_0^t \beta(\tau)d\tau}, \end{aligned}$$

which is inequality (3.5). □

Next we prove that the linearization of the flow φ^t is given by the flow of the linearized system, i.e. e^{tA}. In other words, the operations "solving" and "linearizing" commute.

Lemma 3.4 *Let $\varphi^t(x)$ denote the flow generated by (3.1) and let $M \ge 0$ and $\omega \in \mathbb{R}$ be such that*

$$\|e^{tA}\| \le M e^{\omega t}, \quad t \ge 0. \tag{3.6}$$

Then there exists $\delta_1 = \delta_1(\varepsilon, T)$ such that for any $\varepsilon > 0, T > 0$ the estimate

$$\|\varphi^t(x) - e^{tA}x\| \le M e^{\omega t}(e^{\varepsilon M t} - 1)\|x\| \tag{3.7}$$

holds for all $t \in [0,T]$ and $\|x\| < \delta_1$.

Remark The existence of the constants M and ω for which (3.6) holds follows from Theorem 2.18 in Chap. 2. The lemma implies that

$$\|\varphi^t(x) - e^{tA}x\| = o(\|x\|)$$

for $x \to 0$, uniformly on closed bounded time intervals. We could have given an easier proof of this implication by invoking Theorem 1.5, cf. Lemma 3.6 below. However, since we didn't prove Theorem 1.5, here we prefer to give all details of a more detailed estimate, namely (3.7). Moreover, we shall use this estimate in the proof of the Centre Manifold Theorem in Chap. 6. ♢

Proof of Lemma 3.4 For any $\varepsilon > 0$ there is $\delta = \delta(\varepsilon)$, such that $\|f(x)\| \leq \varepsilon\|x\|$ for $\|x\| \leq \delta$. Define, for given $\varepsilon, \delta, M, \omega$ and $T > 0$,

$$\delta_1 = \begin{cases} \frac{\delta}{M} e^{-(\omega+\varepsilon M)T} & \text{if } \omega + \varepsilon M > 0, \\ \frac{\delta}{M} & \text{if } \omega + \varepsilon M \leq 0. \end{cases}$$

We are going to show that, for any $\varepsilon > 0$ and $t \in [0, T]$,

$$\|\varphi^t(x) - e^{tA}x\| \leq Me^{\omega t}(e^{\varepsilon M t} - 1)\|x\|,$$

provided that $\|x\| < \delta_1$. Since $e^{\varepsilon M t} \to 1$ as $\varepsilon \to 0$, uniformly on closed bounded t-intervals, this implies the lemma statement.

From the variation of constants formula (3.2) and the assumption (3.6) follows the estimate

$$\|\varphi^t(x)\| \leq Me^{\omega t}\|x\| + Me^{\omega t}\int_0^t e^{-\omega\tau}\|f(\varphi^\tau(x))\|d\tau\,.$$

Multiply by $e^{-\omega t}$ and use that $\|f(\varphi^\tau(x))\| \leq \varepsilon\|\varphi^\tau(x)\|$, provided that $\|\varphi^\tau(x)\| \leq \delta$ for $0 \leq \tau \leq T$. Then, under this condition,

$$e^{-\omega t}\|\varphi^t(x)\| \leq M\|x\| + \varepsilon M\int_0^t e^{-\omega\tau}\|\varphi^\tau(x)\|d\tau\,,$$

which by Gronwall Lemma 3.3 yields

$$e^{-\omega t}\|\varphi^t(x)\| \leq M\|x\|e^{\varepsilon M t}$$

or

$$\|\varphi^t(x)\| \leq Me^{(\omega+\epsilon M)t}\|x\|\,. \tag{3.8}$$

If $\omega + \varepsilon M \leq 0$, then

$$\|\varphi^\tau(x)\| \leq M\|x\| < \delta$$

for $\|x\| < \delta_1$. If $\omega + \varepsilon M > 0$, then, for arbitrary $T > 0$

$$\|\varphi^\tau(x)\| \leq Me^{(\omega+\varepsilon M)T}\|x\| < \delta$$

for $\|x\| < \delta_1$ and $0 \leq \tau \leq T$. In both cases the condition $\|\varphi^\tau(x)\| \leq \delta$ (under which the estimate (3.8) is derived) cannot be violated for $0 \leq \tau \leq T$.

From

$$\varphi^t(x) - e^{tA}x = \int_0^t e^{(t-\tau)A} f(\varphi^\tau(x))d\tau$$

it then follows that

$$\|\varphi^t(x) - e^{tA}x\| \leq Me^{\omega t}\varepsilon \int_0^t e^{-\omega\tau} Me^{(\omega+\varepsilon M)\tau}\|x\|d\tau = Me^{\omega t}(e^{\varepsilon Mt} - 1)\|x\|$$

for $\|x\| < \delta_1$ and $0 \leq t \leq T$. □

Theorem 3.5 (Lyapunov, 1892) *Consider a C^1-system* (3.1).
(i) *If $s(A) < 0$ then the equilibrium $x = 0$ is asymptotically stable.*
(ii) *If $s(A) > 0$ then the equilibrium $x = 0$ is unstable.*

Proof

(i) If $s(A) < 0$, the parameter ω in (3.6) can be chosen such that $\omega < 0$ (cf. Theorem 2.21). Next choose $\varepsilon > 0$ so that $\omega + \varepsilon M < 0$. Now recall the estimate (3.8) derived in the proof of Lemma 3.4. It shows that $\varphi^t(x) \to 0$ exponentially fast as $t \to \infty$ for all x with $\|x\| < \delta_1$.

(ii) If $s(A) > 0$, then consider $g(x) = \varphi^1(x)$, where φ^t is the flow generated by (3.1). From the smoothness assumption and Lemma 3.4 it follows that g is a C^1 map and that $g(x) = Lx + o(\|x\|)$ where $L = e^A$. The spectral mapping principle implies $r(L) > 1$ and the instability follows at once from Theorem 3.2 (ii) applied to $g(x)$. □

Remarks

(1) If $s(A) = 0$, one cannot decide about the stability character of $x = 0$ on the basis of information about A only: Depending on the nonlinear terms $f(x)$, the equilibrium can be either stable or unstable (see Chap. 5).

(2) There are efficient algebraic criteria to determine whether $s(A) < 0$ for a given matrix. They can be formulated either in terms of the coefficients of the characteristic polynomial $\det(\lambda I - A)$ (*Routh–Hurwitz Criterion*, see Exercise 3.8) or directly in terms of the elements of A, via the *bialternate matrix product*.

(3) Note that the original proof by Lyapunov was different from the one given above and was based on what is nowadays called a *Lyapunov function* (see Exercise 3.5). ◇

3.2 Local Stability of Periodic Orbits in ODEs

In *linear* ODEs, periodic orbits come in families (see Sect. 2.2 of Chap. 2) and so they are at most Lyapunov stable but never asymptotically stable. In this respect nonlinear systems are really different.

In the discrete-time case, a k-periodic orbit is a fixed point of the k times iterated map and we define stability accordingly, so nothing is needed beyond Theorem 3.2.[1] For continuous-time dynamical systems the situation is not much different but, since a closed orbit (cycle) is a continuous curve of points, the formulation requires quite a bit of care and is, in fact, rather subtle.

In this section we shall formulate a test in terms of the linearized system to decide about the stability character of a generic cycle in an ODE. This test, however, is not as easily carried out as in the case of a steady state. Both conceptually and computationally it is helpful to think about the stability of a cycle in terms of a discrete-time dynamical system generated by the so-called Poincaré map. The key point about this map is that it eliminates the "neutrality" of shifting along the cycle and focuses instead on "recurrent behaviour" after full turns.

Consider a system

$$\dot{x} = f(x), \quad x \in \mathbb{R}^n, \tag{3.9}$$

where $f : \mathbb{R}^n \to \mathbb{R}^n$ is C^1-smooth. Recall that the corresponding flow $\varphi^t(x)$ is at least C^1 jointly in (t, x) and that the mixed partial derivatives with respect to t and (components of) x exist and commute (see, Theorem 1.5).

Lemma 3.6 *The $n \times n$ matrix*

$$Y(t) = D_x\varphi^t(x)\big|_{x=x_0}$$

satisfies the linear differential equation

$$\dot{Y} = Df(\varphi^t(x_0))Y \tag{3.10}$$

and the initial condition $Y(0) = I_n$.

Proof Let $x(t, x_0) := \varphi^t(x_0)$. Then

$$\begin{aligned}\dot{Y}(t) &= D_t D_{x_0} x(t, x_0) = D_{x_0} D_t x(t, x_0) = D_{x_0}(f(x(t, x_0))) \\ &= D_x f(x(t, x_0)) D_{x_0} x(t, x_0) = D_x f(x(t, x_0)) Y(t) = Df(\varphi^t(x_0))Y(t).\end{aligned}$$

Furthermore, $Y(0) = D_x\varphi^0(x)\big|_{x=x_0} = (D_x x)\big|_{x=x_0} = I_n$. □

Nota bene that Y depends on x_0 even though we didn't incorporate this into the notation.

[1] Note, however, that a k-periodic orbit consists of k different points and that, in order to speak about stability of the orbit, we should verify that each point has the same stability character, see Exercise 3.2.

Lemma 3.7 *Suppose that $\varphi^t(x_0)$ is defined for $t \in [0,T]$, where $T > 0$. Let $y_0 = f(x_0)$ and $y_1 = f(\varphi^{t_1}(x_0))$, where $t_1 \in [0,T]$. Then $y_1 = Y(t_1)y_0$.*

Proof Since $x(t) = \varphi^t(x_0)$ is a solution to (3.9), we have

$$\frac{d}{dt}\varphi^t(x_0) = f(\varphi^t(x_0)).$$

Differentiating this equation with respect to t we find

$$\frac{d}{dt}\left(\frac{d}{dt}\varphi^t(x_0)\right) = f_x(\varphi^t(x_0))\frac{d}{dt}\varphi^t(x_0),$$

so

$$y(t) = f(\varphi^t(x_0))$$

is a solution to the linearized problem

$$\dot{y} = Df(\varphi^t(x_0))y, \quad y \in \mathbb{R}^n, \tag{3.11}$$

with the initial condition $y(0) = f(x_0) = y_0$. Since any such solution to (3.11) has the form $y(t) = Y(t)y_0$, we get

$$y_1 = y(t_1) = Y(t_1)y_0.$$ □

Let $\Gamma_0 \subset \mathbb{R}^n$ be a periodic orbit (cycle) of the dynamical system generated by (3.9), i.e. there exists $T > 0$ (the minimal period) such that for every $x_0 \in \Gamma_0$ we have $\varphi^T(x_0) = x_0$ and $\varphi^t(x_0) \neq x_0$ for $t \in (0,T)$. In this case, the orbit $\Gamma_0 = \{x \in \mathbb{R}^n : x = \varphi^t(x_0),\ 0 \le t \le T\}$ is a C^1-smooth closed curve.

Theorem 3.8 *For any $x_0 \in \Gamma_0$, $f(x_0)$ is an eigenvector of $Y(T)$ corresponding to eigenvalue 1.*

Proof By Lemma 3.7, $f(\varphi^T(x_0)) = Y(T)f(x_0)$. The periodicity now yields $f(x_0) = Y(T)f(x_0)$. □

Definition 3.9 *$Y(T)$ is called the* ***monodromy matrix****. Its eigenvalues are called the (****characteristic*** *or)* ***Floquet multipliers****. The multiplier 1 is called* ***trivial****, while all others are called* ***nontrivial*** *multipliers.*

The property found in Theorem 3.8 reflects the fact that, whenever *both* x_0 and $\tilde{x}_0$ belong to Γ_0, the distance between $\varphi^t(x_0)$ and $\varphi^t(\tilde{x}_0)$ does not go to zero for $t \to \infty$, no matter how small we take the distance between x_0 and $\tilde{x}_0$. Actually, the distance varies periodically in time. So it seems a good strategy to focus on the nontrivial multipliers.

Definition 3.10 *A cycle Γ_0 of (3.9) is called* ***simple*** *if $\lambda = 1$ is a simple eigenvalue of $Y(T)$.*

Remarks

(1) The famous *Liouville formula* says that for a linear matrix differential equation $\dot{Y} = A(t)Y$, with $Y \in \mathbb{R}^{n\times n}$ and $A \in C^0(\mathbb{R}, \mathbb{R}^{n\times n})$, the function $\det Y(t)$ satisfies the scalar linear ODE

$$\frac{d}{dt}\det(Y(t)) = \mathrm{Tr}(A(t))\,\det(Y(t)),$$

so that

$$\det(Y(t)) = \exp\left(\int_0^t \mathrm{Tr}(A(\tau))\,d\tau\right)\det(Y(0)).$$

Applying the last formula to (3.10) with the initial condition $Y(0) = I_n$, and taking into account that $\mathrm{Tr}(Df(x)) = \mathrm{div}\, f(x)$, we obtain immediately

$$\det Y(T) = \lambda_1\lambda_2\cdots\lambda_n = \exp\left(\int_0^T \mathrm{div}\, f(\varphi^t(x_0))\,dt\right),$$

where $\lambda_1, \lambda_2, \ldots, \lambda_n$ are the Floquet multipliers. For planar systems ($n = 2$), the combination of this formula with Theorem 3.8 allows us to express the only nontrivial multiplier λ_2 of $Y(T)$ as

$$\lambda_2 = \exp\left(\int_0^T \mathrm{div}\, f(\varphi^t(x_0))\,dt\right).$$

Note that for general $n \geq 2$, we have $\det(Y(t_1)) > 0$ for any $t_1 \in [0, T]$.

(2) *Floquet Theorem*: One may write $Y(t) = Z(t)e^{tD}$ with a constant complex matrix D and a T-periodic complex matrix $Z(t)$ such that $Z(0) = Z(T) = I_n$. Indeed, since $\det Y(T) > 0$, we can find a "logarithm" matrix D, such that

$$Y(T) = e^{TD}.$$

(When $Y(T)$ has real negative eigenvalues, then D must have complex elements!) Now define $Z(t) = Y(t)e^{-tD}$. Then

$$Z(t+T) = Y(t+T)e^{-(t+T)D} = Y(t)Y(T)e^{-TD}e^{-tD} = Y(t)e^{-tD} = Z(t).$$

Obviously, $Z(0) = Z(T) = I_n$. Therefore, the multipliers are the eigenvalues of e^{TD}. It is possible to avoid complex matrices by only requiring that Z is $2T$-periodic and not necessarily T-periodic. ◇

Choose $x_0 \in \Gamma_0$ and define

$$\Sigma_0 = \{\xi \in \mathbb{R}^n : \langle f(x_0), \xi\rangle = 0\}$$

and introduce a *cross section*

$$\Pi_{x_0} = \{x \in \mathbb{R}^n : x = x_0 + \xi,\ \xi \in \Sigma_0\}.$$

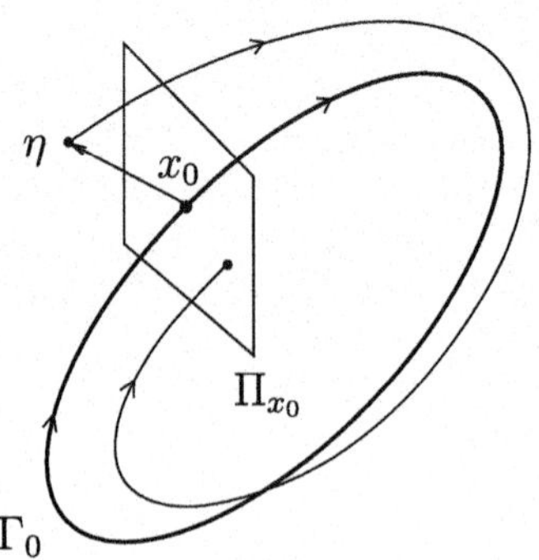

Fig. 3.3 Construction of the Poincaré map

The orbit starting at x_0 hits Π_{x_0} again after T units of time. Our next aim is to show that orbits of (3.9) starting at points on Π_{x_0} near x_0 also hit Π_{x_0} after approximately T units of time. Actually, we show that this is true for all orbits starting near x_0, either on or off Π_{x_0} (see Fig. 3.3).

Lemma 3.11 *There exists a C^1 map $\tau : \mathbb{R}^n \to \mathbb{R}$, $\eta \mapsto \tau(\eta)$, defined in a neighbourhood of $\eta = 0$ and such that*
(i) $\tau(0) = T$;
(ii) $\varphi^{\tau(\eta)}(x_0 + \eta) \in \Pi_{x_0}$.

Moreover, for η in that neighbourhood of $\eta = 0$, if $t \in \mathbb{R}$ is such that $\varphi^t(x_0 + \eta) \in \Pi_{x_0}$ and $|t - T|$ is small enough, then $t = \tau(\eta)$.

Proof Define $F : \mathbb{R} \times \mathbb{R}^n \to \mathbb{R}$ by

$$F(t, \eta) := \langle f(x_0), \varphi^t(x_0 + \eta) - x_0 \rangle$$

and consider the equation

$$F(t, \eta) = 0.$$

Clearly, $F \in C^1$ and

$$F(T, 0) = \langle f(x_0), x_0 - x_0 \rangle = 0,$$

while

$$\frac{\partial}{\partial t} F(T, 0) = \langle f(x_0), f(x_0) \rangle = \|f(x_0)\|^2 \neq 0.$$

The Implicit Function Theorem now yields the statement. □

Definition 3.12 *The map $\mathcal{P} : \Sigma_0 \to \Sigma_0$, defined for $\xi \in \Sigma_0$ near $\xi = 0$ by the formula*

$$\mathcal{P}(\xi) = \varphi^{\tau(\xi)}(x_0 + \xi) - x_0, \tag{3.12}$$

is called a ***Poincaré map*** *of the periodic orbit Γ_0.*

Remarks

(1) $\mathcal{P}$ is a (locally defined) map on the $(n-1)$-dimensional subspace Σ_0. Let $N_i^0 \in \mathbb{R}^n$, $i = 1, 2, \ldots, n-1$, be linearly independent vectors in Σ_0, so that

$$\langle N_i^0, f(x_0) \rangle = 0.$$

Then any $\xi \in \Sigma_0$ can be written as

$$\xi = \zeta_1 N_1^0 + \zeta_2 N_2^0 + \cdots + \zeta_{n-1} N_{n-1}^0.$$

Given such coordinates ζ in Σ_0, the map $\mathcal{P}$ as defined in (3.12) is fully described by a (local) C^1-map

$$P : \mathbb{R}^{n-1} \to \mathbb{R}^{n-1}, \quad \zeta \mapsto P(\zeta),$$

which is often also called the Poincaré map of Γ_0. Note that $\zeta = 0$ is a fixed point of this map: $P(0) = 0$. The eigenvalues of the $(n-1) \times (n-1)$-matrix $DP(0)$ are the eigenvalues of the linear map $D\mathcal{P}(0)$.

(2) Note that $\overline{\mathcal{P}} : \Pi_{x_0} \to \Pi_{x_0}$ defined by $\overline{\mathcal{P}}(x) = \mathcal{P}(x - x_0) + x_0$ takes $x \in \Pi_{x_0}$ to the next point where the orbit through x hits Π_{x_0}. Often, when one speaks about a Poincaré map, $\overline{\mathcal{P}}$ is meant. We prefer to work with $\mathcal{P}$, since it is defined on the linear space Σ_0.

(3) Poincaré maps can be defined using any smooth $(n-1)$-dimensional manifold Π transversal to Γ_0 at any point $x_0 \in \Gamma_0$. All such maps are locally topologically equivalent (C^1-conjugate). The conjugacy is provided by a *correspondence map* defined from one cross section to another along the orbits of the system. Its existence and smoothness can be established by the same arguments as used for constructing the Poincaré map above, see Exercise 3.10. ◇

Note that $\xi = 0$ is a fixed point of the Poincaré map: $\mathcal{P}(0) = 0$. If all $(n-1)$ eigenvalues of its linearization at $\xi = 0$ satisfy $|\lambda| < 1$, then $\xi = 0$ is asymptotically stable as a fixed point of $\mathcal{P}$ in Σ_0. As the following theorem shows, the periodic orbit Γ_0 is then asymptotically stable as well.

Theorem 3.13 (i) *If all $(n-1)$ eigenvalues of the linearization $D\mathcal{P}(0)$ of the Poincaré map $\mathcal{P}$ at $\xi = 0$ satisfy $|\lambda| < 1$, then Γ_0 is asymptotically stable.*

(ii) *If $|\lambda| > 1$ for some eigenvalue λ of the linearization of the Poincaré map $\mathcal{P}$, then Γ_0 is unstable.*

Proof

(i) Introduce an equivalent norm $\|\cdot\|_1$ in $\mathbb{R}^{n-1}$ in which $\mathcal{P}_\xi(0)$ is a linear contraction on Σ_0. There exists $\delta_0 > 0$ such that for all $\xi \in \Sigma_0$ with $\|\xi\|_1 \leq \delta_0$ the inequality

$$\|\mathcal{P}(\xi)\|_1 \leq \rho_1 \|\xi\|_1 \tag{3.13}$$

holds with some $\rho_1 < 1$ (see the proof of part (i) of Theorem 3.2). For any $\delta \leq \delta_0$, construct a neighbourhood U_δ of Γ_0 as follows. Take the ball in Σ_0

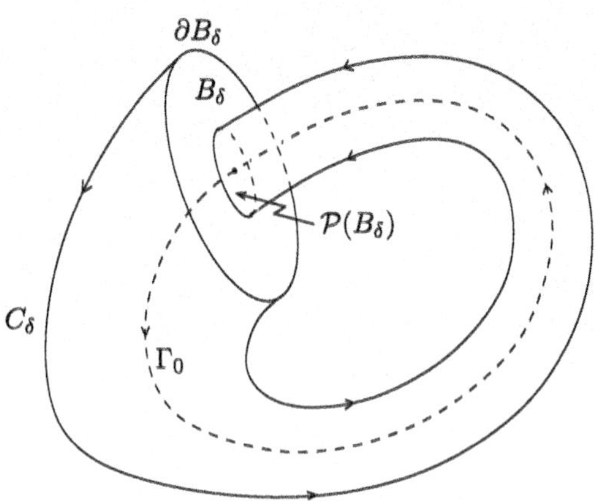

Fig. 3.4 Construction of U_δ

$$B_\delta = \{\xi \in \Sigma_0 : \|\xi\|_1 \le \delta\}$$

and consider all orbits of (3.9) starting at $x_0 + \xi$ with $\xi \in B_\delta$. Any such orbit returns back to Π_0 after $\tau(\xi)$ units of time. Define now $U_\delta \subset \mathbb{R}^n$ as the union of all such orbit segments, i.e.

$$U_\delta = \{x \in \mathbb{R}^n : x = \varphi^t(x_0 + \xi),\ \xi \in B_\delta,\ 0 \le t \le \tau(\xi)\}$$

(see Fig. 3.4). The set U_δ is a closed tubular neighbourhood of Γ_0 that shrinks to Γ_0 as $\delta \to 0$. Since $\mathcal{P}(B_\delta)$ is located strictly in B_δ, U_δ is a *trapping region*, i.e. any orbit starting in U_δ remains in it for all $t \ge 0$. Indeed, the boundary ∂U_δ of U_δ consists of a cylinder C_δ, which is formed by translations of all points of ∂B_δ by the flow until they return to Σ_0, and a set D_δ defined by

$$D_\delta = \text{Int } B_\delta \setminus \text{Int } \mathcal{P}(B_\delta),$$

which is an annulus in Σ_0 between B_δ and $\mathcal{P}(B_\delta)$. Provided δ_0 is sufficiently small, all orbits of the ODE that start in D_δ cross it transversally and then enter U_δ. This implies that any orbit starting in U_δ cannot leave U_δ for $t \ge 0$, if one takes into account that the cylinder C_δ is positively invariant with respect to the system flow.

Consider now any small open neighbourhood U of Γ_0. Making δ sufficiently small, we can guarantee that $U_\delta \subset U$. Since U_δ is a trapping region, this implies Lyapunov stability of Γ_0.

By induction, it follows from (3.13) that

$$\|\mathcal{P}^k(\xi)\|_1 \le \rho_1^k \|\xi\|_1, \quad k = 1, 2, 3, \ldots .$$

This means that the forward half of any orbit starting in U_δ can be divided into finite segments, whose end-points $x_0 + \xi_k$ where $\xi_k \in \Sigma_0$ form a convergent sequence with $\{\xi_k\} \to 0$. Using the continuous dependence of solutions of (3.9) with respect to the initial point (on finite time intervals), we conclude that $\text{dist}(\varphi^t(x_0), \Gamma_0) \to 0$ as $t \to +\infty$, i.e. Γ_0 is asymptotically stable.

(ii) This part is obvious, since instability for the Poincaré map $\mathcal{P}$ immediately implies instability of Γ_0 with respect to the flow generated by the ODE. □

Actually, a result stronger than part (i) of Theorem 3.13 can be established.

Definition 3.14 *A cycle Γ_0 through x_0 is called **exponentially orbitally stable** with asymptotic phase if there exist $c > 0, K > 1$, and $t_0 = t_0(x) \in [0, T)$ such that*

$$\|\varphi^t(x) - \varphi^{t-t_0}(x_0)\| \leq Ke^{-ct}, \quad t \geq 0,$$

for all x with sufficiently small $\text{dist}(x, \Gamma_0)$.

Theorem 3.15 *If all $(n-1)$ eigenvalues of the linearization $D\mathcal{P}(0)$ of the Poincaré map $\mathcal{P}$ at $\xi = 0$ satisfy $|\lambda| < 1$, then Γ_0 is exponentially orbitally stable with asymptotic phase.*

Proof Consider first $x = x_0 + \xi_0, \ \ \xi_0 \in \Sigma_0$, and define for a given ξ_0:

$$\begin{aligned} \xi_k &= \mathcal{P}(\xi_{k-1}), \\ \tau_k &= \tau(\xi_{k-1}) + \tau_{k-1}, \ \tau_0 = 0, \end{aligned}$$

for $k = 1, 2, \ldots$. We know that there exists $\delta > 0$ such that for all ξ_0 with $\|\xi_0\| \leq \delta$ the estimate

$$\|\xi_k\| \leq Me^{-\alpha k}\|\xi_0\|$$

holds for some $M \geq 1$ and some $\alpha > 0$. Hence, $\tau(\xi_k) \to T$ and

$$\frac{\tau_k}{kT} \to 1$$

as $k \to +\infty$. But since $\tau \in C^1$, we can derive a better estimate:

$$|\tau(\xi_{k-1}) - T| \leq C\|\xi_{k-1}\| \leq MCe^{-\alpha(k-1)}\|\xi_0\|.$$

This implies

$$|(\tau_k - kT) - (\tau_{k-1} - (k-1)T)| = |\tau(\xi_{k-1}) - T| \leq MCe^{-\alpha(k-1)}\|\xi_0\|.$$

Thus, $\theta_k = \tau_k - kT$ is a Cauchy sequence, so it has a limit that we denote by t_0. We have

$$|\tau_{k+m} - (k+m)T - (\tau_k - kT)| \leq MC\|\xi_0\| \sum_{j=0}^{m-1} e^{-\alpha(k+j)} \leq MC\|\xi_0\|\frac{e^{-\alpha k}}{1 - e^{-\alpha}}.$$

Taking the limit $m \to +\infty$, we find

$$|\tau_k - kT - t_0| \leq MC\|\xi_0\|\frac{e^{-\alpha k}}{1 - e^{-\alpha}}.$$

Next consider

$$\|\varphi^{t+\tau_k}(x_0 + \xi_0) - \varphi^t(x_0)\| = \|\varphi^t(x_0 + \xi_k) - \varphi^t(x_0)\| \leq C_1\|\xi_k\| \leq MC_1e^{-\alpha k}\|\xi_0\|$$

for $0 \leq t \leq T$, since $\varphi^t(x)$ is a C^1-function of (t, x). Likewise

$$\|\varphi^{t+\tau_k}(x_0 + \xi_0) - \varphi^{t+kT+t_0}(x_0 + \xi_0)\| \leq MC_2e^{-\alpha k}\|\xi_0\|,$$

for $0 \le t \le T$, since the left-hand side is equal to

$$\|\varphi^t(x_0+\xi_k) - \varphi^t(\varphi^{kT+t_0-\tau_k}(x_0+\xi_k))\|.$$

Combining the last two inequalities, we find

$$\|\varphi^{\tau+t_0}(x_0+\xi_0) - \varphi^{\tau}(x_0)\| \le M(C_1+C_2)e^{-\alpha k}\|\xi_0\|$$

for $kT \le \tau \le (k+1)T$, where $\tau = kT + t$ with $t \in [0,T]$. Note that the equality $\varphi^{kT+t}(x_0) = \varphi^t(x_0)$ is used above.

Now take any $x \in \mathbb{R}^n$ near Γ_0. If this point does not belong to Π_0, consider the first intersection of the forward half-orbit starting at x with Π_0 and represent it as $x_0 + \xi_0$. Apply then the proof given above. □

Our next aim is to relate the eigenvalues of the linearization of $\mathcal{P}$ with the eigenvalues of $Y(T)$.

Lemma 3.16 (i) *$D\mathcal{P}(0)$ is the restriction to Σ_0 of the linear map in $\mathbb{R}^n$*

$$\eta \mapsto (D\tau(0)\eta)f(x_0) + Y(T)\eta, \quad \eta \in \mathbb{R}^n. \tag{3.14}$$

(ii) *In terms of a coordinate $c \in \mathbb{R}$ and $\xi \in \Sigma_0$ defined by the decomposition*

$$\mathbb{R}^n = \text{span}\{f(x_0)\} \oplus \Sigma_0,$$

$Y(T)$ maps (c,ξ) to $(c - D\tau(0)\xi, D\mathcal{P}(0)\xi)$.

Proof

(i) By Lemma 3.11 the map $\eta \mapsto \tau(\eta)$ is defined and differentiable in a neighbourhood of the origin in $\mathbb{R}^n$. This implies that the same is true for the map $\eta \mapsto \varphi^{\tau(\eta)}(x_0+\eta) - x_0$. Since $Y(T) = D_x\varphi^T(x_0)$, straightforward differentiation of this map shows that the derivative at $\eta = 0$ is exactly given by (3.14). Next we simply restrict to Σ_0.

(ii) Since $Y(T)f(x_0) = f(x_0)$, the point with coordinates (c,ξ) is mapped to $cf(x_0) + Y(T)\xi$. According to part (i) we may write

$$Y(T)\xi = D\mathcal{P}(0)\xi - (D\tau(0)\xi)f(x_0).$$

Since $D\mathcal{P}(0)\xi \in \Sigma_0$, the image point has coordinates $(c - (D\tau(0)\xi), D\mathcal{P}(0)\xi)$. □

In the (c,ξ)-coordinates, the map $Y(T)$ is represented by the block matrix

$$Y(T) = \begin{pmatrix} 1 & -D\tau(0) \\ 0 & D\mathcal{P}(0) \end{pmatrix}. \tag{3.15}$$

Theorem 3.17 (i) $\lambda \neq 1$ *is an eigenvalue of* $D\mathcal{P}(0)$ *if and only if* λ *is an eigenvalue of* $Y(T)$.

(ii) $\lambda = 1$ *is an eigenvalue of* $D\mathcal{P}(0)$ *if and only if the eigenvalue* 1 *of* $Y(T)$ *has multiplicity bigger than one.*

Proof

(i) If $Y(T)\eta = \lambda\eta$ and η has coordinates (c, ξ), then $D\mathcal{P}(0)\xi = \lambda\xi$ because of Lemma 3.16 (ii). If $\lambda \neq 1$ then $\xi \neq 0$ since $f(x_0)$ is eigenvector of $Y(T)$ corresponding to eigenvalue 1. So the "if" part is proved. On the other hand, if $D\mathcal{P}(0)\xi = \lambda\xi$ and $\lambda \neq 1$, then, as one easily verifies,

$$\eta = \frac{1}{1-\lambda}(D\tau(0)\xi)f(x_0) + \xi$$

is such that $Y(T)\eta = \lambda\eta$.

(ii) Suppose first that $Y(T)\eta = \eta$ where η is not a multiple of $f(x_0)$. Then the Σ_0-component ξ of η is nonzero. And, by Lemma 3.16 (ii), $D\mathcal{P}(0)\xi = \xi$, so 1 is an eigenvalue of $D\mathcal{P}(0)$. If, conversely, $D\mathcal{P}(0)\xi = \xi$ with $\xi \neq 0$, we distinguish the case where $D\tau(0)\xi \neq 0$ from the case where $D\tau(0)\xi = 0$. In the latter case it follows right away that $Y(T)\xi = \xi$, so 1 is an eigenvalue of $Y(T)$. In the former case, we find that the normalized vector

$$\zeta = -\frac{1}{D\tau(0)\xi}\xi$$

satisfies $Y(T)\zeta - \zeta = f(x_0)$ showing that, corresponding to the eigenvalue 1, $Y(T)$ has a higher-than-one-dimensional generalized eigenspace. Finally, suppose that $Y(T)\eta - \eta = f(x_0)$. Then it follows directly from Lemma 3.16 (ii) that $D\mathcal{P}(0)\xi = \xi$, where ξ is the Σ_0-component of η, which necessarily is nonzero, since $Y(T)f(x_0) - f(x_0) = 0$. The proof is complete. □

Theorem 3.17 and the representation (3.15) imply that

$$\det(\lambda I_n - Y(T)) = (\lambda - 1)\det(\lambda I_{n-1} - DP(0)), \tag{3.16}$$

where the map $P : \mathbb{R}^{n-1} \to \mathbb{R}^{n-1}$ is defined in a remark immediately after Definition 3.12. Furthermore, by combining Theorems 3.13, 3.15, and 3.17 we arrive at the following result.

Theorem 3.18 *If all nontrivial Floquet multipliers of a simple cycle have modulus less than one, then the cycle is exponentially orbitally stable with asymptotic phase. If some multiplier lies outside the unit circle, the cycle is unstable.* □

The theorem gives a method to ascertain, at least in principle, the stability of a cycle: Show that the trivial multiplier is simple, while all nontrivial multipliers are located strictly inside the unit circle. In practice, this has to be done numerically, by first computing $Y(T)$ and then its eigenvalues.

3.3 Grobman–Hartman Theorems

Theorems 3.2 and 3.5 do not provide a precise correspondence between orbits of the nonlinear system and orbits of the linearized system. Yet, as we shall show in this section, such a correspondence does indeed exist in the hyperbolic situation: Any nonlinear system is locally topologically equivalent to its linearization at a steady state, *provided that the linearized system is hyperbolic*. This is a strong result which requires more machinery to prove it. The entire next subsection is an interlude. It introduces some notions and estimates that are heavily used in the following subsections, where we return to the relationship between the local behaviour of a nonlinear system and its linearization at a steady state.

3.3.1 Lipschitz Maps

Let X be a Banach space, for example $X = \mathbb{R}^n$ with a norm $\|\cdot\|$, or $X = BC(\mathbb{R}^n, \mathbb{R}^n)$, the Banach space of bounded continuous maps $v : \mathbb{R}^n \to \mathbb{R}^n$ equipped with the supremum norm

$$\|v\| = \sup_{x \in \mathbb{R}^n} \|v(x)\|.$$

Definition 3.19 *A map $f : X \to X$ is called* ***Lipschitz*** *if there is a positive constant K such that*

$$\|f(u) - f(v)\| \leq K\|u - v\|,$$

for all $u, v \in X$. The infimum of the constants K for which the above estimate holds is called the (***exact***) ***Lipschitz constant*** *of f and is denoted by* $\mathrm{Lip}(f)$.

This definition can easily be generalized for a map $f : X \to Y$, where X and Y are two different Banach spaces equipped with their own norms. Clearly, any Lipschitz map is continuous. For a bounded linear operator $L : X \to X$, $\mathrm{Lip}(L) = \|L\|$, the operator norm of L. If $f : X \to X$ and $g : X \to X$ are two Lipschitz maps, then obviously

$$\mathrm{Lip}(f + g) \leq \mathrm{Lip}(f) + \mathrm{Lip}(g)$$

and

$$\mathrm{Lip}(f \circ g) \leq \mathrm{Lip}(f)\,\mathrm{Lip}(g),$$

where $(f \circ g)(u) = f(g(u))$ is the composition of f and g.

The following fundamental result can be found in any good book on differential equations, since the standard proof of Theorem 1.5 formulated in Chap. 1 is based on it. We include this result here for completeness.

Theorem 3.20 (Contraction Mapping Principle) *If $f : X \to X$ is a Lipschitz map with $\theta = \mathrm{Lip}(f) < 1$ (**contraction**), then f has a unique fixed point $p = f(p)$ and*

$$\lim_{k\to\infty} f^{(k)}(x) = p,$$

for all $x \in X$.

Proof

(a) Take any point $x \in X$ and consider its iterates

$$x, f(x), f^2(x), \; f^3(x), \ldots .$$

First we show that $\{f^k(x)\}$ is a Cauchy sequence. Using Definition 3.19, we get by induction

$$\|f^{k+1}(x) - f^k(x)\| \le \theta^k \|f(x) - x\| \tag{3.17}$$

for all $k \in \mathbb{N}$. Applying the triangle inequality we deduce:

$$\|f^{k+m}(x) - f^k(x)\| \le \sum_{j=0}^{m-1} \|f^{k+j+1}(x) - f^{k+j}(x)\|.$$

The right-hand side can be estimated using (3.17) as follows:

$$\sum_{j=0}^{m-1} \|f^{k+j+1}(x) - f^{k+j}(x)\| \le \sum_{j=0}^{m-1} \theta^{k+j} \|f(x) - x\| \le \theta^k \|f(x) - x\| \sum_{j=0}^{m-1} \theta^j .$$

Since $0 < \theta < 1$,

$$\sum_{j=0}^{m-1} \theta^j \le \sum_{j=0}^{\infty} \theta^j = \frac{1}{1-\theta}$$

by the geometric series. Thus, for any $\varepsilon > 0$ and any $m \ge 1$,

$$\|f^{k+m}(x) - f^k(x)\| \le \frac{\theta^k}{1-\theta} \|f(x) - x\| \le \varepsilon,$$

when k is big enough. This implies that $\{f^k(x)\}$ is a Cauchy sequence. Since X is a Banach space, there exists $p \in X$ such that

$$\lim_{k\to\infty} f^k(x) = p.$$

(b) Our next aim is to prove that p is a fixed point of f. As any Lipschitz map, f is continuous, therefore

$$f(p) = f(\lim_{k\to\infty} f^k(x)) = \lim_{k\to\infty} f(f^k(x)) = \lim_{k\to\infty} f^{k+1}(x) = p.$$

(c) Finally, assume that f has another fixed point $q \in X$, so that $f(q) = q$ and $\|p - q\| > 0$. However,

$$\|p - q\| = \|f(p) - f(q)\| \le \theta \|p - q\|$$

implying

$$(1 - \theta)\|p - q\| \le 0.$$

Since $1 - \theta > 0$, we must have $\|p - q\| = 0$, which is a contradiction. Thus the fixed point is unique. □

Remarks

(1) Clearly, one can formulate a variant of the Contraction Mapping Principle in which f maps a closed subset of X into itself, while satisfying the Lipschitz condition with a constant less than one. In this case, f need not be defined on the whole space X.

(2) Replacing everywhere above $\|x - y\|$ by the distance $d(x, y)$ in a complete metric space X, we obtain another variant of Theorem 3.20, where no linear structure on X is assumed. ◇

Theorem 3.21 (Lipschitz Inverse Function Theorem) *Let $L : X \to X$ be a bounded linear operator with bounded inverse L^{-1}, and let $f : X \to X$ be a Lipschitz map with*

$$\mathrm{Lip}(f) < \|L^{-1}\|^{-1}.$$

Then $(L + f) : X \to X$ is invertible with $(L + f)^{-1} = L^{-1} - v$, where $v : X \to X$ is a Lipschitz map with

$$\mathrm{Lip}(v) \leq \frac{\|L^{-1}\| \; \mathrm{Lip}(f)}{\|L^{-1}\|^{-1} - \mathrm{Lip}(f)}, \tag{3.18}$$

so that $(L + f)^{-1}$ is a Lipschitz map with

$$\mathrm{Lip}[(L + f)^{-1}] \leq \frac{1}{\|L^{-1}\|^{-1} - \mathrm{Lip}(f)}. \tag{3.19}$$

Proof We want to show that the equation

$$(L + f)(x) = y \tag{3.20}$$

has a unique solution $x \in X$ for any $y \in X$. Equation (3.20) with $f = 0$ has a solution, namely $L^{-1}y$. Write $x = L^{-1}y - w$ for some unknown $w \in X$ that depends on y. Then (3.20) is equivalent to the fixed-point equation

$$w = g(w)$$

with the map $g : X \to X$ given by

$$g(w) := L^{-1} f(L^{-1}y - w). \tag{3.21}$$

This is a Lipschitz map with $\mathrm{Lip}(g) \leq \|L^{-1}\| \; \mathrm{Lip}(f) < 1$ by hypothesis. By Theorem 3.20 (Contraction Mapping Principle), g has a unique fixed point $v(y) \in X$. It follows that

$$x = L^{-1}y - v(y) \in X$$

is the unique solution of (3.20) for any given $y \in X$.

To obtain the estimate on Lip(v), introduce

$$H := \left\{ h \in X : \text{Lip}(h) \leq \frac{\|L^{-1}\| \, \text{Lip}(f)}{\|L^{-1}\|^{-1} - \text{Lip}(f)} \right\}$$

(cf. (3.18)). The set $H \subset X$ is closed and is—therefore—a complete metric space with respect to the standard distance in X defined via the norm. Moreover, we have $g(H) \subset H$. Indeed, using the condition on Lip(h), we get

$$\text{Lip}(g(h)) \leq \|L^{-1}\| \, \text{Lip}(f) \, [\|L^{-1}\| + \text{Lip}(h)] \leq \frac{\|L^{-1}\| \, \text{Lip}(f)}{\|L^{-1}\|^{-1} - \text{Lip}(f)}.$$

Thus, the restriction of g to H has a unique fixed point $w \in H$ that must coincide with $v(y)$. Thus, $v(y) \in H$ implying (3.18). The estimate (3.19) on $\text{Lip}[(L + f)^{-1}]$ then follows directly from (3.18). □

3.3.2 Grobman–Hartman Theorems for Maps

Let A be a hyperbolic linear map on $\mathbb{R}^n$ and let $\|\cdot\|_1$ be the equivalent norm from Definition 2.14. If necessary, choose another equivalent norm on $\mathbb{R}^n$ such that the projection operators P_s and P_u on, respectively, the stable subspace T^s and the unstable subspace T^u of A both have unit norm. For instance, take $\|x\|_2 = \|P_s x\|_1 + \|P_u x\|_1$. According to Definition 2.14, we have the estimates

$$\|A_s\|_2 \leq \rho, \quad \|A_u^{-1}\|_2 \leq \rho,$$

where the operator norm corresponding to $\|\cdot\|_2$ is used and

$$A_s = A|_{T^s}, \quad A_u = A|_{T^u},$$

while $0 < \rho < 1$.

Let $B = BC(\mathbb{R}^n, \mathbb{R}^n)$ be the Banach space of bounded continuous maps $v : \mathbb{R}^n \to \mathbb{R}^n$ equipped with the supremum norm

$$\|v\| = \sup_{x \in \mathbb{R}^n} \|v(x)\|_2,$$

where $\|\cdot\|_2$ is defined above.[2]

The spectral decomposition $\mathbb{R}^n = T^s \oplus T^u$ induces a decomposition $B = B^s \oplus B^u$ with the corresponding projectors $P_{s,u}$ defined by

$$(P_{s,u} v)(x) = P_{s,u} v(x).$$

[2] By definition: $(c_1 v_1 + c_2 v_2)(x) = c_1 v_1(x) + c_2 v_2(x)$, for all $v_{1,2} \in B$, $c_{1,2} \in \mathbb{R}$, and $x \in \mathbb{R}^n$.

In other words, a bounded continuous map belongs to B^s if and only if its values belong to T^s and similarly with s replaced by u. Thus we write $v = P_s v + P_u v$. In a similar spirit, the map A on $\mathbb{R}^n$ induces a map, which we shall also denote by A, on B according to

$$(Av)(x) = Av(x).$$

In the following, I is the identity map $I(x) = x$ on any space.

To motivate the next lemma, recall from Sect. 1.3 of Chap. 1 that a linear map $x \mapsto Ax$ is topologically conjugate to a nonlinear map $y \mapsto g(y) = Ay + f(y)$ if there is a homeomorphism $x \mapsto y = h(x)$ such that

$$g = h \circ A \circ h^{-1} \tag{3.22}$$

This equality can be rewritten either as an equation for h,

$$h \circ A = (A + f) \circ h \tag{3.23}$$

(take the composition of (3.22) with h from the right), or as an equation for the inverse map h^{-1},

$$h^{-1} \circ (A + f) = A \circ h^{-1} \tag{3.24}$$

(take the composition of (3.22) with h^{-1} from the left). Now notice that both equations (3.23) and (3.24) can be combined into a single one, namely:

$$H \circ (A + R_1) = (A + R_2) \circ H.$$

Indeed, we recover the first equation for $R_1 = 0$ and $R_2 = f$, while we obtain the second equation for $R_1 = f$ and $R_2 = 0$. The following Lemma provides, therefore, the existence and uniqueness of both h and h^{-1}.

Lemma 3.22 *For given $R_1, R_2 \in B$ with sufficiently small* $\mathrm{Lip}(R_i), i = 1, 2$, *there exists a unique solution $v = v(R_1, R_2) \in B$ of the equation*

$$(I + v) \circ (A + R_1) = (A + R_2) \circ (I + v). \tag{3.25}$$

Proof We can rewrite (3.25) in the form

$$Lv + w(v) = R_1, \tag{3.26}$$

where

$$Lv = Av - v \circ (A + R_1), \quad w(v) = R_2 \circ (I + v).$$

Here and in the following we often write $A \circ v = Av$, since A is linear. The map $v \mapsto Lv$ is linear, while $v \mapsto w(v)$ from B into B is nonlinear. We will show with the help of Theorem 3.21 that $(L + w)$ is invertible, so that (3.26) is uniquely solvable in B. Obviously,

$$\|w(v_1) - w(v_2)\| \leq \mathrm{Lip}(R_2)\|v_1 - v_2\|,$$

implying $\mathrm{Lip}(w) \leq \mathrm{Lip}(R_2)$. Since $\mathrm{Lip}(R_2)$ can be taken as small as we want, it is—in view of Theorem 3.21—sufficient to prove that L is invertible.

Take $r \in B$ and consider the equation

$$Lv = r. \tag{3.27}$$

Projecting this equation to T^s and T^u, and taking the commutativity relations $P_s A v = A_s P_s v$ and $P_u A v = A_u P_u v$ into account, we obtain

$$\begin{cases} A_s P_s v - P_s v \circ (A + R_1) = P_s r, \\ A_u P_u v - P_u v \circ (A + R_1) = P_u r. \end{cases} \tag{3.28}$$

Provided that $\mathrm{Lip}(R_1) < \|A^{-1}\|_2^{-1}$, Theorem 3.21 guarantees the invertibility of $(A + R_1)$. Note also that A_u is nonsingular. Taking the composition of the first equation in (3.28) with $(A + R_1)^{-1}$ from the right and multiplying the second equation in (3.28) by A_u^{-1} from the left, we see that (3.28) is equivalent with

$$\begin{cases} P_s v = A_s P_s v \circ (A + R_1)^{-1} - P_s r \circ (A + R_1)^{-1}, \\ P_u v = A_u^{-1} P_u v \circ (A + R_1) + A_u^{-1} P_u r. \end{cases} \tag{3.29}$$

This means that (3.27) is equivalent with a linear equation

$$v = Mv + s, \tag{3.30}$$

where

$$Mv = A_s P_s v \circ (A + R_1)^{-1} + A_u^{-1} P_u v \circ (A + R_1)$$

(the linear part of the right-hand side of (3.29)) and

$$s = -P_s r \circ (A + R_1)^{-1} + A_u^{-1} P_u r$$

(the v-independent part of the right-hand side of (3.29)). Since $\|P_{s,u}\|_2 = 1$, we have

$$\|Mv\| \leq \max\{\|A_s\|_2, \|A_u^{-1}\|_2\}\|v\| \leq \rho\|v\|,$$

with $0 < \rho < 1$. Therefore, the map $v \mapsto Mv + s$ is a contraction in B. According to Theorem 3.20, Eq. (3.30) has a unique solution $v \in B$ for any $s \in B$ and therefore Eq. (3.27) has a unique solution $v \in B$ for any $r \in B$. Thus, L is invertible and we can write the solution v to (3.30) as $v = L^{-1}r$.

Moreover, Theorem 3.21 is also applicable to (3.30) leading to the estimate

$$\|v\| \leq \frac{1}{1-\rho}\|s\| \leq \frac{1}{1-\rho}\|r\|.$$

In other symbols:

$$\|L^{-1}\| \leq \frac{1}{1-\rho}.$$

Taking now $\mathrm{Lip}(R_2) < 1 - \rho$, we can finally apply Theorem 3.21 to (3.26), keeping in mind that our estimates were valid provided $\mathrm{Lip}(R_1) < \|A^{-1}\|_2^{-1}$. The theorem implies the unique solvability of (3.26) and, therefore, of (3.25). □

Theorem 3.23 (Global Grobman–Hartman Theorem for Maps) *A hyperbolic linear map*

$$x \mapsto Ax, \quad x \in \mathbb{R}^n, \tag{3.31}$$

is topologically conjugate to any map

$$y \mapsto g(y) = Ay + f(y), \quad y \in \mathbb{R}^n, \tag{3.32}$$

where $f \in B$ has sufficiently small $\mathrm{Lip}(f)$ *and* $f(0) = 0$.

In fact, there is a unique homeomorphism $h : \mathbb{R}^n \to \mathbb{R}^n$, $h(0) = 0$, in the form $h = I + v$, $v \in B$, such that

$$h \circ A = g \circ h \tag{3.33}$$

Proof Set $H_1 = I + v(0, f)$ and $H_2 = I + v(f, 0)$, where v is defined in Lemma 3.22. These are the unique solutions with $H_k - I \in B$ to the equations

$$(A + f) \circ H_1 = H_1 \circ A, \tag{3.34}$$

$$H_2 \circ (A + f) = A \circ H_2. \tag{3.35}$$

Taking the composition of the first equation with H_2 from the right and using the second one, we find

$$(A + f) \circ H_1 \circ H_2 = H_1 \circ A \circ H_2 = H_1 \circ H_2 \circ (A + f),$$

which is again an equation of the form (3.25), now with $R_1 = R_2 = f$. Notice that $H_1 \circ H_2 = I + w$ with $w \in B$. Since $w = v(f, f) = 0$ by the uniqueness part of Lemma 3.22, $H_1 \circ H_2 = I$. Similarly, by composing (3.34) with H_2 from the left and again using (3.35):

$$H_2 \circ H_1 \circ A = H_2 \circ (A + f) \circ H_1 = A \circ H_2 \circ H_1,$$

from which it follows that $H_2 \circ H_1 = I$ by the same arguments. We conclude that $H_2 = H_1^{-1}$.

Notice now that (3.34) is exactly the condition (3.33) of topological conjugacy of A and g. Therefore, there exists a unique solution $h = H_1$ that is a homeomorphism with $h - I \in B$.

Evaluating the conjugacy relation (3.33) at 0, we see that $h(0) = g(h(0))$, so that $h(0)$ is a fixed point of g. Since 0 is the unique fixed point of g (by conjugacy to the linear hyperbolic map A), we must have $h(0) = 0$. □

Our next aim is to formulate a local version of Theorem 3.23. We first introduce some terminology.

Definition 3.24 *A fixed point x_0 of a C^1-map $G : \mathbb{R}^n \to \mathbb{R}^n$ is called* ***hyperbolic*** *if $A = DG(x_0)$ is hyperbolic. If both the stable and the unstable eigenspace are nontrivial, the hyperbolic fixed point is called a* ***saddle****.*

Shifting the origin to x_0, we can consider the C^1-map g defined by

$$g(y) = G(x_0 + y) - x_0 = Ay + f(y),$$

for which $y = 0$ is a fixed point with $A = Dg(0)$, $f(0) = 0$, and $f(y) = o(\|y\|_2)$.

Theorem 3.25 (Grobman–Hartman Theorem for Maps) *A hyperbolic linear map*

$$A : x \mapsto Ax, \quad x \in \mathbb{R}^n,$$

is locally topologically conjugate near the origin to any C^1-map

$$g : y \mapsto Ay + f(y), \quad y \in \mathbb{R}^n, \tag{3.36}$$

with $f(0) = 0$ and $Df(0) = 0$, i.e. there exists a homeomorphism $x \mapsto y = h(x)$ with $h(0) = 0$ defined on a neighbourhood of the origin for which

$$h \circ A = g \circ h.$$

Proof Denote by $K(\varepsilon)$ the Lipschitz constant of f on the ball of radius $\varepsilon > 0$. Then, since $f_y(0) = 0$, $K(\varepsilon) \to 0$ as $\varepsilon \to 0$. Instead of g consider a map which is modified outside the ball of radius ε:

$$g_\varepsilon : y \mapsto Ay + f_\varepsilon(y), \tag{3.37}$$

where

$$f_\varepsilon(y) = \begin{cases} f(y) & \text{for } \|y\|_2 \leq \varepsilon, \\ f(\varepsilon y \|y\|_2^{-1}) & \text{for } \|y\|_2 > \varepsilon \end{cases} \tag{3.38}$$

The map g_ε coincides with g in the ball $\|y\|_2 \leq \varepsilon$ and is *globally* Lipschitz with $\mathrm{Lip}(f_\varepsilon) \leq 2K(\varepsilon)$. For instance, when $\|y_2\|_2, \|y_1\|_2 \geq \varepsilon$, we have

$$\begin{aligned}
\|f_\varepsilon(y_1) - f_\varepsilon(y_2)\|_2 &= \|f(\varepsilon y_1\|y_1\|_2^{-1}) - f(\varepsilon y_2\|y_2\|_2^{-1})\|_2 \\
&\le K(\varepsilon)\|\varepsilon y_1\|y_1\|_2^{-1} - \varepsilon y_2\|y_2\|_2^{-1}\|_2 \\
&\le K(\varepsilon)\|y_1 - \|y_1\|_2\|y_2\|_2^{-1}y_2\|_2 \\
&= K(\varepsilon)\|y_1 - y_2 + (\|y_2\|_2 - \|y_1\|_2)\|y_2\|_2^{-1}y_2\|_2 \\
&\le K(\varepsilon)(\|y_1 - y_2\|_2 + |\|y_2\|_2 - \|y_1\|_2|) \\
&\le 2K(\varepsilon)\|y_2 - y_1\|_2.
\end{aligned}$$

Application of Theorem 3.23 now proves the statement. Indeed, let $\rho < 1$ be such that

$$\|A_s\|_2, \|A_u^{-1}\|_2 \le \rho$$

and choose ε so small that

$$\mathrm{Lip}(f_\varepsilon) \le 2K(\varepsilon) < 1 - \rho.$$

According to Theorem 3.23, $h = I + v(0, f_\varepsilon)$ is a homeomorphism satisfying

$$h \circ A = g_\varepsilon \circ h,$$

where g_ε is given by (3.37) so that $g_\varepsilon(y) = g(y)$ whenever $\|y\|_2 \le \varepsilon$. □

Let $y = 0$ be a hyperbolic fixed point of a smooth map $g : \mathbb{R}^n \to \mathbb{R}^n$. Using Theorem 3.25, we can immediately characterize the orbit structure of g near the origin. Indeed, let ε and h be as introduced in the proof of the theorem and let

$$B_\varepsilon = \{y \in \mathbb{R}^n : \|y\|_2 \le \varepsilon\}.$$

Choose δ such that $h(B_\delta) \subset B_\varepsilon$. Define two subsets of $\mathbb{R}^n$ as follows:

$$W_\delta^s(0) = h(T^s \cap B_\delta), \quad W_\delta^u(0) = h(T^u \cap B_\delta)$$

(see Fig. 3.5). Since h is a homeomorphism, $\dim W_\delta^s(0) = n_s$ and $\dim W_\delta^u(0) = n_u$. Due to Theorem 3.25, all orbits of g starting in $W_\delta^s(0)$ remain in this set

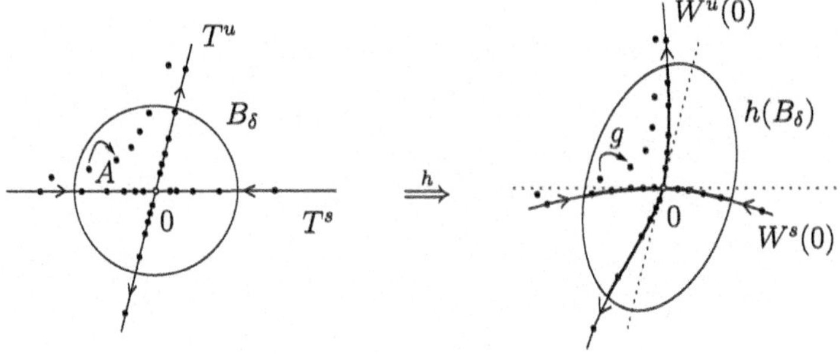

Fig. 3.5 Local orbit structure of g

and converge to $y = 0$ as integer time $k \to +\infty$, while all orbits of g starting in $W^u_\delta(0)$ are defined for all negative integers k and converge to $y = 0$ as $k \to -\infty$, remaining in $W^u_\delta(0)$. Orbits of g, which start in B_δ but neither in $W^s_\delta(0)$ nor in $W^u_\delta(0)$, leave B_δ in both the forward and the backward time direction.

Definition 3.26 *$W^s_\delta(0)$ is called the **local stable manifold** of the hyperbolic fixed point $y = 0$, while $W^u_\delta(0)$ is called the **local unstable manifold** of the hyperbolic fixed point $y = 0$.*

It also holds that $W^s_\delta(0)$ and $W^u_\delta(0)$ can be represented as the graphs of Lipschitz-continuous functions $\psi_s : T^s \to T^u$ and $\psi_u : T^u \to T^s$, respectively.

Remark Actually, $W^s_\delta(0)$ and $W^u_\delta(0)$ are graphs of C^k-functions $\psi_s : T^s \to T^u$ and $\psi_u : T^u \to T^s$, respectively, provided that $g \in C^k$. Moreover, $W^s_\delta(0)$ is tangent to T^s and $W^u_\delta(0)$ is tangent to T^u at $y = 0$. Note that the C^{k-1} smoothness of $W^s_\delta(0)$ and $W^u_\delta(0)$ follows from the results proven in Chap. 6. ♢

Define now global stable and unstable invariant sets of the fixed point $y = 0$ of a map $g : \mathbb{R}^n \to \mathbb{R}^n$.

Definition 3.27 *The **stable set** $W^s(0)$ is the set consisting of all points $y_0 \in \mathbb{R}^n$, such that there exists a sequence $\{y_i\}_{i=1}^\infty$ with the properties:*
(i) $y_i \to 0$ as $i \to +\infty$;
(ii) $y_{i+1} = g(y_i)$ for $i = 0, 1, 2, \ldots$.

Clearly,

$$W^s(0) = \{y \in \mathbb{R}^n : g^k(y) \to 0, \quad k \to \infty\}.$$

Definition 3.28 *The **unstable set** $W^u(0)$ is the set consisting of all points $y_0 \in \mathbb{R}^n$, such that there exists a sequence $\{y_i\}_{i=0}^\infty$ with the properties:*
(i) $y_i \to 0$ as $i \to +\infty$;
(ii) $y_i = g(y_{i+1})$ for $i = 0, 1, 2, \ldots$.

When g is invertible,

$$W^u(0) = \{y \in \mathbb{R}^n : g^{-k}(y) \to 0, \quad k \to \infty\}.$$

Notice, however, that neither invertibility of g nor hyperbolicity of the fixed point $y = 0$ is assumed in Definitions 3.27 and 3.28. Therefore, they are applicable to the fixed point $y = 0$ of any map in $\mathbb{R}^n$ and, mutatis mutandis, to any fixed point, no matter where it is located. Note that g_ε is invertible by Theorem 3.23.

The following result is now obvious.

Theorem 3.29 *Local stable and unstable manifolds of a hyperbolic fixed point of a C^1-map are subsets of the stable and unstable sets of this point, respectively.* □

Remarks

(1) It can be proved that the global stable and unstable sets of a fixed point of a diffeomorphism g are embedded smooth manifolds in $\mathbb{R}^n$.

(2) These manifolds can behave in a very complicated manner. For example, $W^s(x_0)$ of a hyperbolic fixed point x_0 can intersect $W^u(x_0)$ near x_0 in infinitely many points when a so-called *Poincaré homoclinic structure* occurs. This phenomenon will be treated in detail in Chap. 7. ♢

3.3.3 Grobman–Hartman Theorem for ODEs

To conclude this section we show that, just as with the Principle of Linearized Stability, one can prove the corresponding result for ODEs almost directly from the one for maps. We begin with the corresponding global theorem.

Theorem 3.30 (Global Grobman–Hartman Theorem for ODEs) *A hyperbolic linear system*

$$\dot{x} = Ax, \quad x \in \mathbb{R}^n, \tag{3.39}$$

is topologically conjugate to any system

$$\dot{y} = Ay + f(y), \quad y \in \mathbb{R}^n, \tag{3.40}$$

where $f \in B$ has sufficiently small $\mathrm{Lip}(f)$ *and $f(0) = 0$.*

Moreover, there is a unique homeomorphism $h : \mathbb{R}^n \to \mathbb{R}^n,\ h(0) = 0,$ in the form $h = I + v,\ v \in B,$ such that

$$h \circ \mathrm{e}^{tA} = \varphi^t \circ h, \tag{3.41}$$

where φ^t is the flow corresponding to (3.40).

Proof The linear system (3.39) defines the global flow e^{tA}. The Lipschitz system (3.40) also defines a global flow, say φ^t. Since

$$\varphi^t(y) = e^{tA}y + \int_0^t e^{(t-\tau)A} f(\varphi^\tau(y))\, d\tau \tag{3.42}$$

implies that

$$\varphi^1(y) = e^A y + \Phi(y),$$

where $\Phi(0) = 0$ and $\mathrm{Lip}(\Phi)$ is bounded by a constant times $\mathrm{Lip}(f)$, while the matrix e^A has neither eigenvalue zero nor eigenvalues with unit modulus, Theorem 3.23 yields a unique homeomorphism $h : \mathbb{R}^n \to \mathbb{R}^n$ satisfying

$$h \circ e^A = \varphi^1 \circ h, \tag{3.43}$$

for which $h - I \in B$ and $h(0) = 0$.

Now define a family of *homeomorphisms* $H(t) : \mathbb{R}^n \to \mathbb{R}^n$,

$$H(t) = \varphi^t \circ h \circ e^{-tA}, \quad t \in \mathbb{R}.$$

Then

$$H(t) \circ e^A = \varphi^t \circ (h \circ e^A) \circ e^{-tA} = \varphi^t \circ (\varphi^1 \circ h) \circ e^{-tA} = \varphi^1 \circ H(t),$$

so that

$$H(t) \circ e^A = \varphi^1 \circ H(t)$$

for all $t \in \mathbb{R}$. Now compare this identity with (3.43). Provided that we can make sure that $H(t) = I + v(t)$, where $v(t) \in B$, we conclude from the uniqueness part of Theorem 3.23 that $H(t) = h$ and therefore that

$$h \circ e^{tA} = \varphi^t \circ h.$$

Then, by definition, (3.39) and (3.40) are topologically conjugate, and the conjugating homeomorphism h satisfies $h - I \in B$.

To complete the proof, consider

$$H(t) - I = \varphi^t \circ h \circ e^{-tA} - e^{tA} \circ e^{-tA} = (\varphi^t - e^{tA}) \circ h \circ e^{-tA} + e^{tA} \circ (h - I) \circ e^{-tA}.$$

Since $h - I \in B$, the second term belongs to B. If $\varphi^t - e^{tA}$ belongs to B, so does the first term. The variation of constants formula (3.42) shows that indeed $\varphi^t - e^{tA} \in B$. □

The Theorem 3.30 for globally Lipschitz ODEs can also be proved directly, without use of the corresponding theorem for maps.

Let A be hyperbolic (meaning Re $\lambda \neq 0$ for all $\lambda \in \sigma(A)$) and let $P = P_s$ be the projection operator to its stable eigenspace T^s. Then $I - P = P_u$, the projection operator to the unstable eigenspace T^u. From Theorem 2.24, it follows that there are constants $M, \alpha > 0$ such that

$$\|e^{tA}P\| \leq Me^{-\alpha t}, \quad \|e^{-tA}(I - P)\| \leq Me^{-\alpha t}, \quad t \geq 0.$$

Recall that $PA = AP$ and $P^2 = P$.

Lemma 3.31 *If $b : \mathbb{R} \to \mathbb{R}^n$ is a bounded continuous function, then the equation*

$$\dot{y} = Ay + b(t) \tag{3.44}$$

has a unique solution $y(t)$ that is bounded on $(-\infty, +\infty)$, given by

$$y(t) = \int_{-\infty}^{t} e^{(t-s)A}Pb(s)\, ds - \int_{t}^{\infty} e^{(t-s)A}(I - P)b(s)\, ds. \tag{3.45}$$

Moreover, for all $t \in \mathbb{R}$, the estimate

$$\|y(t)\| \leq 2M\alpha^{-1} \sup_{-\infty<t<\infty} \|b(t)\| \tag{3.46}$$

holds.

Proof If y is not unique, i.e. there exist at least two solutions y_1 and y_2 to (3.44), then $y_1 - y_2$ is a bounded nontrivial solution of the autonomous linear system $\dot{y} = Ay$. But since there are no such solutions, this is impossible.

Differentiating (3.45), we get

$$\begin{aligned}\dot{y}(t) &= \int_{-\infty}^{t} Ae^{(t-s)A}Pb(s)\,ds - \int_{t}^{\infty} Ae^{(t-s)A}(I-P)b(s)\,ds + Pb(t) - [-(I-P)b(t)]\\ &= Ay(t) + b(t).\end{aligned}$$

Finally,

$$\begin{aligned}\|y(t)\| &= \left\|\int_{-\infty}^{t} e^{(t-s)A}Pb(s)\,ds - \int_{t}^{\infty} e^{(t-s)A}(I-P)b(s)\,ds\right\|\\ &\leq \int_{-\infty}^{t} \|e^{(t-s)A}Pb(s)\|ds + \int_{t}^{\infty} \|e^{(t-s)A}(I-P)b(s)\|ds\\ &\leq \int_{-\infty}^{t} Me^{-\alpha(t-s)}\|b(s)\|\,ds + \int_{t}^{\infty} Me^{-\alpha(s-t)}\|b(s)\|\,ds\\ &\leq M\left[\int_{-\infty}^{t} e^{-\alpha(t-s)}ds + \int_{t}^{\infty} e^{-\alpha(s-t)}ds\right] \sup_{-\infty<t<\infty}\|b(t)\|\\ &= 2M\alpha^{-1} \sup_{-\infty<t<\infty}\|b(t)\|.\end{aligned}$$

□

Lemma 3.32 *Let A be hyperbolic and let $c : \mathbb{R} \times \mathbb{R}^n \to \mathbb{R}^n$ be a continuous function satisfying*

(*i*) $\|c(t,0)\| \leq \mu$ *for all* $t \in \mathbb{R}$;
(*ii*) $\|c(t,y_1) - c(t,y_2)\| \leq K\|y_1 - y_2\|$ *for all* $t \in \mathbb{R}$, $y_{1,2} \in \mathbb{R}^n$,

with $\mu > 0$ and K such that $0 < K < (2M)^{-1}\alpha$.

Then the equation

$$\dot{y} = Ay + c(t,y) \tag{3.47}$$

has a unique solution $y(t)$ that is bounded on $(-\infty,\infty)$. Moreover, for all $t \in \mathbb{R}$, the estimate

$$\|y(t)\| \leq 2\mu M(\alpha - 2KM)^{-1} \tag{3.48}$$

holds.

Also, if $\tilde{c} : \mathbb{R} \times \mathbb{R}^n \times \mathbb{R}^k \to \mathbb{R}^n$ satisfies conditions (i) and (ii) for each $\eta \in \mathbb{R}^k$ with constants μ and K independent of η, then the equation

$$\dot{u} = Au + \tilde{c}(t,u,\eta) \tag{3.49}$$

has a unique solution $u(t,\eta)$ for each $\eta \in \mathbb{R}^k$ and u is a continuous function of (t,η) satisfying (3.48) for all (t,η).

Proof Denote by BC the Banach space $BC(\mathbb{R},\mathbb{R}^n)$ of bounded continuous functions on $\mathbb{R}$ equipped with the norm

$$\|y\| = \sup_{-\infty<t<\infty}\|y(t)\|.$$

Define $(Tz)(t)$ as the unique bounded solution to

$$\dot{y} = Ay + c(t,z(t)), \tag{3.50}$$

corresponding to a continuous function $z(t)$ that is bounded on $(-\infty, \infty)$. Since

$$\|c(t, z(t))\| \le \|c(t, 0)\| + \|c(t, z(t)) - c(t, 0)\| \le \mu + K\|z\|,$$

this solution exists by Lemma 3.31 and hence $Tz = y$ defines a map $T : BC \to BC$, which satisfies

$$\|Tz\| \le 2M\alpha^{-1}(\mu + K\|z\|).$$

If $z_1, z_2 \in BC$ then $v(t) = (Tz_1)(t) - (Tz_2)(t)$ is the unique solution to

$$\dot{v} = Av + c(t, z_1(t)) - c(t, z_2(t))$$

and by Lemma 3.31 we have

$$\|Tz_1 - Tz_2\| = \|v\| \le 2KM\alpha^{-1}\|z_1 - z_2\|.$$

Thus $T : BC \to BC$ is a contraction. Therefore, there exists a unique $y \in BC$ such that $y = Ty$. This proves the existence of a solution to (3.47).

From the definition of T, y is bounded. The uniqueness follows from the fact that any bounded solution to (3.47) must be in BC and must be a fixed point of T.

Moreover,

$$\|y\| = \|Ty\| \le 2M\alpha^{-1}(\mu + K\|y\|),$$

so that

$$\left(1 - 2KM\alpha^{-1}\right)\|y\| \le 2\mu M\alpha^{-1},$$

from which (3.48) follows.

The continuous dependence on the parameter η of the unique bounded solution $u(t, \eta)$ of (3.49) follows from the continuous dependence of the fixed point of a continuous family of contractions with a uniform contraction rate in BC. We leave details to the reader. □

Theorem 3.33 *Let A by hyperbolic. Let $f : \mathbb{R}^n \to \mathbb{R}^n$ be a function satisfying*

(1) $\|f(y)\| \le \mu$ *for all* $y \in \mathbb{R}^n$;
(2) $\|f(y_1) - f(y_2)\| \le K\|y_1 - y_2\|$ *for all* $y_{1,2} \in \mathbb{R}^n$,

where $\mu > 0$ and K is such that

$$0 < K < \alpha(2M)^{-1}.$$

Then there exists a unique map $H : \mathbb{R}^n \to \mathbb{R}^n$ with the following properties:

(*i*) $\|H(y) - y\|$ *is bounded;*
(*ii*) *whenever $y(t)$ is a solution to*

$$\dot{y} = Ay + f(y), \tag{3.51}$$

then $x(t) = H(y(t))$ is a solution to

$$\dot{x} = Ax. \tag{3.52}$$

Moreover, H is a homeomorphism and for all $y \in \mathbb{R}^n$ the estimate

$$\|H(y) - y\| \le 2\mu M\alpha^{-1}$$

holds.

Proof Let $y(t, \eta)$ be the solution to the initial-value problem

$$\dot{y} = Ay + f(y), \quad y(0) = \eta.$$

Since f is bounded and globally Lipschitz, $y(t, \eta)$ is defined for all t and continuous. Then we introduce $v(t, \eta)$ as the unique bounded solution to

$$\dot{v} = Av + f(y(t, \eta)).$$

This solution exists by Lemma 3.31. Moreover,

$$\|v(t, \eta)\| \leq 2\mu M\alpha^{-1}$$

for all (t, η).

We have $v(t, y(s, \eta)) = v(t + s, \eta)$. This follows from the fact that for fixed (s, η) $v(t, y(s, \eta))$ is the unique solution to

$$\dot{v} = Av + f(y(t, y(s, \eta))) = Au + f(y(t + s, \eta)).$$

Now define

$$H(\eta) = \eta - v(0, \eta).$$

Then

$$H(y(t, \eta)) = y(t, \eta) - v(0, y(t, \eta)) = y(t, \eta) - v(t, \eta)$$

and

$$\frac{d}{dt}H(y(t, \eta)) = A(y(t, \eta) - v(t, \eta)) = AH(y(t, \eta)).$$

The map H so constructed has the property that if $y(t)$ is a solution of (3.51) then $x(t) = H(y(t))$ is the unique solution of the linear system (3.52) such that $\|H(y(t)) - y(t)\| = \|v(t, \eta)\|$ is bounded, namely:

$$\|H(y(t)) - y(t)\| \leq 2\mu M\alpha^{-1}. \tag{3.53}$$

Clearly, any map H satisfying (i) and (ii) will have this property and hence must coincide with the one constructed.

It remains to show that H is a homeomorphism. To this end we construct its inverse. Let $u(t, \eta)$ be the unique bounded solution to

$$\dot{u} = Au + f(e^{tA}\eta + u).$$

Such a solution exists by Lemma 3.32 with $\tilde{c}(t, u, \eta) = f(e^{tA}\eta + u)$. Moreover, for all (t, η) the estimate

$$\|u(t, \eta)\| \leq 2\mu M(\alpha - 2KM)^{-1}$$

holds and $u(t, \eta)$ is a continuous function. We have

$$u(t, e^{sA}\eta) = u(t + s, \eta).$$

Now define

$$J(\eta) = \eta + u(0, \eta).$$

Then J is continuous, satisfies the estimate

$$\|J(\eta) - \eta\| \leq \mu M(\alpha - 2KM)^{-1} \tag{3.54}$$

for all η, and

$$J(e^{tA}\eta) = e^{tA}\eta + u(t, \eta)$$

is a solution to (3.51). Thus, the map J has the property that if $u(t)$ is a solution to the linear system (3.52), then $J(u(t))$ is the unique solution to the nonlinear system (3.51) such that $\|J(u(t)) - u(t)\|$ is bounded.

Now let $y(t)$ be a bounded solution to (3.51). It also has the property that

$$J(H(y(t))) - y(t) = J(H(y(t))) - H(y(t)) + H(y(t)) - y(t)$$

is bounded, see estimates (3.53) and (3.54). But $w(t) = J(H(y(t))) - y(t)$ is a solution to the system

$$\dot{w} = Aw + f(y(t) + w) - f(y(t))$$

of which $w(t) \equiv 0$ is also a bounded solution. By uniqueness in Lemma 3.32, we have $J(H(y(t))) - y(t) \equiv 0$ or

$$y(t) \equiv J(H(y(t))).$$

Since $y(t)$ is an arbitrary solution to (3.51), it follows that $J \circ H = I$. Similarly, one can show that $H \circ J = I$, so that H and J are inverses of each other. Thus, H is a homeomorphism. □

Similar to the map case, the global Theorem 3.30 implies the corresponding local theorem for C^1-ODEs.

Theorem 3.34 (Grobman–Hartman Theorem for ODEs) *A hyperbolic linear system*

$$\dot{x} = Ax, \quad x \in \mathbb{R}^n,$$

is locally topologically conjugate near the origin to any C^1-system

$$\dot{y} = Ay + f(y), \quad y \in \mathbb{R}^n, \tag{3.55}$$

where $f(0) = 0$ and $Df(0) = 0$, i.e. there exists a homeomorphism $x \mapsto y = h(x)$ with $h(0) = 0$ defined on a neighbourhood of the origin for which

$$h \circ e^{tA} = \varphi^t \circ h,$$

where φ^t is the flow generated by (3.55).

Proof Instead of (3.55) consider a modified system

$$\dot{y} = Ay + f_\varepsilon(y), \tag{3.56}$$

where f_ε is defined by (3.38) and so has a Lipschitz constant which can be made as small as we want by choosing ε sufficiently small. Denote the flow generated by (3.56) by φ_ε^t.

Theorem 3.30 yields a homeomorphism $h : \mathbb{R}^n \to \mathbb{R}^n$ satisfying

$$h \circ e^{tA} = \varphi_\varepsilon^t \circ h$$

with $h(0) = 0$. Then, (3.39) and (3.55) are locally topologically conjugate, since φ_ε^t agrees with φ^t near the origin. □

Definition 3.35 *Let $F : \mathbb{R}^n \to \mathbb{R}^n$ be a C^1-map. An equilibrium x_0 of $\dot{x} = F(x)$, is called* **hyperbolic** *if $\dot{x} = Ax$ with $A = DF(x_0)$ is a hyperbolic linear system. If both stable and unstable eigenspaces are nontrivial, the hyperbolic equilibrium is called a* **saddle**.

As for maps, one can define stable and unstable sets of any equilibrium x_0. In fact, since we can always go backward in time, we can use the following definitions.

Definition 3.36 *The* **stable set** *$W^s(x_0)$ is*

$$W^s(x_0) = \{x \in \mathbb{R}^n : \varphi^t(x) \to x_0, \quad t \to \infty\}$$

and the **unstable set** *$W^u(x_0)$ is*

$$W^u(x_0) = \{x \in \mathbb{R}^n : \varphi^t(x) \to x_0, \quad t \to -\infty\}.$$

As a corollary of Theorem 3.34, we obtain a description of $W^s(x_0)$ and $W^u(x_0)$ near a hyperbolic equilibrium. In particular, it follows that their dimensions are n_s and n_u, respectively.

Definition 3.37 *A cycle Γ_0 of $\dot{x} = F(x)$, is called* **hyperbolic** *(saddle) if the corresponding fixed point of its Poincaré map is hyperbolic (saddle).*

Definition 3.38 *The* **stable set** *$W^s(\Gamma_0)$ of a cycle Γ_0 is defined by*

$$W^s(\Gamma_0) = \{x \in \mathbb{R}^n : \varphi^t(x) \to \Gamma_0, \quad t \to \infty\}$$

and the **unstable set** *$W^u(\Gamma_0)$ is defined by*

$$W^u(\Gamma_0) = \{x \in \mathbb{R}^n : \varphi^t(x) \to \Gamma_0, \quad t \to -\infty\}.$$

Applying the Grobman–Hartman Theorem for Maps to the Poincaré maps of a hyperbolic cycle Γ_0, one can construct $W^{s,u}(\Gamma_0)$ near Γ_0. Indeed, take the union of the (un)stable local invariant manifolds of the fixed point of the Poincaré map on various transversal sections to Γ_0.

The following statement, that we give without proof, is valid.

Theorem 3.39 *Suppose a C^1-system*

$$\dot{x} = F(x), \quad x \in \mathbb{R}^n, \tag{3.57}$$

has a cycle Γ_0, while a C^1-system

$$\dot{y} = G(y), \quad y \in \mathbb{R}^n, \tag{3.58}$$

has a cycle Λ_0. If their respective Poincaré maps are locally conjugate near the fixed points corresponding to Γ_0 and Λ_0, then (3.57) is locally topologically equivalent near Γ_0 to (3.58) near Λ_0. □

Together with Theorem 3.25, this theorem allows one to classify hyperbolic cycles in ODEs.

3.4 References

The asymptotic stability of an equilibrium with all eigenvalues satisfying Re $\lambda < 0$ can also be proved directly by constructing a *Lyapunov function*. Such proofs can be found, for example, in Arnol'd (1973), Verhulst (1996) (see also Exercise 3.5).

The stability of limit cycles in ODEs and its relation to the Poincaré maps and the multipliers are analysed in detail in Hartman (2002).

There are many different proofs of the Grobman–Hartman Theorems, e.g. Hartman (2002), Nitecki (1971), Kirchgraber and Palmer (1990), Katok and Hasselblatt (1995). In the latter book, stable and unstable invariant foliations near the saddle are used to construct the conjugating homeomorphism, see also Shilnikov et al. (1998). This approach also provides the existence of local (centre-)stable and (centre-)unstable invariant manifolds, which we will prove using a different method in Chap. 6.

3.5 Exercises

Exercise 3.1 (Delayed Logistic Map) Investigate the stability of a positive fixed point of the map[3]

$$\begin{pmatrix} x \\ y \end{pmatrix} \mapsto \begin{pmatrix} rx(1-y) \\ x \end{pmatrix}$$

for $r > 0$. For which r does the linearization fail to provide the answer?

Exercise 3.2 (Stability of cycles for maps) Let $g : \mathbb{R}^n \to \mathbb{R}^n$, $x \mapsto g(x)$ be a C^1-map. Assume that x_1 is mapped by g to x_2, while x_2 is mapped to x_1, so that $\Gamma = \{x_1, x_2\}$ is a 2-cycle. Both x_1 and x_2 are fixed points of $f = g \circ g$.

(i) Prove that two Jacobian matrices, $A_1 = Df(x_1)$ and $A_2 = Df(x_2)$, have the same stability properties, i.e. $n_s(A_1) = n_s(A_2)$ and $n_u(A_1) = n_u(A_2)$.

(ii) Verify directly from Definitions 1.14 and 1.15 in Chap. 1 that x_1 and x_2 have the same stability character as fixed points of f.

(iii) The cycle Γ is obviously an invariant set for g. Prove that Γ is an (asymptotically) stable cycle of g if and only if x (or y) is an (asymptotically) stable fixed point of f.

Exercise 3.3 (A differential version of Gronwall's Lemma) Prove the following version of Gronwall's Lemma which uses a differential instead of an integral inequality.

Lemma 3.40 *Let $T > 0$ and let α, β be continuous functions and let u be a continuously differentiable function on $[0, T]$ satisfying*

$$u'(t) \le \alpha(t) + \beta(t)u(t), \quad t \in [0, T].$$

[3] Maynard Smith, J. (1968). *Mathematical Ideas in Biology.* London: Cambridge University Press.

Then

$$u(t) \le e^{\int_0^t \beta(\tau)d\tau}\left(u(0) + \int_0^t e^{-\int_0^s \beta(\tau)d\tau}\alpha(s)ds\right), \quad t \in [0,T]. \qquad \square$$

Hint: Note that there is no assumption on the sign of β. Use the Fundamental Theorem of Calculus for the function $v(t) = e^{-\int_0^t \beta(\tau)d\tau}u(t)$.

Exercise 3.4 (Lyapunov Theorem revisited) Prove the stability part of Theorem 3.5 as follows:

(i) Introduce an equivalent norm $\|\cdot\|_2$ in $\mathbb{R}^n$ as in Theorem 2.19.

(ii) Using the C^1-assumption on f, show that for any small $\varepsilon > 0$, there exists a closed ball $\overline{B}_\delta$ of $\|\cdot\|_2$-radius $\delta > 0$, such that for all $x \in \overline{B}_\delta$ we have $\|f(x)\|_2 \le \varepsilon\|x\|_2$.

(iii) Take any μ such that $0 < \mu < \delta$ and introduce another ball $\overline{B}_\mu \subset \overline{B}_\delta$.

(iv) Let $T > 0$ be such that $x(t) = \varphi^t(x_0) \in \overline{B}_\delta$ for $x_0 \in \overline{B}_\mu$ and $t \in [0,T]$, and let $u(t) = e^{\alpha t}\|x(t)\|_2$. Use the variation of constant formula (3.2) and Gronwall's inequality to prove that

$$u(t) \le \|x_0\|_2 e^{\varepsilon t}, \quad t \in [0,T].$$

(v) Prove that $x(t) \in \overline{B}_\mu$ for all $t \ge 0$ and next that $x(t) \to 0$ as $t \to +\infty$. Conclude from this that the equilibrium $x = 0$ is asymptotically stable.

Exercise 3.5 (Lyapunov Theorem via Lyapunov's method) Here we indicate yet another proof of the stability part of Theorem 3.5.

(i) Choose $\beta < 0$ such that $\mathrm{Re}\lambda < \beta$ for all eigenvalues of $A = DF(0)$. Show that the matrix

$$H = \int_0^\infty e^{t(A^{\mathrm{T}} - \beta I_n)} e^{t(A - \beta I_n)} dt$$

is well defined, positive definite and solves *Lyapunov's equation*

$$(A^{\mathrm{T}} - \beta I_n)H + H(A - \beta I_n) = -I_n.$$

Hint: Insert H and use integration by parts.

(ii) Show that

$$\|e^{tA}x\|_H \le e^{\beta t}\|x\|_H, \quad t \ge 0$$

holds for the norm $\|x\|_H = \sqrt{x^T H x}$.
Hint: Use Exercise 3.3 for $u(t) = \|e^{tA}x\|_H^2$.

(iii) Prove again Lyapunov Theorem by using the *Lyapunov function* $V(x) = \|x\|_H^2$. More precisely, determine $\varepsilon > 0$ such that the estimate

$$\frac{d}{dt}V(\varphi^t(x)) \le \beta V(\varphi^t(x))$$

holds for $\|x\|_H \le \varepsilon$ on a maximal interval $[0, \tau_0)$. Prove that $\tau_0 < \infty$ leads to a contradiction. Finally, use Exercise 3.3 to obtain the conclusion.

Exercise 3.6 (Local analysis of equilibria) Consider the following system of two differential equations:

$$\begin{cases} \dot{x} = y, \\ \dot{y} = -x - y + x^2, \end{cases} \tag{3.59}$$

which describes a nonlinear oscillator with friction: $\ddot{x} + \dot{x} + x - x^2 = 0$.

(i) Find all equilibrium points of (3.59).

(ii) Determine their topological type and sketch the phase portrait of (3.59) near each of the equilibrium points.

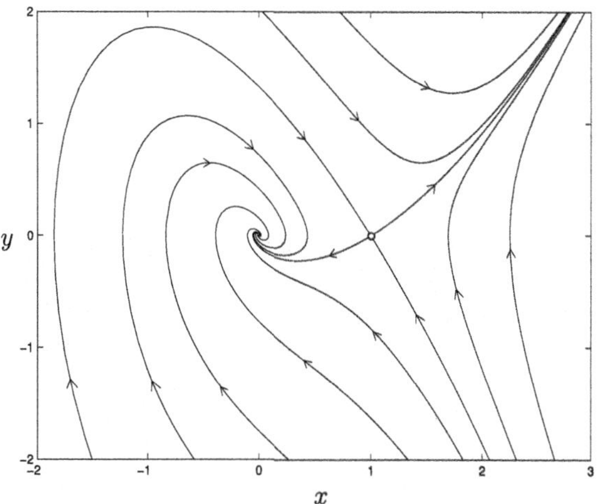

Fig. 3.6 The phase portrait of (3.59)

(iii) Prove that (3.59) has neither periodic nor homoclinic orbits in the (x, y)-plane. *Hint*: The *total mechanical energy* of the oscillator

$$E = \frac{y^2}{2} + U(x), \quad U(x) = \frac{x^2}{2} - \frac{x^3}{3},$$

dissipates due to friction.

(iv) Sketch the global phase portrait of (3.59). See Fig. 3.6.

Exercise 3.7 (Explicit linearization) Compare the phase portraits of the nonlinear system

$$\begin{cases} \dot{x} = -x, \\ \dot{y} = y + x^2, \end{cases} \tag{3.60}$$

with that of its linearization near the origin

$$\begin{cases} \dot{x} = -x, \\ \dot{y} = y. \end{cases} \tag{3.61}$$

(i) Determine α such that $y + \alpha x^2 = 0$ is invariant under (3.60).

(ii) Using Grobman–Hartman Theorem, sketch the phase portraits of these systems near the origin.

(iii) Make a guess for the conjugating homeomorphism (in fact diffeomorphism) h and then verify whether you made the right choice by checking the formula (1.12) in Sect. 1.3. *Remark* It is also possible to solve both systems explicitly and then find h satisfying (1.11), see Perko (2001).

Exercise 3.8 (Stability of equilibria in Lorenz system) We begin by stating an important general result.

Theorem 3.41 (Routh–Hurwitz Criterion)[4] *Consider a polynomial*

$$P_n(z) = p_0 z^n + p_1 z^{n-1} + p_2 z^{n-2} + \cdots + p_{n-1} z + p_n,$$

where $p_0 = 1$, $p_i \in \mathbb{R}$ *for* $i = 1, 2, \ldots, n$ *and set* $p_i = 0$ *for all* $i \notin \{0, 1, 2, \ldots, n\}$. *Introduce* $k \times k$ *matrices* $H^{(k)}$, $k = 1, 2, \ldots, n$, *whose elements are defined by the following formulas*

$$h_{ij}^{(k)} = p_{2j-i}, \quad i, j = 1, 2, \ldots, k,$$

and define $d_k = \det H^{(k)}$ *for* $k = 1, 2, \ldots, n$.

Then all roots λ *of* $P_n(z)$ *satisfy* $\operatorname{Re} \lambda < 0$ *if and only if*

$$d_k > 0, \quad k = 1, 2, \ldots, n. \qquad \square$$

Using this theorem, analyse linear stability of all equilibria in the famous three-dimensional *Lorenz system*[5]

$$\begin{cases} \dot{x} = \sigma(y - x), \\ \dot{y} = rx - y - xz, \\ \dot{z} = -bz + xy, \end{cases}$$

where all parameters σ, r, b are positive.

Exercise 3.9 (Floquet multipliers revisited) Definition 3.9 introduces the monodromy matrix and Floquet multipliers of a periodic orbit Γ_0. These involve the choice of a point $x_0 \in \Gamma_0$ but x_0 is not incorporated in the notation. The aim of this exercise is to show that there are indeed good reasons to down play the role of x_0.

Assume that both x_0 and x_1 lie on the same periodic orbit Γ_0 with period T_0. Define $Y(t)$ as in Lemma 3.6 and let $Z(t)$ be the corresponding matrix when starting in x_1, i.e.

$$Z(t) = D_x \varphi^t(x)(x_1),$$

where φ^t is the flow defined by the ODE.

(i) Use the identity $\varphi^t(x) = \varphi^{-s}(\varphi^t(\varphi^s(x)))$ to derive that $Y(T) = Z(-s_0)Z(T)Y(s_0)$ for a suitable choice of s_0.

(ii) Use the identity $x = \varphi^{-s}(\varphi^s(x))$ to show that $Z(-s_0)Y(s_0) = I = Y(s_0)Z(-s_0)$, where s_0 is as in (i).

(iii) Combine these two observations and deduce that the matrices $Y(T)$ and $Z(T)$ have the same eigenvalues.

Exercise 3.10 (Correspondence maps) The construction of the Poincaré map involved a choice of a point $x_0 \in \Gamma_0$ through which a cross section Π_{x_0} is constructed.

(i) Prove that two Poincaré maps corresponding to different points $x_0, x_1 \in \Gamma_0$ are smoothly conjugate. (*Hint*: For x_0 and x_1 introduce the cross sections Π_{x_0} and Π_{x_1} and define two *correspondence maps*:

$$Q_{10} : \Pi_{x_1} \to \Pi_{x_0}$$

and

$$Q_{01} : \Pi_{x_0} \to \Pi_{x_1}$$

[4] Hurwitz, A. (1964). On the conditions under which an equation has only roots with negative real parts. In R. Bellman & R. Kalaba (Eds.), *Selected Papers on Mathematical Trends in Control Theory*. New York: Dover Publications.

[5] Lorenz, E. (1963). Deterministic non-periodic flow. *Journal of the Atmospheric Sciences, 20*, 130–141.

along the orbits of the system as in Lemma 3.11. Express both Poincaré maps $\mathcal{P}_k : \Pi_{x_k} \to \Pi_{x_k}$, $k = 0, 1$, in terms of $\mathcal{Q}_{10}$ and $\mathcal{Q}_{01}$ and prove their conjugacy.)

(ii) How would you deal with Poincaré maps corresponding to different (and not necessarily planar) cross sections through the same point x_0?

Chapter 4
Planar ODEs

Abstract In this chapter we prove the famous Poincaré–Bendixson Theorem that classifies the asymptotic behaviour of bounded orbits in smooth planar ODEs. We shall see that in a generic planar system any such orbit tends either to an equilibrium or to a cycle (periodic orbit), although in general more complicated limit sets are possible. Then we give some conditions ensuring existence or nonexistence of cycles in planar systems, in particular, when the system is close to a special (Hamiltonian) system. We apply many results and techniques to the analysis of several prey–predator ecological models. Basic facts about multidimensional Hamiltonian systems are collected in the Appendix.

4.1 Limit Sets

Before focusing on the special case $n = 2$, we introduce some notions that are applicable in any finite dimension. So, consider

$$\dot{x} = f(x), \quad x \in \mathbb{R}^n, \tag{4.1}$$

where $f \in C^1(\mathbb{R}^n, \mathbb{R}^n)$. Let $\varphi(t, x) = \varphi^t(x)$, where φ^t is the (local) flow generated by (4.1). Theorem 1.5 implies that the map $(t, x) \mapsto \varphi(t, x)$ is continuous in any point (t, x), where it is defined.

Definition 4.1 *A point $y \in \mathbb{R}^n$ is an **ω-limit point** of $x \in \mathbb{R}^n$, if $\varphi(t, x)$ is defined for all $t \geq 0$ and there exists a sequence $\{t_i\} \to \infty$, such that $\varphi(t_i, x) \to y$ for $i \to \infty$. The set of all ω-limit points of x is called the **ω-limit set** of x and denoted by $\omega(x)$.*

Remarks

(1) We use the expression $\{t_i\} \to \infty$ to indicate that the sequence $\{t_i\}$ converges to infinity, i.e. for every $\varepsilon > 0$ there exists $K(\varepsilon)$ such that $t_i \geq (1/\varepsilon)$ for all $i \geq K(\varepsilon)$. Similarly, $\{p_i\} \to p$ means that the sequence of points $p_i \in \mathbb{R}^n$ converges to $p \in \mathbb{R}^n$.

Yu. Kuznetsov et al., *Dynamical Systems Essentials*, Texts in Applied Mathematics 83, https://doi.org/10.1007/978-3-032-04083-1_4

(2) Considering sequences $\{t_i\} \to -\infty$, we obtain the α-limit points of $x \in \mathbb{R}^n$ and the set $\alpha(x)$.[1]

(3) Since $\omega(\varphi^t(x)) = \omega(x)$, one can use the notation $\omega(\Gamma)$, where $\Gamma = \Gamma(x)$ is the orbit through x, or $\omega(\Gamma^+)$, where $\Gamma^+ = \Gamma^+(x)$ is the *positive half-orbit* through x:

$$\Gamma^+(x) = \{y \in \mathbb{R}^n : y = \varphi^t(x),\ t \geq 0 \text{ such that } \varphi^t(x) \text{ is defined}\}\ .$$

(4) If $x^0 \in \mathbb{R}^n$ is an equilibrium point of (4.1), then $\omega(x^0) = x^0$. If $x \in \Gamma_0$, where Γ_0 is a periodic orbit of (4.1), then $\omega(x) = \Gamma_0$. ◊

For any two subsets $A, B \subset \mathbb{R}^n$ define

$$\text{dist}(A, B) = \inf_{x \in A, y \in B} \|x - y\|\ .$$

If A consists of one point $x \in \mathbb{R}^n$ and B is fixed, the function $x \mapsto \text{dist}(x, B)$ (that has been already used in Chap. 1) is continuous. The following four lemmas are valid for all $n \geq 1$.

Lemma 4.2 (Basic properties) *Let $x \in \mathbb{R}^n$ and assume that $\Gamma^+(x)$ is bounded. Then the set $\omega(x)$ is* (i) *nonempty,* (ii) *bounded,* (iii) *closed, and* (iv) *connected.*

Proof Since $\Gamma^+(x)$ is bounded, $\varphi(t, x)$ is defined for all $t \geq 0$.

(i) Consider $x_i = \varphi(i, x)$, $i = 1, 2, \ldots$. The sequence $\{x_i\}$ is bounded and infinite. So, there is a convergent subsequence: $\{x_{I(k)}\} \to p$. This means $p \in \omega(x)$, so $\omega(x) \neq \emptyset$.

(ii) If $\omega(x)$ is unbounded, then $\Gamma^+(x)$ is also unbounded, a contradiction.

(iii) Let $\{p_i\}$ be a sequence of points $p_i \in \omega(x)$ and $\{p_i\} \to p \in \mathbb{R}^n$. Let us prove that $p \in \omega(x)$.

Indeed, for each p_i there is a sequence of times $\{t_k^{(i)}\} \to \infty$ as $k \to \infty$, such that the corresponding sequence $\{x_k^{(i)}\}$ defined by

$$x_k^{(i)} = \varphi(t_k^{(i)}, x)$$

converges to p_i, i.e. $\|x_k^{(i)} - p_i\| \to 0$ for $k \to \infty$. For each i, take the first $K = K(i)$ such that

$$\|x_{K(i)}^{(i)} - p_i\| \leq \frac{1}{i}.$$

Then

$$\|x_{K(i)}^{(i)} - p\| = \|(x_{K(i)}^{(i)} - p_i) + (p_i - p)\| \leq \|x_{K(i)}^{(i)} - p_i\| + \|p_i - p\| \leq \frac{1}{i} + \|p_i - p\| \to 0$$

as $i \to \infty$. This means that $p \in \omega(x)$, so $\omega(x)$ is closed.

---

[1] Recall that the Greek alphabet begins with α and ends with ω.

(iv) Suppose $\omega(x) = A \cup B$, where $A, B \subset \mathbb{R}^n$ are nonempty, closed, and disjoint, i.e. $A \cap B = \emptyset$. Then, since A, B are closed subsets of the compact set $\omega(x)$ we have

$$\mathrm{dist}(A, B) = \delta > 0.$$

Since A and B are nonempty and composed of ω-limit points, there exist two sequences of times $\{t_i^A\} \to \infty$ and $\{t_i^B\} \to \infty$ such that

$$\mathrm{dist}(\varphi(t_i^A, x), A) < \frac{\delta}{2} \quad \text{and} \quad \mathrm{dist}(\varphi(t_i^B, x), B) < \frac{\delta}{2}.$$

This implies

$$\mathrm{dist}(\varphi(t_i^B, x), A) > \frac{\delta}{2}$$

and, by continuity of dist and φ, there exists a sequence $\{\tau_j\} \to \infty$, such that

$$\mathrm{dist}(\varphi(\tau_j, x), A) = \frac{\delta}{2}.$$

The sequence $\{\varphi(\tau_j, x)\}$ is infinite and bounded. Thus, there exists a subsequence $\{\tau_{J(k)}\}$ such that

$$\{\varphi(\tau_{J(k)}, x)\} \to z, \quad k \to \infty,$$

with $z \notin A, B$. But clearly $z \in \omega(x)$, which is a contradiction. Therefore, $\omega(x)$ is connected. □

Lemma 4.3 (Convergence) *Let $\Gamma^+(x)$ be bounded. Then*

$$\mathrm{dist}(\varphi(t, x), \omega(x)) \to 0 \quad \textit{as} \quad t \to +\infty.$$

Proof Suppose the claim is false, then there is an $\varepsilon > 0$ and a sequence $\{t_i\} \to \infty$ of times such that

$$\mathrm{dist}(\varphi(t_i, x), \omega(x)) \geq \varepsilon > 0 \tag{4.2}$$

for all $i = 1, 2, \ldots$. As in the proof of Lemma 4.2(i), the sequence $\{x_i\}$ with $x_i = \varphi(t_i, x)$ is bounded and infinite. So, it has a convergent subsequence:

$$\{x_{I(k)}\} \to p \in \mathbb{R}^n.$$

Using the continuity of dist, we get from (4.2)

$$\mathrm{dist}(p, \omega(x)) > 0,$$

which is a contradiction, since by definition $p \in \omega(x)$. □

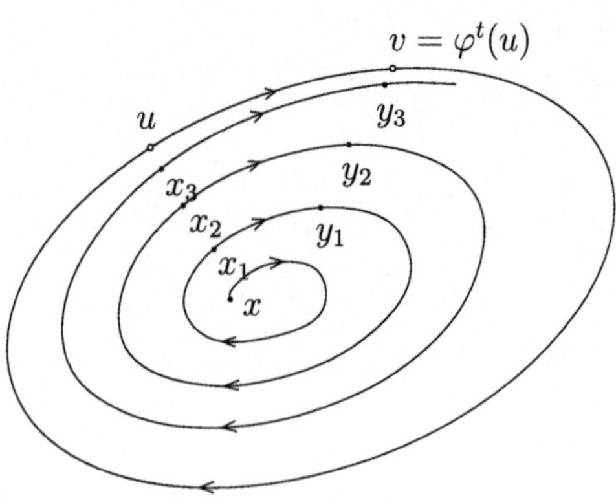

Fig. 4.1 Forward invariance of $\omega(x)$

Lemma 4.4 (Invariance) *Let $\Gamma^+(x)$ be bounded. Then $\omega(x)$ is an invariant set of the flow φ^t, i.e. if $y \in \omega(x)$ then $\varphi^t(y) \in \omega(x)$ for all $t \in \mathbb{R}$.*

Proof First consider forward invariance. Let $u \in \omega(x)$. We want to prove that $\varphi^t(u)$ is defined and $v = \varphi^t(u) \in \omega(x)$ for $t > 0$. By definition, there is a sequence of times $\{t_i\} \to \infty$ such that

$$x_i = \varphi(t_i, x) \to u,$$

as $i \to \infty$. Consider the sequence

$$y_i = \varphi^t(x_i) = \varphi(t, \varphi(t_i, x))$$

(see Fig. 4.1). Since φ is a continuous function of both of its arguments, $\varphi(t, u)$ exists and

$$\{y_i\} \to \varphi(t, u) = v.$$

On the other hand, by the semigroup property,

$$y_i = \varphi(t + t_i, x),$$

so

$$\varphi(\tau_i, x) \to v$$

as $i \to \infty$ for $\tau_i = t + t_i$ and $\{\tau_i\} \to \infty$. Therefore, $v \in \omega(x)$.

For $t < 0$ (backward invariance), the same construction works with a slight modification. Namely, it could happen that $\varphi^t(x_i)$ is not defined for $i \leq N$ for some integer $N > 0$, but $\varphi^t(x_i)$ is well defined for $i > N$ (see Fig. 4.2, where $\mathcal{B}$ represents infinitely remote points). In this case, start from τ_{N+1}, i.e. take $\tilde{\tau}_i = \tau_{N+i}$. □

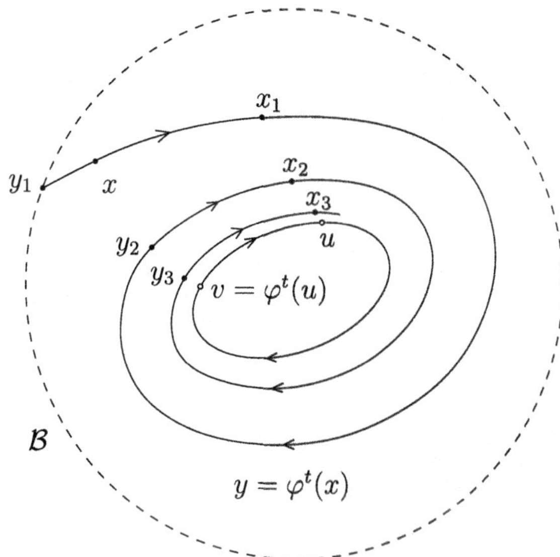

Fig. 4.2 Backward invariance of $\omega(x)$

Lemma 4.5 (Transitivity) *Let $x, y, z \in \mathbb{R}^n$. If $z \in \omega(y)$ and $y \in \omega(x)$, then $z \in \omega(x)$.*

Proof Let $\{x_i\} \to y$, where $x_i = \varphi(t_i, x)$ for some sequence $\{t_i\} \to +\infty$, while $\{y_i\} \to z$, where $y_i = \varphi(\tau_i, y)$ for some sequence $\{\tau_j\} \to +\infty$. Define

$$z_i = \varphi^{\tau_i}(x_i) = \varphi(\tau_i, \varphi(t_i, x)) = \varphi(\tau_i + t_i, x).$$

Then, by continuity of $\varphi(\cdot, \cdot)$,

$$z_i = \varphi(\tau_i, \varphi(t_i, x)) \to z.$$

Therefore, for the sequence $\{\tau_i + t_i\} \to +\infty$,

$$z_i = \varphi(\tau_i + t_i, x) \to z.$$

□

4.2 The Poincaré–Bendixson Theorem

For the rest of this section, fix $n = 2$, i.e. consider a planar system

$$\begin{cases} \dot{x}_1 = f_1(x_1, x_2), \\ \dot{x}_2 = f_2(x_1, x_2), \end{cases} \tag{4.3}$$

where $f_{1,2}$ are C^1-functions of (x_1, x_2). Natural candidates for ω-limit sets in such systems are equilibria and cycles. Recall that a point x^0 is an equilibrium if $f(x^0) = 0$, while a cycle is a closed orbit Γ_0 that corresponds to a nonequilibrium solution for which $\varphi(t, x) = \varphi(t + T, x)$ with some $T > 0$ and $x \in \Gamma_0$. The minimal T is called the *period* of Γ_0.

Definition 4.6 *A closed line segment L is called* ***transverse*** *for* (4.3) *if the vector field f neither vanishes in L, nor is anywhere tangent to L.*

A transverse segment can obviously be constructed through any point $x \in \mathbb{R}^2$ where $f(x) \neq 0$ (for example, take a sufficiently small segment orthogonal to $f(x)$). Consider such a segment L, then there are no equilibria in L and any orbit crosses L at a nonzero angle. Introduce a coordinate s along L.

Definition 4.7 *A continuous image of a circle without self-intersections is called a* ***Jordan curve****.*

According to a famous theorem by Jordan, the complement to a Jordan curve $\Gamma \subset \mathbb{R}^2$ is the union of two disjoint connected open sets, the interior $\text{Int}(\Gamma)$ and the exterior $\text{Ext}(\Gamma)$, the first one of which is bounded while the second is unbounded.

Lemma 4.8 (Monotonicity) *If an orbit of* (4.3) *intersects L for an increasing sequence of times $\{t_i\}$, then the corresponding sequence of intersection points $\{p_i\}$ is either constant or strictly monotone.*

Proof Consider an orbit that crosses a transverse segment L at two *distinct* points

$$p_0 = \varphi(t_0, x), \ p_1 = \varphi(t_1, x),$$

with $t_0 < t_1$, and has no other intersections with L for $t \in [t_0, t_1]$. Denote by Γ the closed (piecewise-smooth) curve composed of the orbit part connecting p_0 and p_1, and the open subsegment $L_0 \subset L$ between these same points. Γ is a Jordan curve. Moreover $\text{Int}(\Gamma)$ is a forward-invariant open set for (4.3), since no orbit can cross its boundary Γ outwards.

This is obvious for the part of the boundary consisting of the orbit that connects p_0 and p_1. Now suppose that orbits starting at points in L_0 enter $\text{Int}(\Gamma)$ for small positive times (see Fig. 4.3). By continuity, the orbit starting at p_1 will also enter $\text{Int}(\Gamma)$ and hence belong to $\text{Int}(\Gamma)$ for all positive times. Thus, all its possible intersections with L, including the next one p_2, must be located in $\text{Int}(\Gamma)$. This implies that the corresponding sequence of intersection points $\{p_i\}$ is monotone.

If $p_0 = p_1$, the orbit is periodic and all its further intersections with L occur at the same point, giving a constant sequence $p_i = p_0$.

The case, when orbits starting at points in L_0 enter $\text{Int}(\Gamma)$ for small negative times, reduces to the case considered above by reversing time. □

The map $p_i \mapsto p_{i+1}$ or, in terms of the coordinate s along L, $s_i \mapsto s_{i+1}$ is obviously a Poincaré map as introduced in Chap. 3.

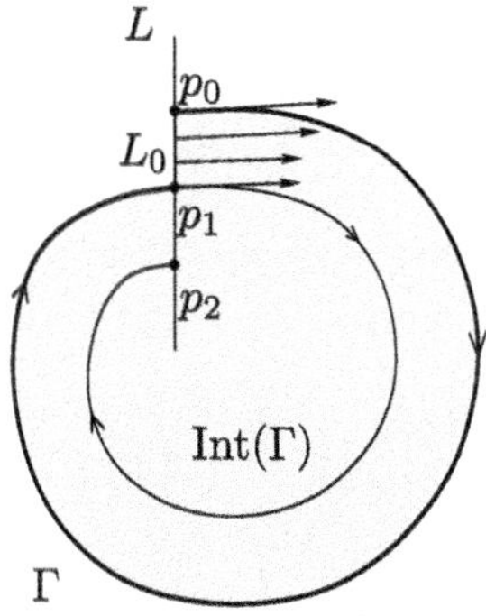

Fig. 4.3 Monotonicity of the intersections

Lemma 4.9 *Let $p \in \omega(x)$ and let L be a transverse segment through p. Then there is an increasing sequence $\{t_i\} \to \infty$ such that for $q_i = \varphi(t_i, x)$ we have $\{q_i\} \to p$ and $q_i \in L$.*

Proof By definition, there is a sequence of points $\{p_i\} \to p$, such that

$$p_i = \varphi(t_i, x)$$

for an increasing sequence of times $\{t_i\} \to +\infty$. All these points belong to the same orbit $\Gamma(x)$ but, in general, $p_i \notin L$. For each point p_i near p, take $q_i = \varphi(\tau(p_i), p_i)$, where $\tau(x)$ is the minimal (in absolute value) time needed to move along $\Gamma(x)$ from p_i to L (see Fig. 4.4). When $\{p_i\} \to p$, we have $\tau(p_i) \to 0$ so that $d(q_i, p_i) \to 0$ and therefore $\{q_i\} \to p$ as $i \to \infty$.

The existence of $\tau(x)$ can be established as in Lemma 3.11 of Chap. 3. Indeed, let $a \in \mathbb{R}^2$ be a vector orthogonal to L, so that for any point $q \in L$ we have

$$\langle a, q - p \rangle = 0 \, .$$

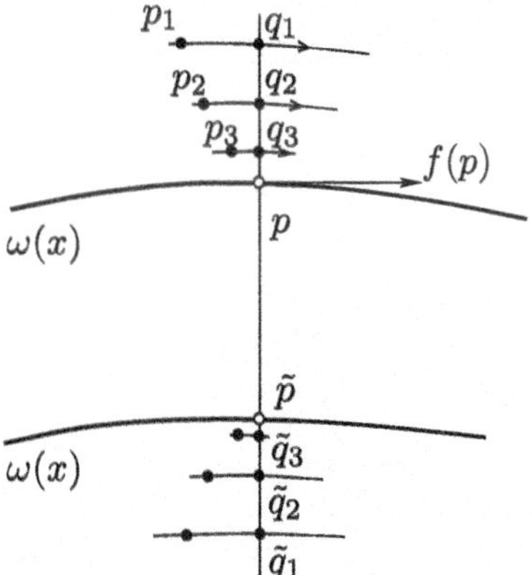

Fig. 4.4 $L \cap \omega(x)$ consists of only one point p

The condition that the orbit starting at x arrives at $t = \tau$ to a point $q \in L$ can be written as

$$g(x,\tau) = \langle a, \varphi(\tau, x) - p\rangle = 0\ .$$

The scalar function g is defined near $(p, 0)$ and smooth. It obviously satisfies $g(p, 0) = 0$. Moreover, since L is a transverse segment,

$$\frac{\partial}{\partial \tau} g(p, 0) = \langle a, f(p)\rangle \neq 0\ .$$

Therefore, the Implicit Function Theorem guarantees the existence of a function $x \mapsto \tau(x)$ having the property $\tau(p) = 0$, defined and smooth in a neighbourhood of p, and satisfying $g(x, \tau(x)) = 0$ in this neighbourhood. □

Lemma 4.10 *Under the assumptions of* Lemma *4.9, the set* $\omega(x)$ *can intersect the transverse segment* L *in at most one point.*

Proof Let $p \in \omega(x) \cap L$. By Lemma 4.9, there is a sequence of points $\{q_i\} \to p$ with $q_i \in L$ and belonging to $\Gamma(x)$ (see Fig. 4.4 again).

According to Lemma 4.8, the sequence $\{q_i\}$ is monotone. Any monotone sequence along a line has at most one limit point. □

The following lemma is often used to prove the existence of a periodic orbit.

Lemma 4.11 *If* $\omega(x)$ *is nonempty and does not contain equilibria, then it contains a periodic orbit* Γ_0.

Proof Take $y \in \omega(x)$ and consider $z \in \omega(y)$ (Fig. 4.5). By Lemma 4.5, $z \in \omega(x)$ and, thus, z is not an equilibrium. Take a transverse segment L through z, and consider a sequence $\{y_i\} \to z$ with

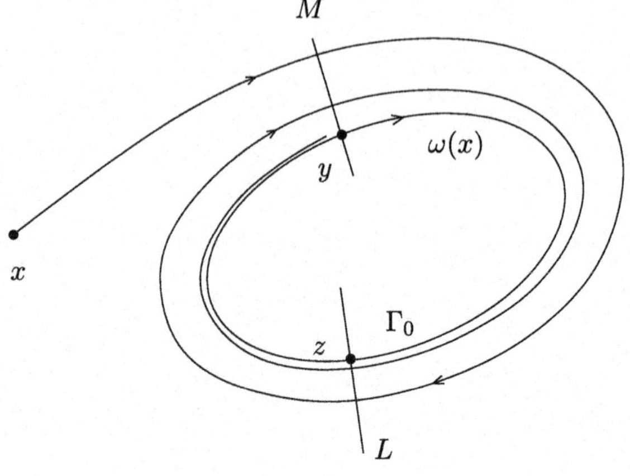

Fig. 4.5 Lemmas 4.11 and 4.12

$$y_i = \varphi(t_i, y) \in L \cap \Gamma(y)$$

(the existence of such a sequence is guaranteed by Lemma 4.9). Notice that by Lemma 4.4 all $y_i \in \omega(x)$. By Lemma 4.8, the sequence $\{y_i\}$ is either strictly monotone or constant. If $\{y_i\}$ is monotone, $\omega(x)$ has more than one intersection with L, a contradiction with Lemma 4.10. Thus, it is constant, i.e.

$$y_i = z,$$

implying that $\Gamma_0 = \Gamma(y)$ is a periodic orbit. □

Lemma 4.12 *If $\omega(x)$ contains a periodic orbit Γ_0, then $\omega(x) = \Gamma_0$.*

Proof There is a transverse segment M through any point $y \in \Gamma_0$ (see Fig. 4.5). By Lemma 4.10, the intersection $\omega(x) \cap M = \{y\}$. Therefore, there is an open annulus in which Γ_0 is the only subset of $\omega(x)$. Since $\omega(x)$ is connected, $\omega(x) \setminus \Gamma_0$ is empty. □

Lemma 4.13 *If $\omega(x)$ contains equilibrium points and $q \in \omega(x)$ is not an equilibrium, then both $\omega(q)$ and $\alpha(q)$ contain only equilibrium points.*

Proof Lemma 4.5 (and its reformulation for α-limit sets) implies

$$\omega(q), \alpha(q) \subset \omega(x).$$

If $\omega(q)$ does not contain an equilibrium, then $\omega(q)$ (and hence $\omega(x)$) contains a periodic orbit (Lemma 4.11). Therefore, by Lemma 4.12, $\omega(x)$ coincides with this periodic orbit. However, $\omega(x)$ contains equilibria, a contradiction. Thus, $\omega(q)$ contains an equilibrium (see Fig. 4.6 for an illustration).

Assume that $\omega(q)$ also contains a nonequilibrium point y. Take a transverse segment L through y. Due to Lemma 4.4, all points of $\Gamma(q)$ belong to $\omega(x)$. By Lemma 4.9, the orbit $\Gamma(q)$ must intersect L infinitely many times near y.

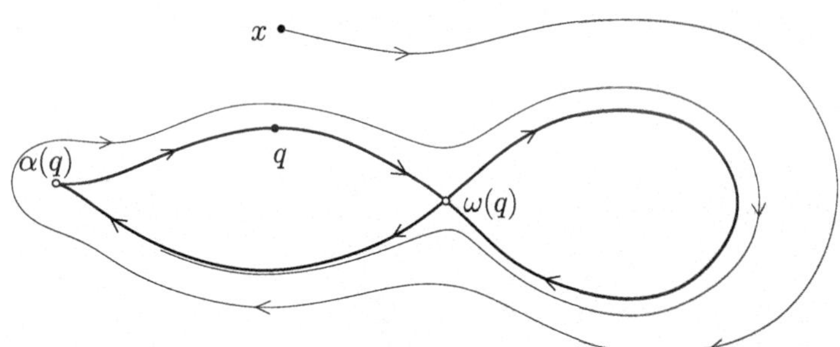

Fig. 4.6 Lemma 4.13. In this figure, $\omega(x)$ consists of two equilibria, two heteroclinic orbits, and one homoclinic orbit

Moreover, all the intersection points must be different (otherwise, $\Gamma(q)$ is a periodic orbit, which is already excluded). Thus, $\omega(x)$ intersects with L more than once, which by Lemma 4.10 is a contradiction. This implies that $\omega(q)$ contains only equilibria.

Essentially the same arguments can be applied to show that $\alpha(q)$ contains only equilibria. □

We are ready now to prove the central result of this section.

Theorem 4.14 (Poincaré–Bendixson, 1901) *Let $x \in \mathbb{R}^2$ and assume that $\Gamma^+(x)$ belongs to a bounded closed subset of $\mathbb{R}^2$ containing only a finite number of equilibria. Then one of the following possibilities holds:*

(i) *$\omega(x)$ is an equilibrium;*
(ii) *$\omega(x)$ is a periodic orbit;*
(iii) *$\omega(x)$ consists of equilibria and orbits having these equilibria as their α- and ω-limit sets.*

Proof

(i) If $\omega(x)$ contains only equilibria, it is a union of a finite number of points. However, according to Lemma 4.2, $\omega(x)$ is connected. Therefore, in this case $\omega(x)$ is just one equilibrium.

(ii) If $\omega(x)$ contains no equilibria, there is a periodic orbit $\Gamma_0 \subset \omega(x)$ (Lemma 4.11). However, by Lemma 4.12 $\omega(x) = \Gamma_0$.

(iii) Suppose $\omega(x)$ contains both equilibrium and nonequilibrium points. Let $q \in \omega(x)$ be a nonequilibrium point. Then, Lemma 4.13 implies that both $\omega(q)$ and $\alpha(q)$ contain only equilibria. Thus, as in step (i), these limit sets are just (possibly coinciding) equilibria. □

Remarks

(1) If f is an analytic vector field, in particular a polynomial vector field, then (4.3) automatically has a finite number of equilibria in any closed bounded subset of the plane.

(2) With Lemma 4.3, option (iii) of the Poincaré–Bendixson Theorem implies that $\omega(x)$ actually consists of a contour formed by equilibria and their connecting (homoclinic or heteroclinic) orbits, i.e. orbits tending to these equilibria as $t \to \pm\infty$. Clearly then, case (iii) is exceptional.

It can be proved that there can be at most one heteroclinic orbit $\Gamma \subset \omega(x)$ connecting an equilibrium $x^1 \in \omega(x)$ to another equilibrium $x^2 \in \omega(x)$ such that $x^1 \neq x^2$. Note that there could simultaneously exist another heteroclinic orbit that connects x^2 to x^1 (see Fig. 4.6). Moreover, any equilibrium $x^0 \in \omega(x)$ of an analytical system (4.3) can have only a finite number of homoclinic orbits $\Gamma_i \in \omega(x)$.

(3) A periodic orbit Γ_0 is called a *limit cycle* if there is an annulus around Γ_0 in which there are no other periodic orbits. A limit cycle can be stable, unstable, or semi-stable (meaning stable from only one side). ◇

4.3 Stability, Existence, and Uniqueness of Planar Cycles

As we have already mentioned in Sect. 3.2 of Chap. 3, for planar systems one can express the only nontrivial multiplier λ_2 of the monodromy matrix corresponding to a periodic solution $\varphi(t)$ of (4.3) as

$$\lambda_2 = \exp\left(\int_0^T \operatorname{div} f(\varphi(t))\, dt\right). \tag{4.4}$$

The cycle is (exponentially) stable if $\lambda_2 < 1$ and unstable if $\lambda_2 > 1$.

Linear stability of a T-periodic cycle of a smooth planar system (4.3) can also be studied directly using the so-called *normal coordinates*. Suppose that (4.3) has a periodic orbit (cycle) Γ_0 and

$$\varphi(t) = \begin{pmatrix} \varphi_1(t) \\ \varphi_2(t) \end{pmatrix} \tag{4.5}$$

is the corresponding periodic solution, $\varphi(t+T) = \varphi(t)$ for all $t \in \mathbb{R}$. Introduce new coordinates (τ, ξ) in a neighbourhood of the cycle by the relations:

$$\begin{cases} x_1 = \varphi_1(\tau) + \xi\dot{\varphi}_2(\tau), \\ x_2 = \varphi_2(\tau) - \xi\dot{\varphi}_1(\tau), \end{cases} \tag{4.6}$$

where $\tau \in [0, T]$ and $\xi \in \mathbb{R}$ has small absolute value (see Fig. 4.7). Note that $\xi = 0$ corresponds to the cycle. The Jacobian matrix of (4.6) is

$$J = \begin{pmatrix} \frac{\partial x_1}{\partial \tau} & \frac{\partial x_1}{\partial \xi} \\ \frac{\partial x_2}{\partial \tau} & \frac{\partial x_2}{\partial \xi} \end{pmatrix} = \begin{pmatrix} \dot{\varphi}_1 + \xi\ddot{\varphi}_2 & \dot{\varphi}_2 \\ \dot{\varphi}_2 - \xi\ddot{\varphi}_1 & -\dot{\varphi}_1 \end{pmatrix},$$

so that

$$\det(J) = -(\dot{\varphi}_1^2 + \dot{\varphi}_2^2) - \xi(\dot{\varphi}_1\ddot{\varphi}_2 - \dot{\varphi}_2\ddot{\varphi}_1).$$

Since $\dot{\varphi}(\tau) = f(\varphi(\tau)) \neq 0$ along the cycle,

$$\det(J)|_{\xi=0} = -(\dot{\varphi}_1^2 + \dot{\varphi}_2^2) = -\|\dot{\varphi}\|^2 \neq 0.$$

This implies that the coordinate transformation (4.6) is regular in a neighbourhood of Γ_0. Now write (4.3) in the (τ, ξ)-coordinates, i.e.

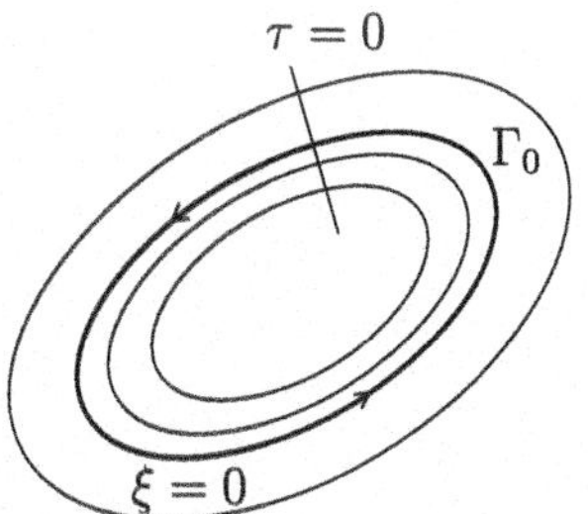

Fig. 4.7 Normal coordinates near the cycle Γ_0

$$\begin{pmatrix} \dot{\tau} \\ \dot{\xi} \end{pmatrix} = J^{-1} f(x),$$

where x is given by (4.6). First, compute the inverse of J:

$$J^{-1} = \frac{1}{\det(J)} \begin{pmatrix} -\dot{\varphi}_1 & -\dot{\varphi}_2 \\ -\dot{\varphi}_2 + \xi\ddot{\varphi}_1 & \dot{\varphi}_1 + \xi\ddot{\varphi}_2 \end{pmatrix}.$$

Since φ satisfies (4.3), we have

$$\begin{cases} \dot{\varphi}_1 = f_1(\varphi_1, \varphi_2), \\ \dot{\varphi}_2 = f_2(\varphi_1, \varphi_2), \end{cases} \tag{4.7}$$

so that

$$f(x) = \begin{pmatrix} \dot{\varphi}_1 + (a_{11}\dot{\varphi}_2 - a_{12}\dot{\varphi}_1)\xi \\ \dot{\varphi}_2 + (a_{21}\dot{\varphi}_2 - a_{22}\dot{\varphi}_1)\xi \end{pmatrix} + O(\xi^2),$$

where

$$a_{ij} = \frac{\partial f_i}{\partial x_i}, \quad i, j = 1, 2.$$

Thus

$$\dot{\tau} = -\frac{1}{\det(J)} \left[\dot{\varphi}_1^2 + \dot{\varphi}_2^2 + (-a_{12}\dot{\varphi}_1^2 + (a_{11} - a_{22})\dot{\varphi}_1\dot{\varphi}_2 + a_{21}\dot{\varphi}_2^2)\xi + O(\xi^2)\right]$$

and

$$\dot{\xi} = -\frac{1}{\det(J)} \left[-(\dot{\varphi}_1\ddot{\varphi}_1 + \dot{\varphi}_2\ddot{\varphi}_2)\xi + (a_{11}\dot{\varphi}_2^2 - (a_{12} + a_{21})\dot{\varphi}_1\dot{\varphi}_2 + a_{22}\dot{\varphi}_1^2)\xi + O(\xi^2)\right],$$

which yields

$$\frac{d\xi}{d\tau} = \left[-\frac{\dot{\varphi}_1\ddot{\varphi}_1 + \dot{\varphi}_2\ddot{\varphi}_2}{\dot{\varphi}_1^2 + \dot{\varphi}_2^2} + \frac{a_{11}\dot{\varphi}_2^2 - (a_{12} + a_{21})\dot{\varphi}_1\dot{\varphi}_2 + a_{22}\dot{\varphi}_1^2}{\dot{\varphi}_1^2 + \dot{\varphi}_2^2}\right] \xi + O(\xi^2).$$

Differentiating both equations of (4.7) with respect to time, multiplying the first by $\dot{\varphi}_1$ and the second by $\dot{\varphi}_2$, and adding the resulting expressions, we find

$$(a_{12} + a_{21})\dot{\varphi}_1\dot{\varphi}_2 = \dot{\varphi}_1\ddot{\varphi}_1 + \dot{\varphi}_2\ddot{\varphi}_2 - a_{11}\dot{\varphi}_1^2 - a_{22}\dot{\varphi}_2^2.$$

Using this formula, we obtain

$$\frac{d\xi}{d\tau} = \left[(a_{11} + a_{22}) - \frac{2(\dot{\varphi}_1\ddot{\varphi}_1 + \dot{\varphi}_2\ddot{\varphi}_2)}{\dot{\varphi}_1^2 + \dot{\varphi}_2^2}\right] \xi + O(\xi^2).$$

Therefore

$$\frac{d\xi}{d\tau} = A_1(\tau)\xi + O(\xi^2), \quad A_1(\tau) = \frac{\partial f_1}{\partial x_1}(\varphi(\tau)) + \frac{\partial f_2}{\partial x_2}(\varphi(\tau)) - \frac{d}{d\tau} \ln \|\dot{\varphi}(\tau)\|^2. \tag{4.8}$$

The solution $\xi(\tau, \xi_0)$ to (4.8) with $\xi(0, \xi_0) = \xi_0$ can be expanded as

$$\xi(\tau, \xi_0) = a_1(\tau)\xi_0 + O(\xi_0^2),$$

where $a_1(0) = 1$. From (4.8) we get the linear equation

$$\dot{a}_1 = A_1(\tau)a_1,$$

which gives, using the periodicity of $\varphi(\tau)$,

$$
\begin{aligned}
a_1(T) &= \exp\left(\int_0^T A_1(\tau)\, d\tau\right) \\
&= \exp\left(\int_0^T \left[\frac{\partial f_1}{\partial x_1}(\varphi(\tau)) + \frac{\partial f_2}{\partial x_2}(\varphi(\tau))\right] d\tau - \ln \|f(\varphi(\tau))\|^2 \Big|_0^T\right) \\
&= \exp\left(\int_0^T \text{div}\ f(\varphi(\tau))\ d\tau\right).
\end{aligned}
$$

So the Poincaré map $\xi_0 \mapsto \xi_1 = \xi(T, \xi_0)$ on the local transverse segment $\tau = 0$ has the expansion

$$
\xi_1 = e^{\int_0^T \text{div}\ f(\varphi(\tau))\ d\tau} \xi_0 + O(\xi_0^2).
$$

Therefore, the cycle is linearly stable if

$$
\int_0^T \text{div}\ f(\varphi(\tau))\ d\tau < 0
$$

and unstable if the opposite inequality is true, in accordance with Sect. 3.2 of Chap. 3.

Using (4.4), one can derive several important results on the existence and uniqueness of periodic orbits in planar systems. Let Γ_0 be a periodic orbit of (4.3) and let $\varphi(t)$ be the corresponding T-periodic solution. Recall that Γ_0 is naturally oriented by the advance of time. Denote by L_0 the same closed orbit but always oriented counterclockwise. Then, if Γ_0 and L_0 have the same (i.e. counterclockwise) orientation, we have

$$
\begin{aligned}
\oint_{L_0} f_1 dx_2 - f_2 dx_1 &= \int_0^T [f_1(\varphi(t))\dot{\varphi}_2(t) - f_2(\varphi(t))\dot{\varphi}_1(t)]dt \\
&= \int_0^T [f_1(\varphi(t))f_2(\varphi(t)) - f_2(\varphi(t))f_1(\varphi(t)]dt = 0.
\end{aligned}
$$

The result remains valid if Γ_0 has opposite (i.e. clockwise) orientation. In this case

$$
\begin{aligned}
\oint_{L_0} f_1 dx_2 - f_2 dx_1 &= \int_0^T [-f_1(\varphi(t))\dot{\varphi}_2(t) + f_2(\varphi(t))\dot{\varphi}_1(t)]dt \\
&= \int_0^T [-f_1(\varphi(t))f_2(\varphi(t)) + f_2(\varphi(t))f_1(\varphi(t)]dt = 0.
\end{aligned}
$$

We call a domain Ω *connected* if any two points from Ω can be linked by a continuous curve lying entirely in the domain. The domain is called *simply connected* if in addition any closed curve in this domain can be continuously shrunk within the domain to a point. In simply connected domains Green's Theorem holds, which asserts that

$$
\oint_{L_0} f_1 dx_2 - f_2 dx_1 = \int_{\text{Int}(L_0)} \text{div} f(x)\ dx, \tag{4.9}
$$

provided that L_0 is a piecewise-smooth counterclockwise-oriented Jordan curve. Here by definition $\operatorname{div} f(x) = \operatorname{Tr} Df(x)$. Combining these two observations we can prove (by contradiction) the following result.

Theorem 4.15 (Bendixson's Criterion) *Let Ω be a simply connected domain in $\mathbb{R}^2$. If*

$$\operatorname{div} f(x) = \frac{\partial f_1(x)}{\partial x_1} + \frac{\partial f_2(x)}{\partial x_2}$$

is not identically zero on any open subset of Ω and does not change sign in Ω, then (4.3) *has no periodic orbits lying entirely in Ω.* □

Since the vector fields $f(x)$ and $\mu(x)f(x)$, where the function $\mu : \mathbb{R}^2 \to \mathbb{R}$ is smooth and everywhere positive, define orbitally equivalent systems of differential equations, one also has the following variant of the above theorem.

Theorem 4.16 (Dulac's Criterion) *Let Ω be a simply connected domain in $\mathbb{R}^2$. Assume that*

$$\operatorname{div}(\mu(x)f(x)) = \frac{\partial}{\partial x_1}(\mu(x)f_1(x)) + \frac{\partial}{\partial x_2}(\mu(x)f_2(x))$$

is not identically zero on any open subset of Ω and does not change sign in Ω for some positive C^1-function $\mu : \Omega \to \mathbb{R}$.

Then (4.3) *has no periodic orbits lying entirely in Ω.* □

An annulus is an example of a domain that is not simply connected. Indeed, closed curves that encircle the "hole" in the domain cannot be shrunk within the domain to a point. Two Jordan curves are called equivalent if they can be deformed into each other, such that each intermediate curve is located within the domain and remains a Jordan curve. In *doubly connected* domains there are exactly two equivalence classes of Jordan curves: Those that can be shrunk to a point and those that cannot.

Theorem 4.17 *Let Ω be a doubly connected domain in $\mathbb{R}^2$ in which* (4.3) *has no equilibria. Assume that*

$$\operatorname{div} f(x) = \frac{\partial f_1(x)}{\partial x_1} + \frac{\partial f_2(x)}{\partial x_2}$$

is not identically zero on any open subset of Ω and does not change sign in Ω.

Then (4.3) *has at most one periodic orbit lying entirely in Ω.*

Proof According to Bendixson's Criterion, there cannot be periodic orbits that can be shrunk within Ω to a point. Thus, if there are periodic orbits of (4.3) in Ω, they must be nested around the "hole" in the domain.

Any such periodic orbit is hyperbolic, since

$$\mu_2 = \exp\left(\int_0^T \text{div } f(\varphi(\tau))\, d\tau\right) \neq 1$$

by the assumption, and thus is *isolated* (i.e. has an annular neighbourhood in which there are no other periodic orbits).

Suppose that there are two or more isolated periodic orbits in Ω and consider a pair of them, such that there are no other periodic orbits in an open annulus bounded by these two periodic orbits. Both periodic orbits are hyperbolic and must have opposite stability. Indeed, the Poincaré–Bendixson theorem implies that, for any point x in this annulus, $\omega(x)$ must be one of these periodic orbits, while $\alpha(x)$ must be the other. (Otherwise, another periodic orbit(s) should exist in the annulus.) Therefore, $\Gamma^+(x)$ must approach one of these periodic orbits as $t \to +\infty$, while $\Gamma^-(x)$ should tend to the other one as $t \to -\infty$. Thus, one periodic orbit is (exponentially) stable with $\mu_2 < 1$, while the other one is unstable with $\mu_2 > 1$ (due to hyperbolicity). One of these inequalities contradicts the assumption. □

As with Dulac's Criterion, applying Theorem 4.17 to an orbitally equivalent system, we get a stronger result.

Theorem 4.18 *Let Ω be a doubly connected domain in $\mathbb{R}^2$ in which* (4.3) *has no equilibria. Assume that μ is a scalar positive C^1-function in Ω. If*

$$\text{div}(\mu(x)f(x)) = \frac{\partial}{\partial x_1}(\mu(x)f_1(x)) + \frac{\partial}{\partial x_2}(\mu(x)f_2(x))$$

is not identically zero on any open subset of Ω and does not change sign in Ω for some positive C^1-function $\mu : \Omega \to \mathbb{R}$.

Then (4.3) *has at most one periodic orbit lying entirely in Ω.* □

4.4 Phase Plane Analysis of Prey–Predator Models

In this section we apply the Poincaré–Bendixson–Dulac theory to analyse planar ODEs appearing in ecological modelling. In addition, we introduce the important notion of zero-isoclines. These divide the phase plane into regions with different "slopes" of the vector field. By regarding qualitative information about these slopes in various regions, one often can obtain strong conclusions about ω-limit sets of the systems. This methodology is called "phase plane analysis".

4.4.1 Lotka–Volterra Model

Consider a simple aquatic ecosystem composed of two fish species, one of which (called the *predator*) has the other (called the *prey*) as its main source of food. Assume that the prey population grows exponentially in the absence of predators, while the predator population decreases exponentially in the absence of prey. Encounters between prey and predator occur according to the Law of Mass Action, i.e. proportional to the densities of both species. Upon encounter, the prey is swallowed with some probability and then instantaneously converted into predator offspring. These assumptions translate into the differential equations (*Lotka–Volterra model*):

$$\begin{cases} \dot{v} = av \;-\; bvp, \\ \dot{p} = -cp \;+\; dvp, \end{cases} \tag{4.10}$$

where v stands for prey (victim) density and p for the predator density. Our first task is to derive qualitative and quantitative information about the behaviour of v and p as functions of time t for various nonnegative initial values and, in particular, how this behaviour depends on the four positive parameters a, b, c and d. As a follow-up, we shall consider several modifications of (4.10).

4.4.2 Zero-Isoclines and Equilibria

It is helpful to start the analysis of (4.10) by drawing some auxiliary curves, the so-called *isoclines* (more precisely, *zero-isoclines*). The v-isoclines are the curves on which $\dot{v} = 0$. Likewise, the p-isoclines are the curves where $\dot{p} = 0$. Clearly, an equilibrium of (4.10) is to be found at the intersection of two such isoclines. For the Lotka–Volterra system, the v-isoclines are the straight lines $v = 0$ and

$$p = \frac{a}{b},$$

while the p-isoclines are the straight lines $p = 0$ and

$$v = \frac{c}{d}.$$

Accordingly, there are two equilibrium points: $(0, 0)$ which is called *trivial*, and

$$(\bar{v}, \bar{p}) = \left(\frac{c}{d}, \frac{a}{b}\right), \tag{4.11}$$

which is called *nontrivial*, *internal*, or *positive*.

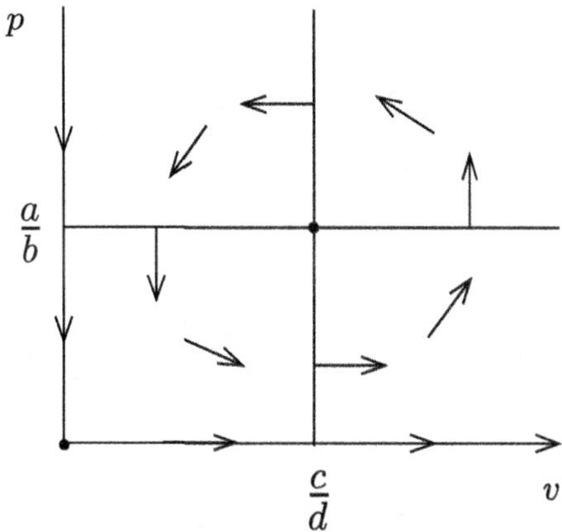

Fig. 4.8 Zero-isoclines of the Lotka–Volterra system (4.10)

Before proceeding, let us address the effect of fishing on the positive equilibrium. Suppose, fishing produces for both prey and predator a per capita probability μ to be caught per unit of time. This means that a should be replaced by $a - \mu$ and $(-c)$ by $(-c - \mu)$ in (4.10). The corresponding equilibrium will become

$$\left(\frac{c+\mu}{d}, \frac{a-\mu}{b}\right).$$

This shows that fishing *increases* $\bar{v}$ while it *decreases* $\bar{p}$. Therefore, in a period in which fishing effort is much reduced, we may expect the $\bar{v}$ to $\bar{p}$ ratio to decrease.[2]

The isoclines divide the nonnegative quarter plane into four regions (see Fig. 4.8). To indicate the direction of the flow, we put vertical arrows on the v-isoclines and horizontal arrows on the p-isoclines. Note that the direction of the arrows flips when passing the nontrivial equilibrium. Placing directional arrows inside the regions by continuity, we see at a glance that orbits of (4.10) have a tendency to spin around the nontrivial equilibrium. However, we cannot see whether they spiral inward or outward.

To determine the local behaviour of orbits of (4.10) near equilibria, one may attempt to apply the Grobman–Hartman Theorem. One might suspect that the trivial equilibrium is a saddle point from the fact that both coordinate axes are invariant. The Jacobian matrix of (4.10) evaluated at the trivial equilibrium,

$$\begin{pmatrix} a & 0 \\ 0 & -c \end{pmatrix},$$

[2] This was Volterra's explanation for an increase of sharks observed by fishermen in the Adriatic Sea when they resumed fishing after World War I. During the war fishing had stopped.

obviously has two real eigenvalues: $\lambda_1 = a > 0,\ \lambda_2 = -c < 0$. Thus, the origin is a hyperbolic *saddle*, as expected. The Jacobi matrix of (4.10) evaluated at the nontrivial equilibrium (4.11),

$$\begin{pmatrix} a - b\bar{p} & -b\bar{v} \\ d\bar{p} & d\bar{v} - c \end{pmatrix} = \begin{pmatrix} 0 & -\frac{bc}{d} \\ \frac{ad}{b} & 0 \end{pmatrix},$$

has purely imaginary eigenvalues $\lambda_{1,2} = \pm i\sqrt{ac}$. Therefore, this equilibrium is nonhyperbolic, and the Grobman–Hartman Theorem is not applicable. To analyse the phase portrait of (4.10) near the nontrivial equilibrium and in the whole positive quarter plane (*quadrant*), one has to apply some other techniques.

Fortunately, the Lotka–Volterra system has a *constant of motion*

$$L(v,p) = dv - c\ln v + bp - a\ln p. \tag{4.12}$$

Indeed, along the orbits

$$\begin{aligned} \frac{d}{dt}L(v(t),p(t)) &= \frac{\partial L}{\partial v}(v(t),p(t))\dot{v}(t) + \frac{\partial L}{\partial p}(v(t),p(t))\dot{p}(t) \\ &= \left(d - \frac{c}{v(t)}\right) v(t)(a - bp(t)) \\ &+ \left(b - \frac{a}{p(t)}\right) p(t)(-c + dv(t)) \ \equiv\ 0, \end{aligned}$$

meaning that L is constant. Consequently, orbits of (4.10) in the positive quadrant correspond to level sets $L(v,p) = L_0$. Actually, all these level sets are closed curves. To see this, notice that (4.12) can be written as

$$L(v,p) = L_1(v) + L_2(p)$$

with functions $L_{1,2}$ that have similar properties. Indeed, $L_1(v) = dv - c\ln v$ has one global quadratic minimum at $v = \frac{c}{d}$, while $L_2(p) = bp - a\ln p$ has one global quadratic minimum at $p = \frac{a}{b}$. Both functions tend to ∞ as $|v|, |p| \to \infty$. In fact, $L(v,p)$ as a function of two variables (v,p) has a unique global quadratic minimum at the nontrivial equilibrium (4.11), and all its level curves are closed and form a nested family surrounding the nontrivial equilibrium. This completes the construction of the phase portrait of the Lotka–Volterra system in the positive quadrant (see Fig. 4.9). The nontrivial equilibrium is a *centre*: It is Lyapunov stable but not asymptotically stable.

It is interesting to calculate the average values of v and p over one period. To this end, we write the first equation in (4.10) as

Fig. 4.9 Phase portrait of the Lotka–Volterra system

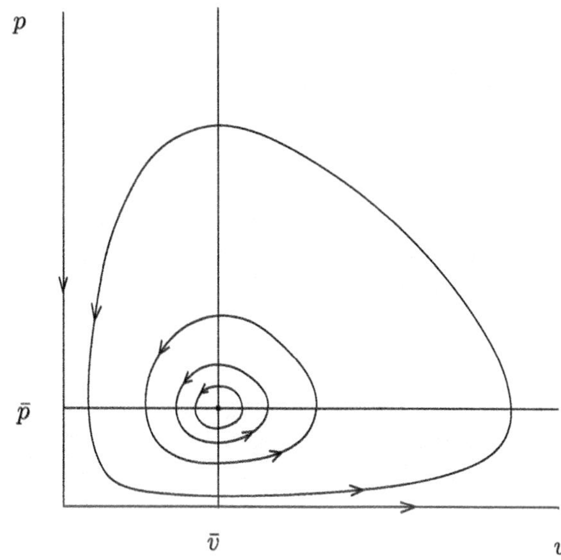

$$\frac{d}{dt}\ln v(t) = \frac{1}{v(t)}\frac{dv(t)}{dt} = a - bp(t)$$

and integrate it to obtain

$$\ln\left(\frac{v(t)}{v(0)}\right) = at - b\int_0^t p(\tau)\, d\tau.$$

When $t = T$ (the period), we have $v(T) = v(0)$, so the left-hand side vanishes and consequently

$$\frac{1}{T}\int_0^T p(\tau)\, d\tau = \frac{a}{b} = \bar{p}.$$

Similarly,

$$\frac{1}{T}\int_0^T v(\tau)\, d\tau = \frac{c}{d} = \bar{v}.$$

In other words, the average of $v(t)$ and $p(t)$ over one period is equal to the equilibrium values $\bar{v}$ and $\bar{p}$, respectively. Thus, the equilibrium values are representative characteristics of the nonequilibrium dynamics of (4.10).

4.4.3 Stabilization by Competition

The Lotka–Volterra system (4.10) gives an oversimplified description of prey–predator interaction. Yet it is a convenient reference point for the incorporation of additional mechanisms. In particular, the neutral stability of the nontrivial equilibrium allows us to classify such mechanisms as either *stabilizing* or *destabilizing*, depending on whether the corresponding equilibrium in a modified model is locally asymptotically stable or unstable.

We begin with modifying the equation for the isolated prey population. The so-called *logistic equation*

$$\dot{v} = av - ev^2 = v(a - ev)$$

describes in a phenomenological manner that, even in the absence of a predator, prey growth may be limited, with the prey population size settling down at the carrying capacity

$$K = \frac{a}{e}.$$

The nonnegative parameter e describes the competition amongst prey for limited external resources. Now introduce the predator as in the original Lotka–Volterra model, i.e. consider

$$\begin{cases} \dot{v} = v(a - ev - bp), \\ \dot{p} = p(-c + dv). \end{cases} \tag{4.13}$$

It is clear that the p-isoclines of (4.13) are the same lines as for the Lotka–Volterra system, i.e. $p = 0$ and

$$v = \frac{c}{d},$$

while the v-isoclines of (4.13) are given by $v = 0$ and

$$p = \frac{a}{b} - \frac{e}{b}v.$$

There are two typical configurations of the isoclines. When

$$\frac{a}{e} < \frac{c}{d},$$

the system (4.13) has two equilibria on the v-axis,

$$(0, 0), \quad \left(\frac{a}{e}, 0\right) = (K, 0),$$

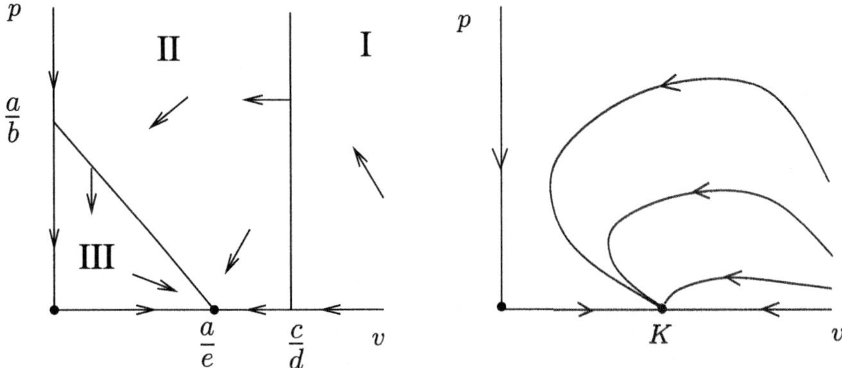

Fig. 4.10 Zero-isoclines and the phase portrait of (4.13): $ad - ce < 0$

the first of which is a saddle, while the second is globally asymptotically stable. This can be seen from Fig. 4.10, where the isoclines and the equilibria are shown together with the flow directions. As one can prove, all orbits starting in region I enter region II, while all orbits starting in II eventually either enter region III or converge to $(0, K)$. The triangular region III is *positively invariant*, i.e. the orbits can only enter it, not leave it. Using monotonicity arguments, we can show that all orbits starting inside this triangle tend to the point $(K, 0)$ as $t \to \infty$.

In case of the opposite inequality

$$\frac{a}{e} > \frac{c}{d},$$

the system (4.13) still has two equilibria on the v-axis,

$$(0, 0), \quad (K, 0),$$

which are both hyperbolic *saddles*, as well as a nontrivial equilibrium

$$(\tilde{v}, \tilde{p}) = \left(\frac{c}{d}, \frac{ad - ec}{bd} \right), \tag{4.14}$$

which is a locally asymptotically *stable node* or *focus*. This follows from the analysis of the eigenvalues of this equilibrium and the Grobman–Hartman Theorem, since the nontrivial equilibrium is hyperbolic. Indeed, the Jacobian matrix of (4.13) evaluated at the nontrivial equilibrium (4.14) is

$$\begin{pmatrix} a - 2e\tilde{v} - b\tilde{p} & -b\tilde{v} \\ d\tilde{p} & d\tilde{v} - c \end{pmatrix} = \begin{pmatrix} -\frac{ec}{d} & -\frac{bc}{d} \\ \frac{ad - ec}{b} & 0 \end{pmatrix}.$$

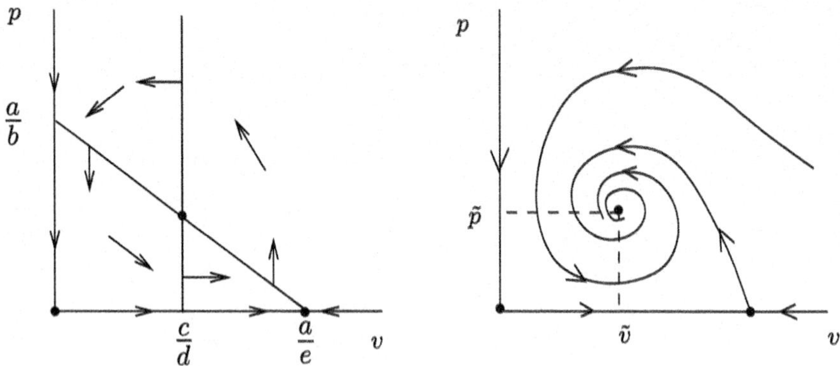

Fig. 4.11 Zero-isoclines and the phase portrait of (4.13): $ad - ce > 0$

Its eigenvalues $\lambda_{1,2}$ satisfy the characteristic equation

$$\lambda^2 - \sigma\lambda + \Delta = 0,$$

where

$$\sigma = \lambda_1 + \lambda_2 = -\frac{ec}{d} < 0, \quad \Delta = \lambda_1\lambda_2 = \frac{c(ad - ec)}{d} > 0.$$

Therefore, Re $\lambda_{1,2} < 0$, since all parameters are positive and $ad - ec > 0$.

Figure 4.11 suggests that the orbits of (4.13) tend to spiral around the nontrivial equilibrium.

In fact, the nontrivial equilibrium (4.14) of (4.13) is not only locally but also *globally* asymptotically stable, i.e. all orbits starting in the positive quarter plane tend to this equilibrium. To prove this, notice first that the constant of motion L for the Lotka–Volterra system (4.10) can be written as

$$L(v, p) = d(v - \bar{v} \ln v) + b(p - \bar{p} \ln p),$$

where $(\bar{v}, \bar{p})$ are the coordinates of the nontrivial equilibrium (4.11). Now introduce a similar function

$$V(v, p) = d(v - \tilde{v} \ln v) + b(p - \tilde{p} \ln p) \tag{4.15}$$

where $(\tilde{v}, \tilde{p})$ are the coordinates of the nontrivial equilibrium (4.14). The function (4.15) is nonincreasing along the orbits of the modified system (4.13):

$$\frac{d}{dt}V(v(t),p(t)) = d\left(1-\frac{d}{dv(t)}\right)v(t)(a-ev(t)-bp(t)) + b\left(1-\frac{ad-ec}{bdp(t)}\right)p(t)(-c+dv(t)) = -de\left(v(t)-\frac{c}{d}\right)^2 \le 0,$$

with equality only on the p-isocline

$$v = \frac{c}{d}.$$

The function V is therefore a *Lyapunov function* for (4.13). It has a unique extremum (minimum) at the nontrivial equilibrium $(\tilde{v},\tilde{p})$ and its closed level curves $V(v,p) = V_0$ surround this point. Moreover, the above formula shows that any orbit of (4.13), starting in the positive quadrant of the (v,p)-plane, passes any level $V(v,p) = V_0$ only once. At such a passage, the orbit goes from the *outer* into the *inner* region delimited by the corresponding level curve. The passage is transversal everywhere except on the p-isocline, where it is tangential. These facts imply that the nontrivial equilibrium in (4.13) is globally asymptotically stable. Thus, the competition amongst the prey is a stabilizing factor.

4.4.4 Destabilization by Saturation of the Predator

The *functional response* is by definition the per predator per unit of time eaten number of prey, usually as a function of prey density. The *numerical response* is the per predator per unit of time produced number of offspring. Usually, one takes the numerical response proportional to the functional response.

In the Lotka–Volterra model the functional response increases linearly with the prey density, which is another oversimplification. In reality, the predator functional response should approach some constant level at high prey densities. The so-called Holling Type II functional response

$$\frac{bv}{1+\beta bv}$$

takes this saturation effect into account. For large v it approaches the maximal digestion/handling capacity $\frac{1}{\beta}$. Taken alone, this is a destabilizing effect.

Indeed, one can show that the following modification of the Lotka–Volterra system

$$\begin{cases} \dot{v} = v\left(a - \frac{bp}{1+\beta bv}\right), \\ \dot{p} = p\left(-c + \frac{dv}{1+\beta bv}\right), \end{cases} \tag{4.16}$$

has a unique nontrivial equilibrium

$$(v_1, p_1) = \left(\frac{c}{d - \beta bc}, \frac{ad}{b(d - \beta bc)}\right)$$

when $d - \beta bc > 0$. The Jacobian matrix of (4.16) evaluated at the equilibrium (v_1, p_1) is reduced to

$$\begin{pmatrix} \frac{\beta abc}{d} & -\frac{bc}{d} \\ \frac{a(d-\beta bc)}{b} & 0 \end{pmatrix}.$$

Therefore, its eigenvalues $\lambda_{1,2}$ satisfy

$$\sigma = \lambda_1 + \lambda_2 = \frac{\beta abc}{d} > 0, \ \Delta = \lambda_1 \lambda_2 = \frac{ac(d - \beta bc)}{d} > 0,$$

and thus the equilibrium (v_1, p_1) is hyperbolic and locally unstable. It can be either an unstable *node* or a *focus*.

Moreover, any orbit of (4.16) starting in the positive quadrant tends to this equilibrium as $t \to -\infty$ (see Fig. 4.12). To prove this global result, we first notice that orbits of (4.16) in the positive quadrant coincide with those of the polynomial system

$$\begin{cases} \dot{v} = v(a + \beta abv - bp), \\ \dot{p} = p(-c + (d - \beta bc)v), \end{cases} \tag{4.17}$$

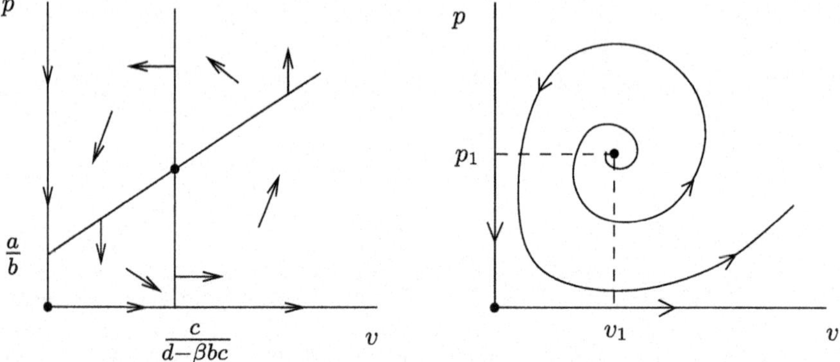

Fig. 4.12 Zero-isoclines and the phase portrait of the system (4.16)

which is obtained by multiplying the right-hand sides of (4.16) by $(1 + \beta bv)$. This factor is obviously positive in the positive quadrant of the (v, p)-plane, so the direction of the motion along the orbits is preserved. According to Chap. 1, the systems (4.17) and (4.16) are *orbitally equivalent* in the positive quadrant.

Introduce new variables in the positive quadrant, namely:

$$\begin{cases} \xi = \ln v, \\ \eta = \ln p. \end{cases}$$

In these variables, (4.17) takes the form

$$\begin{cases} \dot{\xi} = a + \beta ab\, e^{\xi} - b\, e^{\eta} =: P(\xi, \eta), \\ \dot{\eta} = -c + (d - \beta bc)e^{\xi} =: Q(\xi, \eta). \end{cases} \tag{4.18}$$

Since

$$\frac{\partial P}{\partial \xi} + \frac{\partial Q}{\partial \eta} = \beta ab\, e^{\xi} > 0$$

for all ξ, Bendixson's Criterion guarantees the absence of cycles of (4.18) in the whole (ξ, η)-plane. Thus, (4.17) and hence (4.16) have no periodic orbits in the positive quadrant. Moreover, as we will show below, any negative half-orbit of the system is bounded. Together these facts imply that for $d - \beta bc > 0$ any orbit of (4.16) starting in the positive quadrant tends to the nontrivial equilibrium (v_1, p_1) as $t \to -\infty$. Otherwise, according to the Poincaré–Bendixson Theorem for α-limit sets, it must tend to a cycle at $t \to -\infty$, which is impossible. Similar arguments prove that any nonequilibrium orbit of (4.16) is unbounded as $t \to +\infty$.

To prove that all negative half-orbits of (4.17) are bounded, introduce new variables

$$z = \frac{1}{v}, \quad u = \frac{p}{v},$$

so that u is the ratio of the population densities, and $z = 0$ corresponds to "infinitely large" prey density. The dynamics of these new variables is generated by

$$\dot{z} = -\frac{av + \beta abv^2 - bvp}{v^2} = -z\left(a + \frac{\beta ab}{z} - \frac{bu}{z}\right) = -\beta ab - az + bu,$$

$$\dot{u} = \frac{-(a+c)pv + (d - \beta b(a+c))v^2 p + bvp^2}{v^2} = -(a+c)u + (d - \beta b(a+c))\frac{u}{z} + \frac{bu^2}{z}.$$

Multiplying both $\dot{z}$ and $\dot{u}$ by z, we obtain another orbitally equivalent system:

$$\begin{cases} \dot{z} = -\beta abz - az^2 + bzu, \\ \dot{u} = (d - \beta b(a+c))u - (a+c)zu + bu^2, \end{cases} \tag{4.19}$$

that should be studied near the u-axis, i.e. for small $z \geq 0$ and any $u \geq 0$. The results of such an analysis are illustrated in Fig. 4.13. Clearly, both the z-axis and the u-axis are invariant. The origin $(z, u) = (0, 0)$ is always an equilibrium of (4.19). It is a *saddle* if

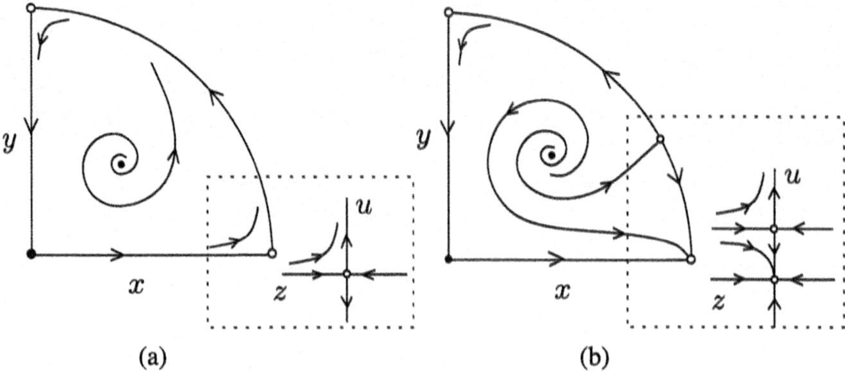

Fig. 4.13 Behaviour of the system (4.16) at the infinity: (**a**) $(d - \beta bc) - \beta ab \geq 0$; (**b**) $(d - \beta bc) - \beta ab < 0$

$$d - \beta b(a + c) = (d - \beta bc) - \beta ab > 0$$

(case (a) in Fig. 4.13) and a *stable node* if

$$d - \beta b(a + c) = (d - \beta bc) - \beta ab < 0$$

(case (b) in Fig. 4.13). In the latter case, there is another equilibrium point on the positive u-axis at

$$u = \frac{\beta b(a + c) - d}{b} > 0,$$

which is a *saddle*.

What does this mean for the system (4.17)? It is useful to consider the image of the (v, p)-portrait of (4.17) in the (x, y)-plane under the map

$$(x, y) = \left(\frac{v}{\sqrt{1 + v^2 + p^2}}, \frac{p}{\sqrt{1 + v^2 + p^2}} \right),$$

which is called the *Poincaré projection*. This map projects the whole (v, p)-plane into the unit disc in the (x, y)-plane, so that the ratio $u = p/v$ is preserved and points on the circle correspond to "infinity". The images of the positive quadrant of the (v, p)-plane are also shown in Fig. 4.13 with the corresponding phase portraits. There are at most three infinitely remote equilibria: Two studied above and one more at the "end" of the p-axis, which is a *saddle* for all $(d - \beta bc) > 0$. From neither of these equilibria emanates an orbit into the positive quadrant: *"Infinity is always stable"*. This implies that all negative half-orbits of (4.17) with positive initial data are bounded.

There is an interesting difference between the two mentioned cases. When $(d - \beta bc) - \beta ab > 0$ (and also when the equality holds), all orbits spiral indefinitely as $t \to +\infty$ while going to infinity. This means that predators gain and lose control over the prey again and again. In contrast, when $(d - \beta bc) - \beta ab < 0$, almost all orbits starting in the positive quadrant spiral for a while around the nontrivial equilibrium but eventually tend to infinity "along the v-axis", i.e. $u(t) = p(t)/v(t) \to 0$ as $t \to +\infty$. There is only one exceptional orbit that connects the unstable positive equilibrium and the saddle at infinity; along this orbit

$$\frac{p(t)}{v(t)} \to \frac{\beta b(a + c) - d}{b}$$

as $t \to +\infty$. So the predator may gain and lose control over the prey a number of times, but eventually the control will be lost and the prey density will grow exponentially, while the predator density also grows exponentially but, generically, at strictly lower rate.

4.4.5 Rosenzweig–MacArthur Model

By combining the stabilizing effect of the competition amongst prey and the destabilizing effect of the predator saturation, one arrives at the model

$$\begin{cases} \dot{v} = v\left(a - ev - \dfrac{bp}{1+\beta bv}\right), \\ \dot{p} = p\left(-c + \dfrac{dv}{1+\beta bv}\right), \end{cases} \tag{4.20}$$

which is a variant of a more general *Rosenzweig–MacArthur model*:

$$\begin{cases} \dot{v} = v(h(v) - p\varphi(v)), \\ \dot{p} = p(-c + \psi(v)). \end{cases} \tag{4.21}$$

The prey isocline in (4.20) is the parabola

$$p(v) = \frac{1}{b}(a - ev)(1 + \beta bv),$$

which has its top for

$$v_0 = \frac{1}{2}\left(\frac{a}{e} - \frac{1}{\beta b}\right).$$

Figure 4.14 illustrates possible isocline configurations. It is not difficult to prove that all positive half-orbits of (4.20) are bounded.

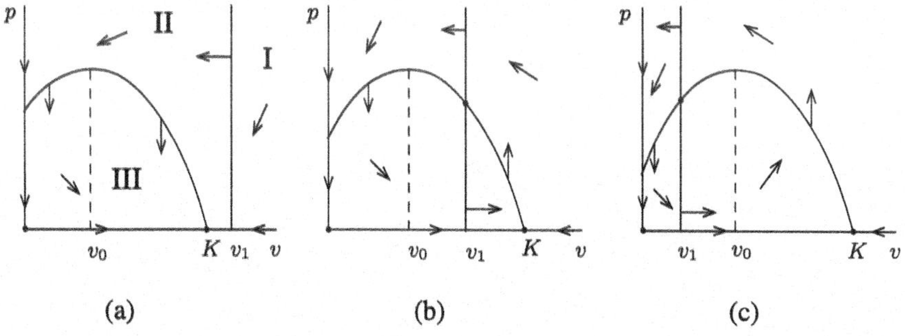

Fig. 4.14 Zero-isoclines of (4.20)

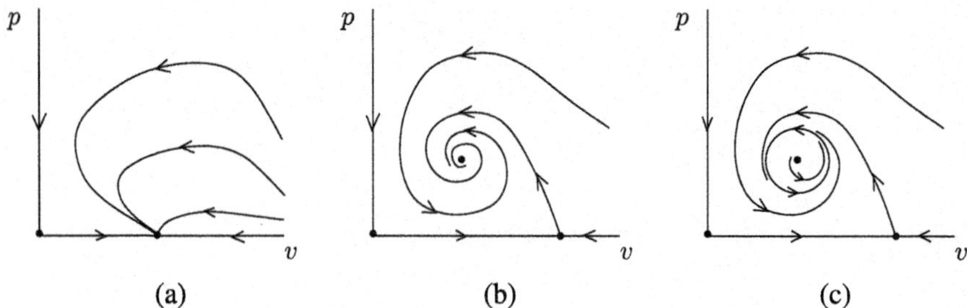

Fig. 4.15 Generic phase portraits of the Rosenzweig–MacArthur system (4.20)

When the predator isocline

$$v = v_1 = \frac{c}{d - \beta bc}$$

lies to the right of the trivial equilibrium

$$(K, 0) = \left(\frac{a}{e}, 0\right)$$

(see Fig. 4.13a), all orbits of (4.20) with positive initial data will tend to this equilibrium as $t \to +\infty$ (see Fig. 4.15a). This fact can be proved just like the corresponding statement in Sect. 4.4.3. The equilibrium remains globally asymptotically stable also when the predator isocline goes through $(K, 0)$.

When the predator isocline goes through the top of the parabola or lies to the right of the top but to the left of the equilibrium $(K, 0)$ (see Fig. 4.14b), the system (4.20) has a unique nontrivial equilibrium $(v_1, p(v_1))$ which is either a *stable node* or a *stable focus*. Moreover, all orbits of (4.20) with positive initial data tend to this equilibrium as $t \to +\infty$ (see Fig. 4.15b). This can be proved with some effort using Dulac's Theorem 4.16.

To simplify further computations, reduce the number of parameters in (4.20) by the linear scaling:

$$v = \frac{a}{e}\, x, \quad p = \frac{ad}{be}\, y, \quad t = \frac{\beta b}{d}\, \tau.$$

This brings (4.20) to the form

$$\begin{cases} \dot{x} = rx(1 - x) - \dfrac{xy}{m + x} & =: \; R(x, y), \\ \dot{y} = -\gamma y + \dfrac{xy}{m + x} & =: \; S(x, y), \end{cases} \tag{4.22}$$

where the dot means now the derivative with respect to τ and

$$r = \frac{\beta ab}{d}, \quad \gamma = \frac{\beta bc}{d}, \quad m = \frac{e}{\beta ab}.$$

After the scaling, $x = 1$ corresponds to the equilibrium of the prey in the absence of the predator, while the coordinates of the nontrivial equilibrium are

$$(x_1, y_1) = \left(\frac{\gamma m}{1-\gamma}, \frac{rm(1-\gamma-\gamma m)}{(1-\gamma)^2} \right). \tag{4.23}$$

This equilibrium is indeed in the positive quadrant $\mathbb{R}^2_+$ if $\gamma < 1$ and $1 - \gamma - \gamma m > 0$. The top of the parabola $\dot{x} = 0$ is now located at

$$x_0 = \frac{1-m}{2}. \tag{4.24}$$

Therefore, the isocline configuration shown in Fig. 4.13b appears when

$$\frac{1-m}{1+m} < \gamma < \frac{1}{1+m}. \tag{4.25}$$

We now look for a function $\mu : \mathbb{R}^2_+ \to \mathbb{R}$ (to use in Dulac's Theorem 4.16) in the form

$$\mu(x, y) = \left(\frac{x}{m+x} \right)^{\alpha} y^{\delta},$$

with some $\alpha, \delta \in \mathbb{R}$. According to Theorem 4.16, system (4.22) will have no cycles in $\mathbb{R}^2_+$ if the expression

$$Q(x, y) = \frac{\partial}{\partial x}(\mu(x, y)R(x, y)) + \frac{\partial}{\partial y}(\mu(x, y)S(x, y))$$

is not identically equal to zero and does not change sign. This expression can be computed as

$$Q(x, y) = \left[-\frac{m(\alpha+1)y}{(m+x)^2} + \frac{r}{m+x} P_{\alpha,\omega}(x) \right] \mu(x, y),$$

where

$$\omega = \frac{\delta+1}{r} \tag{4.26}$$

and

$$P_{\alpha,\omega}(x) = -2x^2 - [(\alpha+2)m + (\gamma-1)\omega - 1]x + (\alpha+1-\gamma\omega)m.$$

Therefore, $Q(x, y)$ will be nonpositive in $\mathbb{R}^2$ if we can select values of α and ω such that $\alpha \geq -1$ and $P_{\alpha,\omega}(x) \leq 0$ for all $x > 0$. The discriminant[3] of the second-degree polynomial $P_{\alpha,\omega}(x)$ is

$$D_{\alpha}(\omega) = (1-\gamma)^2\omega^2 - 2[(\alpha m - 1)(1-\gamma) + 2m(1+\gamma)]\omega + (\alpha+2)^2 m^2 + 2(3\alpha+2)m + 1$$

and in turn the discriminant of the polynomial $D_{\alpha}(\omega)$ is given by

$$D(\alpha) = -32m[(1-\gamma)(1-\gamma-\gamma m)\alpha + 1 - \gamma - 2\gamma m].$$

If there is an α^* such that $D(\alpha^*) > 0$, then $D_{\alpha^*}(\omega) = 0$ has two real solutions $\omega_1 < \omega_2$. If now ω^* is any real number such that $\omega_1 < \omega^* < \omega_2$, then $D_{\alpha^*}(\omega^*) < 0$ and $P_{\alpha^*,\omega^*}(x)$ has no real roots. Since the coefficient of x^2 is negative, $P_{\alpha^*,\omega^*}(x) < 0$ for all x. Choosing $\delta^* = r\omega^* - 1$ according to (4.26) completes the argument.

[3] Recall that for a polynomial $Ax^2 + Bx + C$ the *discriminant* is defined as $D = B^2 - 4AC$.

Consider now three cases when (4.25) holds:

(1) If

$$\frac{1-m}{1+m} < \gamma < \frac{1}{1+2m},$$

choose α^* such that

$$-1 < \alpha^* < -\frac{1-\gamma-2\gamma m}{(1-\gamma)(1-\gamma-\gamma m)} < 0$$

and it follows that $D(\alpha^*) > 0$.

(2) If

$$\gamma = \frac{1}{1+2m},$$

for any $-1 \leq \alpha^* < 0$ we have

$$D(\alpha^*) = -\frac{64m^3\alpha^*}{(1+2m)^2} > 0.$$

(3) If

$$\frac{1}{1+2m} < \gamma < \frac{1}{1+m},$$

set $\alpha^* = 0$. Then, clearly, $D(0) > 0$.

If $\gamma = \frac{1-m}{1+m}$, i.e. when the predator isocline $\dot{y} = 0$ goes through the top of the parabola $\dot{x} = 0$, we can take $\alpha^* = -1$. Then $D(-1) = 0$, implying that

$$D_{-1}(\omega) = \frac{(m^2 + 2\omega m - 1)^2}{(m+1)^2}$$

has the double root

$$\omega^* = \frac{1-m^2}{2m}.$$

With these values of α^* and ω^*, the polynomial P_{α^*,ω^*} takes the form

$$P_{\alpha^*,\omega^*}(x) = -\frac{1}{2}(2x + m - 1)^2 \leq 0.$$

Since it vanishes only along the predator isocline $x = x_0$, Theorem 4.16 guarantees the absence of cycles also in this case.

Finally, when the predator isocline lies to the left of the top (see Fig. 4.14c), the nontrivial equilibrium is an *unstable focus*. Since positive half-orbits of (4.20) could not approach the nontrivial equilibrium due to its repellor character, the Poincaré–Bendixson Theorem implies that there must be at least one *periodic orbit* in the positive quadrant. Since any such orbit must encircle an equilibrium, we conclude that there exists at least one periodic orbit around the nontrivial equilibrium in (4.20). The corresponding solutions $(v(t), p(t))$ are periodic. Thus, the model (4.20) predicts in this case periodic oscillations in the population densities.

With much more effort, one can prove that the system (4.20) has *at most one* periodic orbit. Therefore, when it exists, the nontrivial equilibrium is either globally asymptotically stable, or unstable and is surrounded by the unique stable periodic orbit (*limit cycle*) (see Fig. 4.15c). The proof is sketched below.

Recall that the considered case occurs when $0 < x_1 < x_0$, i.e. when

$$0 < \frac{\gamma m}{1-\gamma} < \frac{1-m}{2}$$

(see (4.24) and (4.23)). The proof of uniqueness is heavily based on the *mirror symmetry* of the prey isocline $y = g(x)$, where

$$g(x) := r(1-x)(m+x), \tag{4.27}$$

with respect to the vertical line $x = x_0$.

Consider in $\mathbb{R}^2_+$ the system

$$\begin{cases} \dot{x} = x[g(x) - y] =: R_1(x,y), \\ \dot{y} = y[-\gamma m + (1-\gamma)x] =: S_1(x,y), \end{cases} \tag{4.28}$$

that is obtained by the multiplication of the right-hand side of (4.22) by $(m+x)$ and hence is *orbitally equivalent* to (4.22). Its advantage is that it is polynomial. The divergence of the vector field $F = (R_1, S_1)$ is given by

$$\text{div } F = \frac{\partial R_1}{\partial x} + \frac{\partial S_1}{\partial x} = g(x) - y + xg'(x) - \gamma m + (1-\gamma)x = \frac{1}{x}\frac{dx}{dt} + xg'(x) + \frac{1}{y}\frac{dy}{dt}.$$

As we have seen, in the considered case (4.22) (and, therefore, (4.28)) has at least one periodic orbit around the nontrivial equilibrium. If

$$I_1 = \int_0^{T_0} \text{div } F(\xi(t), \eta(t))\, dt < 0$$

for *any* periodic solution $(\xi(t), \eta(t))$ with period T_0 of (4.28), then all cycles are hyperbolic and exponentially stable (see Eq. (4.4)). Arguments similar to those used to prove Theorem 4.17 show that there may exist at most one such cycle, i.e. when the cycle exists, it is unique. Notice that I_1 can be computed as

$$I_1 = \int_0^{T_0} \xi(t) g'(\xi(t))\, dt = r\int_0^{T_0} \xi(t)(1-m-2\xi(t))\, dt = 2r\int_0^{T_0} \xi(t)(x_0 - \xi(t))\, dt,$$

since, due to periodicity,

$$\oint_\Gamma \frac{dx}{x} = \oint_\Gamma \frac{dy}{y} = 0,$$

where Γ is the cycle. Note that Γ has counterclockwise orientation. Define

$$I_2 = \int_0^{T_0} \xi(t)(x_0 - \xi(t))\, dt, \tag{4.29}$$

so that $I_1 = 2rI_2$.

Proposition 4.19 *The mirror image Γ' of any cycle Γ of (4.28) with respect to the vertical line $x = x_0$ intersects Γ in precisely six points: A, B, P, Q, P', Q', such that P' and Q' are the mirror images of P and Q, respectively. Moreover,*

(i) $x_A = x_B = x_0$, $y_A < g(x_0) < y_B$;
(ii) $x_{P'} < x_1 < x_0 < x_P$, $x_{Q'} < x_1 < x_0 < x_Q$;
(iii) $y_{P'} = y_P < g(x_P)$, $y_{Q'} = y_Q > g(x_Q)$.

(*see* Fig. *4.16*a).

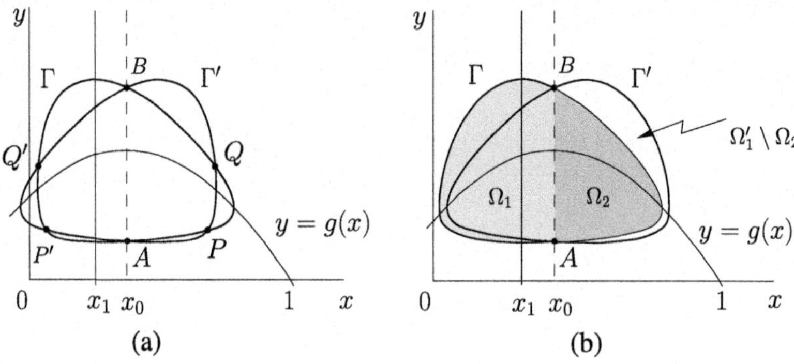

Fig. 4.16 Geometry of the limit cycle: (**a**) real; (**b**) hypothetical

Proof Condition (i) is merely the definition of A and B.

For the rest, it is sufficient to prove that the mirror image of that part of Γ, which is located to the left of the line $x = x_0$, intersects Γ in two points: P and Q. Note that suitable comparison and monotonicity arguments show that there can exist no more than two such points, one of which (say, P) is located below the prey isocline $y = g(x)$, while the other (Q) is above this isocline.

Consider the following auxiliary function

$$V(x, y) = (1 - \gamma) \int_1^x \frac{(\xi - x_1)\, d\xi}{\xi} + y.$$

The total derivative of V along solutions of (4.28) satisfies

$$\dot{V} = (1 - \gamma)\frac{x - x_1}{x}\dot{x} + \dot{y} = (1 - \gamma)g(x)(x - x_1) = \frac{g(x)}{y}\dot{y}.$$

Thus, by periodicity,

$$\oint_\Gamma \frac{g(x)\, dy}{y} = \int_0^{T_0} \dot{V}\, dt = 0.$$

Introduce two open domains Ω_1 and Ω_2, such that Ω_1 is the sub-domain inside Γ located strictly to the left from $x = x_0$, while Ω_2 is the sub-domain inside Γ located strictly to the right from $x = x_0$. Clearly, $\Omega_1 \cap \Omega_2 = \emptyset$.

Now assume that the right part of Γ' does not intersect Γ, i.e. that Ω_2 is located strictly inside the mirror image Ω_1' of Ω_1 (see Fig. 4.16b). Thus, $\Omega_1' \setminus \Omega_2$ is a nonempty simply connected domain.

By Green's Theorem, and using the definition (4.27),

$$0 = \oint_\Gamma \frac{g(x)\, dy}{y} = \int_\Omega \frac{g'(x)}{y}\, dxdy = 2r \int_\Omega \frac{x_0 - x}{y}\, dxdy,$$

where Ω is the domain inside Γ. Since

$$\begin{aligned}
\int_\Omega \frac{x_0 - x}{y}\, dxdy &= \int_{\Omega_1} \frac{x_0 - x}{y}\, dxdy + \int_{\Omega_2} \frac{x_0 - x}{y}\, dxdy \\
&= -\int_{\Omega_1'} \frac{x_0 - x}{y}\, dxdy + \int_{\Omega_2} \frac{x_0 - x}{y}\, dxdy \\
&= -\int_{\Omega_1' \setminus \Omega_2} \frac{x_0 - x}{y}\, dxdy,
\end{aligned}$$

and $x_0 < x$ in $\Omega_1' \setminus \Omega_2$, we have

$$0 = -2r \int_{\Omega_1' \setminus \Omega_2} \frac{x_0 - x}{y}\, dxdy > 0,$$

which is a contradiction. Therefore, Ω_2 cannot be strictly inside the mirror image Ω_1' of Ω_1, meaning that the right part of Γ' must intersect Γ. One can verify that the above arguments also exclude the possibility for Γ' and Γ to have a single common point to the right from $x = x_0$.

Thus, Γ' and Γ intersect at two points to the right of $x = x_0$: P and Q, which, together with their mirror images P' and Q', satisfy conditions (ii) and (iii). □

Proposition 4.20 *Let* $p(\xi) = 2\xi^2 - 4x_0\xi + 2x_0x_1$. *Then*

(*i*) $p(x_{Q'}) > 0$;
(*ii*) $p(x_{P'}) > 0$.

Proof Let $y(x)$ represent the upper part of Γ. Then

$$\frac{dy}{dx} = (1-\gamma)\frac{(x - x_1)y}{(g(x) - y)x}.$$

With a bit of effort one can deduce from Proposition 4.19 that

$$\left.\frac{dy}{dx}\right|_Q > -\left.\frac{dy}{dx}\right|_{Q'}$$

(see Fig. 4.16a) or

$$\frac{(x_Q - x_1)y_Q}{(g(x_Q) - y_Q)x_Q} > -\frac{(x_{Q'} - x_1)y_{Q'}}{(g(x_{Q'}) - y_{Q'})x_{Q'}}.$$

Taking into account $y_Q = y_{Q'}$, $g(x_Q) = g(x_{Q'})$, and $g(x_Q) < y_Q$, we obtain

$$\frac{x_{Q'}}{x_1 - x_{Q'}} < \frac{x_Q}{x_Q - x_1}.$$

Substituting $x_Q = 2x_0 - x_{Q'}$ into this inequality, we immediately get $p(x_{Q'}) > 0$. The proof of (ii) is similar. □

The integral (4.29) can be written as the sum of six integrals (see Fig. 4.16a):

$$I_2 = I_{AP} + I_{PQ} + I_{QB} + I_{BQ'} + I_{Q'P'} + I_{P'A},$$

where each integral is computed over the corresponding time interval. Let

$$0 < t_P < t_Q < t_B < t_{Q'} < t_{P'} < t_A = T_0$$

be the respective times of passage through the points P, Q, B, Q', P', A, while tracing over one period the periodic solution $(\xi(t), \eta(t))$, starting at point A and returning to A.

The next goal is to show that the three sums, $I_{AP} + I_{P'A}, I_{PQ} + I_{Q'P'}$, and $I_{QB} + I_{BQ'}$, are all negative. Let $y = \eta_1(x)$ represent Γ on $[t_{P'}, t_A]$, and let $y = \eta_2(x)$ represent Γ on $[0, t_P]$ (see Fig. 4.17). We can rewrite $I_{P'A} + I_{AP}$ using the mirror symmetry of the prey isocline $y = g(x)$ with respect to the vertical line $x = x_0$, i.e. $g(2x_0 - x) = g(x)$, see (4.27). We get

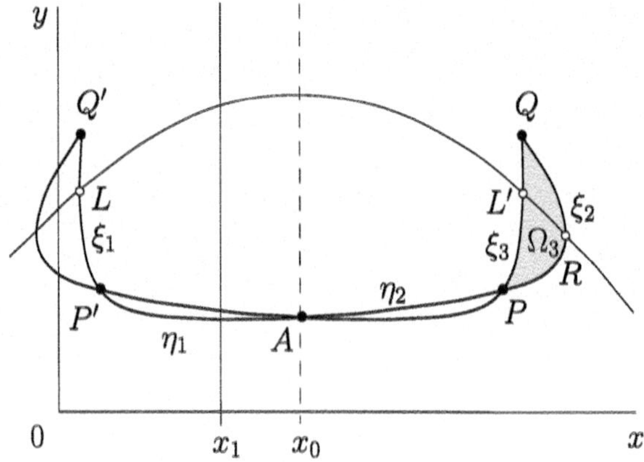

Fig. 4.17 Geometrical constructions for the proof of uniqueness

$$\begin{aligned}
I_{P'A} + I_{AP} &= \left(\int_{t_{P'}}^{t_A} + \int_0^{t_P}\right) \xi(t)(x_0 - \xi(t))\, dt \\
&= \int_{P'A} \frac{(x_0 - x)}{g(x) - \eta_1(x)}\, dx + \int_{AP} \frac{(x_0 - x)}{g(x) - \eta_2(x)}\, dx \\
&= \int_{P'A} \frac{(x_0 - x)}{g(2x_0 - x) - \eta_1(x)}\, dx + \int_{AP} \frac{(x_0 - x)}{g(x) - \eta_2(x)}\, dx \\
&= -\int_{AP} \frac{(x_0 - x)}{g(x) - \eta_1(2x_0 - x)}\, dx + \int_{AP} \frac{(x_0 - x)\, dx}{g(x) - \eta_2(x)} \\
&= \int_{x_A}^{x_P} \frac{(x_0 - x)(\eta_2(x) - \eta_1(2x_0 - x))}{(g(x) - \eta_1(2x_0 - x))(g(x) - \eta_2(x))}\, dx \;<\; 0,
\end{aligned}$$

since from Lemma 4.19 it follows that for all x with $x_A = x_0 < x < x_P$

$$\eta_2(x) - \eta_1(2x_0 - x) > 0, \quad g(x) - \eta_2(x) > 0, \quad g(x) - \eta_1(2x_0 - x) > 0,$$

holds.

Similar arguments imply that

$$I_{QB} + I_{BQ'} = \left(\int_{t_Q}^{t_B} + \int_{t_B}^{t_{Q'}}\right) \xi(t)(x_0 - \xi(t))\, dt < 0.$$

Let now $x = \xi_1(y)$ represent Γ on $[t_{Q'}, t_{P'}]$ and let $x = \xi_2(y)$ represent Γ on $[t_P, t_Q]$ (Fig. 4.17). Then

$$I_{Q'P'} = \int_{t_{Q'}}^{t_{P'}} \xi(t)(x_0 - \xi(t))\, dt = -\frac{1}{1-\gamma} \int_{y_P}^{y_Q} \frac{\xi_1(y)(x_0 - \xi_1(y))}{(\xi_1(y) - x_1)y}\, dy$$

and

$$I_{PQ} = \int_{t_P}^{t_Q} \xi(t)(x_0 - \xi(t))\, dt = \frac{1}{1-\gamma} \int_{y_P}^{y_Q} \frac{\xi_2(y)(x_0 - \xi_2(y))}{(\xi_2(y) - x_1)y}\, dy.$$

The last integral can be rewritten as follows

$$
\begin{aligned}
I_{PQ} &= \frac{1}{1-\gamma}\left[\int_{y_P}^{y_Q} \frac{\xi_2(y)(x_0-\xi_2(y))}{(\xi_2(y)-x_1)y}\,dy + \int_{y_Q}^{y_P} \frac{\xi_3(y)(x_0-\xi_3(y))}{(\xi_3(y)-x_1)y}\,dy \right.\\
&\qquad \left. - \int_{y_Q}^{y_P} \frac{\xi_3(y)(x_0-\xi_3(y))}{(\xi_3(y)-x_1)y}\,dy\right]\\
&= \frac{1}{1-\gamma}\left[\oint_{PRQL'P} \frac{x(x_0-x)}{(x-x_1)y}\,dy + \int_{y_P}^{y_Q} \frac{\xi_3(y)(x_0-\xi_3(y))}{(\xi_3(y)-x_1)y}\,dy\right],
\end{aligned}
$$

where $x=\xi_3(y)=2x_0-\xi_1(y)$ represents the mirror image of $x=\xi_1(y)$, and the first integral is taken over the closed contour $PRQL'P$ (see Fig. 4.17). Denote by Ω_3 the compact domain bounded by this contour. Green's Theorem implies

$$
\oint_{PRQL'P} \frac{x(x_0-x)}{(x-x_1)y}\,dy = \int_{\Omega_3} \frac{(-x^2+2x_1x-x_0x_1)}{(x-x_1)^2y}\,dxdy.
$$

However, since $x_0>x_1$ and

$$
-x^2+2x_1x-x_0x_1 = -[(x-x_1)^2+x_1(x_0-x_1)]<0,
$$

we have

$$
\oint_{PRQL'P} \frac{x(x_0-x)}{(x-x_1)y}\,dy<0.
$$

Therefore

$$
I_{PQ}<\frac{1}{1-\gamma}\int_{y_P}^{y_Q} \frac{\xi_3(y)(x_0-\xi_3(y))}{(\xi_3(y)-x_1)y}\,dy = \frac{1}{1-\gamma}\int_{y_P}^{y_Q} \frac{(2x_0-\xi_1(y))(\xi_1(y)-x_0)}{(2x_0-\xi_1(y)-x_1)y}\,dy.
$$

Now we can estimate

$$
\begin{aligned}
I_{Q'P'}+I_{PQ} &< \frac{1}{1-\gamma}\left[-\int_{y_P}^{y_Q} \frac{\xi_1(y)(x_0-\xi_1(y))}{(\xi_1(y)-x_1)y}\,dy + \int_{y_P}^{y_Q} \frac{(2x_0-\xi_1(y))(\xi_1(y)-x_0)}{(2x_0-\xi_1(y)-x_1)y}\,dy\right]\\
&= \frac{1}{1-\gamma}\int_{y_P}^{y_Q} \frac{(x_0-\xi_1(y))(2\xi_1^2(y)-4x_0\xi_1(y)+2x_0x_1)}{(\xi_1(y)-x_1)(2x_0-\xi_1(y)-x_1)y}\,dy\\
&= \frac{1}{1-\gamma}\int_{y_P}^{y_Q} \frac{(x_0-\xi_1(y))p(\xi_1(y))}{(\xi_1(y)-x_1)(2x_0-\xi_1(y)-x_1)y}\,dy,
\end{aligned}
$$

where

$$
p(\xi)=2\xi^2-4x_0\xi+2x_0x_1
$$

is the polynomial already introduced in Proposition 4.20. This polynomial has two roots:

$$
\xi_\pm = x_0 \pm \sqrt{x_0(x_0-x_1)}>0,
$$

and is positive for $\xi<\xi_-$. If we prove that $\xi_1(y)$ satisfies $\xi_1(y)<\xi_-$ for all $y\in[y_P,y_Q]$, we can conclude that $p(\xi_1(y))>0$, and, since by Proposition 4.19,

$$
\xi_1(y)-x_1<0,\quad 2x_0-\xi_1(y)-x_1>0,
$$

we would get

$$
I_{Q'P'}+I_{PQ}<0.
$$

To complete the proof, notice that by Proposition 4.20 we have $p(x_{Q'}) > 0$ and $p(x_{P'}) > 0$. Moreover, $x_{Q'} < x_0$ and $x_{P'} < x_0$ while $\xi_+ > x_0$. Hence, both $x_{P'}, x_{Q'} < \xi_-$. Since, obviously,

$$\xi_1(y) \leq \max\{x_{P'}, x_{Q'}\},$$

we obtain that $\xi_1(y) < \xi_-$ for all $y \in [y_P, y_Q]$.

Thus, $I_2 < 0$ and therefore $I_1 < 0$ along any periodic orbit Γ.

4.5 Planar Hamiltonian and Related Systems

There is a special class of planar systems, for which a complete characterization of phase portraits is possible and which appear in Classical Mechanics. Moreover, it is possible to draw some conclusions about generic small perturbations of such systems.

4.5.1 Hamiltonian Systems with One Degree of Freedom

Let $H = H(q,p)$ be a smooth (at least C^2) function $H : \mathbb{R} \times \mathbb{R} \to \mathbb{R}$.

Definition 4.21 *A planar system*

$$\begin{cases} \dot{q} = \dfrac{\partial H(q,p)}{\partial p}, \\ \dot{p} = -\dfrac{\partial H(q,p)}{\partial q}, \end{cases} \tag{4.30}$$

is called a ***Hamiltonian system*** *with one degree of freedom and* ***Hamiltonian*** *(or* ***energy function****) H.*

Example 4.22 (Lotka–Volterra system revisited) Consider once again the Lotka–Volterra system from Sect. 4.4.1:

$$\begin{cases} \dot{x} = ax - bxy, \\ \dot{y} = -cy + dxy. \end{cases} \tag{4.31}$$

The linear scaling

$$x = \xi_0 \xi, \; y = \eta_0 \eta, \; t = \tau_0 \tau$$

with

$$\xi_0 = \frac{a}{d}, \; \eta_0 = \frac{a}{b}, \; \tau_0 = \frac{1}{a}$$

reduces (4.31) to

$$\begin{cases} \dot{\xi} = \quad \xi - \xi\eta, \\ \dot{\eta} = -\gamma\eta + \xi\eta, \end{cases} \tag{4.32}$$

where $\dot{\xi}$ and $\dot{\eta}$ are the derivatives of ξ and η with respect to the new time τ, and

$$\gamma = \frac{c}{a}$$

is a new parameter.

Introduce new coordinates in the positive quadrant $\mathbb{R}^2_+$ by the formulas:

$$\begin{cases} q = \ln \xi, \\ p = \ln \eta. \end{cases}$$

One has

$$\dot{q} = \frac{\dot{\xi}}{\xi} = 1 - \eta = 1 - e^p,$$

$$\dot{p} = \frac{\dot{\eta}}{\eta} = -\gamma + \xi = -\gamma + e^q.$$

Thus, (4.32) is smoothly equivalent in $\mathbb{R}^2_+$ to the Hamiltonian system (4.30) with

$$H(q,p) = \gamma q - e^q + p - e^p.$$

The transformation $(\xi,\eta) \mapsto (q,p)$ maps closed orbits of (4.32) in $\mathbb{R}^2_+$ onto closed orbits of (4.30) in $\mathbb{R}^2$, preserving their time parameterization (see Fig. 4.18). ◊

The following result explains why (4.30) is also called a *conservative system* (see also Exercise 4.12).

Theorem 4.23 (Conservation of energy) *The Hamiltonian $H(q,p)$ is a constant of motion for the Hamiltonian system, i.e. $H(q(t),p(t)) = \text{const}$ along solutions.*

Proof $\dfrac{dH}{dt} = \dfrac{\partial H}{\partial q}\dot{q} + \dfrac{\partial H}{\partial p}\dot{p} = \dfrac{\partial H}{\partial q}\dfrac{\partial H}{\partial p} - \dfrac{\partial H}{\partial p}\dfrac{\partial H}{\partial q} = 0.$ □

This theorem has as an immediate implication that any orbit of (4.30) belongs to a level set $\{(q,p) \in \mathbb{R}^2 : H(q,p) = h\}$ for $h \in \mathbb{R}$ of the Hamiltonian function H. Next we can deduce further implications from the geometry of such level sets. As illustrated in the lower half of Fig. 4.19, a connected component of a level set is typically either

- a single point (corresponding to a maximum or a minimum of H);

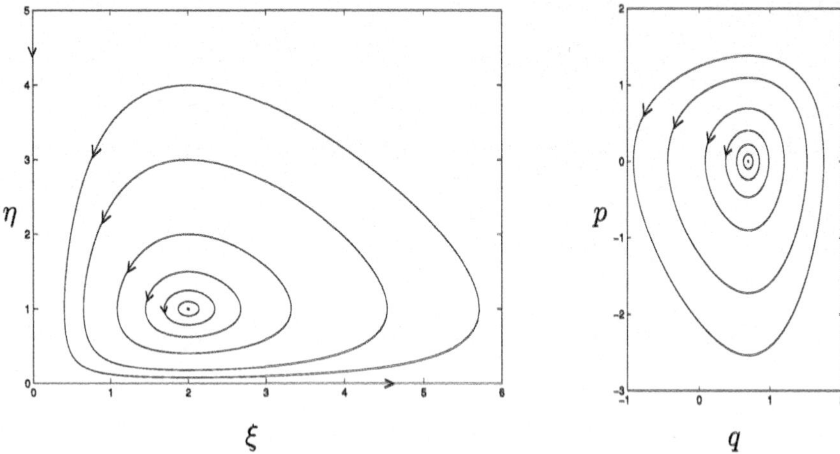

Fig. 4.18 Phase portraits of the Lotka–Volterra system in the (ξ, η)- and (q, p)-coordinates

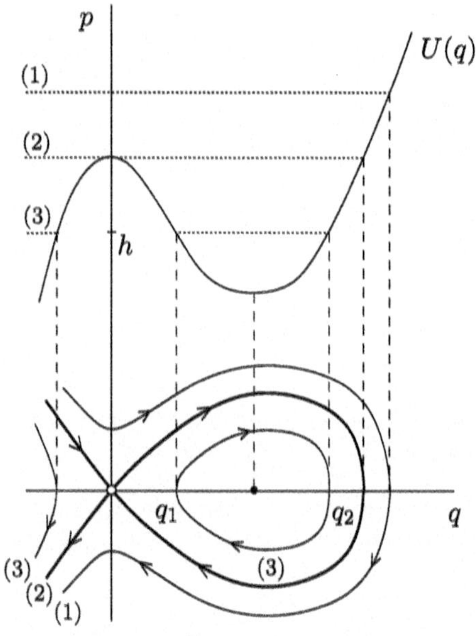

Fig. 4.19 Newton dynamics with one degree of freedom

- a closed curve (these occur in families parametrized by the value of H; a strict maximum or minimum is always surrounded by such a family);
- a number of points connected by curves (more on this case below).

Another direct consequence of (4.30) is that equilibria of a Hamiltonian system correspond exactly to critical points of H. So on the one hand we can use the classification of nondegenerate critical points into extrema (maxima and minima) and saddle points, and on the other hand we can use the classification of equilibria according to their stability properties. How do these two ways of classification correspond to each other?

A first observation is that (4.30) cannot have asymptotically stable equilibria (nor asymptotically stable cycles). Indeed, if all orbits starting in a neighbourhood of an equilibrium converge towards this equilibrium, then H must be constant on this neighbourhood (since H is continuous and constant along orbits). So the entire neighbourhood must consist of equilibria, a contradiction.

As H is assumed to be C^2, the Jacobian matrix at some equilibrium exists and is given by

$$A = \begin{pmatrix} \dfrac{\partial^2 H^0}{\partial q \partial p} & \dfrac{\partial^2 H^0}{\partial p^2} \\ -\dfrac{\partial^2 H^0}{\partial q^2} & -\dfrac{\partial^2 H^0}{\partial q \partial p} \end{pmatrix},$$

where the superscript 0 indicates the evaluation of all quantities at the equilibrium. Thus Tr $A = 0$ and the eigenvalues of A are either real and of opposite sign or purely imaginary, depending on the sign of

$$\det A = \frac{\partial^2 H^0}{\partial q^2}\frac{\partial^2 H^0}{\partial p^2} - \left(\frac{\partial^2 H^0}{\partial q \partial p}\right)^2 \neq 0.$$

If we consider the equilibrium as a critical point of H, we are led to consider the Hessian matrix

$$B = \begin{pmatrix} \dfrac{\partial^2 H^0}{\partial q^2} & \dfrac{\partial^2 H^0}{\partial q \partial p} \\ \dfrac{\partial^2 H^0}{\partial q \partial p} & \dfrac{\partial^2 H^0}{\partial p^2} \end{pmatrix},$$

which has exactly the same determinant as A. When the determinant is positive, H has an extremum (a maximum if tr $B < 0$ and a minimum if tr $B > 0$). Thus, when A has purely imaginary eigenvalues, the corresponding equilibrium is a (nonlinear) centre surrounded by a family of periodic orbits. When the determinant is negative, H has a saddle point and A has one positive and one negative real eigenvalue. Thus, the equilibrium is also a saddle point in the sense of planar dynamical systems (so when locally a level set consists of two intersecting curves, these are exactly the local stable and the local unstable manifolds of the point of intersection, which is a saddle equilibrium point).

In order to formulate our conclusions as a lemma, it is useful to introduce a new term.

Definition 4.24 *An equilibrium is called* ***simple****, if it has no zero eigenvalue.*

Lemma 4.25 *A simple equilibrium of a Hamiltonian system* (4.30) *with one degree of freedom is either a saddle or a centre. The first case applies when*

$$\frac{\partial^2 H^0}{\partial q^2}\frac{\partial^2 H^0}{\partial p^2} - \left(\frac{\partial^2 H^0}{\partial q \partial p}\right)^2 < 0$$

and the second when the opposite inequality holds. □

Remarks

(1) With any smooth function $H : \mathbb{R} \times \mathbb{R} \to \mathbb{R}$ one can also associate the *gradient flow* generated by the system

$$\begin{cases} \dot{q} = -\dfrac{\partial H(q,p)}{\partial q}, \\ \dot{p} = -\dfrac{\partial H(q,p)}{\partial p}, \end{cases} \tag{4.33}$$

For this flow

$$\frac{dH}{dt} = -\left[\left(\frac{\partial H}{\partial q}\right)^2 + \left(\frac{\partial H}{\partial p}\right)^2\right] \leq 0,$$

so orbits converge toward minima of H following steepest descent and crossing level curves of H orthogonally.

(2) Returning to (4.30), we observe that one might wonder how the period varies as a function of the value of H when there is a family of closed orbits? Is the period a monotone function? This is a notoriously difficult problem. With considerable effort one can prove monotonicity for some specific systems. But as far as we know, there is no general result and, in fact, the period may not be monotone for other specific systems.

(3) Finally note that the divergence of the vector field defining a Hamiltonian system (4.30) is zero. As a consequence, planar Hamiltonian flows preserve area (see Theorem 4.38 in the Appendix to this chapter for a general result). ◇

4.5.2 Mechanical Systems with One Degree of Freedom

Definition 4.26 *A planar system* (4.30) *with*

$$H(q,p) = \frac{1}{2}p^2 + U(q),$$

where U is a smooth scalar function, is called a **mechanical system** *with one degree of freedom.*

Here the term $\frac{1}{2}p^2$ is called the *kinetic energy*, while $U(q)$ is called the *potential energy*. The corresponding Hamiltonian system (4.30) has the form:

$$\begin{cases} \dot{q} = p, \\ \dot{p} = -U'(q). \end{cases} \tag{4.34}$$

Eliminating p we get Newton's second law:

$$\ddot{q} = F(q),$$

with the *potential force* $F(q) = -U'(q)$. (Side remark for physicists: Notice that the "mass" is put equal to one, which can always be achieved by scaling.)

The phase portrait of (4.34) in the (q, p)-plane is completely determined by the potential energy $U(q)$ (see Fig. 4.19). The portrait is symmetric with respect to the reflection in the q-axis composed with time reversal. Equilibrium points of (4.34) have the form $(q^0, 0)$, where q^0 is a critical point of $U(q)$, i.e. $U'(q^0) = 0$. If $U''(q^0) < 0$ (implying that q^0 is a local maximum of U) the equilibrium is a saddle; if $U''(q^0) > 0$ (implying that q^0 is a local minimum of U), the equilibrium is a centre. A connected subset of any level set $H(q, p) = h$, where h is a regular value, is diffeomorphic to either a circle (closed curve) or a line. The reader is invited to prove that any closed orbit defines a periodic solution with period

$$T = \int_{q_1}^{q_2} \sqrt{\frac{2}{h - U(q)}}\, dq, \tag{4.35}$$

where $q_1 < q_2$ are the coordinates of the intersections of the orbit with the q-axis and, therefore, $U(q_1) = U(q_2) = h$ (see Exercise 4.11). Critical level sets contain equilibria and homo- or heteroclinic orbits, which are asymptotic to them (see also Exercise 4.11).

Example 4.27 (Famous planar mechanical systems)

(1) *Harmonic oscillator*:

$$H = \frac{1}{2}(p^2 + q^2).$$

(2) *Ideal pendulum*:

$$H = \frac{1}{2}p^2 - \cos q.$$

(3) *Duffing's oscillator* :

$$H = \frac{1}{2}(p^2 - q^2) + \frac{q^4}{4}.$$

You are invited to draw their phase portraits yourself. ◊

4.5.3 Small Perturbations of Planar Hamiltonian Systems

As we have seen in the previous sections, planar Hamiltonian systems have very special phase portraits, which allow for a rather detailed characterization. Can we use this information to analyse planar systems which are not Hamiltonian but close to them? One can think of a potential system subject to a small friction, or about a generalized Lotka–Volterra model that takes into account weak competition amongst prey or predators. As we shall see, the topology of the phase portrait of a Hamiltonian system changes qualitatively under generic (non-Hamiltonian) perturbations: Centres become (stable or unstable) foci, while families of periodic orbits disappear, possibly giving rise to (stable or unstable) limit cycles. Homoclinic orbits also disappear. These qualitative changes are examples of *bifurcations* of dynamical systems, which we will study systematically in Chaps. 5, 6, and 7.

First consider the following one-parameter perturbation of (4.30):

$$\dot{x} = \begin{pmatrix} \dfrac{\partial H(x)}{\partial x_2} \\ -\dfrac{\partial H(x)}{\partial x_1} \end{pmatrix} + \varepsilon f(x), \quad x = (x_1, x_2) \in \mathbb{R}^2, \tag{4.36}$$

where $\varepsilon \in \mathbb{R}$ is a small parameter, and $H : \mathbb{R}^2 \to \mathbb{R}$ and $f : \mathbb{R}^2 \to \mathbb{R}^2$ are smooth functions. Since we are interested in non-Hamiltonian perturbations, $\operatorname{div} f$ is not identically zero.

We begin with a simple result concerning equilibria.

Theorem 4.28 *Consider the system* (4.36) *and assume that* $f(0) = 0$.

(i) *If* $x = 0$ *is a simple saddle at* $\varepsilon = 0$, *then* $x = 0$ *is a saddle of* (4.36) *for all* ε *with sufficiently small* $|\varepsilon| > 0$.

(ii) *If* $x = 0$ *is a simple centre at* $\varepsilon = 0$ *and* $\operatorname{div} f(0) \neq 0$, *then* $x = 0$ *is a hyperbolic focus of* (4.36) *for all* ε *with sufficiently small* $|\varepsilon| > 0$. *Moreover, this focus is stable for* $\varepsilon \operatorname{div} f(0) < 0$ *and unstable for* $\varepsilon \operatorname{div} f(0) > 0$.

Proof Write the Jacobian matrix of (4.36) at the equilibrium $x = 0$ as

$$A(\varepsilon) = \begin{pmatrix} \dfrac{\partial^2 H^0}{\partial x_1 \partial x_2} & \dfrac{\partial^2 H^0}{\partial x_2^2} \\ -\dfrac{\partial^2 H^0}{\partial x_1^2} & -\dfrac{\partial^2 H^0}{\partial x_1 \partial x_2} \end{pmatrix} + \varepsilon Df(0).$$

Its eigenvalues $\lambda_{1,2}(\varepsilon)$ depend smoothly on ε.

Part (i) is then obvious, since if $\lambda_1(\varepsilon)$ and $\lambda_2(\varepsilon)$ have opposite sign at $\varepsilon = 0$, then this property will hold for all sufficiently small ε. Thus, by the Grobman–Hartman Theorem, $x = 0$ is a saddle for such parameter values.

When $x = 0$ is a centre at $\varepsilon = 0$, the matrix $A(\varepsilon)$ has a pair of nonreal eigenvalues $\lambda_1 = \bar{\lambda}_2(\varepsilon)$ for all sufficiently small ε and

$$2\text{Re}\ \lambda_{1,2}(\varepsilon) = \text{Tr}\ A(\varepsilon) = \varepsilon\ \text{Tr}\ Df(0) = \varepsilon\ \text{div} f(0).$$

Applying the Grobman–Hartman Theorem for $\varepsilon \neq 0$, we obtain Part (ii) of the theorem. □

Example 4.29 (Perturbed Lotka–Volterra system) Consider the following small perturbation of the (scaled) Lotka–Volterra system (4.32):

$$\begin{cases} \dot{\xi} = \xi - \xi\eta - \varepsilon\xi^2, \\ \dot{\eta} = -\gamma\eta + \xi\eta, \end{cases} \tag{4.37}$$

where $0 < \varepsilon \ll 1, \gamma > 0$ and the $\varepsilon\xi^2$-term describes weak competition amongst prey. Introducing the same variables as in Example 4.22,

$$\begin{cases} q = \ln \xi, \\ p = \ln \eta, \end{cases}$$

we obtain

$$\begin{cases} \dot{q} = 1 - e^p - \varepsilon e^q, \\ \dot{p} = -\gamma + e^q. \end{cases}$$

This system has for $(1 - \gamma\varepsilon) > 0$ an equilibrium

$$(q^0(\varepsilon), p^0(\varepsilon)) = (\ln \gamma, \ln(1 - \gamma\varepsilon)),$$

which is a centre if $\varepsilon = 0$. Translating the origin of the coordinate system to this equilibrium by the transformation

$$\begin{cases} x_1 = q - \ln \gamma, \\ x_2 = p - \ln(1 - \varepsilon\gamma), \end{cases}$$

we obtain the system

$$\begin{cases} \dot{x}_1 = \quad 1 - e^{x_2} + \varepsilon\gamma(e^{x_2} - e^{x_1}), \\ \dot{x}_2 = -\gamma(1 - e^{x_1}), \end{cases} \tag{4.38}$$

which has the form (4.36) with $H(x) = x_2 - e^{x_2} + \gamma(x_1 - e^{x_1})$ and

$$f(x) = \begin{pmatrix} \gamma(e^{x_2} - e^{x_1}) \\ 0 \end{pmatrix}, \quad f(0) = 0.$$

The system (4.38) satisfies the conditions of Theorem 4.28. Since

$$\varepsilon \operatorname{div} f(0) = -\varepsilon\gamma < 0,$$

the equilibrium $(q^0(\varepsilon), p^0(\varepsilon))$ is a stable focus for sufficiently small $\varepsilon > 0$.

Note that this result perfectly agrees with our analysis in Sect. 4.4.3, where it was shown that system (4.13) has a unique positive globally asymptotically stable equilibrium when $ad - ce > 0$. Indeed, system (4.13) coincides with system (4.37) if we set $a = b = d = 1, c = \gamma$, and $e = \varepsilon$, so that the above condition turns into $1 - \varepsilon\gamma > 0$, which is obviously true for small ε. ◇

Remark Consider a slightly more general system

$$\dot{x} = \begin{pmatrix} \dfrac{\partial H(x)}{\partial x_2} \\ -\dfrac{\partial H(x)}{\partial x_1} \end{pmatrix} + \varepsilon f(x, \varepsilon), \quad x = (x_1, x_2) \in \mathbb{R}^2, \varepsilon \in \mathbb{R}, \tag{4.39}$$

where H is smooth and has a nondegenerate critical point $x = 0$, while f is a smooth function of (x, ε). The Implicit Function Theorem assures that (4.39) has a smooth family $x^0(\varepsilon)$ of equilibria for small $\varepsilon \neq 0$, such that $x^0(0) = 0$. Translating the origin to $x^0(\varepsilon)$, we can assume without loss of generality that $f(0, \varepsilon) = 0$ for all ε with small $|\varepsilon|$. Then, the equilibrium $x = 0$ is either a saddle or, when $\operatorname{div} f(0, 0) \neq 0$, a focus. The stability of the focus is determined by the sign of $\operatorname{div} f(0, 0)$. ◇

Now we want to study limit cycles of the perturbed planar Hamiltonian system (4.36).

Theorem 4.30 (Pontryagin, 1934) *Let L_0 be a clockwise-oriented cycle of (4.36) for $\varepsilon = 0$ corresponding to a periodic solution $\varphi(t)$ with the (minimal) period T_0 and let $\Omega_0 = \operatorname{Int}(L_0) \subset \mathbb{R}^2$ denote the domain inside the cycle L_0. If*

$$M_0 := \int_{\Omega_0} \operatorname{div} f(x)\, dx = 0$$

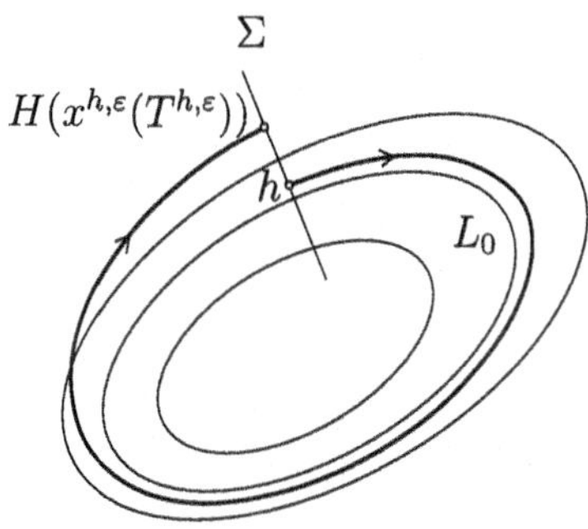

Fig. 4.20 A segment Σ orthogonal to L_0, together with unperturbed and perturbed orbits

while

$$M_1 := \int_0^{T_0} \operatorname{div} f(\varphi(t))\, dt \neq 0,$$

then

(i) *there exists an annulus around* L_0 *in which the system* (4.36) *has, for all* ε *with sufficiently small* $|\varepsilon| > 0$, *a unique cycle* L_ε, *such that* $L_\varepsilon \to L_0$ *as* $\varepsilon \to 0$;

(ii) *this cycle* L_ε *is stable for* $\varepsilon M_1 < 0$ *and unstable for* $\varepsilon M_1 > 0$.

Proof

(i) Consider a cycle L_0 for which the assumptions of the theorem are valid. Without loss of generality, suppose that $H(x) = 0$ for $x \in L_0$. Fix a segment Σ orthogonal to L_0 and consider the value h of the Hamilton function H as a local smooth coordinate along Σ (see Fig. 4.20). This is possible, since L_0 contains no equilibria of the unperturbed system and thus no critical points of H. Denote by $x^{h,\varepsilon} = x^{h,\varepsilon}(t)$ the solution of (4.36) starting at a point in Σ with a small coordinate h, and let $T^{h,\varepsilon}$ be the minimal time needed by the solution to return back to Σ. Due to the smooth dependence of solutions on the initial point and the parameter, $T^{h,\varepsilon}$ is a smooth function of (h, ε) in a neighbourhood of $(0, 0)$ with $T^{0,0} = T_0$, since $x^{0,0} = \varphi$.

Introduce another smooth function

$$\Delta(h, \varepsilon) := H(x^{h,\varepsilon}(T^{h,\varepsilon})) - H(x^{h,\varepsilon}(0)) = H(x^{h,\varepsilon}(T^{h,\varepsilon})) - h,$$

which one calls for obvious reasons the *displacement function*. In general, $\Delta(h, \varepsilon) \neq 0$. However, $\Delta(h, 0) = 0$, since H is constant along orbits of the unperturbed system. Moreover, if $\Delta(h, \varepsilon) = 0$, the perturbed system (4.36) has a closed orbit for the corresponding value of ε passing through the point in Σ with the coordinate h. We will analyse the equation $\Delta(h, \varepsilon) = 0$ with the help of the Implicit Function Theorem.

Along the solutions of (4.36), one has

$$\begin{aligned}
\Delta(h,\varepsilon) &= \int_0^{T^{h,\varepsilon}} \frac{d}{dt} H(x^{h,\varepsilon}(t))\, dt \\
&= \int_0^{T^{h,\varepsilon}} \left[\frac{\partial H(x^{h,\varepsilon}(t))}{\partial x_1} \dot{x}_1^{h,\varepsilon}(t) + \frac{\partial H(x^{h,\varepsilon}(t))}{\partial x_2} \dot{x}_2^{h,\varepsilon}(t) \right] dt \\
&= \varepsilon \int_0^{T^{h,\varepsilon}} \left[\frac{\partial H(x^{h,\varepsilon}(t))}{\partial x_1} f_1(x^{h,\varepsilon}(t)) + \frac{\partial H(x^{h,\varepsilon}(t))}{\partial x_2} f_2(x^{h,\varepsilon}(t)) \right] dt \\
&= \varepsilon \int_0^{T^{h,0}} [-\dot{x}_2^{h,0}(t) f_1(x^{h,0}(t)) + \dot{x}_1^{h,0}(t) f_2(x^{h,0}(t))]\, dt + O(\varepsilon^2),
\end{aligned}$$

where the last integral is computed along solutions of (4.36) with $\varepsilon = 0$. Since these solutions have a clockwise orientation, we can write

$$\Delta(h,\varepsilon) = \varepsilon M(h) + O(\varepsilon^2), \tag{4.40}$$

where

$$M(h) := -\oint_{L_h} f_1 dx_2 - f_2 dx_1 = \int_{\Omega_h} \operatorname{div} f(x)\, dx. \tag{4.41}$$

(The last equality, in which Ω_h is the domain inside the periodic orbit L_h belonging to the level curve $H(x) = h$, is Green's formula (4.9); note the change of sign to incorporate the *clockwise* orientation of L_h.)

By assumption,

$$M(0) = \oint_{H(x)=0} f_1 dx_2 - f_2 dx_1 = \int_{\Omega_0} \operatorname{div} f(x)\, dx = M_0 = 0.$$

We transform the domain $\{(\tau, s) \in \mathbb{R}^2 : 0 \le s \le h,\ 0 \le \tau \le T^{s,0}\}$ into Ω_h by the map $(\tau, s) \mapsto x(\tau, s) = x^{s,0}(\tau)$. Changing the variables in the integral (4.41), we have

$$M(h) = \int_0^h \left(\int_0^{T^{s,0}} \operatorname{div} f(x(\tau,s)) |\det(J_1(\tau,s))|\, d\tau \right) ds,$$

where J_1 is the Jacobian matrix of the transformation. Hence

$$M'(0) = \int_0^{T^{0,0}} \operatorname{div} f(\varphi(\tau))\ |\det(J_1(\tau,0))|\ d\tau.$$

Differentiating the identity

$$H(x^{h,0}(\tau)) = h$$

with respect to h, we obtain

$$\frac{\partial H(x^{h,0}(\tau))}{\partial x_1}\frac{\partial x_1^{h,0}(\tau)}{\partial h}+\frac{\partial H(x^{h,0}(\tau))}{\partial x_2}\frac{\partial x_2^{h,0}(\tau)}{\partial h}=1.$$

Therefore,

$$\det(J_1(\tau,h))=\det\begin{pmatrix}\frac{\partial x_1^{h,0}(\tau)}{\partial \tau} & \frac{\partial x_1^{h,0}(\tau)}{\partial h}\\ \frac{\partial x_2^{h,0}(\tau)}{\partial \tau} & \frac{\partial x_2^{h,0}(\tau)}{\partial h}\end{pmatrix}=\det\begin{pmatrix}\frac{\partial H(x^{h,0}(\tau))}{\partial x_2} & \frac{\partial x_1^{h,0}(\tau)}{\partial h}\\ -\frac{\partial H(x^{h,0}(\tau))}{\partial x_1} & \frac{\partial x_2^{h,0}(\tau)}{\partial h}\end{pmatrix}.$$

Hence $|\det(J_1(\tau,h))|=1$ and thus $|\det(J_1(\tau,0))|=1$, implying that

$$M'(0)=\int_0^{T_0}\operatorname{div} f(\varphi(t))\,dt=M_1\neq 0$$

by the second assumption.

Thus, from (4.40), we have

$$\Delta(h,\varepsilon)=\varepsilon(M_0+hM_1+O(h^2))+O(\varepsilon^2)=\varepsilon F(h,\varepsilon)$$

for some smooth function F. If $\varepsilon=0$, $\Delta=0$ for all small $|h|$ and all orbits are closed (Hamiltonian case). If $\varepsilon\neq 0$, the equation

$$F(h,\varepsilon)=0$$

is such that we can apply the Implicit Function Theorem. Indeed,

$$F(0,0)=M_0=0,\quad \frac{\partial F}{\partial h}(0,0)=M_1\neq 0,$$

by the assumptions we made. Thus, there is a unique smooth function $h=h(\varepsilon)$, $h(0)=0$, such that

$$F(h(\varepsilon),\varepsilon)=0$$

for all small $|\varepsilon|$. This implies that $\Delta(h(\varepsilon),\varepsilon)=0$ for all ε with sufficiently small $|\varepsilon|\neq 0$. Therefore, there exists a unique cycle L_ε through $h(\varepsilon)$.

(ii) Since the map $h\mapsto h+\Delta(h,\varepsilon)$ has derivative

$$1+\frac{\partial\Delta(h(\varepsilon),\varepsilon)}{\partial h}$$

in the fixed point $h(\varepsilon)$ and the sign of

$$\frac{\partial \Delta(h(\varepsilon), \varepsilon)}{\partial h}$$

coincides with the sign of εM_1, the stability assertions follow at once. □

Remarks

(1) If L_0 has counterclockwise orientation, the expression (4.40) will take the form

$$\Delta(h, \varepsilon) = -\varepsilon M(h) + O(\varepsilon^2).$$

The subsequent analysis can then be carried out with obvious modifications. It shows that statement (i) is still valid, but the cycle L_ε is stable when $\varepsilon M_1 > 0$ and unstable when $\varepsilon M_1 < 0$.

(2) When L_0 is oriented counterclockwise,

$$\oint_{L_0} f_1(x)dx_2 - f_2(x)dx_1 = \int_0^{T_0} f(\varphi(t)) \wedge \dot{\varphi}(t)\, dt,$$

where the *wedge product* of two vectors $u, v \in \mathbb{R}^2$ is defined by $u \wedge v := u_1 v_2 - u_2 v_1$. ♢

Example 4.31 (van der Pol equation) The second-order equation

$$\ddot{x} + x = \varepsilon \dot{x}(1 - x^2) \tag{4.42}$$

is called the *van der Pol equation*.[4] It can be rewritten as the equivalent planar system

$$\begin{cases} \dot{x}_1 = x_2, \\ \dot{x}_2 = -x_1 + \varepsilon x_2(1 - x_1^2). \end{cases} \tag{4.43}$$

The system with $\varepsilon = 0$ is Hamiltonian (*harmonic oscillator*) with $H(x_1, x_2) = \frac{1}{2}(x_1^2 + x_2^2)$, which has a family of 2π-periodic solutions

$$\varphi(t) = \begin{pmatrix} r \sin t \\ r \cos t \end{pmatrix}, \quad r > 0.$$

Since $f_1 = 0,\ f_2 = x_2(1 - x_1^2)$, and the periodic orbits are oriented clockwise, we get

[4] van der Pol, B. (1927). Forced oscillations in a circuit with nonlinear resistance (receptance with reactive triode). *London, Edinburgh and Dublin Philosophical Magazine, 3*, 65–80.

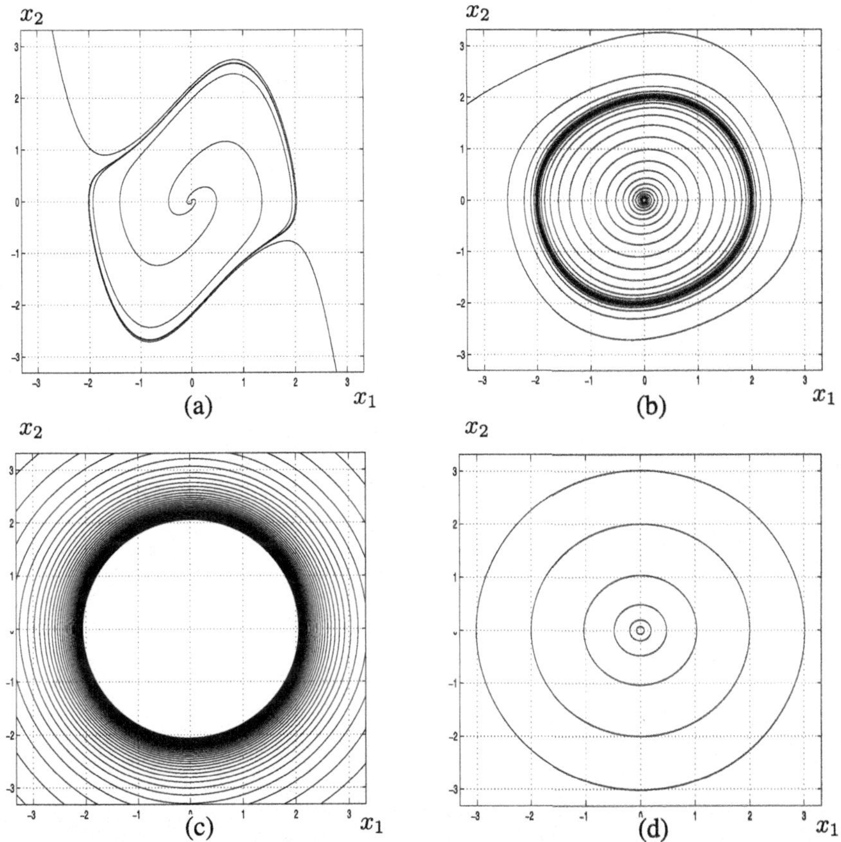

Fig. 4.21 Phase portraits of Van der Pol equation: (**a**) $\varepsilon = 1$; (**b**) $\varepsilon = 0.1$; (**c**) $\varepsilon = 0.01$; (**d**) $\varepsilon = 0$. The (clockwise) orientation of the orbits is not indicated

$$
\begin{aligned}
M(r) &= -\oint_{x_1^2+x_2^2=r^2} f_1(x)dx_2 - f_2(x)dx_1 = \int_0^{2\pi} r^2\cos^2 t(1 - r^2\sin^2 t)\,dt \\
&= \frac{\pi}{4}r^2(4 - r^2).
\end{aligned}
$$

Therefore, $M(r) = 0$ for $r = 2$. Along this solution,

$$
M_1 = \int_0^{2\pi} \operatorname{div} f(\varphi(t))\,dt = \int_0^{2\pi} (1 - 4\sin^2 t)dt = -2\pi < 0.
$$

Thus, by Theorem 4.30, a unique and stable limit cycle bifurcates from the circle $r = 2$ for small $\varepsilon > 0$ (see Fig. 4.21). One can prove that (4.43) has exactly one limit cycle for all $\varepsilon > 0$ (see Exercise 4.6). ◊

Finally, let us consider perturbations of a saddle homoclinic orbit. Suppose that a Hamiltonian planar system

$$\dot{x} = \begin{pmatrix} \dfrac{\partial H(x)}{\partial x_2} \\ -\dfrac{\partial H(x)}{\partial x_1} \end{pmatrix}, \quad x = (x_1, x_2) \in \mathbb{R}^2, \tag{4.44}$$

has an orbit Γ_0 homoclinic to a simple saddle point x_0. Let us denote the corresponding solution by $\gamma(t)$, so that

$$\lim_{t \to \pm\infty} \gamma(t) = x_0.$$

Let $H(x_0) = h_0$, so that $\Gamma_0 \subset \{x \in \mathbb{R}^2 : H(x) = h_0\}$.

Introduce now the following two-parameter perturbation of (4.44):

$$\dot{x} = \begin{pmatrix} \dfrac{\partial H(x)}{\partial x_2} \\ -\dfrac{\partial H(x)}{\partial x_1} \end{pmatrix} + \varepsilon f(x, \mu), \quad x = (x_1, x_2) \in \mathbb{R}^2, \tag{4.45}$$

where $\varepsilon, \mu \in \mathbb{R}$ are parameters, and $f : \mathbb{R}^2 \times \mathbb{R} \to \mathbb{R}^2$ is a smooth function. The reason for introducing the second parameter will become clear later. Finally, suppose that $f(x_0, \mu) = 0$ for $\mu \in \mathbb{R}$. This assumption implies that x_0 is an equilibrium for all values of both parameters.

Consider a segment Σ orthogonal to $\dot{\gamma}(0)$ at $\gamma(0) \in \Gamma_0$. For small values of the parameter ε and any $\mu \in \mathbb{R}$, there exist two solutions of (4.45), say $x_{\pm}^{\mu,\varepsilon}(t)$, such that

$$x_{\pm}^{\mu,\varepsilon}(0) \in \Sigma, \quad \text{and} \quad \lim_{t \to \pm\infty} x_{\pm}^{\mu,\varepsilon}(t) = x_0.$$

Clearly, these solutions belong to the stable and unstable manifolds of the saddle, respectively. Similar to the limit cycle case, parametrize Σ near Γ_0 by the value h of the Hamiltonian function $H(x)$ and introduce the *split function*

$$\Psi(\mu, \varepsilon) := H(x_{+}^{\mu,\varepsilon}(0)) - H(x_{-}^{\mu,\varepsilon}(0))$$

(see Fig. 4.22). This is a smooth function near the origin in the (μ, ε)-plane. Obviously, $\Psi(\mu, 0) = 0$ for all values of μ. Indeed, at $\varepsilon = 0$ the system (4.45) reduces to (4.44) and has the homoclinic orbit Γ_0, so that $x_{+}^{\mu,0}(t) = x_{-}^{\mu,0}(t) = \gamma(t)$ for all $t \in \mathbb{R}$. If $\Psi(\mu, \varepsilon) = 0$ for some $\varepsilon \neq 0$, then the perturbed system (4.45) has a homoclinic orbit Γ to the saddle x_0 at these parameter values. This explains the appearance of the second parameter:

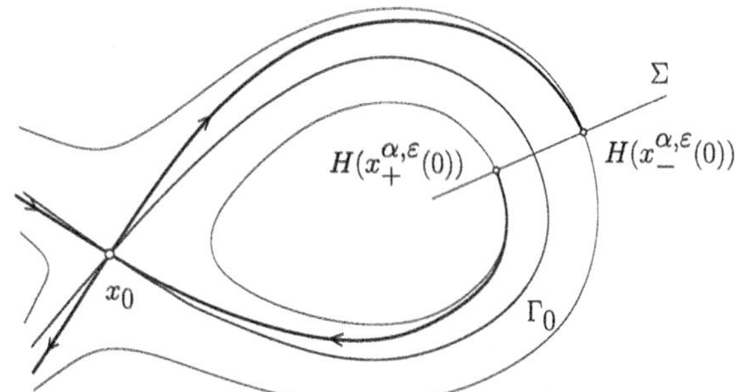

Fig. 4.22 A segment Σ orthogonal to the saddle homoclinic orbitt Γ_0, together with unperturbed and perturbed orbits asymptotic to the saddle

Under a generic perturbation $\varepsilon f(x)$, the invariant manifolds split, so one needs another parameter, e.g. μ, to tune to "compensate" for this in order to preserve the homoclinic orbit in the perturbed system.

We have

$$\begin{aligned}\Psi(\mu,\varepsilon) &= \int_{-\infty}^{0} \frac{d}{dt} H(x_-^{\mu,\varepsilon}(t))\, dt - \int_{\infty}^{0} \frac{d}{dt} H(x_+^{\mu,\varepsilon}(t))\, dt \\ &= \int_{-\infty}^{0} DH(x_-^{\mu,\varepsilon}(t))\dot{x}_-^{\mu,\varepsilon}(t)\, dt + \int_{0}^{\infty} DH(x_+^{\mu,\varepsilon}(t))\dot{x}_+^{\mu,\varepsilon}(t)\, dt \\ &= \varepsilon M(\mu) + O(\varepsilon^2),\end{aligned}$$

where

$$\begin{aligned}M(\mu) &:= \int_{-\infty}^{\infty} [-\dot{\gamma}_2(t) f_1(\gamma(t),\mu) + \dot{\gamma}_1(t) f_2(\gamma(t),\mu)]dt \\ &= \int_{\Gamma_0} f_2(x,\mu)dx_1 - f_1(x,\mu)dx_2. \end{aligned} \tag{4.46}$$

This function is called the *Melnikov homoclinic integral.* The integral converges absolutely, since $f(\gamma(t),\mu)$ and $\dot{\gamma}(t)$ tend exponentially fast to zero as $t \to \pm\infty$.

Fix some $\mu_0 \in \mathbb{R}$ and consider $M_0 = M(\mu_0)$ and

$$M_1 = M'(\mu_0) = \int_{\Gamma_0} \frac{\partial f_2}{\partial \mu}(x,\mu_0)dx_1 - \frac{\partial f_1}{\partial \mu}(x,\mu_0)dx_2,$$

then

$$\Psi(\mu,\varepsilon) = \varepsilon(M_0 + M_1(\mu-\mu_0) + O((\mu-\mu_0)^2)) + O(\varepsilon^2) = \varepsilon\Phi(\mu,\varepsilon).$$

If $M_0 = 0$ but $M_1 \neq 0$, then we can apply the Implicit Function Theorem to the equation $\Phi(\mu,\varepsilon) = 0$. Indeed,

$$\Phi(\mu_0, 0) = M_0 = 0, \quad \frac{\partial \Phi}{\partial \mu}(\mu_0, 0) = M_1 \neq 0.$$

Thus, there is a unique smooth function $\mu = \mu(\varepsilon)$, $\mu(0) = \mu_0$, such that

$$\Phi(\mu(\varepsilon), \varepsilon) = 0$$

for all small $|\varepsilon|$. This implies that $\Psi(\mu(\varepsilon), \varepsilon) \equiv 0$, so that there exists a unique homoclinic orbit Γ to the saddle x_0 in (4.45) for all sufficiently small $|\varepsilon|$, *provided that* $\mu = \mu(\varepsilon)$.

The considerations above can be summarized in the following theorem.

Theorem 4.32 (Melnikov, 1963) *Let Γ_0 be an orbit homoclinic to a saddle equilibrium of* (4.45) *for $\varepsilon = 0$. Suppose that for some $\mu = \mu_0$ holds*

$$\int_{\Gamma_0} f_2(x, \mu_0) dx_1 - f_1(x, \mu_0) dx_2 = 0,$$

while

$$\int_{\Gamma_0} \frac{\partial f_2}{\partial \mu}(x, \mu_0) dx_1 - \frac{\partial f_1}{\partial \mu}(x, \mu_0) dx_2 \neq 0.$$

Then there exists a unique function $\mu(\varepsilon)$ with $\mu(0) = \mu_0$, and an annulus around Γ_0 in which the system (4.45) *has, for all ε with sufficiently small $|\varepsilon|$ and $\mu = \mu(\varepsilon)$, a homoclinic orbit $\Gamma \to \Gamma_0$ as $\varepsilon \to 0$.* □

Example 4.33 (Normal form for Bogdanov–Takens bifurcation) Consider the following planar system,[5]

$$\begin{cases} \dot{x}_1 = x_2, \\ \dot{x}_2 = -1 + x_1^2 + \varepsilon(\mu x_2 + x_1 x_2), \end{cases} \tag{4.47}$$

that appears in the analysis of a generic two-parameter system having an equilibrium with a double zero eigenvalue at some critical parameter values.

For $\varepsilon = 0$ the system (4.47) is Hamiltonian with

$$H(x, y) = \frac{x_2^2}{2} + x_1 - \frac{x_1^3}{3}$$

and has a centre $(-1, 0)$ and a saddle $(x_0, y_0) = (1, 0)$. (Verify!) The saddle is simple and has a homoclinic orbit corresponding with the following explicit solution

$$\gamma_1(t) = 1 - 3\,\mathrm{sech}^2\left(\frac{t}{\sqrt{2}}\right),$$

[5] Guckenheimer, J., & Holmes, Ph. (1983). *Nonlinear Oscillations, Dynamical Systems, and Bifurcations of Vector Fields* (pp. 364–371). New-York: Springer.

$$\gamma_2(t) = 3\sqrt{2}\,\mathrm{sech}^2\left(\frac{t}{\sqrt{2}}\right)\tanh\left(\frac{t}{\sqrt{2}}\right),$$

where

$$\mathrm{sech}(z) = \frac{2}{e^z + e^{-z}}, \quad \tanh(z) = \frac{e^z - e^{-z}}{e^z + e^{-z}}.$$

Notice that (x_0, y_0) is an equilibrium point of (4.47) for all values of (ε, μ). The Melnikov homoclinic integral defined by (4.46) can then be computed explicitly:

$$M(\mu) = \int_{-\infty}^{\infty} \gamma_2(t)[\mu\gamma_2(t) + \gamma_1(t)\gamma_2(t)]\, dt = \frac{24\sqrt{2}}{35}(7\mu - 5).$$

Clearly, for

$$\mu_0 = \frac{5}{7}$$

we have $M(\mu_0) = 0$ and $M_1 = M'(\mu_0) \neq 0$. Thus, Theorem 4.32 implies that, for sufficiently small $|\varepsilon|$ and

$$\mu = \mu(\varepsilon) = \frac{5}{7} + O(\varepsilon),$$

the perturbed system (4.47) has a homoclinic orbit to the saddle equilibrium $(1, 0)$. ◊

4.6 References

The best references for the qualitative theory of autonomous planar ODEs and the theory of their bifurcations are still the classical books (Andronov et al., 1971, 1973). Pontryagin's method to locate limit cycles by perturbing planar Hamiltonian systems is also presented in Andronov et al. (1973). Further results on nonlinear planar ODEs, e.g. the index theory and blow-up techniques to study degenerate equilibrium points, can be found in Arnol'd (1973, 1983), Arnol'd and Il'yashenko (1988), Perko (2001) and, in particular, in Dumortier et al. (2006).

The theory of Hamiltonian systems is a classical and highly developed topic, see Verhulst (1996) for a brief introduction and Arnol'd (1989), Marsden and Ratiu (1999) for advanced presentations.

Phase portraits of various prey–predator models and their bifurcations are systematically studied in Bazykin (1998).

4.7 Exercises

Exercise 4.1 (Bautin's example) Prove that the planar quadratic system

$$\begin{cases} \dot{x} = y, \\ \dot{y} = ax + by + \alpha x^2 + \beta y^2, \end{cases}$$

has no periodic orbits, provided $b \neq 0$. *Hint*: Use Dulac's Criterion with factor $\mu(x, y) = e^{\gamma x}$, where γ is chosen suitably.

Exercise 4.2 (Normal form for Andronov–Hopf bifurcation) Consider the following planar system depending on parameter α:

$$\begin{cases} \dot{x} = \alpha x - y - x(x^2 + y^2) + R(x, y, \alpha), \\ \dot{y} = x + \alpha y - y(x^2 + y^2) + S(x, y, \alpha), \end{cases} \tag{4.48}$$

where R and S are smooth functions of (x, y, α), such that $R, S = \mathcal{O}(\rho^4)$ as $\rho \to 0$ with $\rho^2 = x^2 + y^2$. In Chap. 5, we will show that a generic planar system near the *supercritical Andronov–Hopf bifurcation* is locally smoothly orbitally equivalent to (4.48).

(a) Prove that there exists $\varepsilon > 0$ such that for all $\alpha \leq 0$ with sufficiently small $|\alpha|$ all positive half-orbits of (4.48) starting in the disc

$$D(\varepsilon) = \{(x, y) \in \mathbb{R} : x^2 + y^2 \leq \varepsilon^2\}$$

remain in $D(\varepsilon)$ and converge to $O = (0, 0)$ as $t \to \infty$. *Hint*: Use polar coordinates.

(b) Prove that for all sufficiently small $\alpha > 0$ all positive half-orbits of (4.48) starting in the annulus

$$A(\alpha) = \left\{(x, y) \in \mathbb{R}^2 : \frac{2\alpha}{3} \leq x^2 + y^2 \leq \frac{3\alpha}{2}\right\}$$

remain in $A(\alpha)$. Use the Poincaré–Bendixson Theorem to conclude that (4.48) has at least one periodic orbit in $A(\alpha)$ for all sufficiently small $\alpha > 0$.

(c) Prove that for all sufficiently small $\alpha > 0$ holds: $(\text{div } F)(x, y, \alpha) < 0$ for all $(x, y) \in A(\alpha)$, where F is the vector field defining (4.48). Conclude using Bendixson's Criterion that (4.48) has exactly one periodic orbit in $A(\alpha)$ for all sufficiently small $\alpha > 0$.

(d) Prove that there exists $\varepsilon > 0$ such that for all sufficiently small $\alpha > 0$ simultaneously hold:

- $A(\alpha) \subset D(\varepsilon)$;
- all positive half-orbits of (4.48) starting in $D(\varepsilon)$ remain in $D(\varepsilon)$;
- all positive orbits starting in $D(\varepsilon) \setminus (A(\alpha) \cup O)$ enter $A(\alpha)$.

(e) Combine the results obtained to sketch and describe all possible phase portraits of (4.48) in a fixed small neighbourhood of the origin for all α with sufficiently small $|\alpha|$.

Exercise 4.3 (Chemostat dynamics) A *chemostat* is a laboratory device for growing microorganisms continuously (see Fig. 4.23). Let C_0 denote the concentration of a limiting resource (*substrate*) in the inflow to the chemostat, while C and $N = (N_1, N_2, \ldots)$ denote the concentrations of different types of microorganisms in the chemostat (and in the outflow).

(a) If only one microorganism is present, then, according to some model.[6] and after a scaling, the dynamics of (C, N) are described by

---

[6] Smith, H. L., & Waltman, P. (1995). *The Theory of the Chemostat: Dynamics of Microbial Competition*. Cambridge University Press.

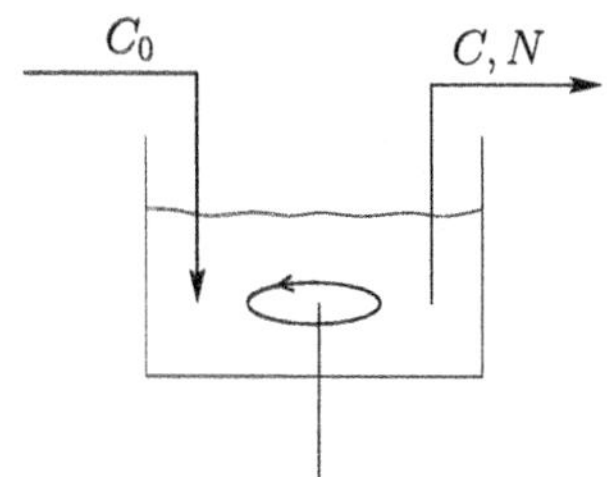

Fig. 4.23 A chemostat

$$\begin{cases} \dot{C} = C_0 - C - K(C)N, \\ \dot{N} = \quad -N + \alpha K(C)N, \end{cases}$$

where K is a positive and increasing smooth function with $K(0) = 0$ and $K(C) \to 1$ as $C \to \infty$, e.g.

$$K(C) = \frac{C}{1+C}.$$

Assume that $\alpha > 1$ (otherwise the microorganisms are washed out, no matter how large is C_0). Prove that

(i) $C(t) + \frac{1}{\alpha}N(t) \to C_0$ for $t \to \infty$;
(ii) If $\alpha K(C_0) < 1$, then the equilibrium $(C_0, 0)$ is globally asymptotically stable;
(iii) If $\alpha K(C_0) > 1$, then the equilibrium $(\overline{C}, \overline{N})$ is asymptotically stable and all orbits with $N(0) > 0$ tend to this equilibrium as $t \to \infty$. Here $\overline{C}$ is characterized by $\alpha K(\overline{C}) = 1$ and

$$\overline{N} = \frac{C_0 - \overline{C}}{K(\overline{C})} = \alpha(C_0 - \overline{C}).$$

(b) If we consider the competition of two different microorganisms for the same substrate, we have to analyse the three-dimensional system

$$\begin{cases} \dot{C} = C_0 - C - K_1(C)N_1 - K_2(C)N_2, \\ \dot{N}_1 = \quad - N_1 + \alpha_1 K_1(C)N_1, \\ \dot{N}_2 = \quad - N_2 \quad + \alpha_2 K_2(C)N_2 \end{cases}$$

where $\alpha_i > 1$, and K_i have the same properties as K in part (a).
(i) Show that $C(t) + \frac{1}{\alpha_1}N_1(t) + \frac{1}{\alpha_2}N_2(t) \to C_0$ as $t \to \infty$;
(ii) Motivated by (i), consider the planar system

$$\begin{cases} \dot{N}_1 = -N_1 + \alpha_1 K_1\left(C_0 - \dfrac{1}{\alpha_1}N_1 - \dfrac{1}{\alpha_2}N_2\right)N_1, \\[2ex] \dot{N}_2 = -N_2 + \alpha_2 K_2\left(C_0 - \dfrac{1}{\alpha_1}N_1 - \dfrac{1}{\alpha_2}N_2\right)N_2. \end{cases} \tag{4.49}$$

We assume that $\alpha_i K_i(C_0) > 1$ for $i = 1, 2$ and define $\overline{C}_i$ by $\alpha_i K_i(\overline{C}_i) = 1$.

Prove that $(\alpha_1(C_0 - \overline{C}_1), 0)$ attracts all orbits with $N_i(0) > 0$ if $\overline{C}_1 < \overline{C}_2$, while $(0, \alpha_2(C_0 - \overline{C}_2))$ attracts all such orbits if $\overline{C}_1 > \overline{C}_2$. (*Hint*: What can you say about the slope of the isoclines?)

Exercise 4.4 (Infectious diseases) (a) The system

$$\begin{cases} \dot{S} = b - \mu S - \beta SI, \\ \dot{I} = \quad -\mu I + \beta SI - \alpha I, \end{cases} \tag{4.50}$$

arises in the context of modelling the spread of an infectious disease.[7] Here S stands for "susceptibles" and I for "infecteds", which are also infectious; b is the population birth rate and all newborns are susceptible; μ is the per capita death rate and α is the per capita rate at which infecteds lose their infectiousness.

(i) Show that the equilibrium

$$\left(\frac{b}{\mu}, 0\right)$$

is globally asymptotically stable when

$$R_0 = \frac{\beta b}{\mu(\mu + \alpha)} < 1.$$

(ii) Show that the equilibrium

$$(\bar{S}, \bar{I}) = \left(\frac{\mu + \alpha}{\beta}, \frac{b}{\mu + \alpha} - \frac{\mu}{\beta}\right)$$

attracts all orbits with $S(0), I(0) > 0$ when $R_0 > 1$. *Hint*: Try to find a Lyapunov function in an already familiar form.

(b) The system

$$\begin{cases} \dot{S} = bS + bR - \mu S - \gamma \frac{SI}{N}, \\ \dot{I} = - \mu I + \gamma \frac{SI}{N} - \alpha I, \\ \dot{R} = - \mu R + f\alpha I, \end{cases}$$

where $N = S + I + R$ describes the spread of an infectious disease in a population that grows with rate $(b - \mu)$ in the absence of the disease. The variables S and I have the same meaning as in part (a), while R denotes the "removed" individuals, which are immune for the rest of their life. After an exponentially distributed (with parameter α) infectious period, individuals die with probability $1 - f$ and become "removed" with probability f. Infected individuals have a zero per capita birth rate. All parameters are positive and we assume that $b > \mu$. In the transmission term $\gamma SI/N$, the N in the denominator makes sure that the contact rate of an individual is independent of the population size (think, for example, of a sexually transmitted disease). In part (a) we did not need to worry about this point, since the population size did stabilize anyhow at b/μ.

(i) Verify that the system is *first-order homogeneous*, i.e. $F(\lambda x) = \lambda F(x)$, where $x = (S, I, R)$ and F is the vector field at the right-hand side. As a consequence, exponential solutions are feasible even though the system is nonlinear.

(ii) Rewrite the system in terms of N, y and z, where $y = I/N$, $z = R/N$. (*Hint*: You should obtain a partially decoupled system: The equations for y and z do not depend on N.)

(iii) Study phase portraits of the (y, z)-subsystem.

Hints: First prove that the triangle $\{(y, z) : y \geq 0, z \geq 0, 1 - y - z \geq 0\}$ is forward-invariant. Then study the cases $\gamma \leq \alpha + b$ and $\gamma > \alpha + b$ separately. In the former case, prove that $y(t) \to 0$ (and, consequently, $z(t) \to 0$) as $t \to \infty$ for all $y(0), z(0) \geq 0$. In the latter case, use Dulac's Criterion with factor $(yz)^{-1}$ to exclude cycles.

(iv) Use the information obtained in (iii) to derive the (asymptotic) growth rate of N. Could the infectious agent slow down, or even stop, the host population growth?

[7] Diekmann, O., Heesterbeek, J. A. P., & Britton, T. (2013). *Mathematical Tools for Understanding Infectious Disease Dynamics*, Princeton University Press.

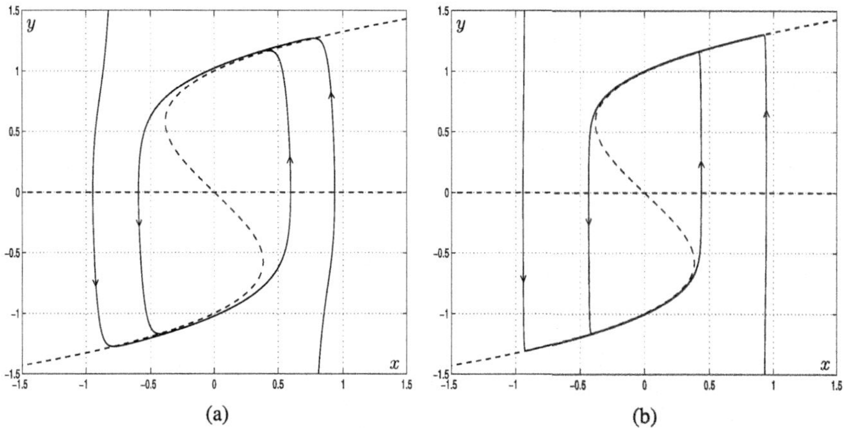

Fig. 4.24 Phase portraits of (4.51): (**a**) $\varepsilon = 0.1$; (**b**) $\varepsilon = 0.01$

Exercise 4.5 (Slow-fast oscillations) Periodic solutions of planar *slow-fast systems*

$$\begin{cases} \dot{\xi} = P(\xi, \eta), \\ \varepsilon\dot{\eta} = Q(\xi, \eta), \end{cases}$$

where $\varepsilon \ll 1$, can be constructed by concatenation of slow motions near the isocline $Q(\xi, \eta) = 0$ with fast "jumps" towards this curve.

(a) Consider the following variant of the *van der Pol oscillator* (cf. Example 4.31):

$$\begin{cases} \dot{x} = -y, \\ \varepsilon\dot{y} = x + y - y^3, \end{cases} \tag{4.51}$$

where $\varepsilon > 0$ is a parameter. Analyse the dynamics of this system for $\varepsilon \ll 1$ (*Hint*: See Fig. 4.24. *Warning*: We realize that "Analyse" is a somehow ambiguous assignment. What we meant is "Make plausible that the phase portraits are as shown in the figure", while leaving it to the reader's own consciousness whether or not to provide rigorous proofs. These proofs require a serious effort.)

(b) Consider the planar system

$$\begin{cases} \varepsilon\dot{x} = x\left(1 - \dfrac{x}{K} - \dfrac{y}{1+x}\right), \\ \dot{y} = y\left(-1 + \theta\dfrac{x}{1+x}\right). \end{cases} \tag{4.52}$$

(i) Show that (4.52) can be obtained from the Rosenzweig–MacArthur model (4.20) by the scaling

$$\tau = ct, \quad x = \beta bv, \quad y = \frac{b}{a}p,$$

and introduction of the new parameters:

$$\varepsilon = \frac{c}{a}, \; \theta = \frac{d}{\beta bc}, \quad K = \frac{\beta ab}{e}.$$

(ii) Assume that

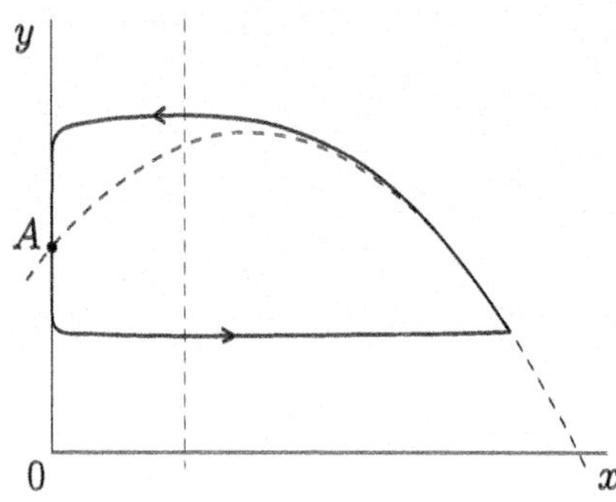

Fig. 4.25 A slow-fast limit cycle in (4.52)

$$\theta > 1, \quad K > 1 + \frac{2}{\theta - 1}.$$

Interpret this assumption in biological terms. Sketch the phase portrait of (4.52) for $\varepsilon \ll 1$.

Warning: A careful and nontrivial analysis shows that the lower part of the slow-fast cycle in (4.52) tends as $\varepsilon \to 0$ to a horizontal line segment located *below* the point A, where the prey isocline

$$y = (1 + x)\left(1 - \frac{x}{K}\right)$$

intersects the y-axis (see Fig. 4.25).[8]

Exercise 4.6 (Cycles in the Liénard system) (a) Prove the following theorem:

Theorem 4.34 *Consider the planar* ***Lienard system***

$$\begin{cases} \dot{x} = y - F(x), \\ \dot{y} = -x, \end{cases} \tag{4.53}$$

where $F : \mathbb{R} \to \mathbb{R}$ is smooth and such that

(i) *$F(-x) = -F(x)$ for all $x \in \mathbb{R}$;*

(ii) *$F(x) \to +\infty$ as $x \to +\infty$ and there exists a constant $\beta > 0$ such that $F(x) > 0$ and $F'(x) > 0$ for $x > \beta$;*

(iii) *$F'(0) < 0$ so that there exists a constant $\alpha > 0$ such that $F(x) < 0$ for all $0 < x < \alpha$.*

Then the system (4.53) has at least one periodic orbit. If moreover $\alpha = \beta$, the periodic orbit is unique and all orbits of (4.53) with $(x(0), y(0)) \neq (0, 0)$ tend to this periodic orbit as $t \to +\infty$.

Hints:

(1) Prove that the origin is an unstable node or focus;

(2) Introduce the function $R(x, y) = \frac{1}{2}(x^2 + y^2)$. Show that $\dot{R} = -xF(x)$.

(3) Consider an orbit of (4.53) starting at a point $A = (0, y_0)$ with sufficiently large $y_0 > 0$ and show that it comes back to the y-axis at a point $E = (0, y_1)$ with $y_1 < 0$ and $|y_1| < y_0$ (see Fig. 4.26a). For this, introduce

$$I(y_0) = R(0, y_1) - R(0, y_0)$$

[8] Kuznetsov, Yu. A., Muratori, S., & Rinaldi, S. (1995). Homoclinic bifurcations in slow-fast second-order systems. *Nonlinear Analysis, 25*, 747–762.

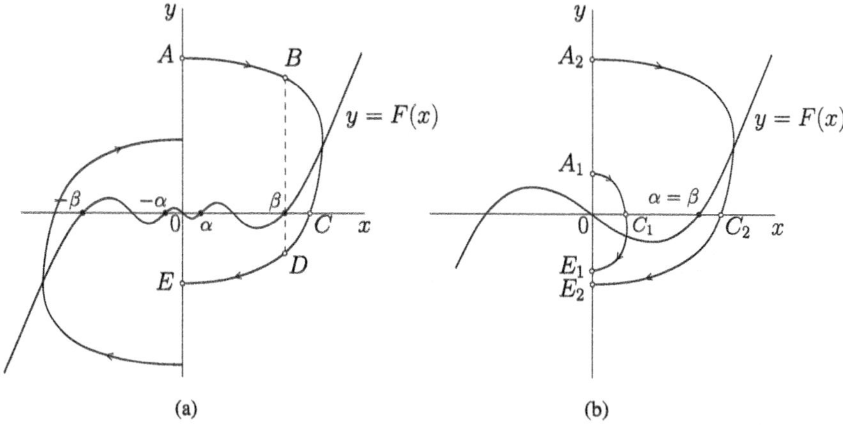

Fig. 4.26 Analysis of limit cycles in Liénard system

and write

$$I(y_0) = \int_{ABCDE} dR = \left(\int_{AB} + \int_{DE}\right) \frac{xF(x)dx}{F(x) - y} + \int_{BCD} F(x)dy$$

and notice that both integrals are negative, implying $I(y_0) < 0$, when $y_0 \to +\infty$.

(4) Then use the symmetry of the system with respect to $(x, y) \mapsto (-x, -y)$ to construct a forward-invariant set composed of the orbit arc connecting A and E, its reflection, and two segments of the y-axis: $[-y_1, y_0]$ and $[-y_0, y_1]$. Apply the Poincaré–Bendixson Theorem.

(5) When $\alpha = \beta$, prove first that for any orbit crossing the interval $0 < x < \alpha$ we have $|y_1| > y_0$ (see Fig. 4.26b). Use the fact that $F(x) < 0$ and, therefore,

$$I(y_0) = \int_{BCD} F(x)dy > 0.$$

(6) Use (3) and the monotonicity of F for $x \geq \beta$ to prove that $I(y_0)$ has only one zero. This implies the uniqueness of the limit cycle and its stability.

(b) Using Theorem 4.34, prove that the cycle in the van der Pol equation (4.42) from Example 4.31 is unique. (*Hint*: To write (4.42) in the form (4.53), introduce a new variable $y = \dot{x} + F(x)$, where

$$F(x) = \int_0^x f(s)\, ds$$

and $f(s) = \varepsilon(s^2 - 1)$. Then the conditions of Theorem 4.34 are satisfied with $\alpha = \beta = \sqrt{3}$.)

Exercise 4.7 (A nontrivial ω-limit set) Consider the system[9]

$$\begin{pmatrix} \dot{x} \\ \dot{y} \end{pmatrix} = \begin{pmatrix} \dfrac{\partial H(x,y)}{\partial y} \\ -\dfrac{\partial H(x,y)}{\partial x} \end{pmatrix} + \mu H(x,y) \begin{pmatrix} \dfrac{\partial H(x,y)}{\partial x} \\ \dfrac{\partial H(x,y)}{\partial y} \end{pmatrix}, \tag{4.54}$$

[9] Batalova, Z. S., & Neimark, Yu. I. (1973). A certain dynamical system with a homoclinic structure. In *Theory of Oscillations, Applied Mathematics and Cybernetics* (vol. 1, pp. 131–147). Gorkii State University. In Russian.

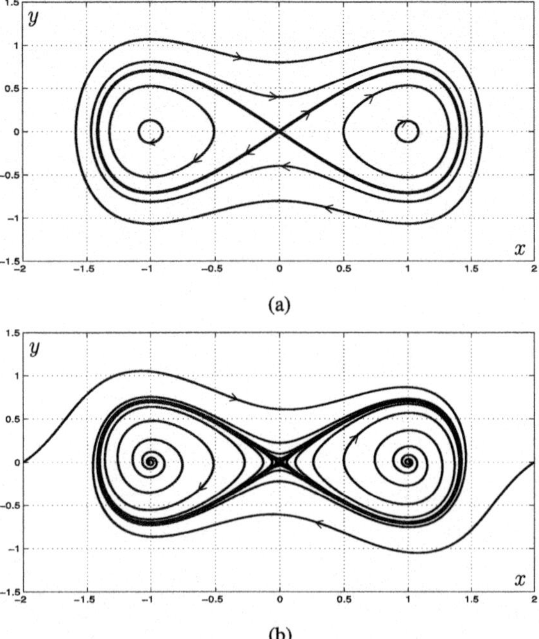

Fig. 4.27 Phase orbits of (4.54): (**a**) $\mu = 0$; (**b**) $\mu = -0.5$

where μ is a parameter,

$$H(x, y) = \frac{y^2}{2} - \frac{x^2}{2} + \frac{x^4}{4}$$

is a scalar function, and

$$\begin{pmatrix} \dfrac{\partial H(x,y)}{\partial x} \\ \dfrac{\partial H(x,y)}{\partial y} \end{pmatrix} = \begin{pmatrix} -x + x^3 \\ y \end{pmatrix}$$

is its gradient.

(a) Plot the phase portrait of (4.54) at $\mu = 0$ with special attention to equilibria and critical level sets. *Answer*: Figure 4.27a.

(b) Prove that for $\mu < 0$ all nonequilibrium orbits of (4.54) tend to the level set $H(x, y) = 0$. *Hint*: $\dot{H} = \mu \left[\left(\frac{\partial H}{\partial x}\right)^2 + \left(\frac{\partial H}{\partial y}\right)^2 \right] H$.

(c) Using (b), prove that the ω-limit set of all points with $H(x, y) > 0$ consists of the saddle point $(x, y) = (0, 0)$ and two orbits homoclinic to this saddle (see Fig. 4.27b).

Exercise 4.8 (A planar Hamiltonian system) Consider the system

$$\begin{cases} \dot{x} = -\mu y + x^2 - y^2, \\ \dot{y} = \mu x - 2xy, \end{cases} \tag{4.55}$$

appearing in the study of bifurcations of a periodic orbit near 1:3 resonance in Hamiltonian systems with two degrees of freedom (Arnol'd, 1989).

(a) Prove that this system is Hamiltonian for all values of the parameter μ and find its Hamiltonian function H.

(b) Draw the phase portrait of (4.55) for $\mu < 0$, $\mu = 0$, and $\mu > 0$. (*Hint*: The level set $H(x, y) = h$ passing through nontrivial equilibria of (4.55) when $\mu \neq 0$ is the union of three straight lines.)

Exercise 4.9 (Rock–scissors–paper dynamics) Consider a planar system

$$\begin{cases} \dot{x}_1 = \left(x_1 + \frac{1}{3}\right)[x_1 + 2x_2 - \varepsilon(x_2 + \Psi(x))], \\ \dot{x}_2 = -\left(x_2 + \frac{1}{3}\right)[2x_1 + x_2 - \varepsilon(x_1 + x_2 - \Psi(x))], \end{cases} \tag{4.56}$$

where $\Psi(x) = x_1^2 + x_1 x_2 + x_2^2$.[10]

(a) Prove that (4.56) is $\mathbb{Z}_3$-symmetric, i.e. it does not change under the transformation $x \mapsto Tx$, where

$$T = \begin{pmatrix} -1 & -1 \\ 1 & 0 \end{pmatrix}, \quad T^2 = \begin{pmatrix} 0 & 1 \\ -1 & -1 \end{pmatrix}, \quad T^3 = \begin{pmatrix} 1 & 0 \\ 0 & 1 \end{pmatrix}.$$

(b) Find all equilibria of (4.56).

(c) Show that (4.56) is Hamiltonian for $\varepsilon = 0$, find its Hamiltonian function $H = H(x)$ and verify that $H(Tx) = H(x)$. Sketch the phase portrait of the system for $\varepsilon = 0$. (*Hint*: Use the trick of Exercise 4.8.)

(d) Prove that $\dot{H} = -3\varepsilon\Psi H$ along orbits of (4.56) and conclude that it has no periodic orbits. Sketch the phase portrait of the system for $\varepsilon \neq 0$.

Exercise 4.10 (Effective Kepler potential) Draw the phase portrait of a mechanical system with one degree of freedom with

$$U(q) = -\frac{k}{q} + \frac{M}{q^2}.$$

Classify all orbits.

Exercise 4.11 (Planar mechanical systems) (a) Show that

$$T(h) = S'(h),$$

where $S(h)$ is the area of the domain bounded by a noncritical closed level curve

$$\frac{p^2}{2} + U(q) = h$$

and where $T(h)$ is the corresponding period.

Hint: Use formula (4.35) for the period $T(h)$ of oscillations of a planar mechanical system.

(b) Find the period of small oscillations near a quadratic minimum of $U(q)$ at q_0, i.e. compute the limit of $T(h)$ as $h \to U(q_0)$.

Answer: $T_0 = 2\pi/\sqrt{U''(q_0)}$.

(c) Find linear approximations to the orbits tending to a simple saddle $(q_0, 0)$ as $t \to \pm\infty$.

Answer: $p = \pm\sqrt{U''(q_0)}(q - q_0)$.

---

[10] Diekmann, O., & van Gils, S. A. (2009). On the cyclic replicator equation and the dynamics of semelparous populations. *SIAM Journal on Applied Dynamical Systems, 8*, 1160–1189.

Exercise 4.12 (Conservative systems)

Definition 4.35 *A smooth system*

$$\dot{x} = f(x), \quad x \in \mathbb{R}^{2m}, \tag{4.57}$$

is called **conservative** *if it is orbitally equivalent to a Hamiltonian system (see Appendix to this chapter):*

$$\dot{x} = J_{2m}[DH(x)]^{\mathrm{T}}, \quad x \in \mathbb{R}^{2m},$$

i.e., there is a positive smooth function $\rho = \rho(x)$ and a smooth function $H = H(x)$, such that $\rho(x)f(x) = J_{2m}[DH(x)]^{\mathrm{T}}$ for all $x \in \mathbb{R}^{2m}$.

Prove the following result.

Theorem 4.36 *If* (4.57) *is a conservative system and φ^t is the corresponding flow, then*

$$W(t) = \int_{\varphi^t(D_0)} \rho(x)\, dx = \text{const}$$

for any domain $D_0 \subset \mathbb{R}^{2m}$, for which $W(0)$ exists. □

Hint: Mimic the proof of Lemma 4.39 in the Appendix and take into account that $\text{div}(\rho f) = \rho \text{ div } f + (D\rho)f$.

Exercise 4.13 (Conservative ecosystems) (a) Prove that the Lotka–Volterra system (4.32) is conservative in any open region of $\mathbb{R}^2_+$ and find the corresponding Hamiltonian function.

Answer: $\rho(\xi, \eta) = (\xi\eta)^{-1}$ and $H(\xi, \eta) = \gamma \ln \xi - \xi + \ln \eta - \eta$.

(b) Consider the following modification of Lotka–Volterra system (4.32) with $\gamma = 1$:

$$\begin{cases} \dot{\xi} = \xi - \dfrac{\xi\eta}{(1+\alpha\xi)(1+\beta\eta)} & =: \ f_1(\xi, \eta), \\ \dot{\eta} = -\eta + \dfrac{\xi\eta}{(1+\alpha\xi)(1+\beta\eta)} & =: \ f_2(\xi, \eta), \end{cases} \tag{4.58}$$

where $\alpha, \beta > 0$ are parameters.[11]

(i) Prove that (4.58) is conservative (see Exercise 4.12) in $\mathbb{R}^2_+$ if $\alpha = \beta$, and has no closed orbits if $\alpha \neq \beta$. (*Hint*: Look for ρ in the form

$$\rho(\xi, \eta) = \xi^a \eta^b (1 + \alpha\xi)(1 + \beta\eta),$$

where a and b are unknown coefficients. By a proper selection of a and b, one can achieve that $\text{div}(\rho f) = (\alpha - \beta)\xi\rho$.)

(ii) Sketch all topologically different phase portraits of (4.58) that exist for different (α, β).

(iii) Can you prove that (4.58) has a family of closed orbits for $\alpha = \beta$, provided that

$$0 < \alpha = \beta < \frac{1}{4},$$

using the symmetry of the system? *Hint*: The transformation $(\xi, \eta, t) \mapsto (\eta, \xi, -t)$ does not change (4.58), i.e. the system is *reversible*.

[11] Bazykin, A. D., Berezovskaya, F. S., Denisov, G. A., & Kuznetsov, Yu. A. (1981). The influence of predator saturation effect and competition amongst predators on predator–prey system dynamics. *Ecological Modelling, 14*, 39–57.

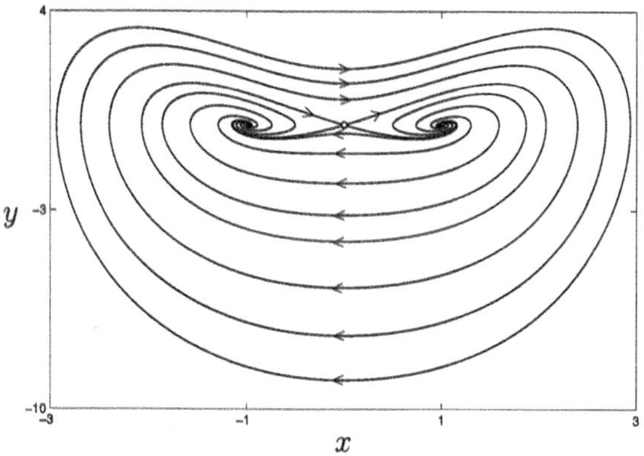

Fig. 4.28 Phase portrait of a reversible system

Exercise 4.14 (Another reversible system) Prove that the phase portrait of the planar system

$$\begin{cases} \dot{x} = y, \\ \dot{y} = x + xy - x^3, \end{cases}$$

is as presented in Fig. 4.28.

Hint: Use the symmetry of the system under a reflection and time reversal: $(x, t) \mapsto (-x, -t)$ or try to prove that it is conservative in (a subset of) $\mathbb{R}^2$.

Exercise 4.15 (Properties of Poisson brackets) (a) Prove properties (P.1)–(P.5) of the Poisson bracket introduced in the Appendix (see Definition 4.41).

(b) Compute $\{q_i, q_j\}$, $\{q_i, p_j\}$, and $\{p_i, p_j\}$.

Exercise 4.16 (An integrable Hamiltonian system with two degrees of freedom) Read Appendix to this chapter in Sect. 4.8 and consider the system

$$\dot{x} = J_4[DH(x)]^{\mathrm{T}}, \quad x \in \mathbb{R}^4, \tag{4.59}$$

with

$$H(q, p) = q_1^2 q_2 + 2q_1 p_1 p_2 - p_1^2 q_2.$$

(a) Prove that not all level sets of H are compact, i.e. that some of them are unbounded. *Hint*: Take $q_2 = 1, p_2 = 0$ and draw the level sets of H in the resulting two-dimensional (q_1, p_1)-space. You will see (unbounded) hyperbolas.

(b) Find a quadratic constant of motion F of (4.59) independent of H. *Hint*: Try

$$F(q, p) = ap_1^2 + bq_1^2 + cp_2^2 + q_2^2$$

with unknown coefficients a, b, c. The condition $\{F, H\} = 0$ is equivalent to a linear system with solution $2a = 2b = c = 1$.

(c) Prove that, although the level sets of H are not bounded, the orbits of (4.59) are bounded. *Hint*: F is definite, i.e. its level sets are compact.

(d) Show that most solutions of (4.59) lie on invariant tori *Hint*: The system is integrable, with only a few degeneracies. *Warning*: Computing on which levels these degeneracies occur is less trivial.

4.8 Appendix: General Properties of Hamiltonian Systems

Consider now a smooth function $H : \mathbb{R}^m \times \mathbb{R}^m \to \mathbb{R},\ H = H(q,p)$, called the *Hamiltonian* or *energy function*.

Definition 4.37 *A system*

$$\begin{cases} \dot{q}_k = \dfrac{\partial H(q,p)}{\partial p_k}, \\ \dot{p}_k = -\dfrac{\partial H(q,p)}{\partial q_k}, \end{cases} \quad k = 1, 2, \ldots, m, \tag{4.60}$$

is called a ***Hamiltonian system*** *with m degrees of freedom.*

The Hamiltonian function is defined modulo an additive constant. The system (4.60) can be written as

$$\dot{x} = J_{2m}[DH(x)]^{\mathrm{T}}, \tag{4.61}$$

where $x = (q,p) \in \mathbb{R}^{2m}$, and

$$J_{2m} = \begin{pmatrix} 0 & I_m \\ -I_m & 0 \end{pmatrix}.$$

Similar to Definition 4.26, (4.60) with

$$H(q,p) = \frac{1}{2}\langle p,p\rangle + U(q),$$

where $q,p \in \mathbb{R}^m$ and U is a smooth scalar function, is called a *mechanical system* with m degrees of freedom. Here the term $\frac{1}{2}\langle p,p\rangle$ is called the *kinetic energy*, while $U(q)$ is called the *potential energy* of the system.

The corresponding $2m$-dimensional Hamiltonian system has the form:

$$\begin{cases} \dot{q} = p, \\ \dot{p} = -[DU(q)]^{\mathrm{T}}. \end{cases} \tag{4.62}$$

Although details of the dynamics generated by (4.60) or (4.62) with $m > 1$ are not known, some general results have been established. The most basic and elementary of these are presented below.

Theorem 4.23, saying that the Hamiltonian H is a constant of the motion, is valid for all $m \geq 1$, since

$$\frac{dH}{dt} = \sum_{k=1}^{m} \left(\frac{\partial H}{\partial q_k}\dot{q}_k + \frac{\partial H}{\partial p_k}\dot{p}_k \right) = \sum_{k=1}^{m} \left(\frac{\partial H}{\partial q_k}\frac{\partial H}{\partial p_k} - \frac{\partial H}{\partial p_k}\frac{\partial H}{\partial q_k} \right) = 0.$$

Therefore, the phase space of a Hamiltonian system is foliated by $(2m-1)$-dimensional (codim 1) invariant level sets $H(q,p) = h$ (*energy levels*). As in the planar case ($m = 1$), this immediately excludes asymptotically stable equilibria and cycles in multidimensional Hamiltonian systems.

Theorem 4.38 (Liouville) *The flow generated by a Hamiltonian system* (4.60) *is volume preserving.*

The proof of this theorem is based on the following lemma valid for general smooth autonomous ODEs.

Lemma 4.39 *Let φ^t be the flow generated by a smooth system*

$$\dot{x} = f(x), \quad x \in \mathbb{R}^n, \tag{4.63}$$

and let $V(t)$ denote the volume of the t-shift $D(t) = \varphi^t(D_0)$ of a measurable domain $D_0 \subset \mathbb{R}^n$. Then

$$\dot{V}(0) = \int_{D(0)} \operatorname{div} f(x)\, dx,$$

where the divergence of the vector field $f(x)$ is defined by

$$\operatorname{div} f(x) = \sum_{i=1}^{n} \frac{\partial f_i(x)}{\partial x_i}.$$

Proof For any diffeomorphism $\Phi : \mathbb{R}^n \to \mathbb{R}^n$, $y = \Phi(x)$, we have

$$\int_{\Phi(D(0))} dy = \int_{D(0)} |\det(D\Phi(x))|\, dx.$$

Thus, for $\Phi = \varphi^t$,

$$V(t) = \int_{\varphi^t(D(0))} dy = \int_{D(0)} \det\left(D_x\varphi^t(x)\right) dx.$$

From the definition of solutions to (4.63), we have

$$\varphi^t(x) = x + tf(x) + o(t).$$

Since

$$D_x\varphi^t(x) = I_n + t\,Df(x) + o(t),$$

we get using the definition of the determinant

$$\begin{aligned}
\det\left(D_x\varphi^t\right) &= \det(I_n + t\,Df) + o(t) \;=\; \prod_{i=1}^{n}(1 + t\lambda_i) + o(t)\\
&= 1 + t\sum_{i=1}^{n}\lambda_i + o(t) \;=\; 1 + t\,\mathrm{Tr}(Df) + o(t)\\
&= 1 + t\,\operatorname{div} f + o(t),
\end{aligned}$$

where λ_i stand for the eigenvalues of the Jacobian matrix Df. Therefore,

$$V(t) \;=\; \int_{D(0)} (1 + t\operatorname{div} f + o(t))\, dx \;=\; V(0) + t\int_{D(0)} \operatorname{div} f\, dx + o(t)$$

and

$$\left.\frac{dV}{dt}\right|_{t=0} = \lim_{t\to 0}\frac{V(t) - V(0)}{t} = \int_{D(0)} \operatorname{div} f\, dx. \qquad \square$$

Proof of Theorem 4.38 For a Hamiltonian vector field f:

$$\operatorname{div}\, f = \sum_{k=1}^{m}\left(\frac{\partial^2 H}{\partial q_k \partial p_k} - \frac{\partial^2 H}{\partial p_k \partial q_k}\right) = 0.$$

Thus $\dot{V}(0) = 0$ and, since for an autonomous system we can put the origin of the time axis wherever we want, $\dot{V}(t) = 0$ for any t, implying $V(t) = \text{const}$. □

Suppose now that the invariant domain

$$D = \{(q, p) \in \mathbb{R}^{2m} : H(q, p) \leq h\}$$

is bounded. Consider the orbits of (4.60) in this domain. Liouville's Theorem 4.38 implies that after some time almost any such orbit will return to an arbitrarily small neighbourhood of its starting point. This is a consequence of the following topological theorem, applied to the unit time-shift $g = \varphi^1$ along the orbits of (4.60).

Theorem 4.40 (Poincaré Recurrence) *Let $D \subset \mathbb{R}^n$ be a bounded domain and let $g : D \to D$ be an invertible, continuous, and volume-preserving mapping. Then in any neighbourhood U of any point in D there is a point x which returns to U after repeated application of the map, i.e. $g^j(x) \in U$ for some integer $j > 0$.*

Proof Take a neighbourhood U of an arbitrary point in D and consider its images:

$$U,\ g(U),\ g^2(U), \ldots, g^j(U), \ldots$$

(see Fig. 4.29). All these sets have the same positive volume. Since the volume of D is finite, these images cannot be all disjoint. Therefore, for some $k \geq 0,\ l \geq 0,\ k > l$,

$$g^k(U) \cap g^l(U) \neq \emptyset.$$

Thus, $g^{k-l}(U) \cap U \neq \emptyset$. Therefore, we can find a point $x \in U$ such that $g^j(x) \in U$ with $j = k - l$. □

Definition 4.41 *Let $F, G : \mathbb{R}^m \times \mathbb{R}^m \to \mathbb{R}$ be two smooth functions. The* ***Poisson bracket*** *is the function $\{F, G\} : \mathbb{R}^m \times \mathbb{R}^m \to \mathbb{R}$ defined by*

$$\{F, G\} := \sum_{i=1}^{m} \left(\frac{\partial F}{\partial q_i} \frac{\partial G}{\partial p_i} - \frac{\partial F}{\partial p_i} \frac{\partial G}{\partial q_i} \right).$$

The functions F and G are said to be in ***involution*** *if $\{F, G\} = 0$.*

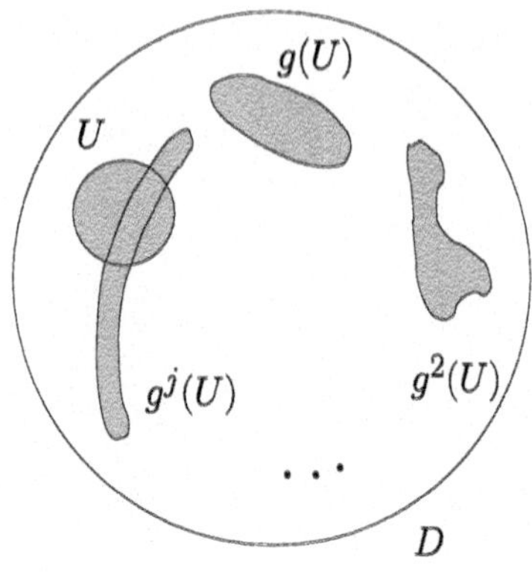

Fig. 4.29 The recurrence phenomenon

The Poisson bracket has the following properties:

(P.1) $\{F, G\} = -\{G, F\}$ (*antisymmetry*);
(P.2) $\{F, G + \lambda H\} = \{F, G\} + \lambda\{F, H\}$ (*bilinearity*);
(P.3) $\{\{F, G\}, H\} + \{\{G, H\}, F\} + \{\{H, F\}, G\} = 0$ (*Jacobi's identity*);
(P.4) $\{FG, H\} = F\{G, H\} + G\{F, H\}$ (*Leibniz' rule*),

where F, G, H are smooth functions and $\lambda \in \mathbb{R}$. The first three properties mean that the set of all smooth functions on $\mathbb{R}^{2m}$ with the usual addition and multiplication by constants and the above-defined Poisson bracket is a *Lie algebra*. Using notations introduced in (4.61), we can write

$$\{F, G\} = DF\ J_{2m}\ [DG]^{\mathrm{T}}.$$

Theorem 4.42 (Poisson) *A smooth function F is a constant of the motion of* (4.61) *if and only if it is in involution with the Hamiltonian H : $\{F, H\} = 0$.*

Proof Along solutions of (4.61), $\dot{F} = DF(x)\dot{x} = DF\ J_{2m}\ [DG]^{\mathrm{T}} = \{F, H\}$. □

Lemma 4.43 *The Poisson bracket $\{F, G\}$ of two constants of the motion F and G of* (4.61) *is also a constant of motion for* (4.61).

Proof Indeed, using the properties of $\{\cdot, \cdot\}$ and Theorem 4.42, we get

$$\{\{F, G\}, H\} = \{F, \{G, H\}\} + \{G, \{H, F\}\} = 0 + 0 = 0.$$ □

Thus, all constants of motion of a Hamiltonian system also form a Lie algebra that is a *subalgebra* of the Lie algebra of all smooth functions.

There is a special class of Hamiltonian systems, called *integrable*, which have m functionally independent constants of the motion in involution (one of them, of course, is the Hamiltonian H). Solutions of such systems can be found analytically by integration. The dynamics of an integrable system is relatively simple: Its phase space is almost completely foliated by m-dimensional invariant tori on which all orbits are either periodic or quasi-periodic. This follows from a fundamental theorem, which we formulate without proof.

Theorem 4.44 (Liouville–Arnold) *Consider m smooth functions*

$$F^{(1)}, F^{(2)}, \ldots, F^{(m)} : \mathbb{R}^m \times \mathbb{R}^m \to \mathbb{R}$$

and introduce the following set:

$$M_f = \{x \in \mathbb{R}^m \times \mathbb{R}^m : F^{(i)}(x) = f_i,\ i = 1, 2, \ldots, m\},$$

where $f = (f_1, f_2, \ldots, f_m) \in \mathbb{R}^m$. Suppose that

(L.1) *The functions $F^{(i)}$ are in involution:* $\{F^{(i)}, F^{(j)}\} = 0,\ i, j = 1, 2, \ldots, m$.
(L.2) *The differentials $DF^{(i)}(x)$, $i = 1, 2, \ldots, m$, are linearly independent in any point $x \in M_f$.*

Then

(i) *M_f is a smooth m-dimensional invariant manifold for the Hamiltonian system $\dot{x} = J_{2m}[DH(x)]^{\mathrm{T}}$ with $H = F^{(1)}$.*

(ii) *If M_f is bounded and connected, then it is diffeomorphic to the m-dimensional torus*

$$\mathbb{T}^m = \{(\varphi_1, \ldots, \varphi_m) : \varphi_i \bmod 2\pi\}.$$

(iii) *The motion on this torus is governed by the equation*

$$\dot{\varphi} = \omega_f, \ \varphi \in \mathbb{T}^m$$

for some $\omega_f \in \mathbb{R}^m$.

(iv) *There is a neighbourhood of* M_f *in which one can introduce new coordinates* $(I, \varphi) \in \mathbb{R}^m \times \mathbb{T}^m$, *such that the Hamiltonian system* $\dot{x} = J_{2m}[DH(x)]^{\mathrm{T}}$ *will take the form:*

$$\begin{cases} \dot{\varphi} = \omega(I), \\ \dot{I} = 0, \end{cases} \tag{4.64}$$

where $\omega(0) = \omega_f$. □

The system (4.64) is Hamiltonian with $H = H(I)$ such that $\omega(I) = [DH(I)]^{\mathrm{T}}$. The coordinates (I, φ) are called the *action-angle variables.*

Obviously, any Hamiltonian system with one degree of freedom ($m = 1$) is integrable. Although several important Hamiltonian systems with several degrees of freedom, for instance, Kepler's two-body problem, are integrable, integrability is exceptional when $m \geq 2$. In general, a nonintegrable Hamiltonian system (4.61) behaves very differently from an integrable one: Invariant tori may exist but are separated by domains of "chaotic motions".

Example 4.45 (Elastic Pendulum) Consider a small ball attached to a massless spring that can both oscillate in the radial direction and swing like a pendulum in the plane (see Fig. 4.30).

This is called the *elastic pendulum.* Let m be the mass of the ball and let l_0 and l be, respectively, the length of the spring in the vertical position in the absence of load and under load. Denote by g the acceleration of gravity, by s the spring constant, and by r the distance between the ball and the suspension point.

Introduce

$$\omega_z = \sqrt{\frac{s}{m}}, \quad \omega_\varphi = \sqrt{\frac{g}{l}}, \quad \sigma = ml^2.$$

Then the Hamiltonian (energy) of the system is given by

$$H = \frac{1}{2\sigma}\left(p_z^2 + \frac{p_\varphi^2}{(z+1)^2}\right) + \frac{\sigma}{2}\omega_z^2\left(z + \frac{\omega_\varphi^2}{\omega_z^2}\right)^2 - \sigma\omega_\varphi^2(z+1)\cos\varphi,$$

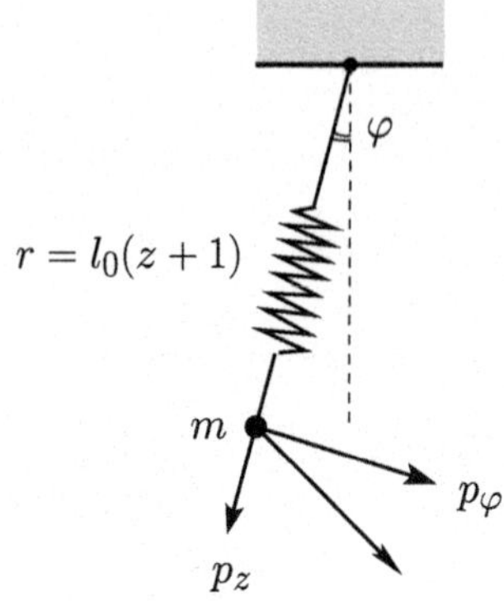

Fig. 4.30 Elastic pendulum

where φ is the angular displacement of the pendulum and

$$z = \frac{r - l_0}{l_0}$$

measures its radial displacement.

Obviously, the corresponding system (4.61) with

$$q = \begin{pmatrix} z \\ \varphi \end{pmatrix}, \quad p = \begin{pmatrix} p_z \\ p_\varphi \end{pmatrix}, \quad x = \begin{pmatrix} q \\ p \end{pmatrix},$$

is Hamiltonian with two degrees of freedom. The energy levels $H = h$ are three-dimensional submanifolds in the four-dimensional phase space. To analyse the dynamics in a level set, introduce the plane

$$z = 0$$

and parametrize it by (φ, p_φ). Given the energy h, consider a map

$$P : \begin{pmatrix} \varphi \\ p_\varphi \end{pmatrix} \mapsto \begin{pmatrix} \varphi' \\ p'_\varphi \end{pmatrix},$$

where (φ', p'_φ) is the point of the *second* intersection with the plane of the orbit starting at the point (φ, p_φ). One can show that P is an area-preserving map. This map can be approximated by numerical integration with high accuracy.[12] Orbits of P computed in this manner are depicted in Fig. 4.31a for the following numerical values of the parameters:

$$h = 0.6125, \quad \sigma = 1, \quad \omega_z = 4, \quad \omega_\varphi = 1.$$

Note that the eigenfrequencies ω_z and ω_φ are in $4 : 1$ resonance.

Several fixed points can be seen in the picture. Amongst them we can clearly distinguish four saddle points. Actually, they belong to one periodic orbit crossing the plane $z = 0$ four times, and so forming two period-two cycles of P. If the system describing the elastic pendulum were integrable, we would see families of closed invariant curves filling the whole plane, separated by coinciding stable and unstable invariant manifolds of saddle fixed points. As Fig. 4.31b, giving a zoom-in of the right upper part, shows, this is not the case. The stable and unstable manifolds of the saddles intersect transversally and form the *heteroclinic structure* for P (see Chap. 7) that is most pronounced near the saddles. Note that, in fact, it is just a *homoclinic structure* for one cycle in the phase space, formed by the intersecting stable and unstable manifolds of this cycle. Inside the homoclinic structure, the dynamics is irregular ("chaotic") and consists of permanent meandering between long-periodic saddle cycles. ◊

Thus, as Poincaré discovered in the 1890s, homoclinic structures prevent integrability. It should be noted that "chaotic domains" in Hamiltonian systems with two degrees of freedom are confined between invariant tori, so that orbits starting inside them cannot leave these domains. Note that this is not true for Hamiltonian systems with $m > 2$, where the chaotic domains can overlap. This phenomenon is called *Arnold's diffusion*.

[12] Tuwankotta, J. M., & Quispel, G. R. W. (2003). Geometric numerical integration applied to the elastic pendulum at higher-order resonance. *Journal of Computational and Applied Mathematics, 154*, 229–242.

Fig. 4.31 Poincare map of the elastic pendulum

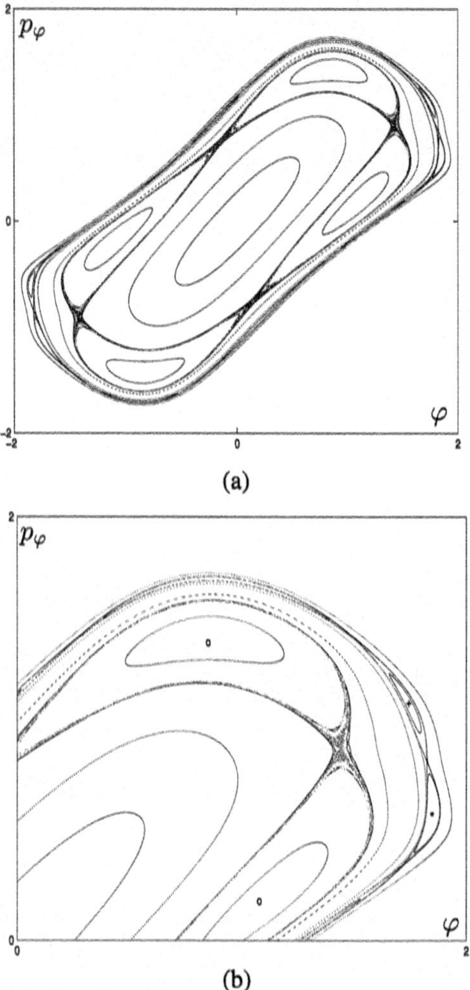

Chapter 5
Local Bifurcations in Minimal Dimensions

Abstract A *bifurcation* is a qualitative (topological) change of the phase portrait under variation of system parameters. For example, the coalescence and disappearance of equilibria or fixed points, or a change of their stability character are bifurcations. In this chapter, we study *local* bifurcations, i.e. those which happen in small neighbourhoods of equilibria or fixed points. We focus on dynamical systems depending on *one* parameter. Local bifurcations happen at the (exceptional) parameter values for which the steady state under consideration lacks hyperbolicity. Our strategy will be to study in this chapter the simplest local bifurcations in generic one-parameter systems in state spaces of the *lowest possible* dimension. The dimension is determined by the number of *critical eigenvalues* at the bifurcating equilibrium (fixed point), i.e. those eigenvalues which make the point nonhyperbolic. The loose meaning of "generic" is "as a rule", while the precise meaning depends on the context. The general idea is that "generic" excludes extra degeneracies that are exceptional in the situation that we consider. In Chap. 6, we will show how to apply the obtained results in general n-dimensional situations.

5.1 Bifurcation Diagrams and Normal Forms

Local bifurcation theory answers two related questions:

(1) What can happen to the phase portrait near an equilibrium (fixed point) when a parameter passes a bifurcation value?

(2) How to determine which of the alternatives occurs in a concrete system in which one is interested?

The answer to the first question is given in the form of a catalogue of *normal forms*, i.e. canonical systems exhibiting certain bifurcations. The second question asks for a recognition algorithm and is answered by listing computable *genericity conditions*, which guarantee that the system under consideration

Yu. Kuznetsov et al., *Dynamical Systems Essentials*, Texts in Applied Mathematics 83, https://doi.org/10.1007/978-3-032-04083-1_5

has phase portraits equivalent to those of a corresponding normal form for corresponding parameter values.[1]

Consider an m-parameter family of dynamical systems:

$$\{\mathbb{T}, \mathbb{R}^n, \varphi_\alpha^t\}, \quad \alpha \in \mathbb{R}^m, \tag{5.1}$$

generated by either an ODE

$$\dot{x} = f(x, \alpha)$$

or a map

$$x \mapsto f(x, \alpha),$$

where $\mathbb{T} = \mathbb{R}$ or $\mathbb{Z}$, respectively, and $f : \mathbb{R}^n \times \mathbb{R}^m \to \mathbb{R}^n$ is smooth. To begin with, assume that—for all $\alpha \in \mathbb{R}^m$—the evolution operator $\varphi_\alpha^t(x)$ is defined for all $x \in \mathbb{R}^n$ and $t \in \mathbb{R}$ or $\mathbb{Z}$, and, in the discrete-time setting, the map f is invertible.

Definition 5.1 *A point $\alpha_0 \in \mathbb{R}^m$ is called a* ***bifurcation point*** *of* (5.1) *if in any neighbourhood of α_0 there is a point α such that the dynamical systems corresponding to parameters α_0 and α are not topologically equivalent.*

A *bifurcation* is a change of the topological class under parameter variation. Its *codimension* is the number of independent equality conditions determining the bifurcation point, in particular, the nature of its degeneracy. These conditions are called the *bifurcation conditions.* One expects that a codim 1 bifurcation happens at isolated parameter values in generic one-parameter families of dynamical systems, on curves in the parameter plane of generic two-parameter families, two-dimensional surfaces in the parameter space of generic three-parameter families, etc. Thus, one would not meet a bifurcation of codimension bigger than m in a generic m-parameter family. Therefore, a bifurcation of codim m should be studied in, at least, m-parameter families. So it is very natural to order the catalogue of bifurcations by codimension.

Example 5.2 (Saddle-node homoclinic bifurcation) Consider the following system on the plane:

$$\begin{cases} \dot{x}_1 = x_1(1 - x_1^2 - x_2^2) - x_2(1 + \alpha + x_1), \\ \dot{x}_2 = \ x_1(1 + \alpha + x_1) + x_2(1 - x_1^2 - x_2^2), \end{cases} \tag{5.2}$$

where α is a parameter. In polar coordinates (ρ, θ) system (5.2) takes the form:

---

[1] We should warn the reader that for one bifurcation considered in this chapter, the so-called Neimark–Sacker bifurcation of fixed points, the constructed normal form does capture essential features of the phase portrait near the bifurcation, but fails to describe its fine details.

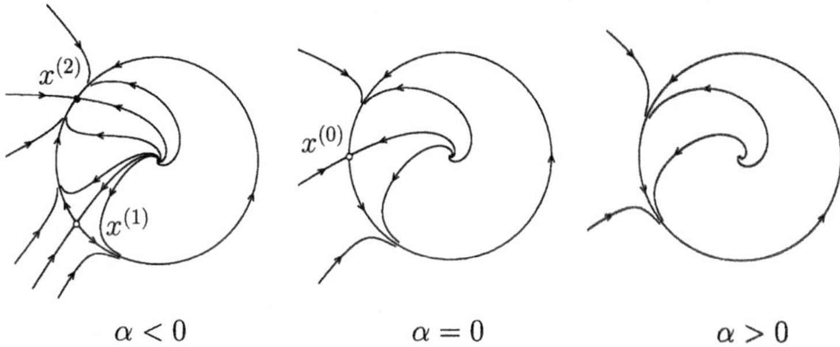

Fig. 5.1 Saddle-node homoclinic bifurcation

$$\begin{cases} \dot{\rho} = \rho(1-\rho^2), \\ \dot{\theta} = 1 + \alpha + \rho\cos\theta. \end{cases} \tag{5.3}$$

The origin is an unstable focus for all $|\alpha| < 1$. Moreover, for $\alpha = 0$, there is a nonhyperbolic equilibrium point of system (5.3), namely:

$$x^{(0)} = (\rho_0, \theta_0) = (1, \pi)$$

(see Fig. 5.1). It has eigenvalues $\lambda_1 = 0, \lambda_2 = -2$ (check!). For small positive values of α there is no equilibrium, while for small negative α there are two, a saddle $x^{(1)}$ and a node $x^{(2)}$. This is a bifurcation which is called a *fold* or *saddle-node* bifurcation. It is a local codimension one bifurcation. We can study it by restricting our attention to the equation for θ with $\rho = 1$.

Notice that for $\alpha > 0$ a stable *limit cycle* appears in (5.2) coinciding with the unit circle $\{(\rho, \theta) : \rho = 1\}$. This circle is always invariant, but for $\alpha \leq 0$ it contains equilibria. Looking at only a small neighbourhood of the nonhyperbolic equilibrium, we miss the global phenomenon of the appearance of the cycle. Notice that at $\alpha = 0$ there is exactly one orbit that is homoclinic to the nonhyperbolic equilibrium $x^{(0)}$. Thus, a local bifurcation can occur simultaneously with a global bifurcation, here the so-called *saddle-node homoclinic bifurcation.* ◊

Definition 5.3 *A* ***bifurcation diagram*** *of* (5.1) *is a partitioning of its parameter space* $\mathbb{R}^m$ *induced by topological equivalence, together with representative phase portraits for each subset.*

Figure 5.1 shows the bifurcation diagram of (5.2) for small $|\alpha|$.

Let

$$\{T, \mathbb{R}^n, \psi_\beta^t\}, \quad \beta \in \mathbb{R}^m, \tag{5.4}$$

be another family of dynamical systems generated by either

$$\dot{y} = g(y, \beta)$$

or

$$y \mapsto g(y, \beta),$$

where $g : \mathbb{R}^n \times \mathbb{R}^m \to \mathbb{R}^n$ is smooth.

Definition 5.4 *The family* (5.1) *is* ***topologically equivalent*** *to the family* (5.4) *if*

(i) *there exists a homeomorphism of the parameter space* $p : \mathbb{R}^m \to \mathbb{R}^m$, $\beta = p(\alpha)$;

(ii) *there is a parameter-dependent homeomorphism of the state space* $h_\alpha : \mathbb{R}^n \to \mathbb{R}^n$, $y = h_\alpha(x)$, *mapping orbits of* (5.1) *at parameter value* α *onto orbits of* (5.4) *at parameter value* $\beta = p(\alpha)$, *preserving the direction of time.*

Bifurcation diagrams of topologically equivalent families are also called equivalent. Notice that the dimension of the state space and the number of parameters in (5.4) must be the same as in (5.1).

For local bifurcations, one can modify Definition 5.4 by considering a neighbourhood Λ of some (usually bifurcation) point $\alpha_0 \in \mathbb{R}^m$ and a (parameter dependent but *not shrinking* to a point) neighbourhood U_α of an equilibrium or a fixed point $x_0 \in \mathbb{R}^n$, and then speak about *local topological equivalence* if h_α is defined on U_α for $\alpha \in \Lambda$. In this case, the map $(x, \alpha) \mapsto (h_\alpha(x), p(\alpha))$ is defined in a neighbourhood of (x_0, α_0) in $\mathbb{R}^n \times \mathbb{R}^m$.

In many cases, we can construct a "nice" representative of an equivalence class of parametrized families having a generic local bifurcation of a certain type. This is called a *topological normal form.* A system satisfying the same bifurcation conditions is locally topologically equivalent to the corresponding normal form, provided certain *genericity conditions* are fulfilled. These genericity conditions will be given explicitly as inequalities involving partial derivatives of the right-hand side of the ODE (or the map) with respect to the state variables and parameters at (x_0, α_0). The genericity conditions involving only derivatives with respect to the state variables are called the *nondegeneracy conditions*, while the genericity conditions that also involve derivatives with respect to the parameters are called the *transversality conditions.*

For codim 1 bifurcations of equilibria in ODEs, the topological normal forms exist and are polynomial. This is not always the case: There is already a codim 1 bifurcation of fixed points of maps—the Neimark–Sacker bifurcation—for which no polynomial (or any other) topological normal form can be constructed.

5.2 One-Parameter Bifurcations of Equilibria

Consider a smooth n-dimensional system smoothly depending on one parameter:

$$\dot{x} = f(x, \alpha), \quad x \in \mathbb{R}^n, \ \alpha \in \mathbb{R} \tag{5.5}$$

and suppose that it has an equilibrium x_0 at some parameter value α_0, i.e. $f(x_0, \alpha_0) = 0$. If x_0 has no zero eigenvalue, i.e.:

$$\det D_x f(x_0, \alpha_0) \neq 0, \tag{5.6}$$

then (5.5) has a unique equilibrium $x(\alpha)$ near x_0 for small $|\alpha - \alpha_0|$ such that $x(\alpha_0) = x_0$. This follows from the Implicit Function Theorem, which also implies that $x(\alpha)$ is smooth and

$$x'(\alpha_0) = -[D_x f(x_0, \alpha_0)]^{-1} D_\alpha f(x_0, \alpha_0). \tag{5.7}$$

If x_0 also has no eigenvalue λ with Re $\lambda = 0$, implying that x_0 is hyperbolic, then α_0 cannot be a bifurcation point related to a local bifurcation of x_0. This follows from the Grobman–Hartman Theorem 3.34. Indeed, the eigenvalues of the Jacobian matrix

$$A(\alpha) = D_x f(x(\alpha), \alpha)$$

depend continuously on α near α_0. Therefore, the equilibrium $x(\alpha)$ remains hyperbolic for all α with sufficiently small $|\alpha - \alpha_0|$. Moreover, the number n_s of eigenvalues of $A(\alpha)$ with Re $\lambda < 0$ (and, therefore, the number n_u of the eigenvalues with Re $\lambda > 0$) remains constant near α_0. Therefore, the local phase portrait of (5.5) does not change qualitatively near α_0, since it is locally topologically equivalent to the phase portrait of the linearized system $\dot{\xi} = A(\alpha)\xi$ (Grobman–Hartman theorem) and, as follows from Theorem 2.26, all such linear ODEs are topologically equivalent to the standard saddle:

$$\begin{cases} \dot{\xi}_s = -\xi_s, & \xi_s \in \mathbb{R}^{n_s}, \\ \dot{\xi}_u = \xi_u, & \xi_u \in \mathbb{R}^{n_u}. \end{cases}$$

These considerations prove that local bifurcations can only happen to *non-hyperbolic* equilibria of (5.5). Therefore, we should consider two *critical cases* (or *singularities*):

(a) the equilibrium x_0 has a zero eigenvalue $\lambda_1 = 0$ (Fig. 5.2a);

(b) the equilibrium x_0 has a pair of complex-conjugate eigenvalues $\lambda_{1,2} = \pm i\omega_0$ with $\omega_0 > 0$ (Fig. 5.2b).

Clearly, in a generic one-parameter system (5.5) we may encounter either the first or the second singularity at some critical parameter value α_0, but not both simultaneously! Moreover, in such systems the corresponding critical eigenvalue(s) will be algebraically simple.

Notice, finally, that case (a) may occur already in one-dimensional systems ($n \geq 1$), while for case (b) we need at least a planar system ($n \geq 2$). Following the strategy outlined above, we first consider these critical cases in the lowest possible dimension.

Fig. 5.2 Critical cases for equilibria

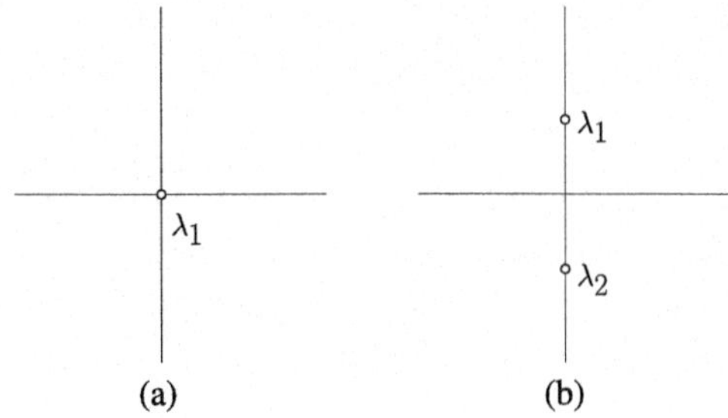

5.2.1 Fold Bifurcation of Scalar ODEs

Consider a dynamical system generated by

$$\dot{x} = f(x, \alpha), \quad x \in \mathbb{R}, \; \alpha \in \mathbb{R},$$

where $f : \mathbb{R} \times \mathbb{R} \to \mathbb{R}$ is smooth. What happens if this system has an equilibrium x_0 at a parameter value α_0 with eigenvalue $\lambda = \frac{\partial f}{\partial x}(x_0, \alpha_0) = 0$? Generically, a *fold* (also called *limit point, turning point, tangent, saddle-node*) bifurcation occurs, as in the following example.

Example 5.5 (Normal form of the fold bifurcation) Consider the following scalar differential equation depending on one parameter:

$$\dot{x} = \alpha + x^2 =: f(x, \alpha), \quad x \in \mathbb{R}, \; \alpha \in \mathbb{R}. \tag{5.8}$$

At $\alpha = 0$ the corresponding dynamical system has a nonhyperbolic equilibrium $x_0 = 0$ with $\lambda = \frac{\partial f}{\partial x}(0, 0) = 0$. The behaviour of the system for all the other values of α is also clear (see Fig. 5.3). For $\alpha < 0$ there are two equilibria in the system: $x_{1,2}(\alpha) = \mp\sqrt{-\alpha}$, one of which is stable, while the other is unstable. For $\alpha > 0$ there are no equilibria. While α crosses zero from negative to positive values, the two equilibria "collide", forming at $\alpha = 0$ a degenerate equilibrium with $\lambda = 0$, and disappear. This is an example of the fold bifurcation.

Figure 5.1 presents the bifurcation diagram of (5.8) in the direct product of the phase and parameter spaces, i.e. the (x, α)-plane. It is customary to put the parameter on the horizontal and the phase variable on the vertical axis. The parabola $\alpha = -x^2$ is the *equilibrium manifold* of the system, determined by the equation $f(x, \alpha) = 0$. Fixing some α, we can easily determine the number and stability of equilibria for this parameter value. The projection of the equilibrium manifold into the parameter axis has a *fold singularity* at $(x, \alpha) = (0, 0)$.

Fig. 5.3 Fold bifurcation in the normal form (5.8)

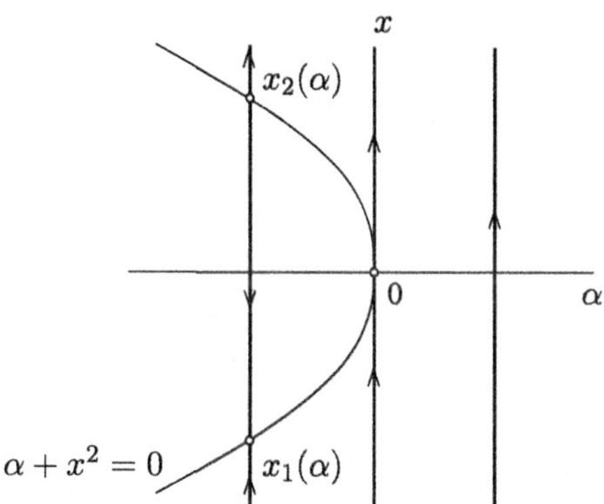

The equation $\dot{x} = \alpha - x^2$ can be considered similarly. Since the reflection $(x, \alpha) \mapsto (-x, -\alpha)$ transforms it into (5.8), it has two equilibria appearing for $\alpha > 0$. ◊

Theorem 5.6 *Any smooth system*

$$\dot{y} = \alpha + y^2 + O(y^3), \tag{5.9}$$

where the $O(y^3)$-term can smoothly depend on α, is locally topologically equivalent near the origin to the system

$$\dot{x} = \alpha + x^2. \tag{5.10}$$

Proof The proof involves two steps. It is based on the fact that for scalar systems a homeomorphism mapping equilibria onto equilibria will automatically map orbits asymptotic to these equilibria onto each other.
Step 1: Analysis of equilibria.

Write system (5.9) as

$$\dot{y} = F(y, \alpha) = \alpha + y^2 + y^3\psi(y, \alpha), \tag{5.11}$$

where ψ is a smooth function of (y, α) near $(0, 0)$. Consider the equilibrium manifold of (5.11) near the origin $(0, 0)$ of the (y, α)-plane:

$$M = \{(y, \alpha) : F(y, \alpha) = \alpha + y^2 + y^3\psi(y, \alpha) = 0\}.$$

The curve M passes through the origin ($F(0, 0) = 0$). By the Implicit Function Theorem, it can be locally parametrized by y (since $\frac{\partial}{\partial\alpha}F(0, 0) = 1 \neq 0$):

$$M = \{(y, \alpha) : \alpha = g(y)\},$$

where g is smooth, defined for small $|y|$, and satisfies $g(0) = 0$. Moreover,

$$g(y) = -y^2 + O(y^3).$$

To see this, expand the function ψ as

$$\psi(y,\alpha) = \psi_{00} + \psi_{10}y + \psi_{01}\alpha + O(\|(y,\alpha)\|^2)$$

and look for the coefficients in the Taylor expansion for g, i.e.:

$$\alpha = g(y) = g_0 + g_1 y + g_2 y^2 + O(y^3).$$

Substituting the above expressions into the equation $F(y,\alpha) = 0$, we get

$$\begin{aligned} g_0 + g_1 y + g_2 y^2 + \cdots = \\ - y^2 - y^3\left(\psi_{00} + \psi_{10}y + \psi_{01}\left(g_1 y + g_2 y^2 + \cdots\right) + \cdots\right), \end{aligned}$$

which implies $g_0 = g_1 = 0$ and $g_2 = -1$.

Thus, for any sufficiently small $\alpha < 0$, there are two equilibria of (5.11) near the origin, $y_1(\alpha)$ and $y_2(\alpha)$, which are close to the equilibria of (5.10), i.e. $x_1(\alpha) = -\sqrt{-\alpha}$ and $x_2(\alpha) = \sqrt{-\alpha}$, for the same parameter value (see Fig. 5.4).

Indeed, write $g(y) = -y^2 + y^3\varphi(y)$, so that the equation for M takes the form:

$$\alpha = -y^2 + y^3\varphi(y), \tag{5.12}$$

where φ is a smooth function that can be written as $\varphi(y) = \varphi_0 + \varphi_1 y + \varphi_2 y^2 + O(y^3)$ with some constants φ_j. For $\alpha \leq 0$, we look for a solution $y = y(\alpha)$ of (5.12) in the form:

$$y = \beta z, \tag{5.13}$$

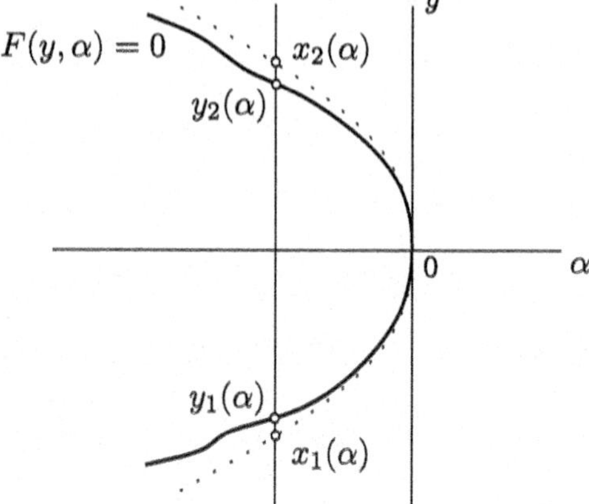

Fig. 5.4 Fold bifurcation in the perturbed normal form

where $\beta = -\sqrt{-\alpha}$ and $z = z(\beta)$ is an unknown smooth function. Substituting (5.13) into (5.12) and factoring out β^2 leads to the equation

$$\Phi(z, \beta) = -1 + z^2 - \beta z^3 \varphi(\beta z) = 0.$$

Since $\Phi(1, 0) = 0$ but $\frac{\partial}{\partial z}\Phi(1, 0) = 2 \neq 0$, the Implicit Function Theorem ensures the local existence and uniqueness of a smooth function $z = z(\beta)$ with $z(0) = 1$ and $\Phi(z(\beta), \beta) \equiv 0$. Therefore, Eq. (5.12) has a solution

$$y_1(\alpha) = -\sqrt{-\alpha}\, z(-\sqrt{-\alpha}).$$

Using the expansion $z(\beta) = z_0 + z_1\beta + O(\beta^2)$, where $z_0 = z(0) = 1$, we see that

$$y_1(\alpha) = -\sqrt{-\alpha}\left(1 - z_1\sqrt{-\alpha} + O(\alpha)\right) = -\sqrt{-\alpha} + O(\alpha) = x_1(\alpha) + O(\alpha).$$

The existence of the branch $y_2(\alpha)$ can be established similarly.

Step 2: Homeomorphism construction.

For small $|\alpha|$, construct a parameter-dependent map $y = h_\alpha(x)$ as follows. For $\alpha \geq 0$ take the identity map

$$h_\alpha(x) = x.$$

For $\alpha < 0$ take an affine transformation

$$h_\alpha(x) = A(\alpha) + B(\alpha)x,$$

where the coefficients A, B are uniquely determined by the conditions

$$h_\alpha(x_j(\alpha)) = y_j(\alpha), \quad j = 1, 2.$$

Namely,

$$A(\alpha) = \frac{y_2(\alpha) + y_1(\alpha)}{2}, \quad B(\alpha) = \frac{y_2(\alpha) - y_1(\alpha)}{2\sqrt{-\alpha}}.$$

Notice that $A(\alpha) \to 0$ and $B(\alpha) \to 1$ as $\alpha \uparrow 0$. The map $h_\alpha : \mathbb{R} \to \mathbb{R}$ thus constructed is a homeomorphism mapping orbits of (5.10) near the origin into the corresponding orbits of (5.11), preserving the direction of time. □

Remarks

(1) Although we do not require for the homeomorphism h_α to depend continuously on α, this property holds here. In particular, h_α tends to the identity map as $\alpha \uparrow 0$.

(2) The equivalence between $\dot{y} = \alpha - y^2 + O(|y|^3)$ and $\dot{x} = \alpha - x^2$ can be established by similar arguments. ◇

Theorem 5.7 *Suppose that a one-dimensional system*

$$\dot{x} = f(x, \alpha), \quad x \in \mathbb{R}, \ \alpha \in \mathbb{R}, \tag{5.14}$$

with smooth f, has at $\alpha = 0$ the equilibrium $x = 0$, and is such that $\lambda = \frac{\partial f}{\partial x}(0,0) = 0$. Assume that the following two genericity conditions are satisfied:

(A.1) $\frac{\partial^2 f}{\partial x^2}(0,0) \neq 0$;
(A.2) $\frac{\partial f}{\partial \alpha}(0,0) \neq 0$.

Then there are invertible smooth coordinate and parameter changes transforming the system into

$$\dot{y} = \beta \pm y^2 + O(y^3). \tag{5.15}$$

Remark One can reformulate the statement of the theorem by saying that generically, near the fold bifurcation at the origin, (5.14) is locally smoothly conjugate to (5.15).

◇

Proof of Theorem 5.7 The proof consists of four steps.

Step 1: Taylor expansion.
Write the right-hand side of (5.14) as

$$f(x, \alpha) = f_0(\alpha) + f_1(\alpha)x + f_2(\alpha)x^2 + O(x^3),$$

where

$$f_0(0) = f(0,0) = 0, \quad f_1(0) = \frac{\partial f}{\partial x}(0,0) = 0, \quad f_2(0) = \frac{1}{2}\frac{\partial^2 f}{\partial x^2}(0,0).$$

Step 2: Shift of the coordinate.
Perform a linear coordinate shift by introducing a new variable ξ:

$$x = \xi + \delta, \tag{5.16}$$

where $\delta = \delta(\alpha)$ is an a priori unknown function that will be defined later. We will see that $\delta = O(\alpha)$ as $\alpha \to 0$. The inverse coordinate transformation is

$$\xi = x - \delta.$$

Substituting (5.16) into (5.14) we find

$$\dot{\xi} = \dot{x} = f_0(\alpha) + f_1(\alpha)(\xi + \delta) + f_2(\alpha)(\xi + \delta)^2 + \cdots .$$

Therefore,

$$\begin{aligned}\dot{\xi} &= \left[f_0(\alpha) + f_1(\alpha)\delta + f_2(\alpha)\delta^2 + O(\delta^3)\right] \\ &+ \left[f_1(\alpha) + 2f_2(\alpha)\delta + O(\delta^2)\right]\xi \\ &+ \left[f_2(\alpha) + O(\delta)\right]\xi^2 \\ &+ O(\xi^3).\end{aligned}$$

Assumption (A.1) implies that

$$f_2(0) = \frac{1}{2}\frac{\partial^2 f}{\partial x^2}(0,0) \neq 0.$$

Then, according to the Implicit Function Theorem, there is a smooth function $\delta(\alpha)$ that annihilates the ξ-term in the equation above for all sufficiently small $|\alpha|$. Indeed, the condition for the ξ-term to vanish can be written as

$$F(\alpha,\delta) \equiv f_1(\alpha) + 2f_2(\alpha)\delta + \delta^2\psi(\alpha,\delta) = 0$$

with some smooth function ψ. We have

$$F(0,0) = 0, \quad \frac{\partial F}{\partial \delta}(0,0) = 2f_2(0) \neq 0, \quad \frac{\partial F}{\partial \alpha}(0,0) = f_1'(0),$$

which implies (local) existence and uniqueness of a smooth function $\delta = \delta(\alpha)$ such that $\delta(0) = 0$ and $F(\alpha,\delta(\alpha)) \equiv 0$. It follows that

$$\delta(\alpha) = -\frac{f_1'(0)}{2f_2(0)}\alpha + O(\alpha^2)$$

(cf. (5.7)). The equation for ξ now no longer contains any ξ-term. Since $f_1(\alpha) = f_1'(0)\alpha + O(\alpha^2)$, we can write

$$\dot{\xi} = [f_0'(0)\alpha + O(\alpha^2)] + [f_2(0) + O(\alpha)]\xi^2 + O(\xi^3). \tag{5.17}$$

Step 3: Introduce a new parameter.

Consider as a new parameter $\mu = \mu(\alpha)$ the constant (ξ-independent) term of (5.17), that we can write in the form:

$$\mu = f_0'(0)\alpha + \alpha^2\phi(\alpha),$$

for some smooth function ϕ. We have

(a) $\mu(0) = 0$;
(b) $\mu'(0) = f_0'(0) = \frac{\partial f}{\partial \alpha}(0,0)$.

Since $\frac{\partial f}{\partial \alpha}(0,0) \neq 0$ due to (A.2), the Inverse Function Theorem implies local existence and uniqueness of a smooth inverse function $\alpha = \alpha(\mu)$ with $\alpha(0) = 0$. Therefore, Eq. (5.17) now reads

$$\dot{\xi} = \mu + a(\mu)\xi^2 + O(\xi^3),$$

where $a(\mu)$ is a smooth function with $a(0) = f_2(0) \neq 0$ due to the first assumption (A.1).

Step 4: Final scaling.
Let $y = |a(\mu)|\xi$ and $\beta = |a(\mu)|\mu$. Then we get

$$\dot{y} = \beta + sy^2 + O(y^3),$$

where $s = \text{sign } a(0) = \pm 1$. This is Eq. (5.15). □

Combining Theorem 5.7 with Theorem 5.6, we can analyse fold bifurcations in generic smooth scalar systems.

Remarks
(1) The genericity conditions (A.1) and (A.2) from Theorem 5.7 play different roles in the analysis: The nondegeneracy condition (A.1) guarantees that the singularity is not too degenerate, while the transversality condition (A.2) assures that the parameter "unfolds" this singularity in a generic way.
(2) What is "generic" depends on the class of systems under consideration. If a scalar ODE $\dot{x} = f(x, \alpha)$ has a trivial equilibrium $x = 0$ for all α, then generically a *transcritical bifurcation* happens when $f_x = 0$. It has the normal form $\dot{x} = x(\alpha \pm x)$. If a scalar system is invariant with respect to the reflection $x \mapsto -x$, then the trivial equilibrium $x = 0$ can undergo a *pitchfork bifurcation* with the normal form $\dot{x} = x(\alpha \pm x^2)$. Figure 5.5 shows the bifurcation diagrams of these normal forms in the "−"-case. Note that branches exchange stability at the bifurcation point where they intersect.
(3) What if the state space has higher dimension? The full answer will be given later in Chap. 6. However, we demonstrate below for some special

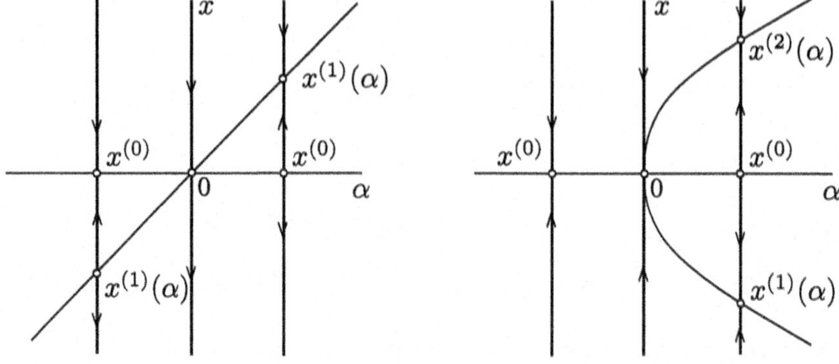

Fig. 5.5 Transcritical (left) and pitchfork (right) bifurcations

case that existence of equilibria can be analysed by the so-called *Lyapunov–Schmidt reduction* to a one-dimensional problem.

If the critical eigenvalue is algebraically simple, we can first reduce to the null space, then apply the scalar theory.

Consider, for example, a nonlinear eigenvalue problem

$$f(x) - \lambda x = 0, \quad f : \mathbb{R}^n \to \mathbb{R}^n, \; f(0) = 0, \tag{5.18}$$

near a bifurcation point corresponding to a simple real eigenvalue λ_0 of the linear eigenvalue problem:

$$Ax - \lambda x = 0, \quad A = Df(0).$$

One can think of analysing equilibria of the n-dimensional system $\dot{x} = -\lambda x + f(x)$ for λ close to λ_0. In the linear case we have a vertical line of equilibria $x = 0, \; \lambda = \lambda_0$ in the (x, λ)-space. In the nonlinear case, as we can expect, the line bends.

Write

$$f(x) = Ax + r(x), \quad \lambda = \lambda_0 + \tau,$$

where λ_0 is a simple real eigenvalue of A. Then, the original problem becomes

$$Lx = h(\tau, x),$$

where $L = A - \lambda_0 I, \; h(\tau, x) = \tau x - r(x)$. Assume simplicity:

$$\mathcal{N} \cap \mathcal{R} = \{0\}, \; \dim \mathcal{N} = 1,$$

for the null space $\mathcal{N}$ and range $\mathcal{R}$ of L. Then

$$\mathbb{R}^n = \mathcal{N} \oplus \mathcal{R}.$$

Let P be the projection operator to $\mathcal{N}$ corresponding to this decomposition. One can explicitly write

$$Px = \langle \psi, x \rangle \varphi,$$

where both $L\varphi = 0$ and $L^{\mathrm{T}}\psi = 0$ and $\langle \psi, \varphi \rangle = 1$.

Write

$$x = \varepsilon\varphi + y, \quad \varepsilon \in \mathbb{R}, \; y \in \mathcal{R}.$$

Define $\mathcal{G} : \mathbb{R} \times \mathcal{R} \times \mathbb{R} \to \mathcal{R}$ by

$$\mathcal{G}(\varepsilon, y, \tau) = (I - P)[Ly - h(\tau, \varepsilon\varphi + y)].$$

Now $(I - P)L : \mathcal{R} \to \mathcal{R}$ is invertible and h contains only higher-order terms. From the Implicit Function Theorem we can conclude that a locally unique solution $y = y(\varepsilon, \tau)$, $y(0, 0) = 0$, to

$$\mathcal{G}(\varepsilon, y, \tau) = 0$$

exists. It remains only to solve

$$Ph(\tau, \varepsilon\varphi + y(\varepsilon, \tau)) = 0$$

with both ε and $\tau \in \mathbb{R}$. This equation is called the *branching* (or *bifurcation*) *equation.* The main approach to study the bifurcation equation is to show that a factor ε can be divided out and that the resulting equation can be solved by the Implicit Function Theorem. We will illustrate this procedure in the following example.

Example 5.8 (Branching of equilibria in CSTR) We illustrate the Lyapunov–Schmidt reduction by a concrete example with $n = 2$ and

$$\begin{cases} f_1(x_1,x_2) = 7x_1 - 2x_2 + 4(x_1-x_2)^2, \\ f_2(x_1,x_2) = -2x_1 + 4x_2 - 4(x_1-x_2)^2 + \frac{21}{4}x_2^2 + \frac{9}{2}x_2^3. \end{cases} \tag{5.19}$$

These functions arise as follows in the modelling of a continuous-flow stirred tank reactor (CSTR).

Consider two chemical reactions, which can go in both directions,

$$2\,NO_2 \underset{k_2}{\overset{k_1}{\rightleftarrows}} N_2O_4$$

$$2\,NO + O_2 \underset{k_4}{\overset{k_3}{\rightleftarrows}} 2\,NO_2$$

and assign a number to each molecule:

molecule	i
O_2	1
NO	2
NO_2	3
N_2O_4	4

and to each atom:

atom	j
O	1
N	2

Let the component c_i of the vector $c = (c_1, c_2, c_3, c_4) \in \mathbb{R}^4$ be the concentration of the ith-molecule. Suppose the reaction occurs in a well-stirred reactor of volume V to which fluid is added at rate w (volume/time), while being removed at the same rate. Such a setting is referred to as a *continuous-flow stirred tank reactor* (*CSTR*). Let c_f denote the vector of molecule concentrations in the inflow. Then

$$\dot{c} = \lambda(c_f - c) + G(c), \tag{5.20}$$

where

$$\lambda = \frac{w}{V}, \quad G(c) = \begin{pmatrix} g_2 \\ 2g_2 \\ -2g_2 + 2g_1 \\ -g_1 \end{pmatrix}$$

with g_j, $j = 1, 2$, the net rate at with the jth-reaction proceeds in the right direction. The Law of Mass Action gives

$$\begin{aligned} g_1(c) &= k_1 c_3^2 - k_2 c_4 \\ g_2(c) &= k_3 c_1 c_2^2 - k_4 c_3^2, \end{aligned}$$

where k_i are the reaction rate constants.

To reduce the number of equations, one can use the conservation of atoms, a_j, $j = 1, 2$. The vector

$$\begin{pmatrix} a_1 \\ a_2 \end{pmatrix} = \begin{pmatrix} 2 & 1 & 2 & 4 \\ 0 & 1 & 1 & 2 \end{pmatrix} \begin{pmatrix} c_1 \\ c_2 \\ c_3 \\ c_4 \end{pmatrix} = Mc$$

satisfies

$$\dot{a} = \lambda(a_f - a),$$

where $a_f = Mc_f$. From the differential equation for a we conclude that $a(t) \to a_f$ as $t \to \infty$. Therefore, c ultimately "lives" on the linear manifold $Mc = a_f$. To parametrize this manifold with $x \in \mathbb{R}^2$, introduce the matrix

$$S = \begin{pmatrix} 0 & 1 \\ 0 & 2 \\ 2 & -2 \\ -1 & 0 \end{pmatrix}$$

describing how many molecules of each type take part in each reaction. Note that

$$MS = 0, \quad G(c) = Sg(c).$$

Now put $c = c_f + Sx$ and require that

$$\dot{x} = -\lambda x + g(c_f + Sx) \tag{5.21}$$

to guarantee that Eq. (5.20) is satisfied.

For the numerical values $k_1 = 1,\ k_2 = 5,\ k_3 = \frac{9}{8},\ k_4 = 1$, and

$$c_f = \left(\frac{1}{2}, \frac{2}{3}, \frac{1}{2}, \frac{1}{20}\right),$$

we get the nonlinear eigenvalue problem (5.18) for equilibria of (5.21) with f given by (5.19). Note that $g(c_f) = 0$, so c_f is an equilibrium for the reaction corresponding to $x = 0$. It is, however, an unstable steady state. The results derived below tell us that, nevertheless, one can find these concentrations in the reactor tank if the replacement rate λ is large enough. But if one slows down the replacement below a critical level, another steady state appears.

The matrix

$$A = Df(0) = \begin{pmatrix} 7 & -2 \\ -2 & 4 \end{pmatrix}$$

has eigenvalues $\lambda_0 = 8$ and $\lambda_1 = 3$. For $\lambda > \lambda_0$, the trivial equilibrium $x = 0$ of (5.21) is stable. We investigate what happens for parameter values λ close to λ_0.

The unit vector

$$\varphi = \frac{1}{\sqrt{5}} \begin{pmatrix} 2 \\ -1 \end{pmatrix}$$

spans the null space $\mathcal{N}$ of

$$L = A - \lambda_0 I = A - 8I = \begin{pmatrix} -1 & -2 \\ -2 & -4 \end{pmatrix}.$$

Since, as it happens, L is symmetric (i.e. $L^{\mathrm{T}} = L$), the projection on $\mathcal{N}$ is given by

$$Px = \langle \varphi, x \rangle \varphi = \frac{1}{\sqrt{5}}(2x_1 - x_2)\varphi.$$

When we rewrite $f(x) - \lambda x = 0$ in the form $Lx = h(\tau, x)$ with $\tau = \lambda - \lambda_0 = \lambda - 8$, we have explicitly

$$h(\tau, x) = \tau x + Ax - f(x) = \begin{pmatrix} \tau x_1 - 4(x_1 - x_2)^2 \\ \tau x_2 + 4(x_1 - x_2)^2 - \frac{21}{4}x_2^2 - \frac{9}{2}x_2^3 \end{pmatrix}.$$

In order to determine the function $\mathcal{G}$ as explicitly as possible, we use a coordinate to describe the range $\mathcal{R}$ of L. Since, clearly, $\mathcal{R}$ consists of vectors for which the second component is twice as large as the first, we choose the unit vector

$$\chi = \frac{1}{\sqrt{5}} \begin{pmatrix} 1 \\ 2 \end{pmatrix}$$

to span $\mathcal{R}$. Since $\mathbb{R}^2 = \mathcal{N} \oplus \mathcal{R}$, we can use $\{\varphi, \chi\}$ as a basis.[2] We shall use ε and δ to denote the coordinates with respect to this basis. If $x = \varepsilon\varphi + \delta\chi$, then

$$\varepsilon = \langle \varphi, x \rangle = \frac{1}{\sqrt{5}}(2x_1 - x_2),$$
$$\delta = \langle \chi, x \rangle = \frac{1}{\sqrt{5}}(x_1 + 2x_2),$$

while, inversely,

$$\begin{cases} x_1 = \dfrac{1}{\sqrt{5}}(2\varepsilon + \delta), \\ x_2 = \dfrac{1}{\sqrt{5}}(-\varepsilon + 2\delta). \end{cases} \tag{5.22}$$

With a slight abuse of notation we define $\mathcal{G} : \mathbb{R} \times \mathbb{R} \times \mathbb{R} \to \mathbb{R}$ by

$$\mathcal{G}(\varepsilon, \delta, \tau) = \langle \chi, \delta L\chi - h(\tau, \varepsilon\varphi + \delta\chi) \rangle$$

and find (using $L\chi = -5\chi$) explicitly

$$\mathcal{G}(\varepsilon, \delta, \tau) = -5\delta - \tau\delta - \frac{4}{\sqrt{5}}\left(\frac{3\varepsilon - \delta}{\sqrt{5}}\right)^2 + \frac{21}{2\sqrt{5}}\left(\frac{-\varepsilon + 2\delta}{\sqrt{5}}\right)^2 + \frac{9}{\sqrt{5}}\left(\frac{-\varepsilon + 2\delta}{\sqrt{5}}\right)^3.$$

According to the Implicit Function Theorem, the equation $\mathcal{G} = 0$ has a unique solution $\delta = \delta(\varepsilon, \tau)$ with $\delta(0, 0) = 0$. By substituting a Taylor expansion for the solution into the equation and equating coefficients, we obtain

$$\delta = -\frac{51\sqrt{5}}{250}\varepsilon^2 + \frac{51\sqrt{5}}{1250}\varepsilon^2\tau + \frac{468}{6250}\varepsilon^3 + \cdots$$

A key point to observe is that $\delta = O(\varepsilon^2)$.

It remains only to solve $Ph(\tau, \varepsilon\varphi + \delta(\varepsilon, \tau)\chi) = 0$. Explicitly this equation reads

$$\varepsilon\tau - \frac{12}{\sqrt{5}}\left(\frac{3\varepsilon - \delta}{\sqrt{5}}\right)^2 + \frac{21}{4\sqrt{5}}\left(\frac{-\varepsilon + 2\delta}{\sqrt{5}}\right)^2 + \frac{9}{2\sqrt{5}}\left(\frac{-\varepsilon + 2\delta}{\sqrt{5}}\right)^3 = 0,$$

where $\delta = \delta(\varepsilon, \tau) = O(\varepsilon^2)$. Therefore, we can eliminate the already known trivial solution $x = 0$ by dividing by ε. Next, the Implicit Function Theorem guarantees that we can express τ as a function of ε. Using Taylor expansion, one finds explicitly

$$\tau = \frac{1}{100}\left(411\sqrt{5}\varepsilon + \frac{5652}{25}\varepsilon^2 + \cdots\right).$$

Finally, the results in terms of ε, δ, and τ can be translated in terms of x_1, x_2, and λ by executing all transformations backwards. This shows that at $\lambda_0 = 8$ a *transcritical bifurcation* occurs: The trivial equilibrium $x = 0$ of (5.21) loses its stability when λ decreases down below λ_0; for $\lambda < \lambda_0$ with small $|\lambda - \lambda_0|$, a stable nontrivial equilibrium coexists with the unstable trivial equilibrium. A qualitative sketch of the bifurcation diagram is depicted in Fig. 5.6. The stability assertions reflect the Principle of Exchange of Stability,

[2] Note, incidentally, that $\langle \varphi, \chi \rangle = 0$ so that, in this particular case of symmetric L, this is actually an orthogonal decomposition.

Fig. 5.6 Transcritical bifurcation in CSTR

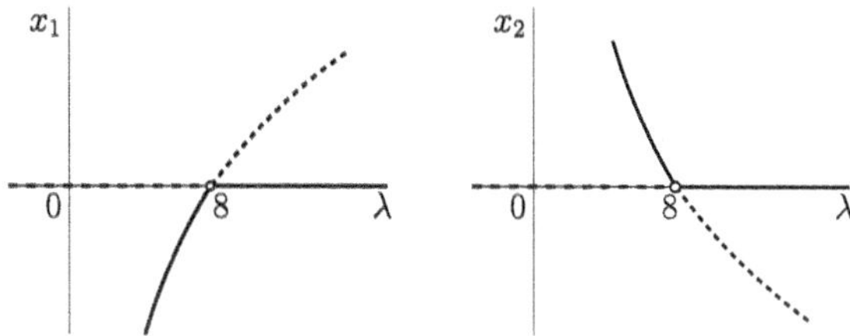

as observed in Remark (2) following the proof of Theorem 5.7. From (5.22) and $\delta = O(\varepsilon^2)$ we obtain $x_1 < 0 < x_2$ for $\lambda < \lambda_0$. Of course, for $\lambda > \lambda_0$ with small $|\lambda - \lambda_0|$, the nontrivial equilibrium also exists but it is unstable and, therefore, not observable in experiments. ◊

5.2.2 Planar Andronov–Hopf Bifurcation

What happens if a planar system has an equilibrium x_0 at some parameter value α_0 with eigenvalues $\lambda_{1,2} = \pm i\omega_0$, $\omega_0 > 0$ (Fig. 5.2b)? Generically, an *Andronov–Hopf* (or *Hopf*) bifurcation happens, as in the following example.

Example 5.9 (Normal form of Andronov–Hopf bifurcation) Consider the following planar system:

$$\begin{cases} \dot{x}_1 = \alpha x_1 - x_2 + s x_1(x_1^2 + x_2^2), \\ \dot{x}_2 = x_1 + \alpha x_2 + s x_2(x_1^2 + x_2^2), \end{cases} \tag{5.23}$$

where $s = \pm 1$. Let $z = x_1 + ix_2$ and $\bar{z} = x_1 - ix_2$. Then (5.23) is equivalent to one complex equation

$$\dot{z} = (\alpha + i)z + sz^2\bar{z}. \tag{5.24}$$

Using $z = re^{i\theta}$, one can write system (5.23) in polar coordinates:

$$\begin{cases} \dot{r} = r(\alpha + sr^2), \\ \dot{\theta} = 1. \end{cases} \tag{5.25}$$

Since the equations for r and θ are separated, one immediately gets the following two bifurcation scenarios.

If $s = -1$, the origin is a globally asymptotically stable equilibrium of (5.25) for $\alpha \le 0$. This equilibrium becomes unstable for $\alpha > 0$ and is surrounded by a stable cycle, which is a circle of radius $\sqrt{\alpha}$. All nonequilibrium orbits tend to this cycle when time advances. This is a *supercritical Andronov–Hopf bifurcation* (see Fig. 5.7). The word “supercritical” means that the cycle

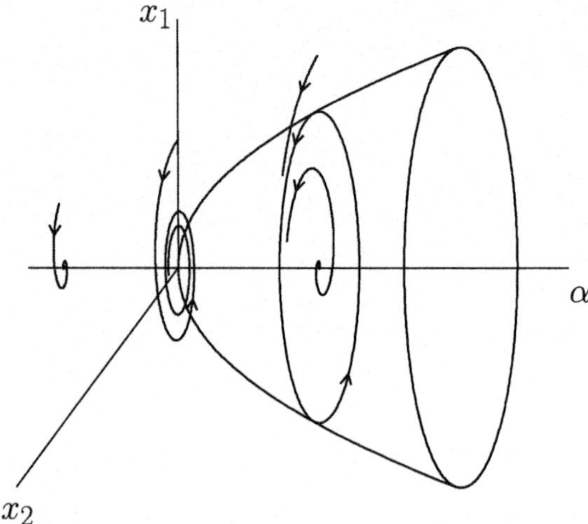

Fig. 5.7 Supercritical Andronov–Hopf bifurcation

exists for parameter values at which the equilibrium is no longer stable, so beyond the critical parameter value corresponding to the stability loss.

If $s = +1$, the origin is still locally asymptotically stable for $\alpha < 0$ but is surrounded by an unstable circular cycle of radius $\sqrt{-\alpha}$. For $\alpha > 0$ the equilibrium at the origin becomes unstable. No cycle is present and all nonequilibrium orbits diverge. This is a *subcritical Andronov–Hopf bifurcation* (see Fig. 5.8). Now indeed the cycle exists for parameter values at which the stability of the equilibrium is not yet lost. The cycle is unstable and separates the domain of attraction of the equilibrium from that of "infinity". When the parameter crosses the critical value, all nontrivial orbits are attracted by infinity. One speaks here about a *hard bifurcation* to contrast with a *soft bifur-*

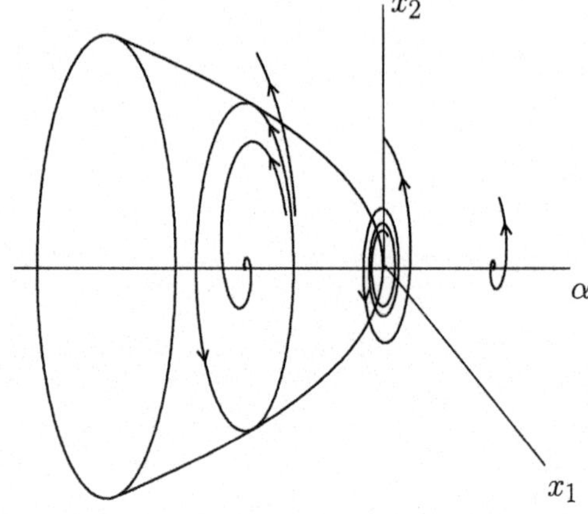

Fig. 5.8 Subcritical Andronov–Hopf bifurcation

cation in the supercritical case, when the amplitude of oscillations increases gradually with the parameter.

Notice that the case $s = +1$ can be reduced to the case $s = -1$ by the substitution $(x_1, x_2, t, \alpha) \mapsto (x_2, x_1, -t, -\alpha)$. Since this transformation involves time reversal, the corresponding systems are *not* topologically equivalent and should be treated separately, because we are interested in stability of their equilibria and cycles. ♢

The use of a complex coordinate z is very convenient, but one should realize that the right-hand side of (5.24) is not differentiable as a function from $\mathbb{C}$ to $\mathbb{C}$ since it depends on $\bar{z}$. So there is a need to define carefully what we mean by "smooth".

Definition 5.10 *A complex-valued function $g = g(z, \bar{z})$ is called* ***smooth*** *if its real and imaginary parts are smooth functions of (x_1, x_2), where $z = x_1 + ix_2$.*

For such functions, one can define the following derivatives:

$$\frac{\partial g}{\partial z} := \frac{1}{2}\left(\frac{\partial G}{\partial x_1} - i\frac{\partial G}{\partial x_2}\right) \quad \text{and} \quad \frac{\partial g}{\partial \bar{z}} := \frac{1}{2}\left(\frac{\partial G}{\partial x_1} + i\frac{\partial G}{\partial x_2}\right),$$

where $G(x_1, x_2) = g(x_1 + ix_2, x_1 - ix_2)$. These operations have standard properties. For example, for two smooth functions $f(z, \bar{z})$ and $g(z, \bar{z})$,

$$\frac{\partial}{\partial z}(fg) = \frac{\partial f}{\partial z}g + f\frac{\partial g}{\partial z} \quad \text{and} \quad \frac{\partial}{\partial \bar{z}}(fg) = \frac{\partial f}{\partial \bar{z}}g + f\frac{\partial g}{\partial \bar{z}}.$$

Clearly,

$$\frac{\partial \bar{z}}{\partial z} = \frac{\partial z}{\partial \bar{z}} = 0 \quad \text{and} \quad \frac{\partial z}{\partial z} = \frac{\partial \bar{z}}{\partial \bar{z}} = 1,$$

implying

$$\frac{\partial z^m}{\partial \bar{z}} = 0 \quad \text{and} \quad \frac{\partial \bar{z}^m}{\partial \bar{z}} = m\bar{z}^{m-1}.$$

Thus, one can differentiate polynomials of z and $\bar{z}$ exactly as in the Real Calculus, considering z and $\bar{z}$ as independent variables. If $z = z(t)$ is a smooth function of time, then the usual chain rule is applicable:

$$\frac{dg}{dt} = \frac{\partial g}{\partial z}\frac{dz}{dt} + \frac{\partial g}{\partial \bar{z}}\frac{d\bar{z}}{dt}.$$

A complex equation

$$\dot{z} = g(z, \bar{z}), \quad z \in \mathbb{C}, \tag{5.26}$$

with smooth g is equivalent to a smooth real system

$$\dot{x} = f(x), \quad x \in \mathbb{R}^2, \tag{5.27}$$

where

$$x = \begin{pmatrix} x_1 \\ x_2 \end{pmatrix}, \quad f(x) = \begin{pmatrix} f_1(x_1, x_2) \\ f_2(x_1, x_2) \end{pmatrix} = \begin{pmatrix} \text{Re}[g(x_1 + ix_2, x_1 - ix_2)] \\ \text{Im}[g(x_1 + ix_2, x_1 - ix_2)] \end{pmatrix},$$

and, vice versa,

$$g(z, \bar{z}) = f_1\left(\frac{z+\bar{z}}{2}, \frac{z-\bar{z}}{2i}\right) + if_2\left(\frac{z+\bar{z}}{2}, \frac{z-\bar{z}}{2i}\right).$$

The phase portrait of (5.26) is by definition the phase portrait of the corresponding real system (5.27).

Definition 5.11 *Two complex smooth equations are called* ***locally topologically equivalent*** *near $z = 0$ if the corresponding planar real systems are locally topologically equivalent near $x = 0$.*

Local topological equivalence of parameter-dependent complex equations can be defined similarly.

Theorem 5.12 *The complex equation*

$$\dot{w} = (\alpha + i)w + sw^2\bar{w} + O(|w|^4), \qquad w \in \mathbb{C}, \tag{5.28}$$

where $s = \pm 1$ and the right-hand side is a smooth function of $(w, \bar{w})$ and α, is locally topologically equivalent near the origin to Eq. (5.24).

Proof The proof involves two steps.

Step 1: Existence and uniqueness of the cycle.

We consider only the case $s = -1$, since the other one can be reduced to it. Our first aim is to construct a Poincaré map for (5.28). Write this equation for $s = -1$ in polar coordinates (ρ, φ) with $w = \rho e^{i\varphi}$. This gives

$$\begin{cases} \dot{\rho} = \rho(\alpha - \rho^2) + \Phi(\rho, \varphi, \alpha), \\ \dot{\varphi} = 1 + \Psi(\rho, \varphi, \alpha), \end{cases} \tag{5.29}$$

where Φ and Ψ are smooth functions of (ρ, φ, α) such that $\Phi = O(\rho^4)$ and $\Psi = O(\rho^3)$. For given α, an orbit of (5.29) starting at $(\rho, \varphi) = (\rho_0, 0)$ has the following representation (see Fig. 5.9): $\rho = \rho(\varphi, \rho_0)$, $\rho(0, \rho_0) = \rho_0$ with ρ satisfying the equation:

$$\frac{d\rho}{d\varphi} = \frac{\rho(\alpha - \rho^2) + \Phi(\rho, \varphi, \alpha)}{1 + \Psi(\rho, \varphi, \alpha)} = \rho(\alpha - \rho^2) + R(\rho, \varphi, \alpha), \tag{5.30}$$

where $R(\rho, \varphi, \alpha) = O(\rho^4)$ is smooth.

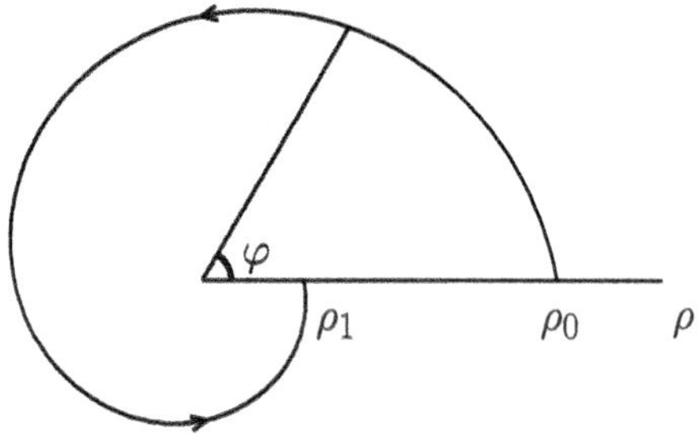

Fig. 5.9 Poincaré (return) map near Andronov–Hopf bifurcation

Since $\rho(\varphi, 0) \equiv 0$ and $\rho(\varphi, \rho_0)$ is a smooth function of its arguments, we can write the Taylor expansion:

$$\rho = u_1(\varphi)\rho_0 + u_2(\varphi)\rho_0^2 + u_3(\varphi)\rho_0^3 + O(\rho_0^4). \tag{5.31}$$

Substituting (5.31) into (5.30) and sorting terms according to powers of ρ_0, we deduce the following *linear* differential equations for the unknown $u_i, i = 1, 2, 3$:

$$\frac{du_1}{d\varphi} = \alpha u_1, \quad \frac{du_2}{d\varphi} = \alpha u_2, \quad \frac{du_3}{d\varphi} = \alpha u_3 - u_1^3,$$

with the initial conditions $u_1(0) = 1, u_2(0) = u_3(0) = 0$ (we also call the equation for u_3 linear, since we can consider u_1 as known when solving for u_3). The solutions are

$$u_1(\varphi) = e^{\alpha\varphi}, \; u_2(\varphi) \equiv 0, \; u_3(\varphi) = \frac{1}{2\alpha} e^{\alpha\varphi}(1 - e^{2\alpha\varphi}).$$

Notice that these expressions are *independent* of the term $R(\rho, \varphi, \alpha)$. Therefore, the Poincaré return map $\rho_0 \mapsto \rho_1 = \rho(2\pi, \rho_0)$ has the form

$$\rho_1 = e^{2\pi\alpha}\rho_0 - e^{2\pi\alpha}[2\pi + O(\alpha)]\rho_0^3 + O(\rho_0^4) \tag{5.32}$$

for all smooth $R = O(\rho^4)$. The map (5.32) can easily be analysed for sufficiently small ρ_0 and $|\alpha|$ with the help of the Implicit Function Theorem. There is a neighbourhood of the origin in which the map has only the trivial fixed point for small $\alpha < 0$ and an additional fixed point, $\rho^{(0)} = \sqrt{\alpha} + O(\alpha)$, for small $\alpha > 0$ (see Fig. 5.10 and note that only nonnegative values of ρ_0 have meaning, so that we can ignore negative roots of the equation $\rho_1(\rho_0, \alpha) = \rho_0$). The stability of the fixed points is also easily obtained from (5.32). Taking into account that a positive fixed point of the map corresponds to a limit cycle of the system, we can conclude that system (5.29) (or Eq. (5.28)) with any $O(|w|^4)$ terms has a unique (stable) limit cycle bifurcating from the origin and existing for $\alpha > 0$ just as Eq. (5.24) for $s = -1$. Therefore, in other

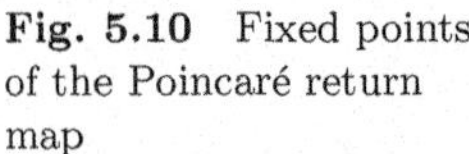

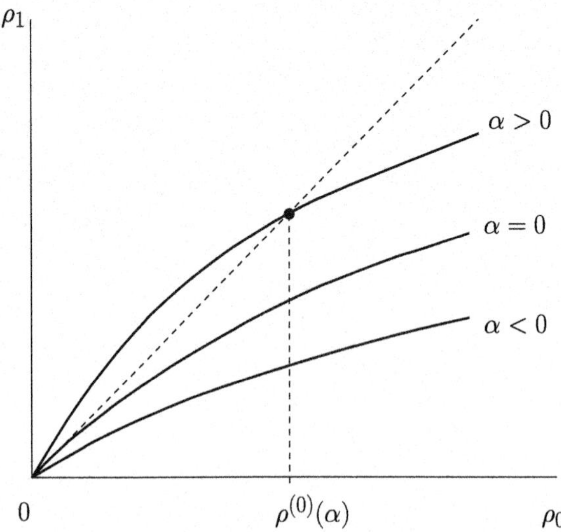

Fig. 5.10 Fixed points of the Poincaré return map

words, higher-order terms do not affect the limit cycle bifurcation in some neighbourhood of $z = 0$ for $|\alpha|$ sufficiently small.

Step 2: Construction of a homeomorphism

The established existence and uniqueness of the limit cycle suffices for most applications. Nevertheless, extra work must be done to prove the topological equivalence of the phase portraits.

Let $w = y_1 + iy_2 = \rho e^{i\varphi}$ and $z = x_1 + ix_2 = re^{i\theta}$. Notice that the transition from (5.29) to (5.30) is equivalent to the introduction of a new time parametrization such that

$$\begin{cases} \dot{\rho} = \rho(\alpha - \rho^2) + R(\rho, \varphi, \alpha), \\ \dot{\varphi} = 1. \end{cases} \tag{5.33}$$

In this system, that is orbitally equivalent to system (5.29), the return time to the half-axis $\varphi = 0$ (mod 2π) is the same for all orbits starting on this axis. Denote the Poincaré map (5.32) for (5.33) by P_α. The normal form (5.24) is equivalent to (5.25), that is (5.33) without the R-term. Denote by Q_α the corresponding Poincaré map. We consider both maps for small nonnegative values of ρ and r and recall that $P_\alpha(0) = Q_\alpha(0) = 0$. Notice that both maps are invertible near the origin.

We want to construct a homeomorphism h_α defined in a neighbourhood of the origin in the phase plane that maps orbits of (5.33) onto orbits of (5.25), preserving their orientation. Such map h_α should conjugate the corresponding Poincaré maps, i.e. its restriction to the half-axis $\rho \geq 0$ for each small parameter value α should be a locally defined homeomorphism $\rho \mapsto H_\alpha(\rho)$

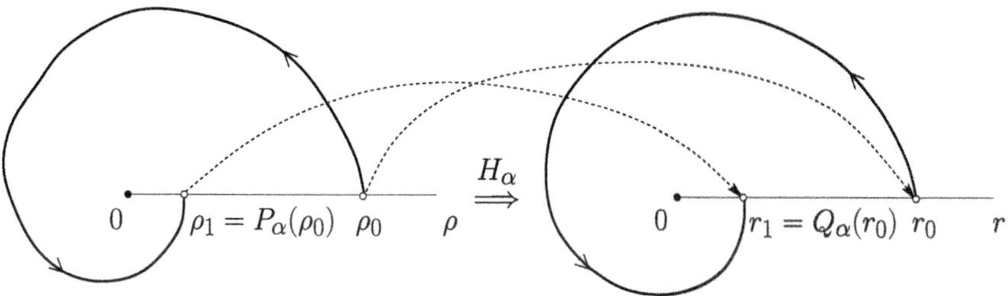

Fig. 5.11 A conjugating homeomorphism H_α near the Hopf bifurcation should conjugate Poincaré maps P_α and Q_α

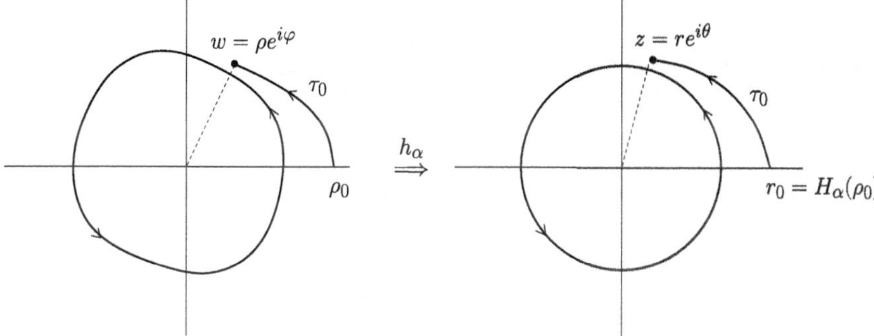

Fig. 5.12 Construction of the homeomorphism near the Hopf bifurcation

that maps orbits of P_α near the origin onto such orbits of Q_α, preserving the direction of time (see Fig. 5.11).

If the map H_α is known, than the planar homeomorphism $y \mapsto x = h_\alpha(y)$ can be constructed by the following procedure (see Fig. 5.12). Take a point $w = \rho e^{i\varphi} = y_1 + iy_2 \neq 0$ and find values (ρ_0, τ_0), where $\tau_0 > 0$ is the minimal time required for an orbit of system (5.33) to reach the point w starting from the horizontal half-axis at $\rho = \rho_0$. Now, take the point on this axis with $r_0 = H_\alpha(\rho_0)$ and construct an orbit of (5.25) on the time interval $[0, \tau_0]$ starting at this point. Denote the resulting point by $z = re^{i\theta} = x_1 + ix_2$. Set $z = 0$ for $w = 0$, i.e. $x = 0$ for $y = 0$.

The map h_α is a homeomorphism that maps orbits of (5.33) (equivalent to (5.28)) in some neighbourhood of the origin onto orbits of (5.25) (equivalent to (5.24)), preserving the time direction.

What is left to do is to construct a scalar homeomorphism H_α by the method of fundamental domains used in Chap. 2, Sect. 2.3.2. We consider separately the cases $\alpha \leq 0$ and $\alpha > 0$, assuming that $|\alpha|$ is sufficiently small.[3]

[3] A small constant $\varepsilon > 0$ appearing below should actually be selected in accordance with the range of α.

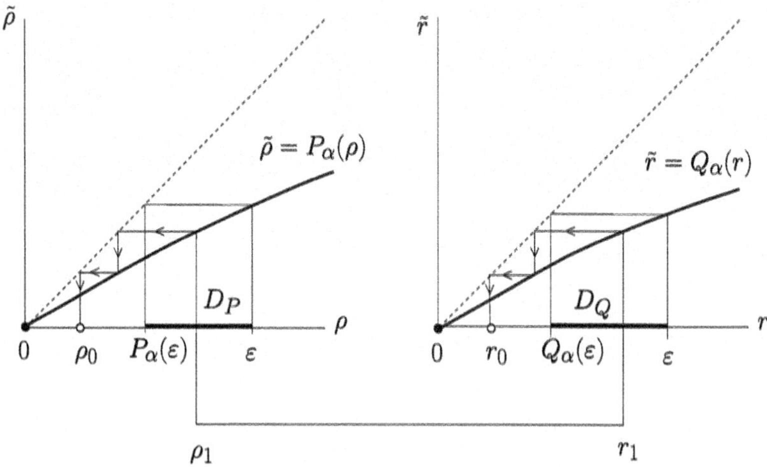

Fig. 5.13 Construction of H_α for $\alpha \leq 0$

Case $\alpha \leq 0$. In this case, the origin is the only fixed point of both P_α and Q_α near the origin, and this point is stable. Take a small $\varepsilon > 0$ and introduce a fundamental domain for P_α:

$$D_P = [P_\alpha(\varepsilon), \varepsilon],$$

and the corresponding domain for Q_α:

$$D_Q = [Q_\alpha(\varepsilon), \varepsilon]$$

(see Fig. 5.13). Let $H_\alpha : D_P \to D_Q$ be an orientation-preserving affine map transforming D_P into D_Q.

For any sufficiently small $\rho_0 > 0$ the sequence $P_\alpha^k(\rho_0)$ converges monotonically to zero. If $\rho_0 > \varepsilon$, there is a positive integer $K = K(\rho_0)$ with $P_\alpha^K(\rho_0) \leq \varepsilon < P_\alpha^{K-1}(\rho_0)$. By the monotonicity of P_α, we obtain $P_\alpha^{K+1}(\rho_0) \leq P_\alpha(\varepsilon) < P_\alpha^K(\rho_0) \leq \varepsilon$. That is, the iteration hits the fundamental domain D_P exactly once. A similar argument holds for $\rho_0 < P_\alpha(\varepsilon)$ with a suitable negative integer $K = K(\rho_0)$ (see Fig. 5.13, where the case $K < 0$ is illustrated).

In other words, for $\rho_0 \notin D_P$ with small $\rho_0 > 0$, there is an integer $K = K(\rho_0)$ with the minimal $|K|$ such that

$$\rho_1 = P_\alpha^K(\rho_0) \in D_P.$$

Set $r_1 = H_\alpha(\rho_1)$ and define

$$r_0 = Q_\alpha^{-K}(r_1).$$

This gives a map $H_\alpha(\rho_0) = r_0$ for small $\rho_0 > 0$. Set $H_\alpha(0) = 0$ by continuity. The map $r = H_\alpha(\rho)$ is defined for small $\rho \geq 0$ and maps orbits of P_α onto orbits of Q_α for small $r \geq 0$. It is continuous and locally invertible, i.e. we have constructed a local homeomorphism conjugating P_α and Q_α at the corresponding $\alpha \leq 0$.

Case $\alpha > 0$. In this case, the origin is the unstable fixed point of P_α and Q_α, but each map has another fixed point that is stable. This fixed point of Q_α is merely $\sqrt{\alpha}$, while the positive fixed point of P_α has the asymptotic expression:

$$\rho^{(0)}(\alpha) = \sqrt{\alpha} + O(\alpha)$$

(see **Step 1**).

Recalling that $0 < \alpha < \sqrt{\alpha}$ for $0 < \alpha < 1$, introduce two pairs of closed intervals:

$$D_P^- = [\alpha, P_\alpha(\alpha)], \quad D_Q^- = [\alpha, Q_\alpha(\alpha)]$$

and

$$D_P^+ = [P_\alpha(\rho^{(0)}(\alpha) + \varepsilon), \rho^{(0)}(\alpha) + \varepsilon], \quad D_Q^+ = [Q_\alpha(\sqrt{\alpha} + \varepsilon), \sqrt{\alpha} + \varepsilon],$$

where $\varepsilon > 0$ is sufficiently small (see Fig. 5.14). Let

$$D_P = D_P^- \cup D_P^+, \quad D_Q = D_Q^- \cup D_Q^+,$$

and let $H_\alpha : D_P \to D_Q$ be a homeomorphism that is defined by the affine maps $D_P^- \to D_Q^-$ and $D_P^+ \to D_Q^+$, so that

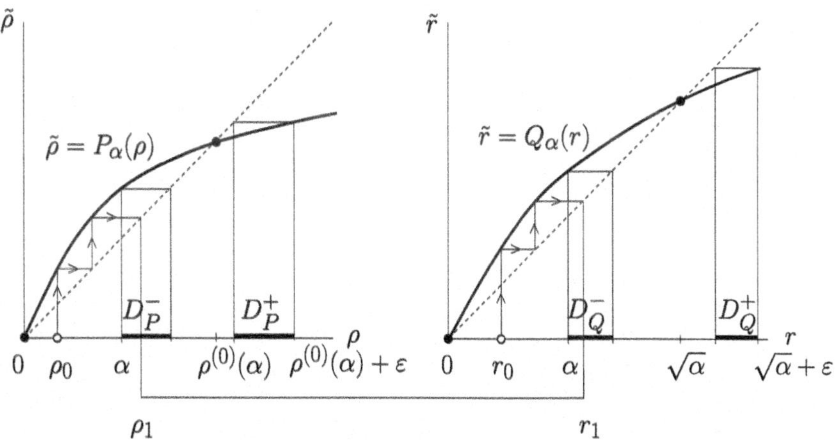

Fig. 5.14 Construction of H_α for $\alpha > 0$; case $K > 0$ with $\rho_1 \in D_P^-$ is illustrated

$$H_\alpha(D_P^{\mp}) = D_Q^{\mp},$$

preserving orientation.

The rest is formally identical to the previous case. For any $\rho_0 \notin D_P$ and such that $r_0 > 0$ and $r_0 \neq \rho^{(0)}(\alpha)$, define an $r_1 > 0$ as follows. Let $K = K(\rho_0)$ be an integer with the minimal $|K|$ such that

$$\rho_1 = P_\alpha^K(\rho_0) \in D_P.$$

Set $r_1 = H_\alpha(\rho_1)$ and define

$$r_0 = Q_\alpha^{-K}(r_1).$$

His gives a map $r_0 = H_\alpha(\rho_0)$ for positive $\rho_0 \neq \rho^{(0)}(\alpha)$ that can be extended to a continuous and locally invertible map $\rho \mapsto H_\alpha(\rho)$ defined for all small $\rho \geq 0$, if we set $H_\alpha(0) = 0$ and $H_\alpha(\rho^{(0)}(\alpha)) = \sqrt{\alpha}$.

The map H_α maps orbits of P_α for small $\rho \geq 0$ onto orbits of Q_α for small $r \geq 0$, i.e. we have constructed a local homeomorphism conjugating P_α and Q_α for small $\alpha > 0$. □

The proof of the following theorem will occupy the rest of this section.

Theorem 5.13 *Suppose a two-dimensional system*

$$\frac{dx}{dt} = f(x, \alpha), \quad x \in \mathbb{R}^2, \ \alpha \in \mathbb{R}, \tag{5.34}$$

with smooth f, has for sufficiently small $|\alpha|$ an equilibrium $x_0(\alpha)$ with eigenvalues

$$\lambda_{1,2}(\alpha) = \mu(\alpha) \pm i\omega(\alpha),$$

where $\mu(0) = 0$, $\omega(0) = \omega_0 > 0$.

Let the following genericity conditions be satisfied:

(B.1) $l_1(0) \neq 0$, *where l_1 is the first Lyapunov coefficient (to be defined below);*

(B.2) $\mu'(0) \neq 0$.

Then there are invertible smooth coordinate and parameter changes and a time reparameterization transforming (5.34) near $x_0(\alpha)$ into the system

$$\frac{d}{d\tau}\begin{pmatrix} y_1 \\ y_2 \end{pmatrix} = \begin{pmatrix} \beta & -1 \\ 1 & \beta \end{pmatrix}\begin{pmatrix} y_1 \\ y_2 \end{pmatrix} + s(y_1^2 + y_2^2)\begin{pmatrix} y_1 \\ y_2 \end{pmatrix} + O(\|y\|^4), \tag{5.35}$$

where $s = \mathrm{sign}\, l_1(0) = \pm 1$.

Remark One can summarize the theorem by saying that generically, near the Hopf bifurcation, (5.34) is locally smoothly orbitally equivalent to (5.35). ◊

Since

$$\det D_x f(x_0(0), 0) = \lambda_1(0)\lambda_2(0) = \omega_0^2 \neq 0,$$

the Implicit Function Theorem guarantees the local uniqueness and smoothness of the equilibrium $x_0(\alpha)$ of (5.34) for small $|\alpha|$. Shifting the origin of the coordinate system to $x_0(\alpha)$, we can then consider a smooth system

$$\dot{x} = A(\alpha)x + F(x, \alpha), \quad x \in \mathbb{R}^2, \alpha \in \mathbb{R}, \tag{5.36}$$

where $F(x, \alpha) = O(\|x\|^2)$ and the matrix $A(\alpha) = D_x f(x_0(\alpha), \alpha)$ has the eigenvalues

$$\lambda_{1,2}(\alpha) = \mu(\alpha) \pm i\omega(\alpha),$$

such that $\mu(0) = 0$, $\omega(0) = \omega_0 > 0$. For small $|\alpha|$,

$$\mu(\alpha) = \frac{1}{2}\mathrm{Tr}A(\alpha), \quad \omega(\alpha) = \frac{1}{2}\sqrt{4\det A(\alpha) - [\mathrm{Tr}A(\alpha)]^2}.$$

Lemma 5.14 *Let $q(\alpha)$ and $p(\alpha)$ be smooth vector-valued functions satisfying*

$$A(\alpha)q(\alpha) = \lambda(\alpha)q(\alpha), \ A^{\mathrm{T}}(\alpha)p(\alpha) = \bar{\lambda}(\alpha)p(\alpha), \ \langle p(\alpha), q(\alpha)\rangle = 1,$$

where $\langle p, q\rangle := \bar{p}^{\mathrm{T}} q$. Introduce

$$z = \langle p(\alpha), x\rangle \in \mathbb{C}.$$

Then, for sufficiently small $|\alpha|$, the system (5.36) is equivalent to one complex equation:

$$\dot{z} = \lambda(\alpha)z + g(z, \bar{z}, \alpha), \tag{5.37}$$

where $g(z, \bar{z}, \alpha) = \langle p(\alpha), F(zq(\alpha) + \bar{z}\bar{q}(\alpha), \alpha)\rangle$ is a smooth function of $(z, \bar{z})$ and α, such that $g = O(|z|^2)$ for small z.

Proof Let $q(\alpha) \in \mathbb{C}^2$ be an eigenvector of $A(\alpha)$ corresponding to the eigenvalue $\lambda(\alpha)$:

$$A(\alpha)q(\alpha) = \lambda(\alpha)q(\alpha),$$

and let $p(\alpha) \in \mathbb{C}^2$ be an eigenvector of the transposed matrix $A^{\mathrm{T}}(\alpha)$ corresponding to its eigenvalue $\overline{\lambda(\alpha)}$:

$$A^{\mathrm{T}}(\alpha)p(\alpha) = \overline{\lambda(\alpha)}p(\alpha).$$

It is always possible to normalize p *with respect to* q:

$$\langle p(\alpha), q(\alpha)\rangle = 1.$$

Any vector $x \in \mathbb{R}^2$ can be uniquely represented for any small α as

$$x = zq(\alpha) + \bar{z}\bar{q}(\alpha) \tag{5.38}$$

for some complex z. Indeed, we have an *explicit* formula to determine z:

$$z = \langle p(\alpha), x\rangle. \tag{5.39}$$

To verify this formula (which results from taking the scalar product with p of both sides of the identity (5.38)), we have to prove that $\langle p(\alpha), \bar{q}(\alpha)\rangle = 0$. This is the case, since

$$\begin{aligned}\langle p(\alpha), \bar{q}(\alpha)\rangle &= \langle p(\alpha), \frac{1}{\bar{\lambda}(\alpha)}A(\alpha)\bar{q}(\alpha)\rangle = \frac{1}{\bar{\lambda}(\alpha)}\langle A^{\mathrm{T}}(\alpha)p(\alpha), \bar{q}(\alpha)\rangle \\ &= \frac{\lambda(\alpha)}{\bar{\lambda}(\alpha)}\langle p(\alpha), \bar{q}(\alpha)\rangle\end{aligned}$$

and therefore

$$\left(1 - \frac{\lambda(\alpha)}{\bar{\lambda}(\alpha)}\right)\langle p(\alpha), \bar{q}(\alpha)\rangle = 0.$$

But $\lambda(\alpha) \neq \bar{\lambda}(\alpha)$ because for all sufficiently small $|\alpha|$ we have $\omega(\alpha) > 0$. Thus, the only possibility is $\langle p(\alpha), \bar{q}(\alpha)\rangle = 0$.

The complex variable z obviously satisfies the equation:

$$\dot{z} = \lambda(\alpha)z + \langle p(\alpha), F(zq(\alpha) + \bar{z}\bar{q}(\alpha), \alpha)\rangle,$$

having the required form (5.37) with

$$g(z, \bar{z}, \alpha) = \langle p(\alpha), F(zq(\alpha) + \bar{z}\bar{q}(\alpha), \alpha)\rangle.$$

□

As we have already mentioned, there is no reason to expect g to be an analytic function of z (i.e. $\bar{z}$- independent). Write g as a formal Taylor series in two independent complex variables (z and $\bar{z}$):

$$g(z, \bar{z}, \alpha) = \sum_{k+l\geq 2} \frac{1}{k!l!} g_{kl}(\alpha) z^k \bar{z}^l,$$

where formally

$$g_{kl}(\alpha) = \frac{\partial^{k+l}}{\partial z^k \partial \bar{z}^l} \langle p(\alpha), F(zq(\alpha) + \bar{z}\bar{q}(\alpha), \alpha)\rangle \bigg|_{z=0},$$

for $k + l \geq 2$, $k, l = 0, 1, \ldots$. In the following, we will call a complex differential equation "smooth" if its right-hand side is a smooth function of $(z, \bar{z})$ and the parameter α.

Lemma 5.15 (Poincaré normal form) *The smooth equation*

$$\dot{z} = \lambda(\alpha)z + \sum_{2 \leq k+l \leq 3} \frac{1}{k!l!} g_{kl}(\alpha) z^k \bar{z}^l + O(|z|^4), \tag{5.40}$$

where $\lambda(\alpha) = \mu(\alpha) + i\omega(\alpha), \mu(0) = 0, \omega(0) = \omega_0 > 0$, *can be transformed, for all sufficiently small* $|\alpha|$, *by an invertible change of complex coordinate*

$$\begin{aligned} z = w &+ \frac{h_{20}(\alpha)}{2} w^2 + h_{11}(\alpha) w\bar{w} + \frac{h_{02}(\alpha)}{2} \bar{w}^2 \\ &+ \frac{h_{30}(\alpha)}{6} w^3 + \frac{h_{12}(\alpha)}{2} w\bar{w}^2 + \frac{h_{03}(\alpha)}{6} \bar{w}^3 + O(|w|^4), \end{aligned}$$

smoothly depending on the parameter, into a smooth equation with only one cubic term:

$$\dot{w} = \lambda(\alpha)w + c_1(\alpha) w^2 \bar{w} + O(|w|^4), \tag{5.41}$$

where

$$c_1(0) = \frac{g_{21}(0)}{2} + \frac{i}{2\omega_0}\left(g_{20}(0)g_{11}(0) - 2|g_{11}(0)|^2 - \frac{1}{3}|g_{02}(0)|^2\right). \tag{5.42}$$

Remark The lemma states that (5.40) is locally smoothly equivalent (diffeomorphic) near the origin to (5.41). The system (5.41) is called the *normal form* for the Hopf bifurcation. The remaining $w^2\bar{w}$-term in (5.41) is called a *resonant term*. ♢

Proof of Lemma 5.15 Write

$$\begin{aligned} \dot{z} &= \lambda z + \frac{1}{2} g_{20} z^2 + g_{11} z\bar{z} + \frac{1}{2} g_{02} \bar{z}^2 \\ &\quad + \frac{1}{6} g_{30} z^3 + \frac{1}{2} g_{21} z^2 \bar{z} + \frac{1}{2} g_{12} z\bar{z}^2 + \frac{1}{6} g_{03} \bar{z}^2 \\ &\quad + O(|z|^4), \end{aligned} \tag{5.43}$$

where the dependence of λ and h_{jk} on α is implicit. Make a substitution

$$
\begin{aligned}
z \;=\; & w + \frac{1}{2}h_{20}w^2 \;+\; h_{11}w\overline{w} \;+\; \frac{1}{2}h_{02}\overline{w}^2 \\
& + \frac{1}{6}h_{30}w^3 \;+\; \frac{1}{2}h_{21}w^2\overline{w} \;+\; \frac{1}{2}h_{12}w\overline{w}^2 \;+\; \frac{1}{6}h_{03}\overline{w}^2
\end{aligned} \tag{5.44}
$$

with unknown complex coefficients $h_{jk} = h_{jk}(\alpha)$. This substitution is invertible in a small neighbourhood of the origin and the inverse is smooth (but not polynomial). The substitution will produce a transformed system

$$
\begin{aligned}
\dot{w} \;=\; & \lambda w + \frac{1}{2}G_{20}w^2 \;+\; G_{11}w\overline{w} \;+\; \frac{1}{2}G_{02}\overline{w}^2 \\
& + \frac{1}{6}G_{30}w^3 \;+\; \frac{1}{2}G_{21}w^2\overline{w} \;+\; \frac{1}{2}G_{12}w\overline{w}^2 \;+\; \frac{1}{6}G_{03}\overline{w}^2 \\
& + O(|w|^4).
\end{aligned} \tag{5.45}
$$

To find G_{jk}, compute $\dot{z}$ in two ways: First, by differentiating (5.44) and replacing $\dot{w}$ using (5.45),

$$
\begin{aligned}
\dot{z} = & \dot{w} + h_{20}w\dot{w} + h_{11}(\overline{w}\dot{w} + w\dot{\overline{w}}) + h_{02}\overline{w}\dot{\overline{w}} \\
& + \frac{1}{2}h_{30}w^2\dot{w} + h_{21}\left(w\overline{w}\dot{w} + \frac{1}{2}h_{21}w^2\dot{\overline{w}}\right) + h_{12}\left(\frac{1}{2}\overline{w}^2\dot{w} + w\overline{w}\dot{\overline{w}}\right) + \frac{1}{2}h_{03}\overline{w}^2\dot{\overline{w}} \\
= & \lambda w + \frac{1}{2}(G_{20} + 2\lambda h_{20})w^2 + (G_{11} + (\lambda + \bar{\lambda}))h_{11})w\overline{w} + \frac{1}{2}(G_{02} + 2\bar{\lambda}h_{02})\overline{w}^2 \\
& + \frac{1}{6}\left(G_{30} + 3G_{20}h_{20} + 3\overline{G}_{02}h_{11} + 3\lambda h_{30}\right)w^3 \\
& + \frac{1}{2}\left(G_{21} + 2G_{11}h_{20} + G_{20}h_{11} + 2\overline{G}_{11}h_{11} + \overline{G}_{02}h_{02} + (2\lambda + \bar{\lambda})h_{21}\right)w^2\overline{w} \\
& + \frac{1}{2}\left(G_{12} + G_{02}h_{20} + 2G_{11}h_{11} + \overline{G}_{20}h_{11} + 2\overline{G}_{11}h_{02} + (\lambda + 2\bar{\lambda})h_{12}\right)w\overline{w}^2 \\
& + \frac{1}{6}\left(G_{03} + 3G_{02}h_{11} + 3\overline{G}_{20}h_{02} + 3\bar{\lambda}h_{03}\right)\overline{w}^3 + O(|w|^4)
\end{aligned}
$$

and second, by directly substituting z in (5.43) using (5.44),

$$
\begin{aligned}
\dot{z} = & \lambda w + \frac{1}{2}(g_{20} + \lambda h_{20})w^2 + (g_{11} + \lambda h_{11})w\overline{w} + \frac{1}{2}(g_{02} + \lambda h_{02})\overline{w}^2 \\
& + \frac{1}{6}\left(g_{30} + \lambda h_{30} + 3g_{20}h_{20} + 3g_{11}\bar{h}_{02}\right)w^3 \\
& + \frac{1}{2}\left(g_{21} + \lambda h_{21} + 2g_{20}h_{11} + 2g_{11}\bar{h}_{11} + g_{11}h_{20} + g_{02}\bar{h}_{02}\right)w^2\overline{w} \\
& + \frac{1}{2}\left(g_{12} + \lambda h_{12} + g_{20}h_{02} + g_{11}\bar{h}_{20} + 2g_{11}h_{11} + 2g_{02}\bar{h}_{11}\right)w\overline{w}^2 \\
& + \frac{1}{6}\left(g_{03} + \lambda h_{03} + 3g_{02}\bar{h}_{20} + 3g_{11}h_{02}\right)\overline{w}^3 + O(|w|^4).
\end{aligned}
$$

By comparing the coefficients in front of $w^2, w\bar{w}$, and $\bar{w}^2$ in both expressions, we find

$$G_{20} = g_{20} - \lambda h_{20}, \tag{5.46}$$
$$G_{11} = g_{11} - \bar{\lambda} h_{11}, \tag{5.47}$$
$$G_{02} = g_{02} - (2\bar{\lambda} - \lambda) h_{02}, \tag{5.48}$$

while the coefficients in front of $w^3, w^2\bar{w}, w\bar{w}^2$, and $\bar{w}^3$ give

$$\begin{aligned} G_{30} &= g_{30} - 2\lambda h_{30} + 3g_{20}h_{20} + 3g_{11}\bar{h}_{02} \\ &\quad - 3G_{20}h_{20} - 3\bar{G}_{02}h_{11}, \\ G_{21} &= g_{21} - (\lambda + \bar{\lambda})h_{21} + 2g_{20}h_{11} + 2g_{11}\bar{h}_{11} + g_{11}h_{20} + g_{02}\bar{h}_{02} \\ &\quad - 2G_{11}h_{20} - G_{20}h_{11} - \bar{G}_{02}h_{02} - 2\bar{G}_{11}h_{11} \\ G_{12} &= g_{12} - 2\bar{\lambda}h_{12} + g_{20}h_{02} + g_{11}\bar{h}_{20} + 2g_{11}h_{11} + 2g_{02}\bar{h}_{11} \\ &\quad - G_{02}h_{20} - 2G_{11}h_{11} - 2\bar{G}_{11}h_{02} - \bar{G}_{20}h_{11} \\ G_{03} &= g_{03} + (\lambda - 3\bar{\lambda})h_{03} + 3g_{02}\bar{h}_{20} + 3g_{11}h_{02} \\ &\quad - 3G_{02}h_{11} - 3\bar{G}_{20}h_{02}, \end{aligned}$$

where G_{20}, G_{11}, and G_{02} are defined by (5.46), (5.47), and (5.48), respectively. Now we can start to simplify the system (5.45) by properly selecting the coefficients h_{jk} of the substitution (5.44).

First, observe that the factors in front of h_{20}, h_{11}, and h_{02} in Eqs. (5.46), (5.47), and (5.48) do not vanish at $\alpha = 0$: $\lambda(0) = i\omega_0 \neq 0$ and

$$2\bar{\lambda}(0) - \lambda(0) = -3i\omega_0 \neq 0.$$

Thus, we can make the quadratic coefficients G_{20}, G_{11}, and G_{02} equal to zero for all sufficiently small $|\alpha|$ by taking

$$h_{20} = \frac{g_{20}}{\lambda}, \quad h_{11} = \frac{g_{11}}{\bar{\lambda}}, \quad h_{02} = \frac{g_{02}}{2\bar{\lambda} - \lambda}. \tag{5.49}$$

In these expressions, the denominators stay away from zero near $\alpha = 0$ by continuity. Thus, the transformed system (5.45) will contain no quadratic terms, which is a considerable simplification.

With $G_{20} = G_{11} = G_{02} = 0$, we get the following equalities:

$$G_{30} = g_{30} - 2\lambda h_{30} + 3g_{20}h_{20} + 3g_{11}\bar{h}_{02}, \tag{5.50}$$
$$G_{21} = g_{21} - (\lambda + \bar{\lambda})h_{21} + 2g_{20}h_{11} + 2g_{11}\bar{h}_{11} + g_{11}h_{20} + g_{02}\bar{h}_{02}, \tag{5.51}$$
$$G_{12} = g_{12} - 2\bar{\lambda}h_{12} + g_{20}h_{02} + g_{11}\bar{h}_{20} + 2g_{11}h_{11} + 2g_{02}\bar{h}_{11}, \tag{5.52}$$
$$G_{03} = g_{03} + (\lambda - 3\bar{\lambda})h_{03} + 3g_{02}\bar{h}_{20} + 3g_{11}h_{02}, \tag{5.53}$$

where now h_{20}, h_{11}, and h_{02} are defined by (5.49).

Then, a closer look at Eqs. (5.50), (5.52), and (5.53) reveals that we can also ensure $G_{30} = G_{12} = G_{03} = 0$ by an appropriate choice of h_{30}, h_{12}, and h_{03}, since the factors in front of these coefficients are nonzero at $\alpha = 0$ and, thus, for all sufficiently small $|\alpha|$. Indeed, $\lambda(0) = i\omega_0 \neq 0$ and also

$$\lambda(0) - 3\bar{\lambda}(0) = 4i\omega_0 \neq 0.$$

However, it is impossible to make $G_{21} = 0$ at $\alpha = 0$ by whatever choice of h_{21} in Eq. (5.51), since $\lambda(0) + \bar{\lambda}(0) = i\omega_0 - i\omega_0 = 0$. We thus will not try to "kill" G_{21} and merely take $h_{21}(\alpha) = 0$ for all $|\alpha|$ small. Note that any choice of h_{21} does not effect the critical value $G_{21}(0)$. Taking into account (5.49), we find from (5.51)

$$G_{21} = g_{21} + \frac{2g_{20}g_{11}}{\bar{\lambda}} + \frac{2g_{11}\bar{g}_{11}}{\lambda} + \frac{g_{11}g_{20}}{\lambda} + \frac{g_{02}\bar{g}_{02}}{2\lambda - \bar{\lambda}}.$$

This gives the system (5.41)

$$\dot{w} = \lambda(\alpha)w + c_1(\alpha)w^2\bar{w} + O(|w|^4),$$

where

$$c_1(\alpha) = \frac{1}{2}G_{21}(\alpha)$$

so that

$$c_1(\alpha) = \frac{g_{21}(\alpha)}{2} + \frac{(2\lambda(\alpha) + \bar{\lambda}(\alpha))g_{20}(\alpha)g_{11}(\alpha)}{2|\lambda(\alpha)|^2} + \frac{|g_{11}(\alpha)|^2}{\lambda(\alpha)} + \frac{|g_{02}(\alpha)|^2}{2(2\lambda(\alpha) - \bar{\lambda}(\alpha))}.$$

Here the dependence of the coefficients on α is indicated explicitly. For $\alpha = 0$, we have $\lambda(0) = i\omega_0$ and thus obtain the formula (5.42) for $c_1(0)$. □

Remark It is instructive to compare the proof of Lemma 5.15 given above with the corresponding proof in terms of real variables. The quadratic part of Eq. (5.43) is equivalent to a real system

$$\begin{cases} \dot{x}_1 = \mu x_1 - \omega x_2 + \frac{1}{2}a_{20}x_1^2 + a_{11}x_1x_2 + \frac{1}{2}a_{02}x_2^2 + O(\|x\|^3), \\ \dot{x}_2 = \omega x_1 + \mu x_2 + \frac{1}{2}b_{20}x_1^2 + b_{11}x_1x_2 + \frac{1}{2}b_{02}x_2^2 + O(\|x\|^3), \end{cases} \tag{5.54}$$

with $\mu = \mu(\alpha), \omega = \omega(\alpha)$ as above and some smooth real functions $a_{ij} = a_{ij}(\alpha), b_{ij} = b_{ij}(\alpha)$. The quadratic part of the transformation (5.44) becomes

$$\begin{cases} x_1 = \xi_1 + \frac{1}{2}g_{20}\xi_1^2 + g_{11}\xi_1\xi_2 + \frac{1}{2}g_{02}\xi_2^2, \\ x_2 = \xi_2 + \frac{1}{2}h_{20}\xi_1^2 + h_{11}\xi_1\xi_2 + \frac{1}{2}h_{02}\xi_2^2, \end{cases}$$

where g_{ij} and h_{ij} are unknown real functions of α. Rewriting (5.54) in terms of the (ξ_1, ξ_2)-coordinates, one gets the system:

$$\left\{\begin{array}{rcl} \dot{\xi}_1 &=& \mu\xi_1 - \omega\xi_2 + \frac{1}{2}(a_{20} - \mu g_{20} - 2\omega g_{11} - \omega h_{20})\xi_1^2 \\ && + (a_{11} + \omega g_{20} - \mu g_{11} - \omega g_{02} - \omega h_{11})\xi_1\xi_2 \\ && + \frac{1}{2}(a_{02} + 2\omega g_{11} - \mu g_{02} - \omega h_{02})\xi_2^2 \\ && + O(\|\xi\|^3), \\ \dot{\xi}_2 &=& \omega\xi_1 + \mu\xi_2 + \frac{1}{2}(b_{20} + \omega g_{20} - \mu h_{20} - 2\omega h_{11})\xi_1^2 \\ && + (b_{11} + \omega g_{11} + \omega h_{20} - \mu h_{11} - \omega h_{02})\xi_1\xi_2 \\ && + \frac{1}{2}(b_{02} + \omega g_{02} + 2\omega h_{11} - \mu h_{02})\xi_2^2 \\ && + O(\|\xi\|^3). \end{array}\right. \tag{5.55}$$

Thus, the requirement to have no quadratic terms in this system is equivalent to the linear algebraic system:

$$\begin{pmatrix} \mu & 2\omega & 0 & \omega & 0 & 0 \\ -\omega & \mu & \omega & 0 & \omega & 0 \\ 0 & -2\omega & \mu & 0 & 0 & \omega \\ -\omega & 0 & 0 & \mu & 2\omega & 0 \\ 0 & -\omega & 0 & -\omega & \mu & \omega \\ 0 & 0 & -\omega & 0 & -2\omega & \mu \end{pmatrix} \begin{pmatrix} g_{20} \\ g_{11} \\ g_{02} \\ h_{20} \\ h_{11} \\ h_{02} \end{pmatrix} = \begin{pmatrix} a_{20} \\ a_{11} \\ a_{02} \\ b_{20} \\ b_{11} \\ b_{02} \end{pmatrix}. \tag{5.56}$$

The matrix in the left-hand side is nonsingular near Hopf bifurcation, since its determinant is equal to

$$(9\omega^2 + \mu^2)(\omega^2 + \mu^2)^2,$$

which reduces to $9\omega_0^6 > 0$ at $\alpha = 0$. Thus, for any given quadratic coefficients a_{ij}, b_{ij} in (5.54), one can find from (5.56) g_{ij}, h_{ij} with $i, j = 0, 1, 2,\ i + j = 2$, such that system (5.55) will have no quadratic terms for all sufficiently small $|\alpha|$.

We leave it to the diligent reader to treat the cubic terms in (5.43) in the real form. Clearly, the complex notation is simpler. ◇

Definition 5.16 *The real number*

$$l_1(0) = \frac{1}{\omega_0}\text{Re } c_1(0)$$

*is called the **first Lyapunov coefficient**.*

Equation (5.42) implies

$$l_1(0) = \frac{1}{2\omega_0^2}\text{ Re}[ig_{20}(0)g_{11}(0) + \omega_0 g_{21}(0)]. \tag{5.57}$$

Remark The absolute value of $l_1(0)$ depends on the choice of the eigenvector q, but the sign of $l_1(0)$, which determines stability of the bifurcating cycle, is independent of this choice. ◊

Lemma 5.17 *Consider the smooth complex equation*

$$\frac{dw}{dt} = (\mu(\alpha) + i\omega(\alpha))w + c_1(\alpha)w|w|^2 + O(|w|^4),$$

where $\mu(0) = 0$ *and* $\omega(0) = \omega_0 > 0$.

Suppose $\mu'(0) \neq 0$ *and* $l_1(0) \neq 0$. *Then, the equation can be transformed by a parameter-dependent linear coordinate transformation, a time rescaling, and a nonlinear time reparametrization into an equation of the form:*

$$\frac{du}{d\tau} = (\beta + i)u + su|u|^2 + O(|u|^4),$$

where u *is a new complex coordinate,* τ, β *are the new time and parameter, respectively, and* $s = \text{sign } l_1(0) = \pm 1$.

Proof
Step 1: Linear time scaling.

Introduce the new time $\theta = \omega(\alpha)t$. The time direction is preserved since $\omega(\alpha) > 0$ for all sufficiently small $|\alpha|$. Then,

$$\frac{dw}{d\theta} = (\beta + i)w + d_1(\beta)w|w|^2 + O(|w|^4),$$

where

$$\beta = \beta(\alpha) = \frac{\mu(\alpha)}{\omega(\alpha)}, \ d_1(\beta) = \frac{c_1(\alpha(\beta))}{\omega(\alpha(\beta))}.$$

We can consider β as a new parameter because

$$\beta(0) = 0, \ \beta'(0) = \frac{\mu'(0)}{\omega(0)} \neq 0,$$

and therefore the Inverse Function Theorem guarantees local existence and smoothness of α as a function of β. Notice that d_1 is *complex*.

Step 2: Nonlinear time reparametrization.

Change the time parametrization along the orbits by introducing a new time τ, where

$$d\tau = (1 + e_1(\beta)|w|^2)\ d\theta$$

with $e_1(\beta) = \text{Im } d_1(\beta)$. We obtain

$$\frac{dw}{d\tau} = (\beta + i)w + l_1(\beta)w|w|^2 + O(|w|^4),$$

where $l_1(\beta) = \text{Re } d_1(\beta) - \beta e_1(\beta)$ is *real* and

$$l_1(0) = \frac{\text{Re } c_1(0)}{\omega(0)}$$

in accordance with Definition 5.16.

Step 3: Linear coordinate scaling.

Finally, introduce a new complex variable u:

$$w = \frac{u}{\sqrt{|l_1(\beta)|}},$$

which is possible since $l_1(0) \neq 0$. The equation then takes the required form:

$$\frac{du}{d\tau} = (\beta + i)u + \frac{l_1(\beta)}{|l_1(\beta)|}u|u|^2 + O(|u|^4) = (\beta + i)u + su|u|^2 + O(|u|^4),$$

with $s = \text{sign } l_1(0) = \text{sign Re } c_1(0)$. □

Lemmas 5.14, 5.15, and 5.17 comprise the proof of Theorem 5.13. Combining Theorem 5.13 with 5.12, we can analyse Andronov–Hopf bifurcations in generic smooth families of planar autonomous ODEs.

Example 5.18 (Andronov–Hopf bifurcation in Brusselator) Let us consider a hypothetical chemical system called *Brusselator*,[4] which is composed of substances that react according to the following irreversible scheme:

$$\begin{aligned} A &\xrightarrow{k_1} X \\ B + X &\xrightarrow{k_2} Y + D \\ 2X + Y &\xrightarrow{k_3} 3X \\ X &\xrightarrow{k_4} E. \end{aligned}$$

Here capital letters denote reactants, while the constants k_i over the arrows indicate the corresponding reaction rate constants. The substances D and E do not re-enter the reaction, while the concentrations of A and B are assumed to remain constant. Thus, the Law of Mass Action gives the following system of two nonlinear equations for the concentrations $[X]$ and $[Y]$:

[4] Prigogine, I., & Lefever, R. (1968). Symmetry breaking instabilities in dissipative systems. II. *The Journal of Chemical Physics, 48* , 1695–1700.

$$\begin{cases} \dfrac{d[X]}{dt} = k_1[A] - k_2[B][X] - k_4[X] + k_3[X]^2[Y], \\ \dfrac{d[Y]}{dt} = k_2[B][X] - k_3[X]^2[Y]. \end{cases}$$

Linear scaling of concentrations and time yields the *Brusselator equations*,

$$\begin{pmatrix} \dot{x}_1 \\ \dot{x}_2 \end{pmatrix} = \begin{pmatrix} a - (b+1)x_1 + x_1^2 x_2 \\ bx_1 - x_1^2 x_2 \end{pmatrix} =: f(x, a, b), \tag{5.58}$$

where a and b are positive parameters.

The system (5.58) has the equilibrium point

$$\left(a, \frac{b}{a}\right).$$

We have

$$\mu = \frac{1}{2} \text{Tr } J = \frac{b - (1 + a^2)}{2},$$

where $J = D_x f$ is the Jacobian matrix of (5.58) evaluated at the equilibrium. Fix a and consider b as a control parameter. Hence, $\mu = 0$ at $b_0 = 1 + a^2$. Moreover, when $b = b_0$, we have

$$\det J = a^2 > 0, \quad \frac{d\mu}{db} = \frac{1}{2} \neq 0.$$

Therefore, at $b_0 = 1 + a^2$ system (5.58) has the equilibrium

$$x_0 = \left(a, \frac{1 + a^2}{a}\right)$$

with purely imaginary eigenvalues $\lambda_{1,2} = \pm i\omega_0$, $\omega_0 = a > 0$. The continuation of these eigenvalues with respect to b crosses the imaginary axis with nonzero velocity at $b = b_0$.

The vectors

$$q \sim \left(-\frac{ia + a^2}{1 + a^2}, 1\right), \quad p \sim \left(\frac{-ia + a^2}{a^2}, 1\right)$$

are the eigenvectors of the Jacobian matrix and its transpose at $b = b_0$,

$$Jq = i\omega_0 q, \quad J^{\mathrm{T}} p = -i\omega_0 p.$$

To achieve the normalization $\langle p, q \rangle = 1$ we take

$$q = \left(-\frac{ia + a^2}{1 + a^2}, \ 1 \right), \quad p = \left(-\frac{i(1 + a^2)}{2a}, \ \frac{1 - ia}{2} \right).$$

Now compose $x = x_0 + zq + \bar{z}\bar{q}$ and compute the function

$$H(z, \bar{z}) = \langle p, f(x_0 + zq + \bar{z}\bar{q}, a, 1 + a^2) \rangle.$$

The first few coefficients of its Taylor expansion at $(z, \bar{z}) = (0, 0)$,

$$H(z, \bar{z}) = i\omega_0 z + \sum_{2 \le j+k \le 3} \frac{1}{j!k!} g_{jk} z^j \bar{z}^k + O(|z|^4),$$

are

$$g_{20} = a - i, \quad g_{11} = \frac{(a - i)(a^2 - 1)}{1 + a^2}, \quad g_{21} = -\frac{a(3a - i)}{1 + a^2}.$$

Therefore, the first Lyapunov coefficient is given by

$$l_1 = \frac{1}{2\omega_0^2} \mathrm{Re}(i g_{20} g_{11} + \omega_0 g_{2,1}) = -\frac{2 + a^2}{2a(1 + a^2)} < 0.$$

Thus, a unique and stable limit cycle bifurcates in (5.58) from the equilibrium for $b > b_0$: Andronov–Hopf bifurcation is supercritical (see Fig. 5.15). ◊

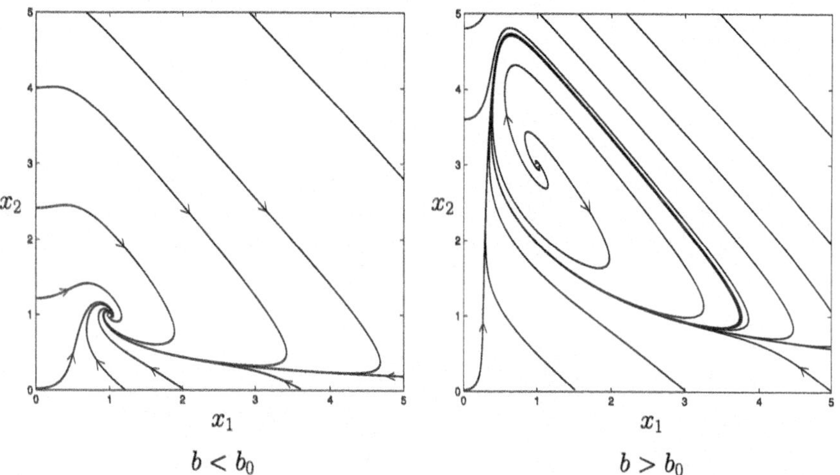

Fig. 5.15 Supercritical Andronov–Hopf bifurcation in Brusselator at $b_0 = 1 + a^2$

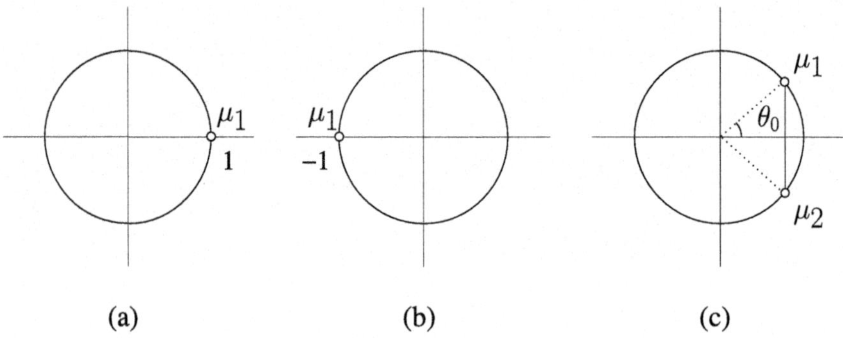

Fig. 5.16 Critical cases for fixed points

5.3 One-Parameter Bifurcations of Fixed Points

Consider a generic smooth one-parameter family of maps

$$x \mapsto f(x, \alpha) =: f_\alpha(x), \quad x \in \mathbb{R}^n, \alpha \in \mathbb{R}.$$

As for equilibria in ODEs, we expect that local bifurcations happen to *non-hyperbolic* fixed points. Thus, we should consider three *critical cases* (or *singularities*, see Fig. 5.16):

(a) the fixed point x_0 has eigenvalue $\mu_1 = 1$;
(b) the fixed point x_0 has eigenvalue $\mu_1 = -1$;
(c) the fixed point x_0 has a pair of complex-conjugate eigenvalues $\mu_{1,2} = e^{\pm i\theta_0}$ with $0 < \theta_0 < \pi$.

Moreover, we can assume that in each case no other eigenvalue satisfies $|\mu| = 1$ and the critical eigenvalues of x_0 are algebraically simple. The codim 1 bifurcation associated to the critical case (a) is called a *fold* (or *limit point*) bifurcation of maps. The codim 1 bifurcation occurring at the critical case (b) is called a *flip* (or *period-doubling*) bifurcation, while the codim 1 bifurcation related to case (c) is referred to as a *Neimark–Sacker* bifurcation. Since local bifurcations of period-k cycles can be treated as bifurcations of fixed points of f_α^k, we only consider local bifurcations of fixed points here.

First we study these bifurcations in the lowest possible phase-space dimension, i.e. $n = 1$ for the fold and flip and $n = 2$ for the Neimark–Sacker bifurcation.

5.3.1 Fold Bifurcation of Scalar Maps

Example 5.19 (Normal form of the fold bifurcation of maps) Consider the one-dimensional dynamical system generated by the following map

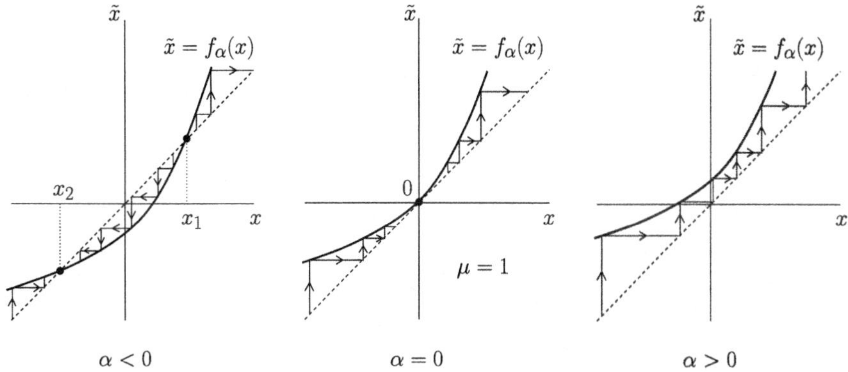

Fig. 5.17 Fold bifurcation

depending on one parameter:

$$x \mapsto f_\alpha(x) = \alpha + x + x^2. \tag{5.59}$$

Let $f(x, \alpha) = \alpha + x + x^2$. The map f_α is invertible in a neighbourhood of the origin for $|\alpha|$ small. The system has at $\alpha = 0$ a nonhyperbolic fixed point $x_0 = 0$ with $\mu = \frac{\partial f}{\partial x}(0, 0) = 1$. The behaviour of the system near $x = 0$ for small $|\alpha|$ is shown in Fig. 5.17. For $\alpha < 0$ there are two fixed points in the system: $x_{1,2}(\alpha) = \pm\sqrt{-\alpha}$, the left of which is stable, while the right one is unstable. For $\alpha > 0$ there are no fixed points in the system. While α crosses zero from negative to positive values, the two fixed points (stable and unstable) "collide", forming at $\alpha = 0$ a fixed point with $\mu = 1$, and disappear. This is a fold bifurcation in the discrete-time dynamical system. ◊

Theorem 5.20 *The smooth map*

$$y \mapsto \alpha + y + y^2 + O(y^3),$$

where the $O(|y|^3)$-term can smoothly depend on α, is locally topologically equivalent near the origin to the map

$$x \mapsto \alpha + x + x^2.$$

Proof Write the first map as

$$y \mapsto g_\alpha(y) = \alpha + y + y^2 + y^3\varphi(y, \alpha), \tag{5.60}$$

where φ is a smooth function of (y, α). The number and stability of the fixed points of (5.60) in a neighbourhood of $y = 0$ for small $|\alpha|$ is the same as in

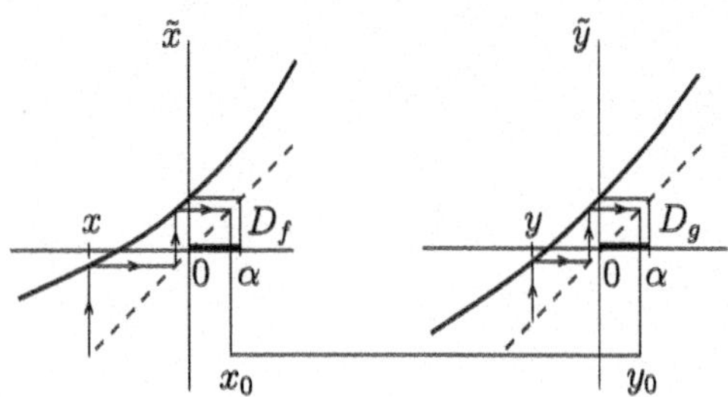

Fig. 5.18 Conjugating homeomorphism for $\alpha > 0$

the normal form (5.59). This can be proved exactly as in the continuous-time fold case (Theorem 5.6).

The construction of a conjugating homeomorphism is more involved, since it has to map not only fixed points of (5.60) onto fixed points of (5.59) but map all orbits onto orbits. Note that both f_α and g_α are invertible near the origin. We consider separately the cases $\alpha > 0$, $\alpha = 0$, and $\alpha < 0$.

Case $\alpha > 0$. Define a fundamental domain for f_α:

$$D_f = [0, f_\alpha(0)] = [0, \alpha],$$

and the corresponding domain for g_α:

$$D_g = [0, g_\alpha(0)] = [0, \alpha].$$

(See Fig. 5.18).

Let the homeomorphism $h : \mathbb{R} \to \mathbb{R}$ under construction be $y = x$ for $x \in D_f$. This identifies D_g and D_f.

Any point $x \notin D_f$ from a small neighbourhood of $x = 0$ falls after a finite number of iterations (forward or backward) with f_α into D_f, provided $\alpha > 0$ is small. Let $K = K(x, \alpha) \neq 0$ be the integer with the minimal $|K|$ such that

$$x_0 = f_\alpha^K(x) \in D_f.$$

Take $y_0 = x_0$ and then set

$$y = g_\alpha^{-K}(y_0).$$

The map $x \mapsto y = h_\alpha(x)$ is a local homeomorphism defined for small $\alpha > 0$. Moreover, it maps orbits of f_α onto orbits of g_α in a non-shrinking neighbourhood of the origin for all small $\alpha > 0$.

Case $\alpha = 0$. In this case we fix a small $\varepsilon > 0$ and introduce two fundamental domains for f_0:

$$D_f^{\pm} = [\pm\varepsilon, f_0(\pm\varepsilon)]$$

and two similar domains for g_0:

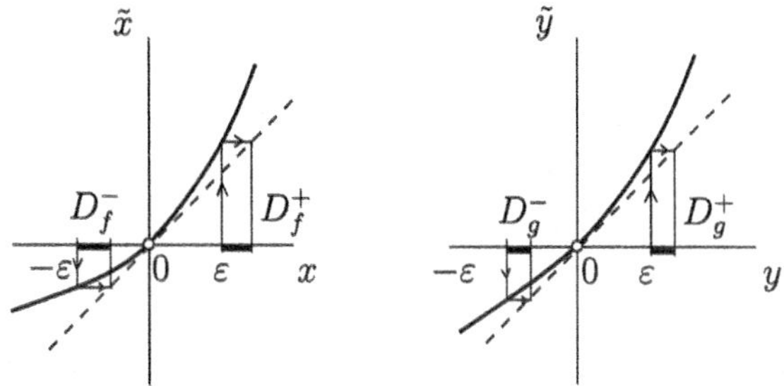

Fig. 5.19 Fundamental domains for $\alpha = 0$

$$D_g^{\pm} = [\pm\varepsilon, g_0(\pm\varepsilon)].$$

(See Fig. 5.19). Next we identify these domains pairwise by affine (orientation-preserving) maps defined for small $x, y > 0$ and $x, y < 0$, respectively. This gives a homeomorphism $h : D_f^{\pm} \to D_g^{\pm}$.

For $x \notin D_f^{\pm}$ with small $|x| \neq 0$, let $K^{\pm} = K^{\pm}(x, \alpha)$ be an integer with the minimal $|K|$ such that

$$x_0 = f_0^{K^{\pm}}(x) \in D_f^{\pm}.$$

Set $y_0 = h(x_0)$, where h is the homeomorphism mentioned above. Then take

$$y = g_0^{-K^{\pm}}(y_0)$$

and define $h_\alpha(x) = y$ for $x \neq 0$. Set $h_\alpha(0) = 0$ by continuity. The map h_α is defined in a small neighbourhood of $x = 0$ and maps orbits of f_0 onto orbits of g_0 in a small neighbourhood of $y = 0$.

Case $\alpha < 0$. Denote by $x_{1,2}(\alpha) = \mp\sqrt{-\alpha}$ the fixed points of f_α and by $y_{1,2}(\alpha)$ the corresponding fixed points of g_α. Similar to the case $\alpha > 0$, define and identify the fundamental domains

$$D_f = [f_\alpha(0), 0] = [\alpha, 0]$$

and

$$D_g = [g_\alpha(0), 0] = [\alpha, 0]$$

(see Fig. 5.20). For any $x \notin D_f$ and such that $x \in (x_1(\alpha), x_2(\alpha))$, define an $y \in (y_1(\alpha), y_2(\alpha))$ as in the case $\alpha > 0$. Extend the map thus obtained by continuity to a map $h : [x_1(\alpha), x_2(\alpha)] \to [y_1(\alpha), y_2(\alpha)]$.

Finally, let

$$D_f^{\mp} = [x_{1,2}(\alpha) \mp \varepsilon, f_\alpha(x_{1,2}(\alpha) \mp \varepsilon)],$$

and

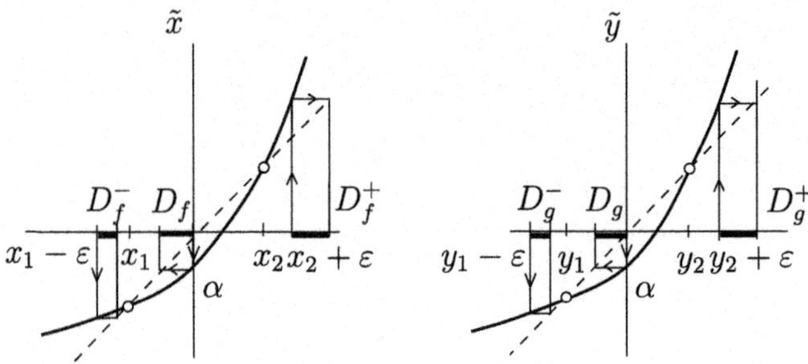

Fig. 5.20 Fundamental domains for $\alpha < 0$

$$D_g^{\mp} = [y_{1,2}(\alpha) \mp \varepsilon, g_\alpha(y_{1,2}(\alpha) \mp \varepsilon)],$$

where $\varepsilon > 0$ is sufficiently small (see Fig. 5.20). Take for the map $h : D_f^{\mp} \to D_g^{\mp}$ an affine (orientation-preserving) transformation of $D_f^{\mp}$ onto $D_g^{\mp}$. Then, repeat the standard procedure to complete the construction of a local homeomorphism. □

Theorem 5.21 *Suppose that a one-dimensional map*

$$x \mapsto f(x, \alpha), \quad x \in \mathbb{R}, \ \alpha \in \mathbb{R}, \tag{5.61}$$

where f is smooth, has at $\alpha = 0$ the fixed point $x_0 = 0$, and let

$$\mu = \frac{\partial f}{\partial x}(0, 0) = 1.$$

Assume that the following generic conditions are satisfied:

(A.1) $\frac{\partial^2 f}{\partial x^2}(0, 0) \neq 0$;
(A.2) $\frac{\partial f}{\partial \alpha}(0, 0) \neq 0$.

Then there are smooth invertible coordinate and parameter changes transforming the map into

$$y \mapsto \beta + y \pm y^2 + O(y^3),$$

where the sign in front of y^2 is given by the sign of the expression in (A.1).

Proof Expand $f(x, \alpha)$ in a Taylor series with respect to x at $x = 0$:

$$f(x, \alpha) = f_0(\alpha) + f_1(\alpha)x + f_2(\alpha)x^2 + O(x^3).$$

Two conditions are satisfied: $f_0(0) = f(0, 0) = 0$ (*fixed-point condition*) and $f_1(0) = \frac{\partial f}{\partial x}(0, 0) = 1$ (*fold bifurcation condition*). Since $f_1(0) = 1$, we may write

$$f(x,\alpha) = f_0(\alpha) + [1+g(\alpha)]x + f_2(\alpha)x^2 + O(x^3),$$

where g is smooth and $g(0) = 0$.

Perform a coordinate shift by introducing a new variable ξ:

$$x = \xi + \delta, \tag{5.62}$$

where $\delta = \delta(\alpha)$ is to be chosen suitably later on. The transformation (5.62) yields

$$\tilde{\xi} = \tilde{x} - \delta = f(x,\alpha) - \delta = f(\xi+\delta,\alpha) - \delta.$$

Therefore,

$$\begin{aligned}\tilde{\xi} &= [f_0(\alpha) + g(\alpha)\delta + f_2(\alpha)\delta^2 + O(\delta^3)] \\ &+ \ \xi + [g(\alpha) + 2f_2(\alpha)\delta + O(\delta^2)]\xi \\ &+ \ [f_2(\alpha) + O(\delta)]\xi^2 + O(\xi^3).\end{aligned}$$

According to (A.1),

$$f_2(0) = \frac{1}{2}\frac{\partial^2 f}{\partial x^2}(0,0) \neq 0.$$

Hence there is a smooth function $\delta(\alpha)$, which annihilates the parameter-dependent linear term in the above map for all sufficiently small $|\alpha|$. Indeed, the condition for that term to vanish can be written as

$$F(\alpha,\delta) = g(\alpha) + 2f_2(\alpha)\delta + \delta^2\varphi(\alpha,\delta) = 0$$

for some smooth function φ. We have

$$F(0,0) = 0, \quad \frac{\partial F}{\partial \delta}(0,0) = 2f_2(0) \neq 0,$$

which implies the (local) existence and uniqueness of a smooth function $\delta = \delta(\alpha)$ such that $\delta(0) = 0$ and $F(\alpha,\delta(\alpha)) \equiv 0$ (use the Implicit Function Theorem). It follows that

$$\delta(\alpha) = -\frac{g'(0)}{2f_2(0)}\alpha + O(\alpha^2).$$

The map written in terms of ξ is now given by

$$\tilde{\xi} \ = \ [f_0'(0)\alpha + \alpha^2\psi(\alpha)] \ + \ \xi \ + \ [f_2(0) + O(\alpha)]\,\xi^2 \ + \ O(\xi^3), \tag{5.63}$$

where ψ is some smooth function.

Consider as a new parameter $\mu = \mu(\alpha)$ the constant (ξ-independent) term of (5.63):

$$\mu = f_0'(0)\alpha + \alpha^2\psi(\alpha).$$

We have

(a) $\mu(0) = 0$;

(b) $\mu'(0) = f_0'(0)$.

Since

$$f_0'(0) = \frac{\partial f}{\partial \alpha}(0,0) \neq 0$$

due to (A.2), the Inverse Function Theorem implies the local existence and uniqueness of a smooth inverse function $\alpha = \alpha(\mu)$ with $\alpha(0) = 0$. Therefore, Eq. (5.63) now reads

$$\tilde{\xi} = \mu + \xi + a(\mu)\xi^2 + O(\xi^3),$$

where $a(\mu)$ is a smooth function with $a(0) = f_2(0) \neq 0$ due to the first assumption (A.1).

Let $y = |a(\mu)|\xi$ and $\beta = |a(\mu)|\mu$. Then we get

$$\tilde{y} = \beta + y + sy^2 + O(y^3),$$

where $s = \text{sign } a(0) = \pm 1$. □

Example 5.22 (Fold bifurcation in a simple population model) Consider the following simple population model

$$x_{k+1} = \frac{\alpha x_k^2}{1 + x_k^2},$$

where x_k is the population density in year k, and $\alpha > 0$ is a parameter determining both the growth rate for small population size and the maximum population level. This recurrence relation corresponds to iterating the map

$$x \mapsto f(x, \alpha) = \frac{\alpha x^2}{1 + x^2} \tag{5.64}$$

which has a trivial fixed point $x = 0$ for all values of the parameter α. At $\alpha_0 = 2$, however, two nontrivial positive fixed points

$$x_{1,2}(\alpha) = \frac{\alpha \pm \sqrt{\alpha^2 - 4}}{2}$$

appear at $x = 1$ via a fold bifurcation. We leave it to the reader to check the genericity conditions (A.1) and (A.2) of Theorem 5.21. ◊

5.3.2 *Flip Bifurcation of Scalar Maps*

Example 5.23 (Normal form of the flip bifurcation) Consider the one-parameter family of scalar dynamical systems generated by the following map:

$$x \mapsto f_\alpha(x) = -(1+\alpha)x + x^3. \tag{5.65}$$

For small $|\alpha|$, the map f_α is invertible in a neighbourhood of the origin. System (5.65) has for all α the fixed point $x_0 = 0$ with multiplier $\mu = -(1+\alpha)$. The fixed point is linearly stable for small $\alpha < 0$ and is linearly unstable for $\alpha > 0$. At $\alpha = 0$ the point is not hyperbolic, since the multiplier $\mu = \frac{df_0}{dx}(0) = -1$, but is nevertheless (nonlinearly) stable. There are no other fixed points near the origin for $|\alpha|$.

Consider now the *second iterate* $f_\alpha^2(x)$ of the map (5.65):

$$f_\alpha^2(x) = -(1+\alpha)[-(1+\alpha)x + x^3] + [-(1+\alpha)x + x^3]^3.$$

The map f_α^2 obviously has the trivial fixed point $x_0 = 0$ for all values of α. It also has *two* nontrivial fixed points for small $\alpha > 0$:

$$x_{1,2} = f_\alpha^2(x_{1,2}),$$

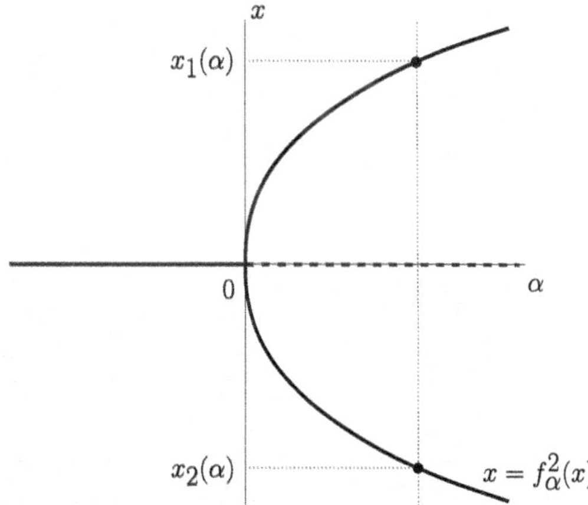

Fig. 5.21 Fixed points of f_α^2 near a flip bifurcation

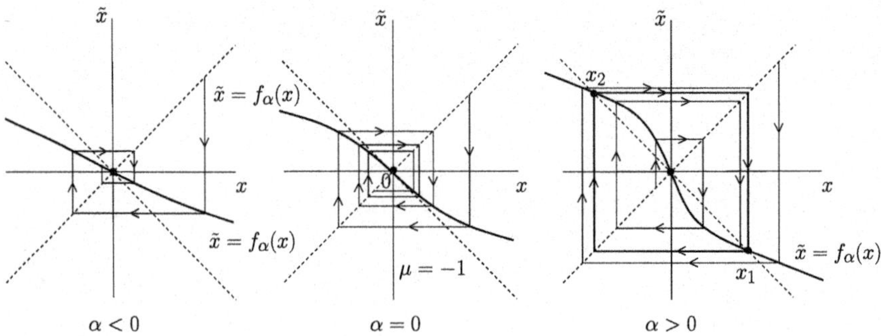

Fig. 5.22 Supercritical flip bifurcation

where $x_{1,2} = \pm\sqrt{\alpha}$ (see Fig. 5.21). These two points are stable and constitute a *cycle of period two* for the original map f_α, since

$$x_2 = f_\alpha(x_1), \;\; x_1 = f_\alpha(x_2),$$

with $x_2 \neq x_1$.

Figure 5.22 shows the complete bifurcation diagram of the map (5.65) with the help of a staircase diagram. As α approaches zero from above, the period-two cycle "shrinks" and disappears. This is a *supercritical flip* (or *period-doubling*) *bifurcation.*

The map $x \mapsto -(1+\alpha)x - x^3$ exhibits a *subcritical flip bifurcation*, as one can easily verify by a similar analysis. ◊

Theorem 5.24 *The smooth map*

$$y \mapsto -(1+\alpha)y + y^3 + O(y^4),$$

where the $O(y^4)$*-term can smoothly depend on* α*, is locally topologically equivalent near the origin to the map*

$$x \mapsto -(1+\alpha)x + x^3.$$

Proof
Step 1: Fixed point analysis.

Write the first map as

$$y \mapsto g_\alpha(y) = -(1+\alpha)y + y^3 + y^4\psi(y,\alpha), \tag{5.66}$$

where ψ is a smooth function. Its second iterate has the form

$$g_\alpha^2(y) = y + (2\alpha + \alpha^2)y - (1+\alpha)(2 + 2\alpha + \alpha^2)y^3 + y^4\Psi(y,\alpha)$$

for some smooth function Ψ. It has a trivial fixed point $y_0 = 0$, as well as two nontrivial fixed points satisfying the equation

$$\Phi(\alpha, y) = 2\alpha + \alpha^2 - (1+\alpha)(2+2\alpha+\alpha^2)y^2 + y^3\Psi(y,\alpha) = 0. \qquad (5.67)$$

Since $\Phi(0,0) = 0$ and $\frac{\partial \Phi}{\partial \alpha}(0,0) = 2 \neq 0$, the Implicit Function Theorem gives the existence of a smooth local solution to (5.67):

$$\alpha = \Lambda(y), \quad \Lambda(0) = 0.$$

Moreover, $\Lambda(y) = y^2 + O(y^3)$, and therefore g_α^2 has two fixed points

$$y_{1,2}(\alpha) = \pm\sqrt{\alpha} + O(\alpha) = x_{1,2}(\alpha) + O(\alpha),$$

for small $\alpha > 0$, corresponding to a period-2 cycle of g_α. Thus, the number and (as is easy to check) stability of the fixed points and period-2 cycles is the same for the map (5.66) and the normal form (5.65).

Step 2: Construction of the homeomorphism.

We use the same method of fundamental domains as in the fold case. Namely, for any sufficiently small $|\alpha|$, we construct two domains D_f and D_g, such that, apart from at most finitely many exceptions, any point x in a neighbourhood of the origin $x = 0$ is mapped by f_α^K into D_f, where $K = K(x,\alpha)$ is the unique integer with the minimal $|K|$ having this property. Both D_f and D_g will be either closed intervals or the union of two closed intervals. Then we identify D_f and D_g with a (piecewise-affine) homeomorphism $h : D_f \to D_g$, and set as before

$$y = g_\alpha^{-K}(h(f_\alpha^K(x)).$$

This gives a (parameter-dependent) map $x \mapsto y = h_\alpha(x)$ that can be extended by continuity to the exceptional points. The extended map is a local homeomorphism mapping orbits of f_α onto orbits of g_α.

For $\alpha \leq 0$ with small $|\alpha|$, the domains (see Fig. 5.23 for $\alpha < 0$; case $\alpha = 0$ is completely analogous)

$$D_f = [f_\alpha^2(\varepsilon), \varepsilon] \text{ and } D_g = [g_\alpha^2(\varepsilon), \varepsilon]$$

have the required properties for sufficiently small $\varepsilon > 0$ (the exceptional point is $x = 0$).

Now consider the case $\alpha > 0$. When $x \in (x_2(\alpha), x_1(\alpha))$, we shall work with D_f^- and D_g^- defined by

$$D_f^- = [\alpha, f_\alpha^2(\alpha)] \text{ and } D_g^- = [\alpha, g_\alpha^2(\alpha)]$$

(see Fig. 5.24).

When $x \notin [x_1(\alpha), x_2(\alpha)]$ we shall work with D_f^+ and D_g^+ defined by

$$D_f^+ = [f_\alpha^2(x_1(\alpha)+\varepsilon), x_1(\alpha)+\varepsilon], \quad D_g^+ = [g_\alpha^2(y_1(\alpha)+\varepsilon), y_1(\alpha)+\varepsilon]$$

for small $\varepsilon > 0$ (see Fig. 5.25). We define $D_f = D_f^- \cup D_f^+$ and $D_g = D_g^- \cup D_g^+$. The exceptional points are the fixed point $x = 0$ and the period-two points $x_1(\alpha)$ and $x_2(\alpha)$. □

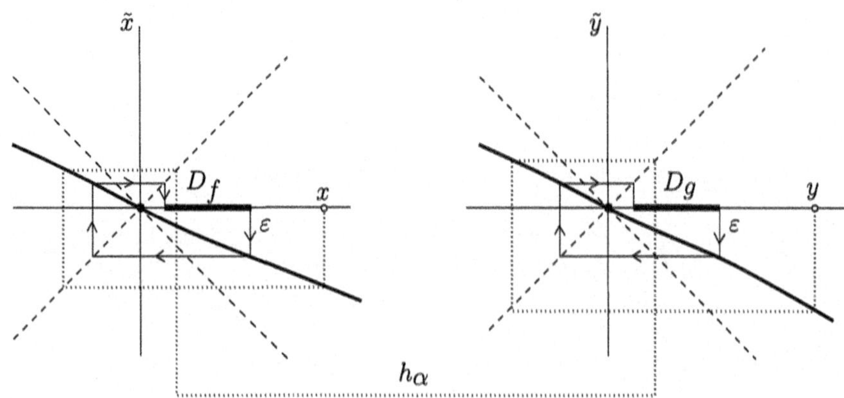

Fig. 5.23 Fundamental domains D_f and D_g for $\alpha < 0$

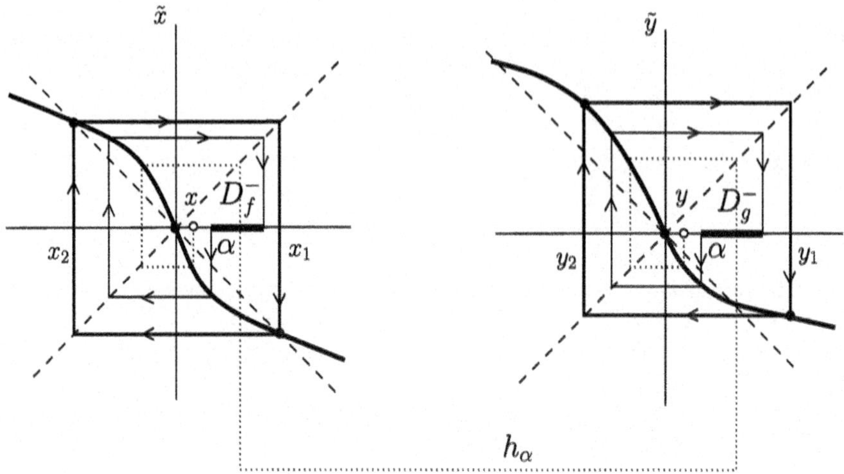

Fig. 5.24 Fundamental domains D_f^- and D_g^- for $\alpha > 0$

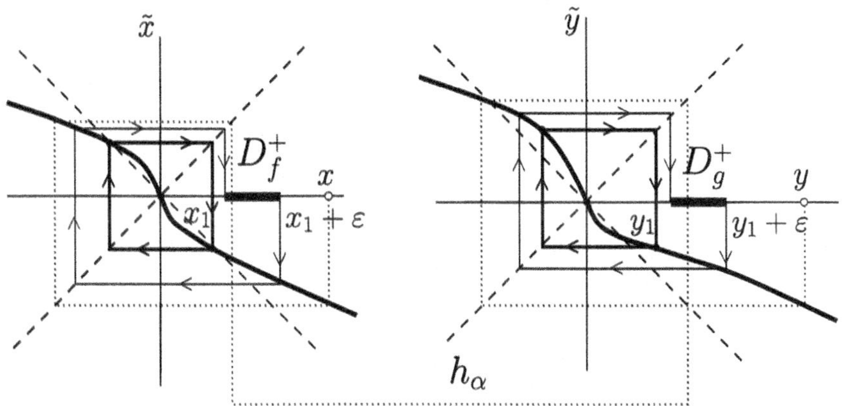

Fig. 5.25 Fundamental domains D_f^+ and D_g^+ for $\alpha > 0$

Theorem 5.25 *Suppose that a one-dimensional smooth map*

$$x \mapsto f(x, \alpha), \quad x \in \mathbb{R}, \; \alpha \in \mathbb{R},$$

has at $\alpha = 0$ *the fixed point* $x_0 = 0$. *Assume that* $\mu = \frac{\partial f}{\partial x}(0,0) = -1$. *Let the following genericity conditions be satisfied:*

(B.1) $\frac{1}{2}\left[\frac{\partial^2 f}{\partial x^2}(0,0)\right]^2 + \frac{1}{3}\frac{\partial^3 f}{\partial x^3}(0,0) \neq 0$;
(B.2) $\frac{\partial^2 f}{\partial x \partial \alpha}(0,0) + \frac{1}{2}\frac{\partial f}{\partial \alpha}(0,0)\frac{\partial^2 f}{\partial x^2}(0,0) \neq 0$.

Then there are smooth invertible coordinate and parameter changes transforming the map into

$$y \mapsto -(1+\beta)y \pm y^3 + O(y^4),$$

where the sign in front of y^3 *is given by the sign of the expression in* (B.1).

Proof Since $\frac{\partial f}{\partial x}(0,0) \neq 1$, the Implicit Function Theorem guarantees that the map has a unique fixed point $x_0(\alpha)$ in some neighbourhood of the origin for all sufficiently small $|\alpha|$. Moreover,

$$x_0(0) = 0, \quad x_0'(0) = \frac{\frac{\partial f}{\partial \alpha}(0,0)}{1 - \frac{\partial f}{\partial x}(0,0)} = \frac{1}{2}\frac{\partial f}{\partial \alpha}(0,0).$$

We can perform a parameter-dependent coordinate shift,

$$x = x_0(\alpha) + \xi,$$

placing the fixed point at $\xi = 0$ for $|\alpha|$ sufficiently small.

The map can then be expressed as

$$\xi \mapsto \tilde{\xi} = g(\xi, \alpha),$$

where $g(\xi, \alpha) := f(x_0(\alpha) + \xi, \alpha) - x_0(\alpha)$ with $g(0, \alpha) \equiv 0$, so that

$$g(\xi, \alpha) = g_1(\alpha)\xi + g_2(\alpha)\xi^2 + g_3(\alpha)\xi^3 + O(\xi^4), \tag{5.68}$$

with smooth functions $g_k, k = 1, 2, 3$ given by

$$g_1(\alpha) = \frac{\partial f}{\partial x}(x_0(\alpha), \alpha), \quad g_2(\alpha) = \frac{1}{2}\frac{\partial^2 f}{\partial x^2}(x_0(\alpha), \alpha), \quad g_3(\alpha) = \frac{1}{6}\frac{\partial^3 f}{\partial x^3}(x_0(\alpha), \alpha).$$

The function g_1 can be represented as $g_1(\alpha) = -[1 + r(\alpha)]$ for some smooth function r. Since $r(0) = 0$ and

$$\begin{aligned} r'(0) &= -\frac{\partial^2 f}{\partial x^2}(0,0)x_0'(0) - \frac{\partial^2 f}{\partial x \partial \alpha}(0,0) \\ &= -\left[\frac{\partial^2 f}{\partial x \partial \alpha}(0,0) + \frac{1}{2}\frac{\partial f}{\partial \alpha}(0,0)\frac{\partial^2 f}{\partial x^2}(0,0)\right] \neq 0 \end{aligned}$$

according to Assumption (B.2), the function g is locally invertible and can be used to introduce a new parameter:

$$\beta = r(\alpha).$$

Our map (5.68) now takes the form

$$\tilde{\xi} = \mu(\beta)\xi + a(\beta)\xi^2 + b(\beta)\xi^3 + O(\xi^4), \tag{5.69}$$

where $\mu(\beta) = -(1 + \beta)$, and the functions $a(\beta)$ and $b(\beta)$ are smooth. We have

$$a(0) = g_2(0) = \frac{1}{2}\frac{\partial^2 f}{\partial x^2}(0,0), \quad b(0) = g_3(0) = \frac{1}{6}\frac{\partial^3 f}{\partial x^3}(0,0).$$

Let us perform a polynomial change of coordinate:

$$\xi = \eta + \delta\eta^2, \tag{5.70}$$

where $\delta = \delta(\beta)$ is a smooth function to be defined. The transformation (5.70) is close to the identity in some neighbourhood of the origin. Written in the new coordinate η, the map (5.69) takes the form:

$$\tilde{\eta} = \mu(\beta)\eta + d(\beta)\eta^2 + c(\beta)\eta^3 + O(\eta^4), \tag{5.71}$$

where d and c are smooth functions of β. Equations (5.70) and (5.71) imply

$$\begin{aligned}\tilde{\xi} = \tilde{\eta} + \delta\tilde{\eta}^2 &= \mu\eta + d\eta^2 + c\eta^3 + \delta(\mu\eta + d\eta^2 + c\eta^3)^2 + O(\eta^4) \\ &= \mu\eta + (d + \delta\mu^2)\eta^2 + (c + 2\delta\mu d)\eta^3 + O(\eta^4).\end{aligned}$$

Substituting (5.70) into (5.69), we get

$$\begin{aligned}\tilde{\xi} &= \mu\xi + a\xi^2 + b\xi^3 + O(\xi^4) \\ &= \mu(\eta + \delta\eta^2) + a(\eta + \delta\eta^2)^2 + b(\eta + \delta\eta^2)^3 + O(\eta^4) \\ &= \mu\eta + (a + \delta\mu)\eta^2 + (b + 2a\delta)\eta^3 + O(\eta^4).\end{aligned}$$

Therefore

$$d(\beta) = a(\beta) + \delta(\beta)\mu(\beta) - \delta(\beta)\mu^2(\beta), \quad c(\beta) = b(\beta) + 2a(\beta)\delta(\beta) - 2\delta(\beta)\mu(\beta)d(\beta).$$

Thus, the quadratic term in (5.71) can be "killed" for all sufficiently small $|\beta|$ by setting

$$\delta(\beta) = \frac{a(\beta)}{\mu^2(\beta) - \mu(\beta)},$$

since $\mu^2(0) - \mu(0) = 2 \neq 0$. With $\delta(\beta)$ thus selected, $d(\beta) \equiv 0$ and the cubic coefficient in (5.71) becomes

$$c(\beta) = b(\beta) + 2a(\beta)\delta(\beta) = b(\beta) + \frac{2a^2(\beta)}{\mu^2(\beta) - \mu(\beta)},$$

so that the map (5.71) takes the form:

$$\tilde{\eta} = -(1 + \beta)\eta + c(\beta)\eta^3 + O(\eta^4).$$

Here the function $c = c(\beta)$ is smooth and

$$c(0) = a^2(0) + b(0) = \frac{1}{4}\left[\frac{\partial^2 f}{\partial x^2}(0,0)\right]^2 + \frac{1}{6}\frac{\partial^3 f}{\partial x^3}(0,0). \tag{5.72}$$

Notice that $c(0) \neq 0$ by Assumption (B.1).

Apply the rescaling

$$\eta = \frac{y}{\sqrt{|c(\beta)|}}.$$

In the y-coordinate the map takes the desired form:

$$\tilde{y} = -(1 + \beta)y + sy^3 + O(y^4),$$

where $s = \text{sign } c(0) = \pm 1$. □

Example 5.26 (Flip bifurcation in Ricker's map) Consider the following simple population model due to the fishery biologist Ricker[5]:

$$x_{k+1} = \alpha x_k e^{-x_k}.$$

Here x_k is the population density in year k, and $\alpha > 0$ is the per capita growth rate at very low density. The factor e^{-x} on the right-hand side takes into account the negative effect of competition at high population densities (originally interpreted by Ricker in terms of cannibalism). The above recurrence relation corresponds to the map

$$x \mapsto f(x, \alpha) = \alpha x e^{-x}. \tag{5.73}$$

This map has the trivial fixed point $x_0 = 0$ for all values of the parameter α. At $\alpha_0 = 1$, however, a nontrivial positive fixed point

$$x_1(\alpha) = \ln \alpha$$

appears via a transcritical bifurcation. The eigenvalue of this point is given by the expression

$$\mu(\alpha) = 1 - \ln \alpha.$$

Thus, x_1 is stable for $1 < \alpha < \alpha_1$ and unstable for $\alpha > \alpha_1$, where $\alpha_1 = e^2 = 7.38907\ldots$. At the critical parameter value $\alpha = \alpha_1$, the fixed point has multiplier $\mu(\alpha_1) = -1$. Therefore, a flip bifurcation takes place. To apply Theorem 5.25, we need to verify the genericity conditions (B.1) and (B.2) in which all the derivatives must be computed at the fixed point $x_1(\alpha_1) = 2$ and at the critical parameter value $\alpha_1 = e^2$.

One can check that

$$\frac{1}{4}\left[\frac{\partial^2 f}{\partial x^2}(2, e^2)\right]^2 + \frac{1}{6}\frac{\partial^3 f}{\partial x^3}(2, e^2) = \frac{1}{6} > 0$$

and

$$\frac{\partial^2 f}{\partial x \partial \alpha}(2, e^2) + \frac{1}{2}\frac{\partial f}{\partial \alpha}(2, e^2)\frac{\partial^2 f}{\partial x^2}(2, e^2) = -\frac{1}{e^2} < 0.$$

Thus, Theorem 5.24 implies that a unique and stable period-two cycle bifurcates from x_1 for $\alpha > \alpha_1$.

The fate of this period-two cycle can be traced further. It can be verified numerically that this cycle loses stability at $\alpha_2 = 12.50925\ldots$ via a supercrit-

[5] Ricker, W. E. (1954). Stock and recruitment. *Journal of the Fisheries Research Board of Canada, 11*, 559–623.

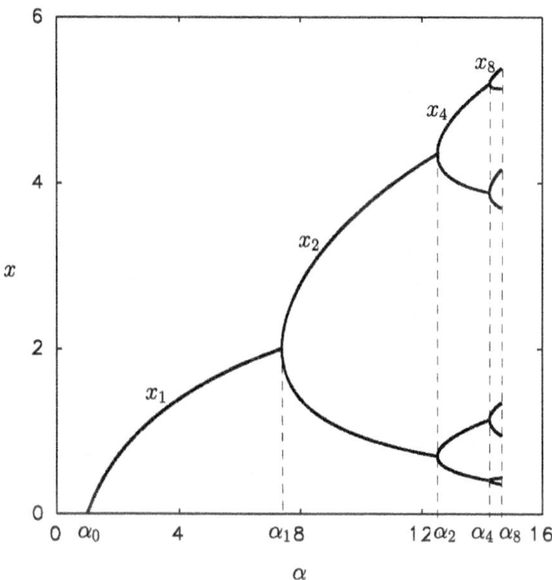

Fig. 5.26 Period-doubling (flip) bifurcations in Ricker's map

ical flip bifurcation, giving rise to a stable period-four cycle. This period-four cycle bifurcates again at $\alpha_4 = 14.24425\ldots$, generating a stable period-eight cycle that loses its stability at $\alpha_8 = 14.65267\ldots$. The next period doubling takes place at $\alpha_{16} = 14.74212\ldots$ (see Fig. 5.26, where several doublings are presented).

In view of Sharkovsky's Theorem (see Sect. 7.1.1), it is natural to expect that there is an *infinite* sequence of bifurcation values: $\alpha_{m(k)}$, $m(k) = 2^k$, $k = 1, 2, \ldots$ ($m(k)$ is the period of the cycle before the kth doubling). Moreover, one can check numerically that at least the first few elements of this sequence closely resemble a *geometric progression*. In fact, the quotient

$$\frac{\alpha_{m(k)} - \alpha_{m(k-1)}}{\alpha_{m(k+1)} - \alpha_{m(k)}}$$

tends to $\mu_F = 4.6692\ldots$ as k increases. This phenomenon is called *Feigenbaum's cascade* of period doublings, and the constant μ_F is referred to as the *Feigenbaum constant*. The most surprising fact is that this constant is the same for many different systems exhibiting a cascade of flip bifurcations. This universality has a deep underlying structure, see references in Sect. 5.4. $\diamondsuit$

Remark The critical normal form coefficient $c(0)$ can be expressed in terms of the second iterate of the map $x \mapsto f_\alpha(x) = f(x, \alpha)$ by the formula:

$$c(0) = -\frac{1}{12} \frac{\partial^3}{\partial x^3} f(f(x, \alpha), \alpha)\bigg|_{(x,\alpha)=(x_0,\alpha_0)},$$

where x_0 is the critical fixed point at the bifurcation parameter value α_0. ♢

5.3.3 *Planar Neimark–Sacker Bifurcation*

In continuous time, we can scale the time variable. In the context of Andronov–Hopf bifurcation, this allowed us to normalize the frequency ω to 1. In discrete time, this is impossible and the position on the unit circle where an eigenvalue crosses the unit circle naturally becomes an extra parameter in the normal form. Moreover, nonlinear terms contribute in a more complicated manner to the dynamics. As a result, the Neimark–Sacker bifurcation is essentially more involved than Andronov–Hopf bifurcation.

Example 5.27 (A model map for the Neimark–Sacker bifurcation) Consider the two-dimensional discrete-time system generated by the following map depending on the parameter α and its smooth functions $\theta(\alpha), a(\alpha)$, and $b(\alpha)$:

$$\begin{aligned}\begin{pmatrix} x_1 \\ x_2 \end{pmatrix} &\mapsto (1+\alpha)\begin{pmatrix} \cos\theta(\alpha) & -\sin\theta(\alpha) \\ \sin\theta(\alpha) & \cos\theta(\alpha) \end{pmatrix}\begin{pmatrix} x_1 \\ x_2 \end{pmatrix} \\ &+ (x_1^2+x_2^2)\begin{pmatrix} \cos\theta(\alpha) & -\sin\theta(\alpha) \\ \sin\theta(\alpha) & \cos\theta(\alpha) \end{pmatrix}\begin{pmatrix} a(\alpha) & -b(\alpha) \\ b(\alpha) & a(\alpha) \end{pmatrix}\begin{pmatrix} x_1 \\ x_2 \end{pmatrix},\end{aligned} \tag{5.74}$$

where $a(0) \neq 0$ and $0 < \theta(0) < \pi$.

This system has the fixed point $x_1 = x_2 = 0$ for all α with Jacobian matrix

$$A = (1+\alpha)\begin{pmatrix} \cos\theta(\alpha) & -\sin\theta(\alpha) \\ \sin\theta(\alpha) & \cos\theta(\alpha) \end{pmatrix}.$$

The matrix has eigenvalues $\mu_{1,2} = (1+\alpha)e^{\pm i\theta(\alpha)}$, which makes the map (5.74) invertible near the origin for all small $|\alpha|$. As can be seen, the fixed point at the origin is nonhyperbolic at $\alpha = 0$ due to a complex-conjugate pair of eigenvalues on the unit circle. To analyse the corresponding bifurcation, introduce the complex variable $z = x_1 + ix_2, \bar{z} = x_1 - ix_2, |z|^2 = z\bar{z} = x_1^2 + x_2^2$. The map (5.74) in the z-coordinate reads as

$$z \mapsto e^{i\theta(\alpha)} z(1+\alpha+d(\alpha)|z|^2), \tag{5.75}$$

where $d(\alpha) = a(\alpha) + ib(\alpha)$ is a smooth complex function of the parameter α.

Using the representation $z = \rho e^{i\varphi}$, we obtain for $\rho = |z|$

$$\rho \mapsto \rho|1+\alpha+d(\alpha)\rho^2|.$$

Since

$$\left|1+\alpha+d(\alpha)\rho^2\right| = \sqrt{(1+\alpha+a(\alpha)\rho^2)^2+b^2(\alpha)\rho^4} = 1+\alpha+a(\alpha)\rho^2+O(\rho^4)$$

and

$$\arg(1+\alpha+d(\alpha)\rho^2) = \arctan\left(\frac{b(\alpha)\rho^2}{1+\alpha+a(\alpha)\rho^2}\right) = \frac{b(\alpha)}{1+\alpha}\rho^2+O(\rho^4),$$

we arrive at the following *polar* form of the map (5.74):

$$\begin{cases} \rho \mapsto \rho(1+\alpha+a(\alpha)\rho^2)+\rho^4 R(\rho,\alpha), \\ \varphi \mapsto \varphi+\theta(\alpha)+\rho^2 Q(\rho,\alpha), \end{cases} \tag{5.76}$$

for functions R and Q, which are smooth functions of (ρ,α) near $(\rho,\alpha) = (0,0)$.[6] Bifurcations of the phase portrait of (5.74) as α passes through zero can easily be analysed using the latter form, since the mapping for ρ is *independent* of φ. The first equation in (5.76) defines a one-dimensional dynamical system that has the fixed point $\rho = 0$ for all values of α. The point is linearly stable if $\alpha < 0$; for $\alpha > 0$ the point becomes linearly unstable. The stability of the fixed point at $\alpha = 0$ is determined by the sign of the coefficient $a(0)$. Suppose that $a(0) < 0$; then the origin is (nonlinearly) stable at $\alpha = 0$. Moreover, the ρ-map of (5.76) has one additional stable fixed point

$$\rho_0(\alpha) = \sqrt{-\frac{\alpha}{a(\alpha)}}+O(\alpha)$$

for small $\alpha > 0$. The φ-map of (5.76) describes a rotation by an angle depending on ρ and α; it is approximately equal to $\theta(\alpha)$. Thus, by superposition of the mappings defined by (5.76), we obtain the bifurcation diagram for the original two-dimensional system (5.74) (see Fig. 5.27).

The system always has a fixed point at the origin. This point is stable for $\alpha < 0$ and unstable for $\alpha > 0$. At the critical parameter value $\alpha = 0$ the point is nonlinearly stable. The fixed point is surrounded for small $\alpha > 0$ by an isolated *closed invariant curve* that is unique and stable. The curve is a circle of radius $\rho_0(\alpha)$. All orbits starting outside or inside the closed invariant curve, but not at the origin, tend to the curve under iterations of (5.76). This is a *supercritical* Neimark–Sacker bifurcation.

When $a(0) > 0$, we get a *subcritical* Neimark–Sacker bifurcation, where an unstable closed invariant curve exists for $\alpha < 0$ with small $|\alpha|$ (see Fig. 5.28).

The structure of orbits of (5.74) on the invariant circle depends on whether the ratio between the rotation angle and 2π is rational or irrational. Define

$$\nu_0(\alpha) = \frac{\theta(\alpha)+\rho_0^2(\alpha)Q(\rho_0(\alpha),\alpha)}{2\pi}.$$

[6] Although $R(\rho,\alpha) = O(\rho)$ for the map (5.74), we factor out only ρ^4, to prepare for a more general case ahead.

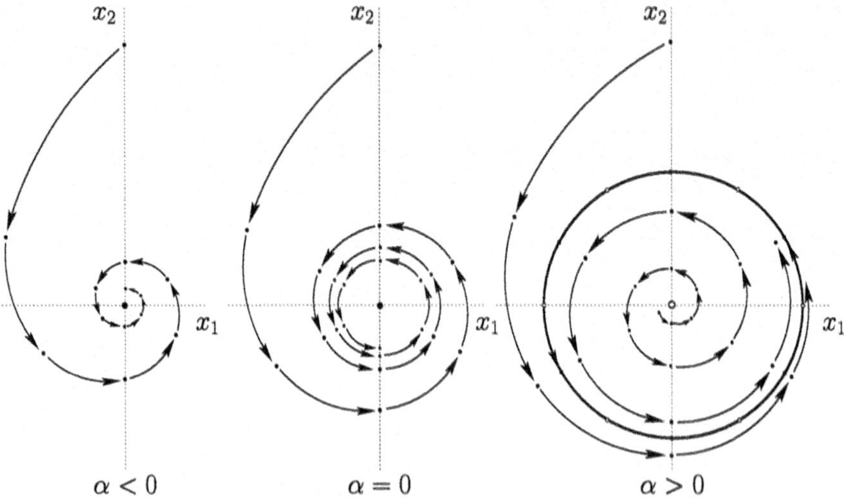

Fig. 5.27 Supercritical Neimark–Sacker bifurcation

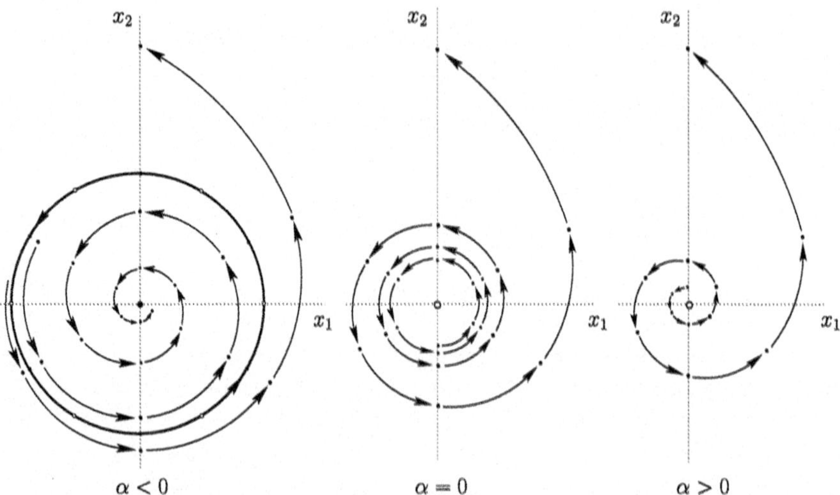

Fig. 5.28 Subcritical Neimark–Sacker bifurcation

If $\nu_0(\alpha)$ is rational, i.e. $\nu_0(\alpha) = \frac{n}{m}$ with $\gcd(n, m) = 1$, then every point on the circle is m-periodic and after m steps it winds n times around the origin. In this case, all orbits on the invariant curve are periodic. If $\nu_0(\alpha)$ is irrational, there are no periodic orbits and all orbits are dense in the circle. ◇

Theorem 5.28 *Consider the map*

$$z \mapsto \tilde{z} = g(z, \bar{z}, \alpha) = e^{i\theta(\alpha)} z(1 + \alpha + d(\alpha)|z|^2) + G(z, \bar{z}, \alpha), \tag{5.77}$$

where $z = x_1 + ix_2$, $d(\alpha) = a(\alpha) + ib(\alpha)$; $a(\alpha)$, $b(\alpha)$, *and* $\theta(\alpha)$ *are smooth real-valued functions;* $a(0) < 0, 0 < \theta(0) < \pi$, *and* G *is a smooth complex-valued function of* $(z, \bar{z}, \alpha)$ *such that* $G = O(|z|^4)$ *for small* z.

The real planar map $x \mapsto \tilde{x} = f(x, \alpha)$ *corresponding to the map* (5.77) *has in a neighbourhood of the origin a unique stable closed invariant curve for sufficiently small* $\alpha > 0$. *Moreover, any nontrivial orbit starting in this neighbourhood converges to the closed invariant curve under iteration of this map.*

Proof The proof involves a number of steps.
Step 1: Constructing an attracting annulus.

The point with polar coordinates (ρ, φ) is mapped by (5.77) to the point with polar coordinates $(\tilde{\rho}, \tilde{\varphi})$ with

$$\begin{cases} \tilde{\rho} = \rho(1 + \alpha + a(\alpha)\rho^2) + \rho^4 R(\rho, \varphi, \alpha), \\ \tilde{\varphi} = \varphi + \theta(\alpha) + \rho^2 Q(\rho, \varphi, \alpha), \end{cases} \tag{5.78}$$

where R and Q are 2π-periodic in φ smooth functions of (ρ, φ, α) near $(\rho, \alpha) = (0, 0)$ and

$$Q(\rho, \varphi, \alpha) = \frac{b(\alpha)}{1 + \alpha} + O(\rho^2)$$

(cf. (5.76) but note the difference: now R and Q depend also on φ and, in particular, the ρ-equation is not decoupled from the φ-equation).

Let $1 > \varepsilon > 0$. Introduce an annulus $A_{\alpha,\varepsilon}$ in the plane by the formula:

$$A_{\alpha,\varepsilon} = \left\{ (\rho, \varphi) : \sqrt{-\frac{\alpha}{a(\alpha)}}(1 - \varepsilon) \le \rho \le \sqrt{-\frac{\alpha}{a(\alpha)}}(1 + \varepsilon),\ \varphi \in [0, 2\pi] \right\}.$$

Consider a point (ρ, φ) and its image $(\tilde{\rho}, \tilde{\varphi})$ under the map (5.78). Then

$$\Delta\rho = \tilde{\rho} - \rho = \rho(\alpha + a(\alpha)\rho^2 + \rho^3 R(\rho, \varphi, \alpha))$$

and (recall that $a(\alpha) < 0$)

$$\Delta\rho \ge \rho(\alpha(2\varepsilon - \varepsilon^2) + O(\alpha^{3/2})) \text{ if } 0 \le \rho \le \sqrt{-\frac{\alpha}{a(\alpha)}}(1 - \varepsilon),$$

$$\Delta\rho \le \rho(\alpha(-2\varepsilon - \varepsilon^2) + O(\alpha^{3/2})) \text{ if } \varepsilon_0 \ge \rho \ge \sqrt{-\frac{\alpha}{a(\alpha)}}(1 + \varepsilon),$$

where the $O(\alpha^{3/2})$-terms are smooth functions of ε and $\varepsilon_0 > 0$ is a constant independent of α. Now take

$$\varepsilon = \alpha^{1/4}. \tag{5.79}$$

The above inequalities then imply

$$\Delta\rho > 0 \text{ if } 0 < \rho \leq \sqrt{-\frac{\alpha}{a(\alpha)}}(1 - \alpha^{1/4}),$$
$$\Delta\rho < 0 \text{ if } \varepsilon_0 \geq \rho \geq \sqrt{-\frac{\alpha}{a(\alpha)}}(1 + \alpha^{1/4}),$$

for all sufficiently small $\alpha > 0$. (Indeed, for any constant C, $\alpha^{5/4} > C\alpha^{3/2}$ provided $\alpha > 0$ is sufficiently small, so that the $\pm 2\alpha^{5/4}$-terms dominate.)

This proves that any nontrivial orbit of (5.77) starting in a small neighbourhood of the origin containing the annulus $A_{\alpha,\alpha^{1/4}}$ enters this annulus after a finite number of iterations. Notice that $A_{\alpha,\alpha^{1/4}}$ has width of order $O(\alpha^{3/4})$ and contains the circle

$$S_0(\alpha) = \left\{(\rho, \varphi) : \rho = \sqrt{-\frac{\alpha}{a(\alpha)}}\right\}$$

having radius $O(\alpha^{1/2})$.

Step 2: Rescaling and shifting.

To analyse the dynamics inside the annulus $A_{\alpha,\alpha^{1/4}}$, introduce first a new radial variable s by the formula:

$$\rho = \sqrt{-\frac{\alpha}{a(\alpha)}}(1 + s). \tag{5.80}$$

Substitution of (5.80) into (5.78) gives

$$\begin{cases} \tilde{s} = (1 - 2\alpha)s - \alpha(3s^2 + s^3) + \alpha^{3/2} r(s, \varphi, \alpha^{1/2}), \\ \tilde{\varphi} = \varphi + \theta(\alpha) + \alpha\nu(\alpha)(1 + s)^2 + \alpha^2 q(s, \varphi, \alpha^{1/2}), \end{cases} \tag{5.81}$$

where

$$\nu(\alpha) = -\frac{b(\alpha)}{a(\alpha)},$$

while r and q are smooth real-valued functions of $(s, \varphi, \alpha^{1/2})$ that are 2π-periodic in φ.

Next define u by

$$s = \alpha^{1/4} u. \tag{5.82}$$

Notice that the band $\{(u, \varphi) : |u| \leq 1\}$ corresponds exactly to the annulus $A_{\alpha,\alpha^{1/4}}$ (see (5.79)). After rescaling according to (5.82), the map (5.81) takes the form:

$$\begin{pmatrix} u \\ \varphi \end{pmatrix} \mapsto \begin{pmatrix} \tilde{u} \\ \tilde{\varphi} \end{pmatrix} = F \begin{pmatrix} u \\ \varphi \end{pmatrix},$$

where

$$F : \begin{cases} \tilde{u} = (1 - 2\alpha)u + \alpha^{5/4} H_\alpha(u, \varphi), \\ \tilde{\varphi} = \varphi + \omega(\alpha) + \alpha^{5/4} K_\alpha(u, \varphi). \end{cases} \tag{5.83}$$

Here

$$\omega(\alpha) = \theta(\alpha) + \alpha\nu(\alpha)$$

is smooth and

$$\begin{aligned} H_\alpha(u, \varphi) &= -(3u^2 + \alpha^{1/4} u^3) + r(\alpha^{1/4} u, \varphi, \alpha^{1/2}), \\ K_\alpha(u, \varphi) &= \nu(\alpha)(2u + \alpha^{1/4} u^2) + \alpha^{3/4} q(\alpha^{1/4} u, \varphi, \alpha^{1/2}) \end{aligned}$$

are smooth functions of $(u, \varphi, \alpha^{1/4})$ that are 2π-periodic in φ.

It is convenient to introduce

$$\lambda = \lambda(\alpha) = \sup_{|u| \le 1, \varphi \in [0, 2\pi]} \left\{ |H_\alpha|, |K_\alpha|, \left| \frac{\partial H_\alpha}{\partial u} \right|, \left| \frac{\partial K_\alpha}{\partial u} \right|, \left| \frac{\partial H_\alpha}{\partial \varphi} \right|, \left| \frac{\partial K_\alpha}{\partial \varphi} \right| \right\} \tag{5.84}$$

and note that $\lambda(\alpha)$ remains bounded as $\alpha \to 0$.

Step 3: Definition of the function space.

We will represent closed curves by elements of a function space U. By definition, $u \in U$ is a 2π-periodic continuous function $u = u(\varphi)$ satisfying the following two conditions:

(U.1) $|u(\varphi)| \le 1$ for all φ;
(U.2) $|u(\varphi_1) - u(\varphi_2)| \le |\varphi_1 - \varphi_2|$ for all φ_1, φ_2.

The first property means that $u(\varphi)$ is *absolutely bounded* by unity, while the second means that $u(\varphi)$ is *Lipschitz continuous* with Lipschitz constant less than or equal to one. The space U is a complete metric space with respect to the distance induced by the supremum norm:

$$\|u\| = \sup_{\varphi \in [0, 2\pi]} |u(\varphi)|.$$

Recall that a map $\mathcal{F} : U \to U$ (transforming a function $u(\varphi) \in U$ into some other function $\tilde{u}(\varphi) = (\mathcal{F}u)(\varphi) \in U$) is a *contraction* if there is a number ϵ, $0 < \epsilon < 1$, such that

$$\|\mathcal{F}(u_1) - \mathcal{F}(u_2)\| \le \epsilon \|u_1 - u_2\|$$

Fig. 5.29 Accumulating closed curves

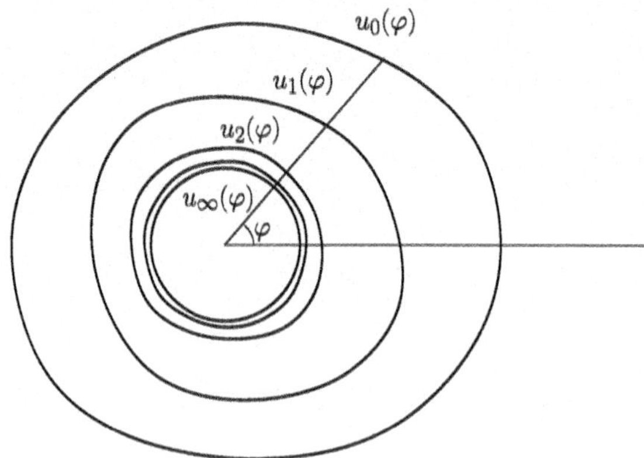

for all $u_{1,2} \in U$. According to Theorem 3.20, a contraction in a complete metric space has a unique fixed point $u^{(\infty)} \in U$:

$$\mathcal{F}(u^{(\infty)}) = u^{(\infty)}.$$

Moreover, the fixed point $u^{(\infty)}$ is globally asymptotically stable as a fixed point of the infinite-dimensional dynamical system $\{\mathbb{N}, U, \mathcal{F}^k\}$, i.e.:

$$\lim_{k \to +\infty} \|\mathcal{F}^k(u) - u^{(\infty)}\| = 0,$$

for all $u \in U$ (see Fig. 5.29).

Step 4: Construction of the map $\mathcal{F}$.

We will consider a map $\mathcal{F}$ on U *induced* by F. This means that if u represents a closed curve, then $\tilde{u} = \mathcal{F}(u)$ represents its image under the map F defined by (5.83). Such a map is called a *Hadamard Graph Transform*. It is clear that the continuous map F transforms a closed curve into a closed curve, but it need not be true that the image can be represented by a function of φ. We will now verify that for α small such a representation is actually possible.

Suppose that a function $u = u(\varphi)$ from U is given. To construct the map $\mathcal{F}$, we have to specify a procedure that for each given φ finds the corresponding $\tilde{u}(\varphi) = (\mathcal{F}u)(\varphi)$. Notice that F is nearly a *rotation* by the angle $\omega(\alpha)$ in φ. So a point $(\tilde{u}(\varphi), \varphi)$ in the resulting curve is the image of a point $(u(\hat{\varphi}), \hat{\varphi})$ in the original curve with a *different* angle coordinate $\hat{\varphi}$ (see Fig. 5.30).

To show that $\hat{\varphi}$ is *uniquely* defined, we have to prove that the equation

$$\varphi = \hat{\varphi} + \omega(\alpha) + \alpha^{5/4} K_\alpha(u(\hat{\varphi}), \hat{\varphi}) \tag{5.85}$$

has a unique solution $\hat{\varphi} = \hat{\varphi}(\varphi)$ for any given $u \in U$.

We solve (5.85) in $X = \mathbb{R}$ for any $\varphi \in \mathbb{R}$ and $u \in U$ by the Lipschitz Inverse Function Theorem 3.21 with the setting

$$L\tilde{\varphi} = \tilde{\varphi}, \quad f(\tilde{\varphi}) = f(\tilde{\varphi}, u) = \omega(\alpha) + \alpha^{5/4} K_\alpha(u(\tilde{\varphi}), \tilde{\varphi}).$$

The function f depends on α and on the function u which will be made explicit only when needed. Combining (U.2) with the definition (5.84) we deduce

$$|f(\varphi_1) - f(\varphi_2)| \le \lambda\alpha^{5/4}(|u(\varphi_1) - u(\varphi_2)| + |\varphi_1 - \varphi_2|) \le 2\lambda\alpha^{5/4}|\varphi_1 - \varphi_2|. \tag{5.86}$$

Hence we have $\mathrm{Lip} f \le 2\ell < 1$ where $\ell = \lambda\alpha^{5/4}$ and α is sufficiently small. Therefore, Eq. (5.85) has a unique solution $\hat{\varphi}(\varphi) = \hat{\varphi}(\varphi, u)$. From (3.19) we obtain the Lipschitz estimate

$$|\hat{\varphi}(\varphi_1) - \hat{\varphi}(\varphi_2)| \le (1 - 2\ell)^{-1}|\varphi_1 - \varphi_2|. \tag{5.87}$$

Further, the solution satisfies $\hat{\varphi}(\varphi + 2\pi) = \hat{\varphi}(\varphi) + 2\pi$, since $f(\varphi)$ is 2π-periodic and the equation

$$(L + f)(\hat{\varphi}(\varphi) + 2\pi) = (L + f)(\hat{\varphi}(\varphi)) + 2\pi = \varphi + 2\pi$$

has the unique solution $\hat{\varphi}(\varphi + 2\pi)$.

We now define the map $u \mapsto \tilde{u} = \mathcal{F}(u)$ by the formula:

$$\tilde{u}(\varphi) = [\mathcal{F}(u)](\varphi) = (1 - 2\alpha)u(\hat{\varphi}(\varphi)) + \alpha^{5/4} H_\alpha(u(\hat{\varphi}(\varphi)), \hat{\varphi}(\varphi)). \tag{5.88}$$

This defines a 2π-periodic function $\tilde{u}$ due to the periodicity properties of u, H_α, and $\hat{\varphi}$.

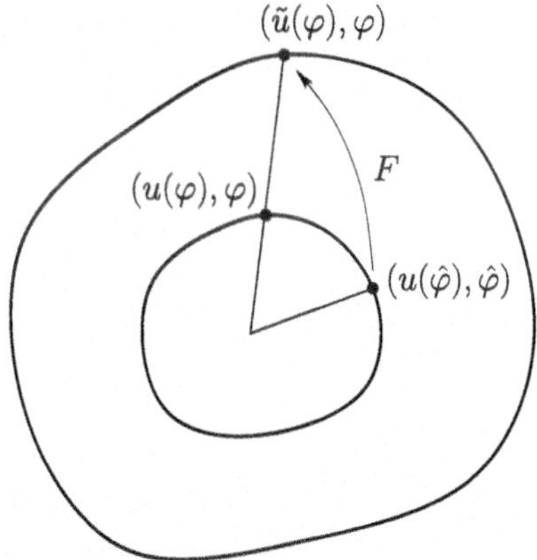

Fig. 5.30 Definition of the map $\mathcal{F}$

Next we verify $\tilde{u} = \mathcal{F}(u) \in U$ for $u \in U$ by checking the conditions (U.1) and (U.2). Setting $\hat{\varphi} = \hat{\varphi}(\varphi)$, condition (U.1) follows from (5.88) and (5.84)

$$|\tilde{u}(\varphi)| \leq (1 - 2\alpha)|u(\hat{\varphi})| + \alpha^{5/4}|H_\alpha(u(\hat{\varphi}), \hat{\varphi})| \leq 1 - 2\alpha + \ell \leq 1$$

if $\alpha > 0$ is small enough. Then we combine the estimates (5.87), (U.2) and the definitions (5.84), (5.88) to obtain

$$\begin{aligned} |\tilde{u}(\varphi_1) - \tilde{u}(\varphi_2)| &\leq (1 - 2\alpha)|u(\hat{\varphi}(\varphi_1)) - u(\hat{\varphi}(\varphi_2))| \\ &\quad + \ell(|u(\hat{\varphi}(\varphi_1)) - u(\hat{\varphi}(\varphi_2))| + |\hat{\varphi}(\varphi_1) - \hat{\varphi}(\varphi_2)|) \\ &\leq (1 - 2\alpha + 2\ell)|\hat{\varphi}(\varphi_1) - \hat{\varphi}(\varphi_2)| \\ &\leq (1 - 2\alpha + 2\ell)(1 - 2\ell)^{-1}|\varphi_1 - \varphi_2|. \end{aligned} \tag{5.89}$$

Thus, (U.2) also holds for $\tilde{u}$ for all sufficiently small positive α. Therefore, the map $\tilde{u} = \mathcal{F}(u)$ is well defined.

Step 5: Verification of the contraction property.

Now suppose two functions $u_1, u_2 \in U$ are given. We need to estimate $\|\tilde{u}_1 - \tilde{u}_2\| = \|\mathcal{F}(u_1) - \mathcal{F}(u_2)\|$ in terms of $\|u_1 - u_2\|$. For $\varphi \in \mathbb{R}$ let us set $\hat{\varphi}_1 = \hat{\varphi}(\varphi, u_1)$, $\hat{\varphi}_2 = \hat{\varphi}(\varphi, u_2)$ where we recall that $\hat{\varphi}$ and f in (5.86) depend on u. By the solution property of $\hat{\varphi}_1$ we have

$$L\hat{\varphi}_1 + f(\hat{\varphi}_1, u_2) = \varphi + f(\hat{\varphi}_1, u_2) - f(\hat{\varphi}_1, u_1).$$

Then the Lipschitz estimates (5.86) and (5.87) yield

$$|\hat{\varphi}_1 - \hat{\varphi}_2| \leq (1 - 2\ell)^{-1}|f(\hat{\varphi}_1, u_2) - f(\hat{\varphi}_1, u_1)| \leq \frac{\ell}{1 - 2\ell}\|u_1 - u_2\|. \tag{5.90}$$

With this and the Lipschitz properties of H_α and u_2 we finally obtain

$$\begin{aligned} |\tilde{u}_1(\varphi) - \tilde{u}_2(\varphi)| &\leq (1 - 2\alpha)|u_1(\hat{\varphi}_1) - u_2(\hat{\varphi}_2)| + \ell\,(|u_1(\hat{\varphi}_1) - u_2(\hat{\varphi}_2)| \\ &\quad + |\hat{\varphi}_1 - \hat{\varphi}_2|) \\ &\leq (1 - 2\alpha + \ell)\,(|u_1(\hat{\varphi}_1) - u_2(\hat{\varphi}_1)| + |u_2(\hat{\varphi}_1) - u_2(\hat{\varphi}_2)|) \\ &\quad + \ell|\hat{\varphi}_1 - \hat{\varphi}_2| \\ &\leq (1 - 2\alpha + \ell)\|u_1 - u_2\| + (1 - 2\alpha + 2\ell)|\hat{\varphi}_1 - \hat{\varphi}_2| \\ &\leq \frac{1 - 2\alpha + 2\ell\alpha}{1 - 2\ell}\|u_1 - u_2\|. \end{aligned}$$

Thus the map $\mathcal{F}$ is a contraction in U if we choose $\alpha > 0$ such that $\ell(1 + \alpha) = \lambda\alpha^{5/4}(1 + \alpha) < \alpha$. Then the map has a unique fixed point $u^{(\infty)} \in U$ such that

$$\lim_{k \to \infty} \mathcal{F}^k(u) = u^{(\infty)}$$

for any $u \in U$.

Step 6: Asymptotic stability of the invariant curve.

The established contraction of $\mathcal{F}$ implies that the closed invariant curve corresponding to $u^{(\infty)}$ is a *stable invariant set* for the map F defined by (5.83). Indeed, images of any band located in $A_{\alpha,\alpha^{1/4}}$ and containing $u^{(\infty)}$ will contain this curve and lie inside the band, for all sufficiently high iterates of F.

Now take a point (u_0, φ_0) within the band $\{(u,\varphi) : |u| \leq 1\}$. If the point belongs to the curve given by $u^{(\infty)}$, it remains on this curve under iteration of F, since the map F maps this curve into itself. If the point does not lie on the invariant curve, take some (noninvariant) closed curve passing through it represented by $u^{(0)} \in U$, say $u^{(0)}(\varphi) = u_0$. Let us apply the iterations of the map F to this point. We get a sequence of points

$$\{(u_k, \varphi_k)\}_{k=0}^{\infty}.$$

It is clear that each point from this sequence belongs to the corresponding iterate of the curve $u^{(0)}$ under the map $\mathcal{F}$. We have just shown that the iterations of the curve converge to the invariant curve given by $u^{(\infty)}$. Therefore, the point sequence must also converge to the curve. This proves asymptotic stability of the closed invariant curve as the only nontrivial invariant set of the map (5.77) in the annulus $A_{\alpha,\alpha^{1/4}}$. Recalling **Step 1** completes the proof. □

Remarks

(1) The orbit structure on the closed invariant curve and the variation of this structure when the parameter changes are generically different in the maps (5.77) and (5.75). In (5.77), there exist generically only a finite number of periodic orbits on the closed invariant curve. When the parameter changes, these orbits collide and disappear/appear via fold bifurcations.

(2) The bifurcating invariant curve in (5.77) generically has only finite smoothness, which increases as $\alpha \to 0$. ◊

We now shall prove that any generic two-dimensional system undergoing a Neimark–Sacker bifurcation can be transformed into the form (5.77).

Consider a system

$$x \mapsto f(x, \alpha), \quad x \in \mathbb{R}^2, \ \alpha \in \mathbb{R},$$

with a smooth function f, which has at $\alpha = 0$ the fixed point $x = 0$ with simple eigenvalues $\mu_{1,2} = e^{\pm i\theta_0}$, $0 < \theta_0 < \pi$. By the Implicit Function Theorem, the system has a unique fixed point $x_0(\alpha)$ in some neighbourhood of the origin for all sufficiently small $|\alpha|$, since $\mu = 1$ is not an eigenvalue of the Jacobian matrix. We can perform a parameter-dependent coordinate shift, placing this fixed point at the origin. Therefore, we may assume without loss of generality that $x = 0$ is the fixed point of the system for $|\alpha|$ sufficiently small. Thus, the system can be written as

$$x \mapsto A(\alpha)x + F(x, \alpha), \tag{5.91}$$

where F is a smooth vector function whose components $F_{1,2}$ have Taylor expansions in x starting with quadratic (or higher-order) terms, $F(0, \alpha) = 0$ for all sufficiently small $|\alpha|$. The Jacobian matrix $A(\alpha)$ has two eigenvalues

$$\mu_{1,2}(\alpha) = r(\alpha)e^{\pm i\varphi(\alpha)},$$

where $r(0) = 1, \varphi(0) = \theta_0$. Thus, $r(\alpha) = 1 + \beta(\alpha)$ for some smooth function $\beta(\alpha), \beta(0) = 0$. Suppose that $\beta'(0) \neq 0$. Then, we can use β as a new parameter and express the multipliers in terms of $\beta : \mu_1(\beta) = \mu(\beta), \mu_2(\beta) = \bar{\mu}(\beta)$, where

$$\mu(\beta) = (1 + \beta)e^{i\theta(\beta)}$$

with a smooth function $\theta(\beta)$ such that $\theta(0) = \theta_0$.

Lemma 5.29 *By the introduction of a complex variable and a new parameter, the system* (5.91) *can be transformed for all sufficiently small* $|\alpha|$ *into the following form:*

$$z \mapsto \mu(\beta)z + g(z, \bar{z}, \beta), \tag{5.92}$$

where $\beta \in \mathbb{R}, z \in \mathbb{C}, \mu(\beta) = (1 + \beta)e^{i\theta(\beta)}$, *and* g *is a complex-valued smooth function of* $(z, \bar{z}, \beta)$ *whose formal Taylor expansion with respect to* $(z, \bar{z})$ *contains only quadratic and higher-order terms:*

$$g(z, \bar{z}, \beta) = \sum_{k+l \geq 2} \frac{1}{k!l!} g_{kl}(\beta) z^k \bar{z}^l,$$

with $k, l = 0, 1, \ldots$. □

The proof of the lemma is completely analogous to that from the Andronov–Hopf bifurcation analysis and is omitted. In particular, we have

$$g(z, \bar{z}, 0) = \langle p, F(zq + \bar{z}\bar{q}, 0) \rangle,$$

where the vectors $q, p \in \mathbb{C}^2$ satisfy

$$A(0)q = e^{i\theta_0}q, \;\; A^{\mathrm{T}}(0)p = e^{-i\theta_0}p, \;\; \langle p, q \rangle = 1.$$

As in the Andronov–Hopf case, we start by making *nonlinear* (complex) coordinate transformations that will simplify the map (5.92).

Lemma 5.30 (Poincaré normal form for NS bifurcation) *The smooth map*

$$z \mapsto z' = \mu(\beta)z + \sum_{2\le j+l\le 3} \frac{1}{j!l!} g_{jl}(\beta) z^j \bar{z}^l + O(|z|^4),$$

where $\mu(\beta) = (1+\beta)e^{i\theta(\beta)}$ *and* $\theta_0 = \theta(0)$ *is such that* $e^{ik\theta_0} \neq 1$ *for* $k = 1,2,3,4$, *can be transformed, for all sufficiently small* $|\beta|$, *by an invertible change of complex coordinate*

$$z = w + \frac{h_{20}(\beta)}{2}w^2 + h_{11}(\beta)w\bar{w} + \frac{h_{02}(\beta)}{2}\bar{w}^2 + \frac{h_{30}(\beta)}{6}w^3 + \frac{h_{12}(\beta)}{2}w\bar{w}^2 + \frac{h_{03}(\beta)}{6}\bar{w}^3,$$

depending smoothly on the parameter, into a map

$$w \mapsto w' = \mu(\beta)w + c_1(\beta)w^2\bar{w} + O(|w|^4), \tag{5.93}$$

where

$$c_1(0) = \frac{g_{20}(0)g_{11}(0)(1-2\mu_1)}{2(\mu_1^2-\mu_1)} + \frac{|g_{11}(0)|^2}{1-\bar{\mu}_1} + \frac{|g_{02}(0)|^2}{2(\mu_1^2-\bar{\mu}_1)} + \frac{g_{21}(0)}{2} \tag{5.94}$$

with $\mu_1 := \mu(0) = e^{i\theta_0}$.

Remarks

(1) The nondegeneracy conditions on θ_0 used in the lemma are said to *exclude the strong resonances*. They can be written as

$$\mu_1^k \neq 1, \quad k = 1,2,3,4.$$

Notice that $\mu_1 \neq 1$ and $\mu_1^2 \neq 1$ hold automatically due to our initial assumption $0 < \theta_0 < \pi$.

(2) The coordinate transformation $w \mapsto z$ is polynomial with coefficients that are smoothly dependent on β. In a small neighbourhood of the origin, the transformation is *nearly the identity* and thus invertible. The inverse transformation is smooth, but not polynomial.

(3) The lemma says that generically the original map $z \mapsto z'$ is locally smoothly conjugate (*diffeomorphic*) to the map (5.93). ◊

Proof of Lemma 5.30 Write

$$\begin{aligned} z' = {} & \mu z + \frac{1}{2}g_{20}z^2 + g_{11}z\bar{z} + \frac{1}{2}g_{02}\bar{z}^2 \\ & + \frac{1}{6}g_{30}z^3 + \frac{1}{2}g_{21}z^2\bar{z} + \frac{1}{2}g_{12}z\bar{z}^2 + \frac{1}{6}g_{03}\bar{z}^2 \\ & + O(|z|^4), \end{aligned} \tag{5.95}$$

where the dependence of μ and g_{jk} on β is implicit. Make a substitution

$$\begin{aligned} z \;=\; & w + \frac{1}{2}h_{20}w^2 \;+\; h_{11}w\overline{w} \;+\; \frac{1}{2}h_{02}\overline{w}^2 \\ & + \frac{1}{6}h_{30}w^3 \;+\; \frac{1}{2}h_{21}w^2\overline{w} \;+\; \frac{1}{2}h_{12}w\overline{w}^2 \;+\; \frac{1}{6}h_{03}\overline{w}^2 \end{aligned} \tag{5.96}$$

with unknown complex coefficients $h_{jk} = h_{jk}(\beta)$. Notice that the $w^2\overline{w}$-term is yet present. This substitution is invertible in a small neighbourhood of the origin and the inverse is smooth (but not polynomial). The substitution will produce a transformed map

$$\begin{aligned} w' \;=\; & \mu w + \frac{1}{2}G_{20}w^2 \;+\; G_{11}w\overline{w} \;+\; \frac{1}{2}G_{02}\overline{w}^2 \\ & + \frac{1}{6}G_{30}w^3 \;+\; \frac{1}{2}G_{21}w^2\overline{w} \;+\; \frac{1}{2}G_{12}w\overline{w}^2 \;+\; \frac{1}{6}G_{03}\overline{w}^2 \\ & + O(|w|^4). \end{aligned} \tag{5.97}$$

To find G_{jk}, compute z' in two ways: First, by substituting w' and its complex conjugate from (5.97) into (5.96)

$$\begin{aligned} z' &= w' + \frac{1}{2}h_{20}w'^2 + h_{11}w'\overline{w}' + \frac{1}{2}h_{02}\overline{w}'^2 \\ &\quad + \frac{1}{6}h_{30}w'^3 \;+\; \frac{1}{2}h_{21}w'^2\overline{w}' \;+\; \frac{1}{2}h_{12}w'\overline{w}'^2 \;+\; \frac{1}{6}h_{03}\overline{w}'^2 \\ &= \mu w + \frac{1}{2}(G_{20} + \mu^2 h_{20})w^2 + (G_{11} + \mu\bar{\mu}h_{11})w\overline{w} + \frac{1}{2}(G_{02} + \bar{\mu}^2 h_{02})\overline{w}^2 \\ &\quad + \frac{1}{6}(G_{30} + \mu^2 h_{30} + 3\mu G_{20}h_{20} + 3\mu\overline{G}_{02}h_{11})w^3 \\ &\quad + \frac{1}{2}(G_{21} + \mu^2\bar{\mu}h_{21} + 2\mu G_{11}h_{20} + 2\mu\overline{G}_{11}h_{11} + \bar{\mu}G_{20}h_{11} + \bar{\mu}\overline{G}_{02}h_{02})w^2\overline{w} \\ &\quad + \frac{1}{2}(G_{12} + \mu\bar{\mu}^2 h_{12} + \mu G_{02}h_{20} + \mu\overline{G}_{20}h_{11} + 2\bar{\mu}G_{11}h_{11} + 2\bar{\mu}\overline{G}_{11}h_{02})w\overline{w}^2 \\ &\quad + \frac{1}{6}(G_{03} + \bar{\mu}^3 h_{30} + 3\bar{\mu}G_{02}h_{11} + 3\bar{\mu}\overline{G}_{20}h_{02})\overline{w}^3 + O(|w|^4) \end{aligned}$$

and second, by substituting z given by (5.96) and its complex conjugate in the original map (5.95)

$$\begin{aligned}
z' = \mu w &+ \frac{1}{2}(g_{20} + \mu h_{20})w^2 + (g_{11} + \mu h_{11})w\overline{w} + \frac{1}{2}(g_{02} + \mu h_{02})\overline{w}^2 \\
&+ \frac{1}{6}(g_{30} + \mu h_{30} + 3g_{11}\bar{h}_{02} + 3g_{20}h_{20})w^3 \\
&+ \frac{1}{2}(g_{21} + \mu h_{21} + 2g_{20}h_{11} + 2g_{11}\bar{h}_{11} + g_{11}h_{20} + g_{02}\bar{h}_{02})w^2\overline{w} \\
&+ \frac{1}{2}(g_{12} + \mu h_{12} + 2g_{02}\bar{h}_{11} + 2g_{11}h_{11} + g_{11}\bar{h}_{20} + g_{20}h_{02})w\overline{w}^2 \\
&+ \frac{1}{6}(g_{03} + \mu h_{03} + 3g_{02}\bar{h}_{20} + 3g_{11}h_{02})\overline{w}^3 + O(|w|^4).
\end{aligned}$$

By comparing the coefficients in front of $w^2, w\overline{w}$, and $\overline{w}^2$ in both expressions, we find

$$G_{20} = g_{20} - (\mu^2 - \mu)h_{20}, \tag{5.98}$$

$$G_{11} = g_{11} - (|\mu|^2 - \mu)h_{11}, \tag{5.99}$$

$$G_{02} = g_{02} - (\bar{\mu}^2 - \mu)h_{02}, \tag{5.100}$$

while the comparison of the coefficients in front of $w^3, w^2\overline{w}, w\overline{w}^2$, and $\overline{w}^3$ gives

$$\begin{aligned}
G_{30} &= g_{30} + (\mu - \mu^3)h_{30} + 3g_{11}\bar{h}_{02} + 3g_{20}h_{20} \\
&\quad - 3\mu G_{20}h_{20} - 3\mu\overline{G}_{02}h_{11}, \\
G_{21} &= g_{21} + (\mu - \mu^2\bar{\mu})h_{21} + g_{02}\bar{h}_{02} + g_{11}h_{20} + 2g_{11}\bar{h}_{11} + 2g_{20}h_{11} \\
&\quad - 2\mu G_{11}h_{20} - 2\mu\overline{G}_{11}h_{11} - \bar{\mu}G_{20}h_{11} - \bar{\mu}\overline{G}_{02}h_{02}, \\
G_{12} &= g_{12} + (\mu - \mu\bar{\mu}^2)h_{12} + 2g_{02}\bar{h}_{11} + 2g_{11}h_{11} + g_{11}\bar{h}_{20} + g_{20}h_{02} \\
&\quad - \mu G_{02}h_{20} - \mu\overline{G}_{20}h_{11} - 2\bar{\mu}G_{11}h_{11} - 2\bar{\mu}\overline{G}_{11}h_{02} \\
G_{03} &= g_{03} + (\mu - \bar{\mu}^3)h_{03} + 3g_{02}\bar{h}_{20} + 3g_{11}h_{02} \\
&\quad - 3\bar{\mu}G_{02}h_{11} - 3\bar{\mu}\overline{G}_{20}h_{02}.
\end{aligned}$$

As in the Hopf case, we can try to make the quadratic coefficients G_{20}, G_{11}, and G_{02} equal to zero by taking

$$h_{20} = \frac{g_{20}}{\mu^2 - \mu},\ h_{11} = \frac{g_{11}}{|\mu|^2 - \mu},\ h_{02} = \frac{g_{02}}{\bar{\mu}^2 - \mu}. \tag{5.101}$$

This is possible, since according to our assumptions the denominators stay away from zero for all sufficiently small $|\beta|$. It is sufficient to check this at $\beta = 0$ and then use the continuity of the multipliers w.r.t. β. Recall that $\mu_1 = \mu(0) = e^{i\theta_0}$. We have $\mu_1^k = e^{ik\theta_0} \neq 1$ for $k = 1$ and 3, so that

$$\begin{aligned}\mu_1^2-\mu_1 &= e^{i\theta_0}(e^{i\theta_0}-1)\neq 0,\\ |\mu_1|^2-\mu_1 &= 1-e^{i\theta_0}\neq 0,\\ \bar{\mu}_1^2-\mu_1 &= e^{i\theta_0}(e^{-3i\theta_0}-1)\neq 0.\end{aligned}$$

With $G_{20}=G_{11}=G_{02}=0$, we have

$$G_{30}=g_{30}+(\mu-\mu^3)h_{30}+3g_{11}\bar{h}_{02}+3g_{20}h_{20}, \tag{5.102}$$

$$G_{21}=g_{21}+(\mu-\mu^2\bar{\mu})h_{21}+g_{02}\bar{h}_{02}+g_{11}h_{20}+2g_{11}\bar{h}_{11}+2g_{20}h_{11}, \tag{5.103}$$

$$G_{12}=g_{12}+(\mu-\mu\bar{\mu}^2)h_{12}+2g_{02}\bar{h}_{11}+2g_{11}h_{11}+g_{11}\bar{h}_{20}+g_{20}h_{02}, \tag{5.104}$$

$$G_{03}=g_{03}+(\mu-\bar{\mu}^3)h_{03}+3g_{02}\bar{h}_{20}+3g_{11}h_{02}, \tag{5.105}$$

where h_{20}, h_{11}, and h_{02} are given by (5.101). Further, we see that according to our assumption on the critical multipliers

$$\begin{aligned}\mu_1-\mu_1^3 &= e^{i\theta_0}(1-e^{2i\theta_0})\neq 0,\\ \mu_1-\mu_1\bar{\mu}_1^2 &= e^{i\theta_0}(1-e^{-2i\theta_0})\neq 0,\\ \mu_1-\bar{\mu}_1^3 &= e^{i\theta_0}(1-e^{-4i\theta_0})\neq 0.\end{aligned}$$

Therefore, using Eqs. (5.102), (5.104), and (5.105), we can assure for all sufficiently small $|\alpha|$ that $G_{30}=G_{12}=G_{03}=0$ by properly selecting the coefficients h_{30}, h_{12}, and h_{03}.

However, since

$$\mu_1-\mu_1^2\bar{\mu}_1=\mu_1(1-|\mu_1|^2)=e^{i\theta_0}(1-1)=0,$$

the coefficient in front of h_{21} in Eq. (5.103) vanishes at the critical parameter value $\beta=0$, so than no choice of h_{21} would make $G_{21}=0$. Thus the $w^2\bar{w}$-term in the transformed map (5.97) cannot be "killed" at $\alpha=0$. As in the Hopf case, we do not try to remove this *resonant term* and merely set $h_{21}=0$ in the substitution (5.96).

The transformed map (5.97) reduces to (5.93):

$$w'=\mu w+c_1w|w|^2+O(|w|^4),$$

where $c_1=\frac{1}{2}G_{21}$ with

$$G_{21}=g_{21}+\frac{2g_{20}g_{11}}{|\mu|^2-\mu}+\frac{2g_{11}\bar{g}_{11}}{|\mu|^2-\bar{\mu}}+\frac{g_{11}g_{20}}{\mu^2-\mu}+\frac{g_{02}\bar{g}_{02}}{\mu^2-\bar{\mu}}.$$

Here, (5.101) in taken into account. Thus

$$c_1 = \frac{g_{20}g_{11}(\bar{\mu} - 3 + 2\mu)}{2(\mu^2 - \mu)(\bar{\mu} - 1)} + \frac{|g_{11}|^2}{|\mu|^2 - \bar{\mu}} + \frac{|g_{02}|^2}{2(\mu^2 - \bar{\mu})} + \frac{g_{21}}{2}, \tag{5.106}$$

where $\mu = \mu(\beta)$ and $g_{jk} = g_{jk}(\beta)$ are smooth functions of β. At the critical parameter value $\beta = 0$, we get (5.94), since $\mu(0) = e^{i\theta_0}$ and thus $(1 - 2\mu(0))(\bar{\mu}(0) - 1) = \bar{\mu}(0) - 3 + 2\mu(0)$. □

We now summarize the obtained results in the following theorem.

Theorem 5.31 *Suppose a two-dimensional discrete-time system generated by the map*

$$x \mapsto f(x, \alpha), \quad x \in \mathbb{R}^2, \ \alpha \in \mathbb{R},$$

with smooth f has, for all sufficiently small $|\alpha|$, the fixed point $x = 0$ with multipliers

$$\mu_{1,2}(\alpha) = r(\alpha)e^{\pm i\varphi(\alpha)},$$

where $r(0) = 1, \varphi(0) = \theta_0$. Let the following genericity conditions be satisfied:

(C.1) $r'(0) \neq 0$;
(C.2) $e^{ik\theta_0} \neq 1$ *for* $k = 1, 2, 3, 4$.

Then there are smooth invertible coordinate and parameter changes transforming the system into

$$\begin{aligned}\begin{pmatrix} y_1 \\ y_2 \end{pmatrix} &\mapsto (1 + \beta) \begin{pmatrix} \cos\theta(\beta) & -\sin\theta(\beta) \\ \sin\theta(\beta) & \cos\theta(\beta) \end{pmatrix} \begin{pmatrix} y_1 \\ y_2 \end{pmatrix} \\ &+ (y_1^2 + y_2^2) \begin{pmatrix} \cos\theta(\beta) & -\sin\theta(\beta) \\ \sin\theta(\beta) & \cos\theta(\beta) \end{pmatrix} \begin{pmatrix} a(\beta) & -b(\beta) \\ b(\beta) & a(\beta) \end{pmatrix} \begin{pmatrix} y_1 \\ y_2 \end{pmatrix} \\ &+ O(\|y\|^4)\end{aligned} \tag{5.107}$$

with $\theta(0) = \theta_0$ and

$$a(0) = \text{Re}\left(\frac{e^{-i\theta_0}g_{21}^0}{2}\right) - \text{Re}\left(\frac{(1 - 2e^{i\theta_0})e^{-2i\theta_0}}{2(1 - e^{i\theta_0})}g_{20}^0 g_{11}^0\right) - \frac{1}{2}|g_{11}^0|^2 - \frac{1}{4}|g_{02}^0|^2, \tag{5.108}$$

where g_{kl}^0 are defined by the formal expansion

$$\langle p, f(zq + \bar{z}\bar{q}, 0)\rangle = e^{i\theta_0}z + \sum_{2 \le k+l \le 3} \frac{1}{k!l!} g_{kl}^0 z^k \bar{z}^l + O(|z|^4),$$

in which $q, p \in \mathbb{C}^2$ satisfy $D_x f(0,0)q = e^{i\theta_0}q$, $[D_x f^{\mathrm{T}}(0,0)]p = e^{-i\theta_0}p$, and $\langle p, q\rangle = 1$.

Proof The only thing left to verify is the formula (5.108) for $a(0)$. Indeed, by Lemmas 5.29 and 5.30, the system can be reduced to the complex Poincaré normal form

$$w \mapsto \mu(\beta)w + c_1(\beta)w|w|^2 + O(|w|^4), \quad w \in \mathbb{C},$$

where $\mu(\beta) = (1+\beta)e^{i\theta(\beta)}$. This map can be rewritten as

$$w \mapsto e^{i\theta(\beta)}(1 + \beta + d(\beta)|w|^2)w + O(|w|^4)$$

with

$$d(\beta) := e^{-i\theta(\beta)}c_1(\beta) \ = \ a(\beta) + ib(\beta)$$

for some smooth real functions $a(\beta)$ and $b(\beta)$. A return to real $y = (y_1, y_2) \in \mathbb{R}^2$ by writing $w = y_1 + iy_2$, gives system (5.107). Finally,

$$a(\beta) = \text{Re } d(\beta) = \text{Re}(e^{-i\theta(\beta)}c_1(\beta)).$$

Thus $a(0) = \text{Re}(e^{-i\theta_0}c_1(0))$ and, taking into account (5.94), we get (5.108). □

Example 5.32 (NS bifurcation of the delayed logistic map) Consider the following recurrence equation[7]:

$$u_{k+1} = ru_k(1 - u_{k-1}).$$

This is yet another simple population dynamics model, where u_k stands for the density of a population in year k, and r is the growth rate at low densities. It is assumed that the growth is determined not only by the current population density but also by its density a year ago.

If we introduce $v_k = u_{k-1}$, the equation can be rewritten as

$$\begin{cases} u_{k+1} = ru_k(1 - v_k), \\ v_{k+1} = u_k, \end{cases}$$

which, in turn, defines the two-dimensional map,

$$\begin{pmatrix} x_1 \\ x_2 \end{pmatrix} \mapsto \begin{pmatrix} rx_1(1 - x_2) \\ x_1 \end{pmatrix} =: \begin{pmatrix} F_1(x, r) \\ F_2(x, r) \end{pmatrix}, \tag{5.109}$$

where $x = (x_1, x_2)$. The map (5.109) has the fixed point $(0, 0)^T$ for all values of r. For $r > 1$, a nontrivial positive fixed point x^0 appears, with the coordinates

[7] Aronson, D. G., Chory, M. A., Hall, G. R., & McGehee, R. P. (1982). Bifurcations from an invariant circle for two-parameter families of maps of the plane: A computer-assisted study. *Communications in Mathematical Physics, 83*, 303–354.

$$x_1^0(r) = x_2^0(r) = 1 - \frac{1}{r}.$$

The Jacobian matrix of the map (5.109) evaluated at the nontrivial fixed point is given by

$$A(r) = \begin{pmatrix} 1 & 1-r \\ 1 & 0 \end{pmatrix}$$

and has eigenvalues

$$\mu_{1,2}(r) = \frac{1}{2} \pm \sqrt{\frac{5}{4} - r}.$$

If $r > \frac{5}{4}$, the eigenvalues are complex and $|\mu_{1,2}|^2 = \mu_1\mu_2 = r - 1$. Therefore, at $r = r_0 = 2$ the nontrivial fixed point loses stability and we have a Neimark–Sacker bifurcation: The critical multipliers are

$$\mu_{1,2} = e^{\pm i\theta_0}, \ \theta_0 = \frac{\pi}{3}.$$

It is clear that conditions (C.1) and (C.2) are satisfied.

We have to verify that the nondegeneracy condition $a(0) \neq 0$ also holds. The critical Jacobian matrix $A_0 = A(r_0)$ has the eigenvectors

$$A_0 q = e^{i\theta_0} q, \ A_0^{\mathrm{T}} p = e^{-i\theta_0} p,$$

where

$$q \sim \left(\frac{1}{2} + i\frac{\sqrt{3}}{2}, \ 1 \right), \ p \sim \left(-\frac{1}{2} + i\frac{\sqrt{3}}{2}, \ 1 \right).$$

To achieve the normalization $\langle p, q \rangle = 1$, we can take, for example,

$$q = \left(\frac{1}{2} + i\frac{\sqrt{3}}{2}, \ 1 \right), \ p = \left(i\frac{\sqrt{3}}{3}, \ \frac{1}{2} - i\frac{\sqrt{3}}{6} \right).$$

Now we compose

$$x = x^0 + zq + \bar{z}\bar{q}$$

and evaluate the function

$$H(z, \bar{z}) = \langle p, F(x^0 + zq + \bar{z}\bar{q}, r_0) - x^0 \rangle.$$

Computing its Taylor expansion at $(z, \bar{z}) = (0, 0)$,

$$H(z, \bar{z}) = e^{i\theta_0} z + \sum_{2 \le j+k \le 3} \frac{1}{j!k!} g_{jk}^0 z^j \bar{z}^k + O(|z|^4),$$

gives

$$g^0_{20} = -2 + i\frac{2\sqrt{3}}{3}, \quad g^0_{11} = i\frac{2\sqrt{3}}{3}, \quad g^0_{02} = 2 + i\frac{2\sqrt{3}}{3}, \quad g^0_{21} = 0,$$

which allows us to evaluate $a(0)$ via (5.108) and see that

$$a(0) = -2 < 0.$$

Therefore, by Theorems 5.31 and 5.28, a unique and stable closed invariant curve bifurcates from the nontrivial fixed point for $r > 2$ (see Fig. 5.31). ◊

5.4 References

Bifurcations of stationary points and periodic orbits in one- and two-parameter families of 1D and 2D ODEs and maps are treated in many textbooks, including (Arnol'd, 1983; Guckenheimer & Holmes, 1983; Arrowsmith & Place, 1990; Shilnikov et al., 2001; Wiggins, 2003; Kuznetsov, 2023). A useful summary is given in Arnol'd et al. (1994), while many technical issues are clarified in Iooss (1979), Vanderbauwhede (1989), Iooss and Adelmeyer (1992). For a detailed theory of circle homeomorphisms and the related ergodic theory we refer to Chaps. 4 and 11 of Katok & Hasselblatt (1995). This chapter is based on Chaps. 3 and 4 from Kuznetsov (2023).

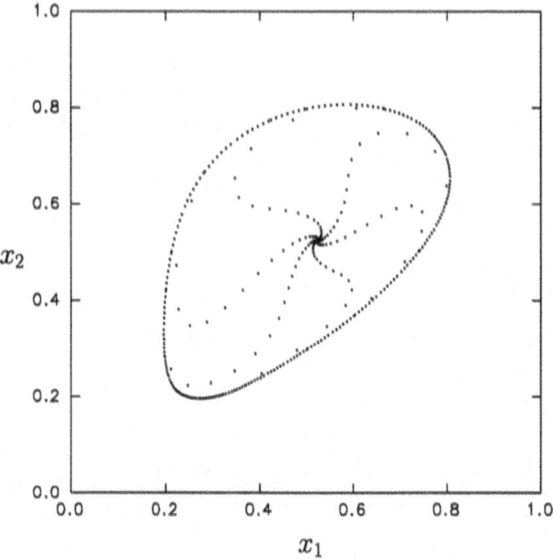

Fig. 5.31 Stable invariant curve in the delayed logistic map

Cascades of period-doubling bifurcations in various maps exhibit universal accumulation properties discovered by Feigenbaum (1978) and explained in Coullet and Eckmann (1980). The relevant theorems were first proved with the help of a computer and delicate error estimates in Lanford (1980), see also Babenko and Petrovich (1983), Petrovich (1990). In Lyubich (2000), an overview of available analytical results is given, while introductory expositions of this topic can be found, e.g., in Guckenheimer and Holmes (1983), Kuznetsov (2023).

5.5 Exercises

Exercise 5.1 (Fold bifurcation in a simple ecological model) Consider the following differential equation, which models a single population under a constant harvest:

$$\dot{x} = rx\left(1 - \frac{x}{K}\right) - \alpha,$$

where x is the population number; r and K are the *intrinsic growth rate* and the *carrying capacity* of the population, respectively; and α is the *harvest rate*, which is a control parameter. Find a parameter value α_0 at which the system has a fold bifurcation, and check the genericity conditions of Theorem 5.7. Based on the analysis, explain what might be a result of overharvesting on the ecosystem dynamics.

Exercise 5.2 (Andronov–Hopf bifurcation in planar systems) Check that each of the following systems has an equilibrium that exhibits an Andronov–Hopf bifurcation at some value of α, and compute the first Lyapunov coefficient:

(a) *Rayleigh's equation:*

$$\ddot{x} + \dot{x}^3 - 2\alpha\dot{x} + x = 0;$$

(*Hint:* Introduce $y = \dot{x}$ and rewrite the equation as a system of two differential equations.)

(b) *Van der Pol's oscillator:*

$$\ddot{y} - (\alpha - y^2)\dot{y} + y = 0;$$

(c) *Bautin's example*[8]:

$$\begin{cases} \dot{x} = y, \\ \dot{y} = -x + \alpha y + x^2 + xy + y^2; \end{cases}$$

(d) *Feichtinger's model*[9]:

$$\begin{cases} \dot{x}_1 = \alpha[1 - x_1x_2^2 + A(x_2 - 1)], \\ \dot{x}_2 = x_1x_2^2 - x_2. \end{cases}$$

[8] Bautin, N. N., & Leontovich, E. A. (1976). *Methods and Techniques for the Qualitative Study of Dynamical Systems on the Plane*, Nauka, Moscow. In Russian.

[9] Feichtinger, G. (1992). Hopf bifurcation in an advertising diffusion model. *Journal of Economic Behavior and Organization, 17*, 401–411.

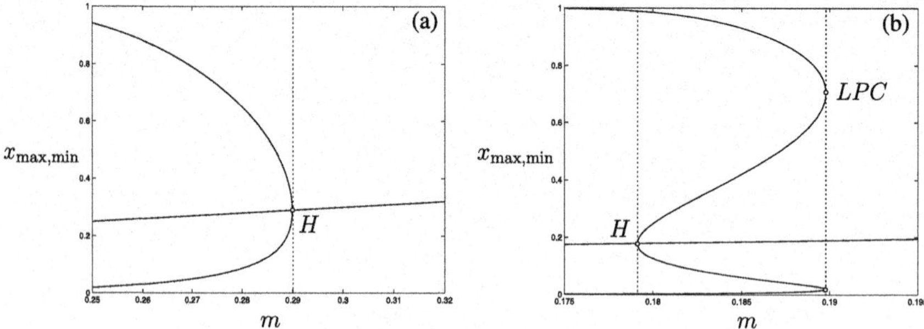

Fig. 5.32 Maximal and minimal values of the x-variable along equilibrium and cycle branches near the Hopf bifurcation of (5.110) for $\gamma = 1$ and (**a**) $n = 0.2$; (**b**) $n = 0.05$

Exercise 5.3 (Sub- and supercritical Hopf bifurcation in a prey–predator model) Analyse Andronov–Hopf bifurcations in the prey–predator model[10]:

$$\begin{cases} \dot{x} = \dfrac{x^2(1-x)}{n+x} - xy, \\ \dot{y} = -\gamma y(m-x), \end{cases} \tag{5.110}$$

where $\gamma > 0$ is fixed, while $n \geq 0$ and $0 \leq m < 1$ are control parameters.

(a) Find a relationship between parameters m and n corresponding to a Hopf bifurcation of a positive equilibrium in the system.

(b) Compute the first Lyapunov coefficient l_1 along the Hopf bifurcation curve H in the (m, n)-plane found at step (a) and show that it vanishes at

$$B = (m, n) = \left(\frac{1}{4}, \frac{1}{8}\right).$$

(c) Fix $\gamma = 1$ and compute numerically families of limit cycles bifurcating at the Hopf bifurcation from for $n = 0.2$ and $n = 0.05$ and find a fold bifurcation of limit cycles (LPC) in one of them. (*Hint*: Fig. 5.32.)

(d) Continue numerically the fold bifurcation curve LPC in two control parameters and convince yourself that it connects the origin of the (m, n)-plane with the point B found at step (b). (*Hint*: Fig. 5.33.)

(e) Produce all typical phase portraits of the system.

Exercise 5.4 (Flip bifurcation in scalar population models) Prove that the genericity conditions (B.1) and (B.2) for the flip bifurcation at $(x, \alpha) = (0, 0)$ of the map

$$x \mapsto xg(x, \alpha)$$

with $g(0, 0) = -1$ are equivalent to

(D.1) $\frac{1}{2}\frac{\partial^2 g}{\partial x^2}(0, 0) + \left[\frac{\partial g}{\partial x}(0, 0)\right]^2 \neq 0$;
(D.2) $\frac{\partial g}{\partial \alpha}(0, 0) \neq 0$.

---

[10] Bazykin, A. D., & Khibnik, A. I. (1981). On sharp excitation of self-oscillations in a Volterra-type model. *Biophysika, 26*, 851–853. In Russian.

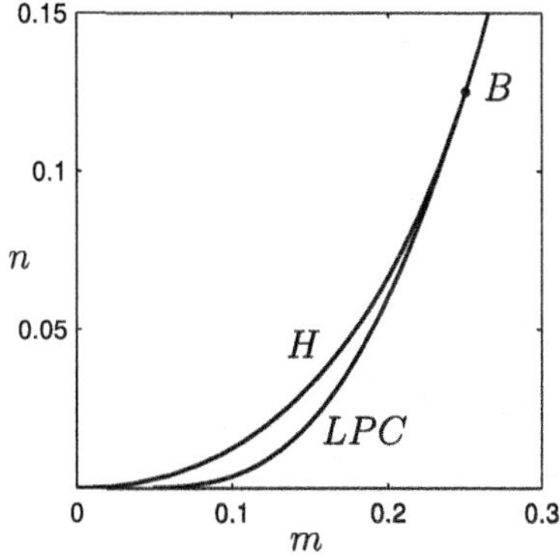

Fig. 5.33 Bifurcation curves of (5.110) for $\gamma = 1$: H—Hopf bifurcation, LPC—fold bifurcation of limit cycles

Exercise 5.5 (Fixed points and periodic orbits of the 2-Ricker population model) Consider the map[11]

$$x \mapsto rxe^{-\nu(x,p)} \quad \text{with} \quad \nu(x,p) = x(1 + pe^{-x}), \tag{5.111}$$

where $r > 1$ and $p > 0$.

(a) Find fixed points of (5.111) and study their fold and period-doubling bifurcations analytically.

(b) Study numerically bifurcations of period-two, -three, and -four cycles of (5.111) within the region $1 \le r \le 500, 0 \le p \le 40$. *Hint*: Introduce new variable and parameters: $y = \ln x, R = \ln r, P = \ln p$.

(c) Present a combined bifurcation diagram using the original (r, p)-parameters. Where could multiple attractors and chaos be expected?

Exercise 5.6 (Discrete-time prey–predator model by Maynard Smith) Consider the following recurrence equations[12]:

$$\begin{cases} x_{k+1} &= \alpha x_k(1 - x_k) - x_k y_k, \\ y_{k+1} &= \dfrac{1}{\beta} x_k y_k, \end{cases}$$

which is a discrete-time version of the Lotka–Volterra model. Here x_k and y_k are the prey and predator numbers, respectively, in year (generation) k, and it is assumed that in the absence of prey the predators become extinct in one generation. Parameters α and β are positive.

(a) Introduce the map

$$\begin{pmatrix} x \\ y \end{pmatrix} \mapsto \begin{pmatrix} \alpha x(1 - x) - xy \\ \dfrac{1}{\beta} xy \end{pmatrix}, \tag{5.112}$$

and prove that a nontrivial fixed point of this map undergoes a Neimark–Sacker bifurcation on a curve in the (α, β)-plane.

---

[11] Davydova, N. V. (2004). Dynamics and bifurcations of single year class maps, Chap. 5 of *Old and Young. Can they coexist?* Ph.D. Thesis, Utrecht University.

[12] Maynard Smith, J. (1968). *Mathematical Ideas in Biology*. London: Cambridge University Press.

(b) Determine the direction of the closed invariant-curve bifurcation by computing the corresponding normal form coefficient.

(c) Iterate the map (5.112) numerically for various parameter values to see what happens to the closed invariant curve away from the Neimark–Sacker bifurcation.

Chapter 6
Centre Manifold Reduction

Abstract The previous chapter gave a rather detailed description of bifurcations of equilibria and fixed points in generic one-parameter families of ODEs and maps with minimal possible dimension of the state space. These results are also applicable to general n-dimensional systems because of the existence of a low-dimensional invariant manifold for parameter values near the bifurcation point, on which all interesting local dynamics in the state space is concentrated. The present chapter is devoted to the constructive definition of this invariant *Centre Manifold* and the efficient computation of normal forms on it yielding all the relevant information. Our proof will also establish the existence and smoothness of the local *stable* and *unstable invariant manifolds* of hyperbolic saddles.

6.1 Centre Manifolds for Maps

Consider a C^k-map with $k \geq 0$,

$$u \mapsto F(u) = Au + R(u), \quad u \in \mathbb{R}^n, \tag{6.1}$$

where $R(0) = 0$. If $k \geq 1$ then we assume that $DR(0) = 0$, i.e. all first-order partial derivatives of R with respect to the components of u evaluated at $u = 0$ vanish. Suppose that $n = n_{cu} + n_s$ and

$$u = \begin{pmatrix} x \\ y \end{pmatrix} \in \mathbb{R}^{n_{cu}} \times \mathbb{R}^{n_s}, \quad A = \begin{pmatrix} B & 0 \\ 0 & C \end{pmatrix}, \quad R(u) = \begin{pmatrix} f(x,y) \\ g(x,y) \end{pmatrix},$$

where $f : \mathbb{R}^{n_{cu}} \times \mathbb{R}^{n_s} \to \mathbb{R}^{n_{cu}}$ and $g : \mathbb{R}^{n_{cu}} \times \mathbb{R}^{n_s} \to \mathbb{R}^{n_s}$ are C^k-functions with $f(0,0) = 0$ and $g(0,0) = 0$. Then we can rewrite (6.1) as

$$\begin{pmatrix} x \\ y \end{pmatrix} \mapsto F(x,y) = \begin{pmatrix} B & 0 \\ 0 & C \end{pmatrix} \begin{pmatrix} x \\ y \end{pmatrix} + \begin{pmatrix} f(x,y) \\ g(x,y) \end{pmatrix}. \tag{6.2}$$

Yu. Kuznetsov et al., *Dynamical Systems Essentials*, Texts in Applied Mathematics 83, https://doi.org/10.1007/978-3-032-04083-1_6

Assume that all eigenvalues of the $n_{cu} \times n_{cu}$ matrix B satisfy $|\lambda| \geq 1$ (implying that B is invertible), while all eigenvalues of the $n_s \times n_s$ matrix C satisfy $|\lambda| < 1$. The spaces $\mathbb{R}^{n_{cu}}$ and $\mathbb{R}^{n_s}$ are the *centre-unstable* and *stable* eigenspaces of A, respectively. A general map (6.1) takes the form (6.2) in the *eigenbasis* composed of all eigenvectors and generalized eigenvectors of A, when one first takes n_{cu} vectors corresponding to the eigenvalues with $|\lambda| \geq 1$ and then n_s vectors corresponding to the eigenvalues with $|\lambda| < 1$. The eigenvalues of B with $|\lambda| = 1$ are often called *critical eigenvalues.* If B has critical eigenvalues, the fixed point $u = 0$ of (6.1) is nonhyperbolic.

We also write (6.2) as

$$\begin{pmatrix} x \\ y \end{pmatrix} \mapsto F(x, y) = \begin{pmatrix} P(x, y) \\ Q(x, y) \end{pmatrix},$$

where

$$P(x, y) := Bx + f(x, y), \quad Q(x, y) := Cy + g(x, y). \tag{6.3}$$

Given any integer j, we can introduce norms for $x \in \mathbb{R}^{n_{cu}}$ and $y \in \mathbb{R}^{n_s}$, such that the corresponding operator norms,

$$\alpha := \|C\|, \quad \beta := \|B^{-1}\|,$$

satisfy

$$\alpha < 1, \tag{6.4}$$

$$\alpha\beta^{j+1} < 1. \tag{6.5}$$

Indeed, from Theorem 2.7 it follows that there are equivalent norms such that $\alpha < 1$ (this is (6.4)) and $\beta < 1 + \varepsilon$, where $\varepsilon > 0$ is arbitrarily small. For any given integer $j \geq 0$, we can achieve that the inequality (6.5) holds by selecting ε sufficiently small.

On $\mathbb{R}^n = \mathbb{R}^{n_{cu}} \times \mathbb{R}^{n_s}$ we use the norm

$$\|u\| = \max\{\|x\|, \|y\|\},$$

where $\|x\|$ and $\|y\|$ are the norms on $\mathbb{R}^{n_{cu}}$ and on $\mathbb{R}^{n_s}$ introduced above. With this norm on $\mathbb{R}^n$, we define the norm on the linear space of bounded functions $R : \mathbb{R}^n \to \mathbb{R}^n$ having the above form:

$$\|R\| = \sup_{u \in \mathbb{R}^n} \|R(u)\| = \max\{\|f\|, \|g\|\},$$

where

$$\|f\| = \sup_{u \in \mathbb{R}^n} \|f(u)\| \quad \text{and} \quad \|g\| = \sup_{u \in \mathbb{R}^n} \|g(u)\|.$$

Clearly, $\|f\| \leq \|R\|$ and $\|g\| \leq \|R\|$.

For a k-linear function

$$E : \underbrace{\mathbb{R}^n \times \cdots \times \mathbb{R}^n}_{k \text{ times}} \to \mathbb{R}^n$$

defined on the product of k copies of $\mathbb{R}^n$, the *operator norm* is

$$\|E\| := \sup_{v^{(k)} \neq 0} \frac{\|E(v^{(1)}, v^{(2)}, \cdots, v^{(n)})\|}{\|v^{(1)}\| \cdot \|v^{(2)}\| \cdots \|v^{(n)}\|} = \sup_{\|v^{(k)}\|=1} \|E(v^{(1)}, v^{(2)}, \cdots, v^{(n)})\|.$$

It is remarkable that for a *symmetric* k-linear function E one has:

$$\|E\| = \sup_{\|v\|=1} \|E(v, \cdots, v)\|.$$

This norm will be used below for $D^k R(u)$, i.e. the k-th derivative of R evaluated at $u \in \mathbb{R}^n$. Recall, that $D^0 R(u) \equiv R(u)$ and for $k \geq 1$, $j = 1, \ldots, n$:

$$D^k R_j(u)(v^{(1)}, \ldots, v^{(k)}) = \sum_{i_1, i_2, \ldots, i_k \in \{1,2,\ldots,n\}} \frac{\partial^k R_j(u)}{\partial u_{i_1} \cdots \partial u_{i_k}} v_{i_1}^{(1)} \cdots v_{i_k}^{(k)}, \quad (6.6)$$

where $v^{(i)} \in \mathbb{R}^n$ for $i = 1, 2, \ldots, k$.

Definition 6.1 *We call $D^k R$* ***Lipschitz on*** *$\mathbb{R}^n$ if there is a constant $L \geq 0$ such that*

$$\|D^k R(u) - D^k R(v)\| \leq L\|u - v\|$$

for all $u, v \in \mathbb{R}^n$. The infimum of the constants L is denoted by $\mathrm{Lip}(D^k R)$.

Definition 6.2 *We say that R* ***belongs to class $C^{k,1}$*** *for $k \geq 0$ if $R \in C^k$ and $D^k R$ is Lipschitz on $\mathbb{R}^n$.*

Similarly, we will write $h \in C^{k,1}$ for some function $h : \mathbb{R}^{n_{cu}} \to \mathbb{R}^{n_s}$ if $h \in C^k$ and $D^k h$ is Lipschitz on $\mathbb{R}^{n_{cu}}$.

As a final preparation, let us define $\|R\|_{C^k} := \|R\| + \|DR\| + \cdots + \|D^k R\|$ and note that $\mathrm{Lip}(D^k R) = \|D^{k+1} R\|$ if $R \in C^{k+1}$.

Theorem 6.3 (Global Lipschitz Centre-Unstable Manifold for Maps) *There exists $\varepsilon > 0$ such that when $R \in C^{0,1}$ with $\|R\| \leq \varepsilon$ and* $\mathrm{Lip}(R) \leq \varepsilon$ *then the map (6.2) has a globally attracting invariant manifold*

$$W^{cu} = \{(x, h(x)) : x \in \mathbb{R}^{n_{cu}}\},$$

where $h : \mathbb{R}^{n_{cu}} \to \mathbb{R}^{n_s}$ is a $C^{0,1}$-map satisfying $h(0) = 0$.

If $R \in C^{k,1}$ with $k \geq 1$, then the above assertion holds and, moreover, h is a $C^{k,1}$-map satisfying $Dh(0) = 0$.

Definition 6.4 W^{cu} *is called a* ***centre-unstable manifold*** *of the fixed point* $(0,0)$ *of* (6.2).

Proof of Theorem 6.3
Step 1: The inequalities (6.4) and (6.5) with $j = 0$ imply that there is $\varepsilon > 0$ such that

$$\varepsilon\beta < 1, \tag{6.7}$$

$$(\varepsilon + \alpha)\frac{\beta}{1 - \varepsilon\beta} < 1, \tag{6.8}$$

$$\alpha + 2\varepsilon < 1. \tag{6.9}$$

hold simultaneously. From now on, assume that with this value of ε we have $\|R\| \leq \varepsilon$ and $\mathrm{Lip}(R) \leq \varepsilon$.

Introduce a set M^0 of maps $H \in C^{0,1}(\mathbb{R}^{n_{cu}}, \mathbb{R}^{n_s})$, satisfying the following conditions:

(i) $\|H\| < \infty$, where $\|H\| := \sup_{x \in \mathbb{R}^{n_{cu}}} \|H(x)\|$;
(ii) $\mathrm{Lip}(H) \leq 1$;
(iii) $H(0) = 0$.

The set M^0 is a complete metric space with respect to the distance $\|H_1 - H_2\|$.

The invariance under the map (6.2) of the (Lipschitz) manifold

$$W = \{(x, H(x)) : x \in \mathbb{R}^{n_{cu}}\}$$

for given $H \in M^0$ means that

$$Q(\xi, H(\xi)) = H(P(\xi, H(\xi))) \tag{6.10}$$

should hold for all $\xi \in \mathbb{R}^{n_{cu}}$. Recall, that P and Q are defined by (6.3). We want to consider (6.10) as a fixed-point equation

$$T(H) = H,$$

where T is such that the graph of H is mapped to a new graph by the mapping (6.2), see Fig. 6.1. Accordingly, the graph of a fixed point of T is invariant under (6.2).

The formal definition of T is as follows. For each $H \in M^0$, introduce the map

$$\hat{H} : \mathbb{R}^{n_{cu}} \to \mathbb{R}^{n_{cu}} \times \mathbb{R}^{n_s}, \quad \xi \mapsto \hat{H}(\xi) = (\xi, H(\xi)).$$

First, we have to prove that for each $x \in \mathbb{R}^{n_{cu}}$ and each $H \in M^0$, there is a unique $\xi = S_H(x)$, such that

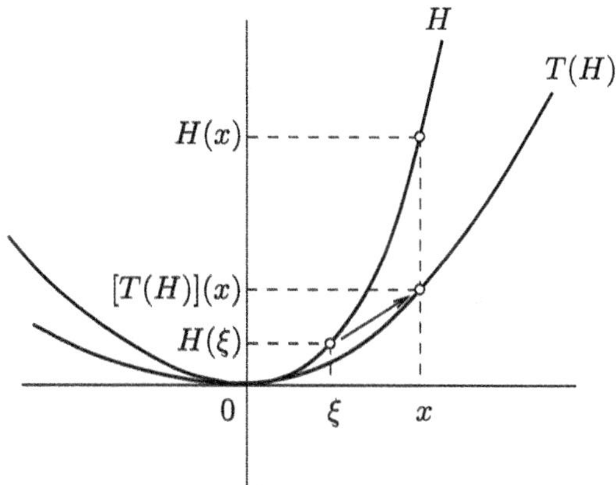

Fig. 6.1 Hadamard Graph Transform: Solve $x = P(\xi, H(\xi))$ for ξ and set $[T(H)](x) = Q(\xi, H(\xi))$

$$x = P(\xi, H(\xi)) = B\xi + f(\xi, H(\xi)). \tag{6.11}$$

Equivalently, we have to prove that the map on $\mathbb{R}^{n_{cu}}$ defined as

$$\xi \mapsto (P \circ \hat{H})(\xi) = P(\xi, H(\xi)) = B\xi + (f \circ \hat{H})(\xi),$$

is invertible. For $H \in M^0$ we have

$$\text{Lip}(f \circ \hat{H}) \leq \text{Lip}(f)\ \text{Lip}(\hat{H}) \leq \text{Lip}(R) \leq \varepsilon.$$

Thus, Theorem 3.21 applies, since due to (6.7)

$$\varepsilon < \beta^{-1} = \|B^{-1}\|^{-1},$$

so that $P \circ \hat{H}$ is indeed invertible with Lipschitz inverse:

$$(P \circ \hat{H})^{-1} = B^{-1} - V_H =: S_H, \tag{6.12}$$

where $V_H : \mathbb{R}^{n_{cu}} \to \mathbb{R}^{n_s}$ is Lipschitz with

$$\text{Lip}(V_H) \leq \frac{\varepsilon\beta^2}{1 - \varepsilon\beta}$$

and S_H is Lipschitz with

$$\text{Lip}(S_H) \leq \frac{\beta}{1 - \varepsilon\beta}. \tag{6.13}$$

Define now the *Hadamard Graph Transform* $H \mapsto T(H)$ by the formula

$$T(H) = Q \circ \hat{H} \circ S_H \;, \tag{6.14}$$

where

$$P \circ \hat{H} \circ S_H = I \tag{6.15}$$

(see Fig. 6.1). Clearly, the equation $H = T(H)$ is equivalent to (6.10), since they are related by the transformation $x = P(\xi, H(\xi))$ where $\xi = S_H(x)$.

Step 2: First we prove that $T(M^0) \subset M^0$. Clearly, $T(H) \in C^{0,1}(\mathbb{R}^{n_{cu}}, \mathbb{R}^{n_s})$. Then we have, using (6.14) and the expression for Q in (6.3),

$$\begin{aligned}\|T(H)\| &\le \|C \circ H \circ S_H\| + \|g \circ \hat{H} \circ S_H\| \le \|C\|\|H \circ S_H\| + \|g \circ \hat{H} \circ S_H\| \\ &\le \|C\| \, \|H\| + \|g\| \le \alpha\|H\| + \|R\| < \infty.\end{aligned}$$

Next we obtain the estimate

$$\begin{aligned}\mathrm{Lip}(T(H)) &\le \mathrm{Lip}(C \circ H \circ S_H) + \mathrm{Lip}(g \circ \hat{H} \circ S_H) \\ &\le \|C\| \,\mathrm{Lip}(H)\,\mathrm{Lip}(S_H) + \mathrm{Lip}(g)\,\mathrm{Lip}(S_H) \\ &\le (\|C\| + \mathrm{Lip}(R))\,\mathrm{Lip}(S_H) \\ &\le (\alpha + \varepsilon)\,\mathrm{Lip}(S_H).\end{aligned}$$

From the second inequality in (6.13) it follows that

$$\mathrm{Lip}(T(H)) \le (\varepsilon + \alpha)\frac{\beta}{1 - \varepsilon\beta}.$$

Then (6.8) directly implies $\mathrm{Lip}(T(H)) < 1$. Finally, it is obvious that $S_H(0) = 0$ and hence $[T(H)](0) = 0$. Thus, $T(M^0) \subset M^0$.

Step 3: Now we verify that T is a contraction in M^0. For each $H \in M^0$ and $(\xi, y) \in \mathbb{R}^{n_{cu}} \times \mathbb{R}^{n_s}$, we have

$$\begin{aligned}&\|Q(\xi,y) - [T(H)](P(\xi,y))\| \\ &\quad\le \|Q(\xi,y) - Q(\xi,H(\xi)) + Q(\xi,H(\xi)) - [T(H)](P(\xi,y)\| \\ &\quad\le \|Q(\xi,y) - Q(\xi,H(\xi))\| + \|Q(\xi,H(\xi)) - [T(H)](P(\xi,y)\| \\ &\quad\le \|Q(\xi,y) - Q(\xi,H(\xi))\| + \|[T(H)](P(\xi,H(\xi))) - [T(H)](P(\xi,y))\| \\ &\quad\le \|Cy - CH(\xi)\| + \|g(\xi,y) - g(\xi,H(\xi))\| \\ &\qquad\qquad + \mathrm{Lip}(T(H))\|P(\xi,y) - P(\xi,H(\xi))\| \\ &\quad\le \|C\| \, \|y - H(\xi)\| + \|g(\xi,y) - g(\xi,H(\xi))\| + \|f(\xi,y) - f(\xi,H(\xi))\| \\ &\quad\le (\|C\| + 2\,\mathrm{Lip}(R))\|y - H(\xi)\|.\end{aligned}$$

Here, we used $\mathrm{Lip}(T(H)) \le 1$. Thus

$$\|Q(\xi,y) - [T(H)](P(\xi,y))\| \le (\alpha + 2\varepsilon)\|y - H(\xi)\|. \tag{6.16}$$

For any two $H_{1,2} \in M^0$ and any $x \in \mathbb{R}^{n_{cu}}$, we obtain using (6.15) that

$$\begin{aligned}\|[T(H_2)](x) - [T(H_1)](x)\| &= \|(Q \circ \hat{H}_2 \circ S_{H_2})(x) - [T(H_1)]((P \circ \hat{H}_2 \circ S_{H_2})(x))\| \\ &= \|Q(\xi, y) - [T(H_1)](P(\xi, y))\|,\end{aligned}$$

where $\xi = S_{H_2}(x)$ and $y = H_2(S_{H_2}(x))$. Taking into account (6.16) for $H = H_1$, we see that

$$\begin{aligned}\|[T(H_2)](x) - [T(H_1)](x)\| &\leq (\alpha + 2\varepsilon)\, \|H_2(S_{H_2}(x)) - H_1(S_{H_2}(x))\| \\ &\leq (\alpha + 2\varepsilon)\|H_2 - H_1\|\end{aligned}$$

for all $x \in \mathbb{R}^{n_{cu}}$, implying

$$\|T(H_2) - T(H_1)\| \leq (\alpha + 2\varepsilon)\|H_2 - H_1\|. \tag{6.17}$$

Therefore, T is a contraction in M^0, since $\alpha + 2\varepsilon < 1$ due to (6.9).

The Contraction Mapping Principle (see Theorem 3.20 in Chap. 3) guarantees the existence of a global Lipschitz centre-unstable manifold given by the graph of h, where $h \in M^0$ is the fixed point of T.

This manifold is attracting for the map (6.2), since $\mathrm{dist}(F^k(x,y), W^{cu}) \to 0$ as $k \to \infty$ for all $(x, y) \in \mathbb{R}^{n_{cu}} \times \mathbb{R}^{n_s}$. Indeed, the above estimates show that T is also a contraction in the set of maps defined as M^0 but without the requirement (iii). The iterates of the constant function $(\xi, y) \mapsto y$ for all $\xi \in \mathbb{R}^{n_{cu}}$ converge then to a fixed point of T in this bigger space. This fixed point must coincide with h due to uniqueness. Consider now an arbitrary point (x, y) and its images $F^k(x, y)$. These images belong to the graphs of the transforms of the constant function and thus converge to W^{cu}.

Step 4: To establish that $h \in C^{k,1}(\mathbb{R}^{n_{cu}}, \mathbb{R}^{n_s})$ when $k \geq 1$, one can show that there is $r = (r_2, r_3, \ldots, r_{k+1}) \in \mathbb{R}^k$ with $r_j > 0$ such that T maps M_r^k into itself, where M_r^k is the set of maps $H \in C^{k,1}(\mathbb{R}^{n_{cu}}, \mathbb{R}^{n_s})$ satisfying the following conditions:

(i) $\|H\| < \infty$, where $\|H\| = \sup_{x \in \mathbb{R}^{n_{cu}}} \|H(x)\|$;
(ii) $\|DH\| \leq 1, \|D^2H\| \leq r_2, \ldots, \|D^kH\| \leq r_k, \mathrm{Lip}(D^kH) \leq r_{k+1}$;
(iii) $H(0) = 0$.

We show that the set M_r^k is a closed subset of M^0 w.r.t. the norm from (i) and $T(M_r^k) \subset M_r^k$. Starting the iteration with $H = 0$ implies that the fixed point $h \in M^0$ of T lies actually in M_r^k, so that $h \in C^{k,1}(\mathbb{R}^{n_{cu}}, \mathbb{R}^{n_s})$. Rather involved details of the proof are sketched below.

The following Lemma is a special case of the so-called Gagliardo–Nirenberg estimates.

Lemma 6.5 *For any function $G \in C^{k,1}(\mathbb{R}^p, \mathbb{R}^q)$ with $k \geq 1$ the following estimate holds*

$$\|D^kG\| \leq 2^k \|G\|^{\frac{1}{k+1}} \left(\mathrm{Lip}(D^kG)\right)^{\frac{k}{k+1}}. \tag{6.18}$$

Let $(R_\ell)_{\ell\in\mathbb{N}}$ be a sequence in M_r^k which converges w.r.t. $\|\cdot\|$ in M_0 to some R. Apply (6.18) to $G = R_\ell - R_m$ and conclude that the derivatives $D^j R_\ell$, $j = 1, \ldots, k$ are Cauchy sequences w.r.t. $\|\cdot\|$. Hence the uniform limit R lies in C^k and, by passing to the limit, one finds that R satisfies the inequalities (i)-(iii) above. Therefore, M_r^k is closed with respect to the norm in M^0.

Proof of Lemma 6.5 The proof proceeds by induction. First observe that

$$DG(u)v = G(u+v) - G(u) + \int_0^1 \big(DG(u) - DG(u+sv)\big)v\, ds$$

for all $u, v \in \mathbb{R}^p$. Taking norms leads to

$$\|DG(u)v\| \le 2\|G\| + \frac{1}{2}\mathrm{Lip}(DG)\|v\|^2 \tag{6.19}$$

for all $u, v \in \mathbb{R}^p$. We select a vector $v_0 \in \mathbb{R}^p$ with $\|v_0\| = 1$ and $\|DG(u)v_0\| = \|DG(u)\|$ and apply (6.19) to $v = cv_0$ with $c > 0$

$$\|DG(u)\| = \frac{1}{c}\|DG(u)(cv_0)\| \le \frac{2}{c}\|G\| + \frac{c}{2}\mathrm{Lip}(DG).$$

Minimizing the right-hand side over $c > 0$ leads to

$$\|DG(u)\| \le 2\sqrt{\|G\|\mathrm{Lip}(DG)},$$

which is our assertion for $k = 1$. For the induction step we apply this estimate to D^kG and obtain

$$\begin{aligned}\|D^{k+1}G\| &\le 2\|D^kG\|^{\frac{1}{2}}\left(\mathrm{Lip}(D^{k+1}G)\right)^{\frac{1}{2}}\\ &\le 2^{\frac{k+2}{2}}\|G\|^{\frac{1}{2(k+1)}}\left(\mathrm{Lip}(D^kG)\right)^{\frac{k}{2(k+1)}}\left(\mathrm{Lip}(D^{k+1}G)\right)^{\frac{1}{2}}\\ &= 2^{\frac{k+2}{2}}\|G\|^{\frac{1}{2(k+1)}}\|D^{k+1}G\|^{\frac{k}{2(k+1)}}\left(\mathrm{Lip}(D^{k+1}G)\right)^{\frac{1}{2}}.\end{aligned}$$

Collecting $\|D^{k+1}G\|$ on the left side and raising to the power $\frac{2(k+1)}{k+2}$ yields (6.18) for $k+1$:

$$\|D^{k+1}G\| \le 2^{k+1}\|G\|^{\frac{1}{k+2}}\left(\mathrm{Lip}(D^{k+1}G)\right)^{\frac{k+1}{k+2}}.$$

□

If $k \ge 1$, then the inequalities (6.4) and (6.5) imply that there is $\varepsilon > 0$ such that

$$(\varepsilon + \alpha)\frac{\beta^{k+1}}{(1-\varepsilon\beta)^{k+2}} < 1. \tag{6.20}$$

Since R is continuously differentiable and Lipschitz, we can assume that $\|R\|_{C^1} \le \varepsilon$ holds with this value of ε.

The following lemma follows directly from the chain rule.

Lemma 6.6 *For any C^l-functions $G_k : \mathbb{R}^{n_k} \to \mathbb{R}^{n_{k-1}}, k = 1, 2, 3$, with $l \ge 1$, holds*

$$\begin{aligned}\|D^l(G_1\circ G_2\circ G_3)\| \le\ &\|D^lG_1\|\,\|DG_2\|^l\,\|DG_3\|^l\\ &+ \|DG_1\|\,\|D^lG_2\|\,\|DG_3\|^l\\ &+ \|DG_1\|\,\|DG_2\|\,\|D^lG_3\| + \mathcal{R}_l,\end{aligned} \tag{6.21}$$

where $\mathcal{R}_l$ depends only on $\|D^jG_m\|$ with $j = 1, 2, \ldots, l-1$ and $m = 1, 2, 3$. □

Lemma 6.7 *Provided all derivatives below exist for each $l \geq 2$, we have*

$$\|D^l S_H\| = \|D^l V_H\| \leq \varepsilon \left(\frac{\beta}{1-\varepsilon\beta}\right)^{l+1} \|D^l H\| + C_l, \tag{6.22}$$

where C_l is some constant that depends only on A, $\|R\|_{C^l}$, and $\|D^j H\|$ with $j = 1, 2, \ldots, l-1$.

Proof From (6.12) and (6.15) it follows that $V_H = B^{-1}(f \circ \hat{H} \circ S_H)$. The estimate (6.21) implies that

$$\|D^l V_H\| \leq \|B^{-1}\| \left(\|D^l f\| \, \|D\hat{H}\|^l \|DS_H\|^l + \|Df\| \, \|D^l \hat{H}\| \, \|DS_H\|^l + \|Df\| \, \|D\hat{H}\| \, \|D^l S_H\| + \mathcal{R}_l\right).$$

Since $\|R\|_{C^1} \leq \varepsilon$, we have $\|Df\| \leq \varepsilon$. Moreover, $\|D\hat{H}\| \leq 1$ and $\|D^l \hat{H}\| = \|D^l H\|$. Thus, (6.13) implies

$$\|D^l V_H\| \leq \beta \|D^l f\| \left(\frac{\beta}{1-\varepsilon\beta}\right)^l + \varepsilon\beta \|D^l H\| \left(\frac{\beta}{1-\varepsilon\beta}\right)^l + \varepsilon\beta \|D^l V_H\| + \tilde{\mathcal{R}}_l,$$

where $\tilde{\mathcal{R}}_l$ has the same properties as $\mathcal{R}_l$. Since $1 - \varepsilon\beta > 0$ due to (6.7), we obtain

$$\begin{aligned} \|D^l V_H\| &\leq \frac{1}{1-\varepsilon\beta} \left[\beta \|D^l f\| \left(\frac{\beta}{1-\varepsilon\beta}\right)^l + \varepsilon\beta \|D^l H\| \left(\frac{\beta}{1-\varepsilon\beta}\right)^l + \tilde{\mathcal{R}}_l\right] \\ &= \varepsilon \left(\frac{\beta}{1-\varepsilon\beta}\right)^{l+1} \|D^l H\| + C_l, \end{aligned}$$

where C_l depends only on A, $\|R\|_{C^l}$, and $\|D^j H\|$ with $j = 1, 2, \ldots, l-1$, but not on $\|D^l H\|$.

The equality $\|D^l S_H\| = \|D^l V_H\|$ for $l \geq 2$ follows directly from (6.12) since B is a constant matrix. □

We can now estimate the derivatives of the graph transform T.

Lemma 6.8 *For $H \in M^0 \cap C^l$ and $l \geq 2$ we have*

$$\|D^l T(H)\| \leq \frac{(\alpha + \varepsilon)\beta^l}{(1-\varepsilon\beta)^{l+1}} \|D^l H\| + \tilde{C}_l \tag{6.23}$$

for some constant $\tilde{C}_l$ that depends only on A, $\|R\|_{C^l}$, and $\|D^j H\|$ with $j = 1, 2, \ldots, l-1$.

Proof The graph transform is defined by Eq. (6.14), i.e. $T(H) = Q \circ \hat{H} \circ S_H$. The estimate (6.21) now gives

$$\|D^l T(H)\| \leq \|D^l Q\| \, \|D\hat{H}\|^l \|DS_H\|^l + \|DQ\| \, \|D^l \hat{H}\| \, \|DS_H\|^l + \|DQ\| \, \|D\hat{H}\| \, \|D^l S_H\| + \mathcal{R}_l.$$

Using $\|D\hat{H}\| \leq \|DH\| \leq 1$, the estimates (6.13) and (6.22), as well as $\|R\|_{C^1} \leq \varepsilon$ implying $\|DQ\| \leq \|C\| + \varepsilon$, we get

$$\begin{aligned} \|D^l T(H)\| \leq\; & \|D^l Q\| \left(\frac{\beta}{1-\varepsilon\beta}\right)^l + \|DQ\| \, \|D^l H\| \left(\frac{\beta}{1-\varepsilon\beta}\right)^l \\ & + \varepsilon \|DQ\| \|D^l H\| \left(\frac{\beta}{1-\varepsilon\beta}\right)^{l+1} + C_l + \mathcal{R}_l \end{aligned}$$

$$= \frac{\|DQ\|\beta^l}{(1-\varepsilon\beta)^{l+1}} \|D^l H\| + \tilde{C}_l \leq \frac{(\alpha+\varepsilon)\beta^l}{(1-\varepsilon\beta)^{l+1}} \|D^l H\| + \tilde{C}_l,$$

where $\tilde{C}_l$ depends only on A, $\|R\|_{C^l}$, and $\|D^j H\|$ for $j = 1, 2, \ldots, l-1$. □

Lemma 6.9 *There exists* $r = (r_2, r_3, \ldots, r_{k+1}) \in \mathbb{R}^k$ *with* $r_j > 0$ *such that* $T(M_r^k) \subset M_r^k$.

Proof We proceed by induction. For $k = 1$ we have

$$M_{r_2}^1 = \{H \in C^{1,1} : \|H\| < \infty, \|DH\| \leq 1, \mathrm{Lip}(DH) \leq r_2\}.$$

First note that $\|DT(H)\| = \mathrm{Lip}(T(H)) \leq 1$ has been shown in *Step 2.* Then, for each $H \in M_{r_2}^1 \cap C^2$, the estimate (6.23) implies that

$$\mathrm{Lip}(DT(H)) \leq \frac{(\alpha+\varepsilon)\beta^2}{(1-\varepsilon\beta)^3} \mathrm{Lip}(DH) + \tilde{C}_2 \leq \frac{(\alpha+\varepsilon)\beta^2}{(1-\varepsilon\beta)^3} r_2 + \tilde{C}_2.$$

Using (6.20) and observing that $\tilde{C}_2$ does not depend on r_2, we see that the right-hand side is less than or equal to r_2, provided we choose r_2 such that

$$r_2 \geq \tilde{C}_2 \left[1 - \frac{(\alpha+\varepsilon)\beta^2}{(1-\varepsilon\beta)^3}\right]^{-1}.$$

Since $M_{r_2}^1 \cap C^2$ is dense in $M_{r_2}^1$, this estimate ensures that $T(M_{r_2}^1) \subset M_{r_2}^1$. The same argument can now be used to show that we can recursively find k constants $r_j > 0, j = 2, 3, \ldots, k+1$, such that $T(M_r^k) \subset M_r^k$ with $r = (r_2, r_3, \ldots, r_{k+1})$. □

Step 5: It is left to demonstrate that $Dh(0) = 0$. For this, observe that from (6.10) now follows

$$C\,Dh(0) - Dh(0)B = 0.$$

We claim that this implies that the linear map $Dh : \mathbb{R}^{n_{cu}} \to \mathbb{R}^{n_s}$ equals zero. Indeed, consider any eigenvector $v \in \mathbb{C}^{n_{cu}}$ of matrix B corresponding to some eigenvalue $\lambda \in \mathbb{C}$. Then $Dh(0)v$ is either zero or an eigenvector of C corresponding to the same eigenvalue λ, since

$$C\,Dh(0)v = Dh(0)Bv = \lambda Dh(0)v.$$

However, $\sigma(B) \cap \sigma(C) = \emptyset$, which implies that $Dh(0)v$ cannot be an eigenvector of C corresponding to the eigenvalue λ. We conclude that $Dh(0)v = 0$. If B has n_{cu} linearly independent eigenvectors, this gives $Dh(0) = 0$. Otherwise, we have to consider a basis including generalized eigenvectors as well. By using the same argument as above, one easily proves that if v belongs to the null-space of $(\lambda I_{n_{cu}} - B)^k$ then $Dh(0)v$ is in the null-space of $(\lambda I_{n_s} - C)^k$. As the nontriviality of the null-space of $(\lambda I_{n_{cu}} - B)^k$ implies triviality of that of $(\lambda I_{n_s} - C)^k$, we find that necessarily $Dh(0)v = 0$. We see that also in this case $Dh(0)$ annihilates n_{cu} linearly independent vectors, and hence must vanish. □

Similar to our approach for the Grobman–Hartman Theorem in Chap. 3, we now set up a local version of Theorem 6.3.

Theorem 6.10 (Existence of a Local Centre-Unstable Manifold) *Assume that the function R in (6.1) is of class C^{k+1} for some $k \geq 1$ and satisfies $R(0) = 0$ and $DR(0) = 0$.*

Then there exists a C^k-map $h : \mathbb{R}^{n_{cu}} \to \mathbb{R}^{n_s}$ and $\delta > 0$ such that

$$W_\delta^{cu} = \{(x, h(x)) : x \in \mathbb{R}^{n_{cu}}, \|x\| \leq \delta\}$$

is conditionally invariant for the map (6.1). Moreover, $h(0) = 0$ and $Dh(0) = 0$.

Remark Recall from Chap. 3 that conditional invariance means that any point $(x, h(x))$ in W_δ^{cu} with image (ξ, η) such that $\|\xi\| \leq \delta$ satisfies $(\xi, \eta) \in W_\delta^{cu}$, i.e. $\eta = h(\xi)$. ♢

Definition 6.11 *The set W_δ^{cu} is called a* ***local centre–unstable manifold*** *of the fixed point 0 of (6.1).*

Proof of Theorem 6.10 Instead of (6.1), consider the map

$$u \mapsto Au + \chi\left(\frac{1}{\delta}u\right) R(u) \tag{6.24}$$

where $\delta > 0$ and $\chi \in C^\infty(\mathbb{R}^n, \mathbb{R})$ is a standard "cut-off" function with $\chi(u) = 1$ for $0 \leq \|u\| \leq 1$ and $\chi(u) = 0$ for $\|u\| \geq 2$. The map (6.24) coincides with (6.1) for $u \in \mathbb{R}^n$ satisfying $\|u\| \leq \delta$. With the scaling $u \mapsto \delta u$ the mapping (6.24) transforms into

$$u \mapsto Au + R_\delta(u), \tag{6.25}$$

where

$$R_\delta(u) = \chi(u)\frac{1}{\delta}R(\delta u).$$

To apply Theorem 6.3 to this map, we first verify that

$$\|R_\delta(u)\| \leq K_\delta, \quad \|DR_\delta(u)\| \leq L_\delta,$$

where K_δ, L_δ can be made arbitrarily small by sending $\delta \to 0$. Note that, by continuity of $DR(u)$ and $DR(0) = 0$, we have $\ell_\delta := \sup\{\|DR(v)\| : \|v\| \leq 2\delta\} \to 0$ as $\delta \to 0$. Since

$$R(\delta u) = R(\delta u) - R(0) = \int_0^1 \frac{d}{dt}R(\delta t u)\, dt = \delta \int_0^1 DR(\delta t u)u\, dt,$$

we have

$$\frac{1}{\delta}\|R(\delta u)\| \le 2\ell_\delta$$

if $\|u\| \le 2$. This yields

$$\|R_\delta(u)\| \le 2\|\chi\|\ell_\delta =: K_\delta,$$
$$\|DR_\delta(u)\| = \left\| D\chi(u)\frac{1}{\delta}R(\delta u) + \chi(u)DR(\delta u)\right\| \le (2\|D\chi\| + \|\chi\|)\ell_\delta =: L_\delta.$$

Therefore, the conditions

$$\|R_\delta\| \le \varepsilon,\ \mathrm{Lip}(R_\delta) \le \varepsilon$$

are satisfied for an arbitrary small $\varepsilon > 0$ provided $\delta = \delta(\varepsilon) > 0$ is sufficiently small. With h_δ being the fixed point from Theorem 6.3 corresponding to R_δ set

$$h(x) = \delta h_\delta\left(\frac{1}{\delta}x\right). \tag{6.26}$$

Then global invariance of $\{(x, h_\delta(x)) : x \in \mathbb{R}^{n_{cu}}\}$ under the map (6.25) yields conditional invariance of W^{cu}_δ under the map (6.1). Finally, $Dh(0) = 0$ directly follows from $Dh_\delta(0) = 0$. □

Remarks

(1) While the global centre-unstable manifold in Theorem 6.3 is unique due to its global attractivity, this is no longer true for the local centre-unstable manifold. In general W^{cu}_δ depends on ε and on the cut-off function χ. However, one can show that the derivatives of the possible functions h agree at the origin up to the guaranteed order of differentiability.

(2) It can happen that $\delta \to 0$ as $k \to \infty$. Thus, there are C^∞-maps having no C^∞ centre manifolds. Moreover, there are analytic mappings admitting C^∞ centre manifolds which are not analytic.

(3) Actually, one can avoid the loss of smoothness in Theorem 6.10, i.e. prove that h has the same smoothness as the map (6.1). For that purpose we say that R is in $C^{k,1}_{\mathrm{loc}}$, $k \ge 0$ if R is of class C^k in some δ-neighbourhood of 0 and D^kR is Lipschitz there. By $\mathrm{Lip}_\delta(D^kR)$ we denote the smallest constant L satisfying

$$\|D^kR(u) - D^kR(v)\| \le L\|u - v\| \quad u, v \in \mathbb{R}^n \text{ with } \|u\|, \|v\| \le \delta.$$

Theorem 6.12 *Assume that the function R in (6.1) is of class $C^{0,1}_{\mathrm{loc}}$ and satisfies $R(0) = 0$ and $\mathrm{Lip}_\delta(R) \to 0$ as $\delta \to 0$.*

Then there exist a map $h \in C^{0,1}_{\mathrm{loc}}(\mathbb{R}^{n_{cu}}, \mathbb{R}^{n_c})$ and $\delta > 0$ such that $h(0) = 0$ and

$$W_\delta^{cu} = \{(x, h(x)) : x \in \mathbb{R}^{n_{cu}}, \|x\| \leq \delta\}$$

is conditionally invariant for the map (6.1). *Moreover, if R is of class $C^{k,1}_{\mathrm{loc}}$ for some $k \geq 1$ and $DR(0) = 0$ then h is of class $C^{k,1}_{\mathrm{loc}}$ and satisfies $Dh(0) = 0$.*

Proof As in the proof of Theorem 6.10, instead of (6.1), consider the map

$$u \mapsto Au + \chi\left(\frac{1}{\delta}u\right) R(u), \tag{6.27}$$

where $\delta > 0$ and $\chi \in C^\infty(\mathbb{R}^n, \mathbb{R})$ is a standard "cut-off" function with $\chi(u) = 1$ for $0 \leq \|u\| \leq 1$ and $\chi(u) = 0$ for $\|u\| \geq 2$. We can assume that the map (6.27) is Lipschitz continuous in the ball of radius 2δ. The map (6.27) coincides with (6.1) for $u \in \mathbb{R}^n$ satisfying $\|u\| \leq \delta$ and it is zero for $\|u\| \geq 2\delta$. With the scaling $u \mapsto \delta u$ the mapping (6.27) transforms into

$$u \mapsto Au + R_\delta(u), \tag{6.28}$$

where

$$R_\delta(u) = \chi(u)\frac{1}{\delta}R(\delta u).$$

To apply Theorem 6.3 to this map, we verify that $\|R_\delta\|$ and $\mathrm{Lip}(R_\delta)$ can be made arbitrarily small by sending $\delta \to 0$. For $\|u\|, \|v\| \leq 2$ we estimate

$$\begin{aligned} \|R_\delta(u) - R_\delta(v)\| &\leq \frac{1}{\delta}\left[|\chi(u)|\|R(\delta u) - R(\delta v)\| + |\chi(u) - \chi(v)|\|R(\delta v) - R(0)\|\right] \\ &\leq \mathrm{Lip}_{2\delta}(R)\|u - v\|\left[\|\chi\| + \mathrm{Lip}(\chi)\|v\|\right]. \end{aligned} \tag{6.29}$$

Using this estimate with $v = 0$ yields $\|R_\delta\| \leq 2\|\chi\|\mathrm{Lip}_{2\delta}(R)$ in view of $R_\delta(0) = 0$ and $R_\delta(u) = 0$ for $\|u\| \geq 2$. Next we claim that

$$\mathrm{Lip}(R_\delta) \leq \mathrm{Lip}_{2\delta}(R)(\|\chi\| + 2\mathrm{Lip}(\chi)).$$

For $\|u\|, \|v\| \leq 2$ the Lipschitz estimate follows from (6.29) while it is trivial for $\|u\|, \|v\| \geq 2$. In case $\|u\| \leq 2 < \|v\|$ we apply (6.29) to u and $\tilde{v} = u + c(v - u)$ where $c \in [0, 1]$ is the unique value such that $\|\tilde{v}\| = 2$. This proves our assertion since $R_\delta(v) = R_\delta(\tilde{v}) = 0$ and $\|u - \tilde{v}\| \leq \|u - v\|$.

With h_δ being the fixed point from Theorem 6.3 corresponding to R_δ set

$$h(x) = \delta h_\delta\left(\frac{1}{\delta}x\right). \tag{6.30}$$

Then global invariance of $\{(x, h_\delta(x)) : x \in \mathbb{R}^{n_{cu}}\}$ under the map (6.28) yields global invariance of $W_\delta^{cu} = \{(x, h(x)) : x \in \mathbb{R}^{n_{cu}}\}$ under the map (6.27) and hence conditional invariance of W_δ^{cu} under the map (6.1). Moreover, if R is in $C^{k,1}_{\mathrm{loc}}$ and $k \geq 1$ then $DR(0) = 0$ implies

$$\mathrm{Lip}_\delta(R) = \sup_{\|u\| \leq \delta} \|DR(u) - DR(0)\| \to 0 \text{ as } \delta \to 0.$$

By Theorem 6.3 the map h_δ is of class $C^{k,1}$ and hence h is of class $C^{k,1}_{\mathrm{loc}}$. □

(4) If the matrix B in (6.2) has no critical eigenvalues, i.e. all its eigenvalues satisfy $|\lambda| > 1$, Theorem 6.10 provides the existence of a smooth local *unstable manifold* of the hyperbolic saddle point $u = 0$ of (6.1). ◇

Consider now the C^{k+2}-map with $k \geq 1$

$$\begin{pmatrix} \xi \\ \eta_1 \\ \eta_2 \end{pmatrix} \mapsto \begin{pmatrix} B_1\xi + f_1(\xi, \eta_1, \eta_2) \\ C_1\eta_1 + g_1(\xi, \eta_1, \eta_2) \\ C_2\eta_2 + g_2(\xi, \eta_1, \eta_2) \end{pmatrix}, \tag{6.31}$$

where $(\xi, \eta_1, \eta_2) \in \mathbb{R}^{n_c} \times \mathbb{R}^{n_s} \times \mathbb{R}^{n_u}$. We now assume that the $n_c \times n_c$ matrix B_1 has n_c eigenvalues with $|\lambda| = 1$, all eigenvalues of the $n_u \times n_u$ matrix C_1 satisfy $|\lambda| > 1$, all eigenvalues of the $n_s \times n_s$ matrix C_2 satisfy $|\lambda| < 1$, while the functions f_1, g_1, and g_2 have neither constant nor linear terms.

Theorem 6.13 (Existence of a Centre Manifold) *The map* (6.31) *has a locally defined invariant manifold*

$$W^c = \{(\xi, h^c(\xi)) : \xi \in \mathbb{R}^{n_c}, \ \|\xi\| \leq \varepsilon\},$$

where $\varepsilon > 0$ *is sufficiently small and* $h^c : \mathbb{R}^{n_c} \longrightarrow \mathbb{R}^{n_s+n_u}$ *is a* C^k*-map satisfying*

$$h^c(0) = 0, \ Dh^c(0) = 0.$$

Proof Theorem 6.10 guarantees the existence of the centre-unstable manifold

$$W^{cu}_{\varepsilon_1} = \{(\xi, \eta_1, h(\xi, \eta_1)) : (\xi, \eta_1) \in \mathbb{R}^{n_c} \times \mathbb{R}^{n_u}, \|(\xi, \eta_1)\| \leq \varepsilon_1\},$$

where $h : \mathbb{R}^{n_c} \times \mathbb{R}^{n_u} \longrightarrow \mathbb{R}^{n_s}$ is a C^{k+1}-function that vanishes at $(\xi, \eta_1) = (0, 0)$ together with its first derivative. The restriction of (6.31) to $W^{cu}_{\varepsilon_1}$ is given by the C^{k+1}-map

$$\begin{pmatrix} \xi \\ \eta_1 \end{pmatrix} \mapsto \begin{pmatrix} B_1\xi + f_1(\xi, \eta_1, h(\xi, \eta_1)) \\ C_1\eta_1 + g_1(\xi, \eta_1, h(\xi, \eta_1)) \end{pmatrix}, \tag{6.32}$$

Applying Theorem 6.10 to a map that is locally inverse to (6.32), we get a manifold W^c with all properties mentioned above. □

Clearly, the manifold W^c is tangent to the generalized eigenspace $\mathbb{R}^{n_c}$ of the matrix

$$\begin{pmatrix} B_1 & 0 & 0 \\ 0 & C_1 & 0 \\ 0 & 0 & C_2 \end{pmatrix}$$

corresponding to its critical eigenvalues, i.e. the eigenvalues of B_1.

Definition 6.14 W^c *is called a* ***centre manifold*** *of the fixed point* $(0, 0)$ *of* (6.2).

We formulate without proof the following theorem.

Theorem 6.15 (Reduction Principle) *Consider a sufficiently smooth map*

$$\begin{pmatrix} \xi \\ \eta \end{pmatrix} \mapsto \begin{pmatrix} B\xi + f(\xi, \eta), \\ C\eta + g(\xi, \eta) \end{pmatrix}, \tag{6.33}$$

where $\xi \in \mathbb{R}^{n_c}$, $\eta \in \mathbb{R}^{n_s+n_u}$, *the* $n_c \times n_c$ *matrix* B *has* n_c *eigenvalues with* $|\lambda| = 1$, *while all eigenvalues of the invertible* $(n_s + n_u) \times (n_s + n_u)$ *matrix* C *satisfy* $|\lambda| \neq 1$, *and the functions* f *and* g *have neither constant nor linear terms.*

The map (6.33) *is locally topologically conjugate near* $(0, 0)$ *to the map*

$$\begin{pmatrix} \xi \\ \eta \end{pmatrix} \mapsto \begin{pmatrix} B\xi + f(\xi, h^c(\xi)), \\ C\eta \end{pmatrix}, \tag{6.34}$$

where h^c *represents the centre manifold* W^c *given by* Theorem 6.13. □

The maps for ξ and η are decoupled in (6.34). Therefore, the map (6.34) is locally topologically conjugate near $(0, 0)$ to a map

$$\begin{pmatrix} \xi \\ \eta \end{pmatrix} \mapsto \begin{pmatrix} b(\xi) \\ C\eta \end{pmatrix},$$

where $\xi \mapsto b(\xi)$ is any map that is locally topologically conjugate near $\xi = 0$ to $\xi \mapsto B\xi + f(\xi, h^c(\xi))$. Indeed, the conjugating homeomorphism can be constructed as the direct product of a conjugating homeomorphism in the ξ-space and the identity map in the η-space.

Corollary 6.16 *If* $\xi = 0$ *is a stable fixed point of the restriction of* (6.33) *to its centre manifold*

$$\xi \mapsto B\xi + f(\xi, h^c(\xi)), \quad \xi \in \mathbb{R}^{n_c},$$

and $n_u = 0$ (*i.e. all eigenvalues of* C *satisfy* $|\lambda| < 1$), *then* $(\xi, \eta) = (0, 0)$ *is a stable fixed point of* (6.33). □

6.2 Centre Manifolds for ODEs

Consider a C^k-smooth system with $k \geq 0$,

$$\dot{u} = Au + r(u), \quad u \in \mathbb{R}^n, \tag{6.35}$$

with $n = n_{cu} + n_s$ and

$$u = \begin{pmatrix} x \\ y \end{pmatrix} \in \mathbb{R}^{n_{cu}} \times \mathbb{R}^{n_s}, \quad A = \begin{pmatrix} B & 0 \\ 0 & C \end{pmatrix}, \quad r(u) = \begin{pmatrix} f(u) \\ g(u) \end{pmatrix},$$

where f and g have neither constant nor linear terms. Suppose that the $n_{cu} \times n_{cu}$ matrix B has n_c *critical eigenvalues* (i.e. eigenvalues with $\operatorname{Re} \lambda = 0$) and n_u eigenvalues with $\operatorname{Re} \lambda > 0$, while all n_s eigenvalues of the $n_s \times n_s$ matrix C satisfy $\operatorname{Re} \lambda < 0$.

We can equivalently rewrite (6.35) as

$$\begin{cases} \dot{x} = Bx + f(x, y), \\ \dot{y} = Cy + g(x, y). \end{cases} \tag{6.36}$$

Theorem 6.17 (Global Centre-Unstable Manifold for ODEs) *There exists $\nu > 0$ such that when $r \in C^{0,1}$ with $\|r\| \leq \nu$ and $\operatorname{Lip}(r) \leq \nu$ then the system (6.35) has a globally attracting invariant manifold*

$$W^{cu} = \{(x, h(x)) : x \in \mathbb{R}^{n_{cu}}\},$$

where $h : \mathbb{R}^{n_{cu}} \to \mathbb{R}^{n_s}$ is a $C^{0,1}$-map satisfying $h(0) = 0$.

If $r \in C^{k,1}$ with $k \geq 1$, then the above assertion holds and, moreover, h is a $C^{k,1}$-map satisfying $Dh(0) = 0$.

Proof

Step 1: Since r satisfies a global Lipschitz estimate, (6.35) generates a global solution flow $\Phi^t(u)$. Define $R^t(u)$ as the nonlinear part of the flow

$$\Phi^t(u) = e^{tA}u + R^t(u) = \begin{pmatrix} e^{tB} & 0 \\ 0 & e^{tC} \end{pmatrix} u + R^t(u). \tag{6.37}$$

We show that Theorem 6.3 applies to this map $\Phi^t(u)$ for all $0 < t \leq 2$.

Step 2: By Theorem 2.19 there exist Lyapunov norms $\|\cdot\|_1$ on $\mathbb{R}^{n_{cu}}$ and $\|\cdot\|_2$ on $\mathbb{R}^{n_s}$ and numbers $0 < b < a$ such that

$$\alpha(t) = \|e^{tC}\|_2 \leq e^{-at}, \quad \beta(t) = \|e^{-tB}\|_1 \leq e^{bt}, \quad t \geq 0. \tag{6.38}$$

Clearly, $\alpha(t) < 1$ and $\alpha(t) < \beta(t)^{-1}$ for all $t > 0$. Thus, the basic conditions (6.4) and (6.5) are satisfied with $j = 0$ for all $t > 0$. In what follows we use these adapted norms and define

$$\|u\| = \max(\|x\|_1, \|y\|_2), \quad u = \begin{pmatrix} x \\ y \end{pmatrix}.$$

Step 3: From the variation of constants formula, we obtain

$$R^t(u) = \int_0^t e^{(t-s)A} r(\Phi^s(u))ds \tag{6.39}$$

so that

$$\|R^t(u)\| = \left\| \int_0^t e^{(t-s)A} r(\Phi^s(u))ds \right\| \leq \nu \int_0^t e^{(t-s)\|A\|} ds = \nu \frac{e^{t\|A\|} - 1}{\|A\|}.$$

Observe that, for $\nu > 0$ sufficiently small,

$$\|R^t(u)\| \leq K_\nu(t) := \nu \frac{e^{t\|A\|} - 1}{\|A\|} \leq \varepsilon, \quad 0 < t \leq 2,$$

where $\varepsilon > 0$ is arbitrarily small. Further note that the variation of constants formula also implies

$$\begin{aligned} \|\Phi^t(u) - \Phi^t(v)\| &\leq \|e^{tA}(u - v)\| + \left\| \int_0^t e^{(t-s)A} (r(\Phi^s(u)) - r(\Phi^s(v)))ds \right\| \\ &\leq e^{t\|A\|} \|u - v\| + \nu \int_0^t e^{(t-s)\|A\|} \|\Phi^s(u) - \Phi^s(v)\| ds \end{aligned}$$

or

$$e^{-t\|A\|} \|\Phi^t(u) - \Phi^t(v)\| \leq \|u - v\| + \nu \int_0^t e^{-s\|A\|} \|\Phi^s(u) - \Phi^s(v)\| ds \ ,$$

which by Gronwall's Lemma 3.3 leads to

$$\|\Phi^t(u) - \Phi^t(v)\| \leq e^{t(\|A\| + \nu)} \|u - v\|. \tag{6.40}$$

In this way we obtain the Lipschitz estimate for R^t

$$\begin{aligned} \|R^t(u) - R^t(v)\| &= \left\| \int_0^t e^{(t-s)A} (r(\Phi^s(u)) - r(\Phi^s(v)))ds \right\| \\ &\leq \int_0^t e^{\|A\|(t-s)} \nu \|\Phi^s(u) - \Phi^s(v)\| ds \\ &\leq e^{t\|A\|} (e^{t\nu} - 1) \|u - v\| = L_\nu(t) \|u - v\|, \end{aligned}$$

where

$$L_\nu(t) := e^{t\|A\|} (e^{t\nu} - 1).$$

Then, if we take $\nu > 0$ sufficiently small, we can satisfy the condition

$$\text{Lip}(R^t) \leq L_\nu(t) \leq \varepsilon$$

for all $0 < t \leq 2$ for arbitrary small $\varepsilon > 0$. Thus, the first part of Theorem 6.3 applies.

Step 4: By the proof of Theorem 6.3, the Hadamard graph transform T_t corresponding to Φ^t has a unique fixed point H_t in M^0 for $0 < t \leq 2$. Now we show that $H_t = H_s$ for all $0 < t, s \leq 1$. From the flow property of Φ^t one finds $T_t \circ T_s = T_{t+s} = T_s \circ T_t$ for $0 < t, s \leq 1$. From $T_s(H_s) = H_s$ we now deduce

$$T_t(H_s) = T_t(T_s(H_s)) = T_s(T_t(H_s)).$$

This means that $T_t(H_s)$ is a fixed point of T_s so that, by uniqueness, $T_t(H_s) = H_s$. Then the uniqueness of the fixed point of T_t gives $H_s = H_t$. Therefore, all the functions $H_t, 0 < t \leq 1$ coincide. This proves that the global centre-unstable manifold $W = W^{cu}$ defined by $h = H_t$ is invariant under the flow $\Phi^s, 0 < s \leq 1$. Since

$$\Phi^n = \underbrace{\Phi^1 \circ \Phi^1 \circ \cdots \circ \Phi^1}_{n \text{ times}},$$

the graph of h is invariant under Φ^n. For arbitrary t, we write $\Phi^t = \Phi^{[t]} \circ \Phi^{t-[t]}$ and note that the graph of h is invariant under both $\Phi^{t-[t]}$ and $\Phi^{[t]}$. Here $[t]$ is the integer part of t. We conclude that we have the invariance under Φ^t without any restriction on $t \geq 0$.

Step 5: To complete the proof of Theorem 6.17, first notice that the conditions (6.4) and (6.5) would also be satisfied for $\alpha = \alpha(t)$ and $\beta = \beta(t)$ with any $j \geq 1$ for $t > 0$. The second part of Theorem 6.17 will then follow from the second part of Theorem 6.3. Indeed, one can show that $r \in C^{k,1}$ with $k \geq 1$ implies that R^t has the same properties for $0 < t < 2$.

Since $r \in C^{k,1}$ with $k \geq 1$, it follows from Theorem 1.5 that $R^t \in C^k$. Below we only indicate how to obtain estimates for the first derivative of $R^t(u)$ with respect to u. The bound

$$\|DR^t(u)\| \leq L_\nu(t) = e^{t\|A\|}(e^{t\nu} - 1)$$

follows from the Lipschitz estimate for R^t above. Differentiating (6.37), we obtain

$$D\Phi^t(u) = e^{tA} + DR^t(u)$$

and thus $\|D\Phi^t(u)\| \leq e^{t\|A\|} + \|DR^t(u)\| \leq e^{t\|A\|} + e^{t\|A\|}(e^{t\nu} - 1)$, so that

$$\|D\Phi^t(u)\| \leq e^{t(\|A\|+\nu)}. \tag{6.41}$$

It remains to establish that DR^t is Lipschitz on $\mathbb{R}^n$ according to Definition 6.1. Differentiating (6.39) gives

$$DR^t(u) = \int_0^t e^{(t-s)A} Dr(\Phi^s(u)) D\Phi^s(u) ds. \tag{6.42}$$

Let ν_1 be simultaneously a bound for Dr and a Lipschitz constant for Dr. Then

$$
\begin{aligned}
\|D\Phi^t(u) - D\Phi^t(v)\| &= \|DR^t(u) - DR^t(v)\| \\
&= \left\| \int_0^t e^{(t-s)A} \left[Dr(\Phi^s(u))D\Phi^s(u) - Dr(\Phi^s(v))D\Phi^s(v) \right] \right\| \\
&\leq \int_0^t e^{(t-s)\|A\|} \|Dr(\Phi^s(u))D\Phi^s(u) - Dr(\Phi^s(v))D\Phi^s(u)\| \, ds \\
&\quad + \int_0^t e^{(t-s)\|A\|} \|Dr(\Phi^s(v))D\Phi^s(u) - Dr(\Phi^s(v))D\Phi^s(v)\| ds \\
&\leq \int_0^t e^{(t-s)\|A\|} \|Dr(\Phi^s(u)) - Dr(\Phi^s(v))\| \|D\Phi^s(u)\| ds \\
&\quad + \int_0^t e^{(t-s)\|A\|} \|Dr(\Phi^s(v))\| \|D\Phi^s(u) - D\Phi^s(v)\| ds
\end{aligned}
$$

$$
\leq \nu_1 \|u - v\| e^{t\|A\|} \int_0^t e^{s(\|A\|+2\nu)} ds + \nu_1 \int_0^t e^{(t-s)\|A\|} \|D\Phi^s(u) - D\Phi^s(v)\| ds.
$$

Here we have used bounds (6.41) and (6.40). Thus

$$
\begin{aligned}
e^{-t\|A\|} \|D\Phi^t(u) - D\Phi^t(v)\| &\leq \nu_1 \|u - v\| \frac{e^{t(\|A\|+2\nu)} - 1}{\|A\| + 2\nu} \\
&\quad + \nu_1 \int_0^t e^{-s\|A\|} \|D\Phi^s(u) - D\Phi^s(v)\| ds
\end{aligned}
$$

Using Gronwall's Lemma 3.3, we get the estimate for $D\Phi^t$:

$$
\|D\Phi^t(u) - D\Phi^t(v)\| \leq \nu_1 e^{t(\|A\|+\nu_1)} \frac{e^{t(\|A\|+2\nu)} - 1}{\|A\| + 2\nu} \|u - v\|.
$$

Thus DR^t is Lipschitz, since $\|DR^t(u) - DR^t(v)\| = \|D\Phi^t(u) - D\Phi^t(v)\|$. □

Theorem 6.18 (Existence of a Local Centre-Unstable Manifold) *Assume that the function r in (6.35) is of class C^{k+1} for some $k \geq 1$ and satisfies $r(0) = 0$ and $Dr(0) = 0$. Then the system (6.35) has a conditionally invariant local centre-unstable manifold*

$$
W_\delta^{cu} = \{(x, h(x)) : x \in \mathbb{R}^{n_{cu}}, \ \|x\| \leq \delta\},
$$

where $\delta > 0$ is sufficiently small and h is a C^k-function.

Proof Take the cut-off function χ from the proof of Theorem 6.10 and replace (6.35) by the scaled system

$$
\dot{u} = Au + r_\delta(u),
$$

where

$$
r_\delta(u) = \chi(u) \frac{1}{\delta} r(\delta u).
$$

The proof of Theorem 6.10 shows that r_δ, Dr_δ have small global bounds and Lipschitz constants as well. Again the global centre-unstable manifold of the cut-off system leads to a conditionally invariant local centre-unstable manifold of the original system (6.35). □

As in the discrete-time case, consider next the C^{k+2}-system

$$\begin{cases} \dot{\xi} = B_1\xi + f_1(\xi, \eta_1, \eta_2) \\ \dot{\eta}_1 = C_1\eta_1 + g_1(\xi, \eta_1, \eta_2) \\ \dot{\eta}_2 = C_2\eta_2 + g_2(\xi, \eta_1, \eta_2) \end{cases} \tag{6.43}$$

where $(\xi, \eta_1, \eta_2) \in \mathbb{R}^{n_c} \times \mathbb{R}^{n_s} \times \mathbb{R}^{n_u}$. We now assume that the $n_c \times n_c$ matrix B_1 has n_c eigenvalues with $\operatorname{Re} \lambda = 0$, all eigenvalues of the $n_u \times n_u$ matrix C_1 satisfy $\operatorname{Re} \lambda > 0$, all eigenvalues of the $n_s \times n_s$ matrix C_2 satisfy $\operatorname{Re} \lambda < 0$, while the functions f_1, g_1, and g_2 have neither constant nor linear terms.

Theorem 6.19 (Existence of a Centre Manifold) *The system* (6.43) *has a locally defined invariant manifold*

$$W^c = \{(\xi, h^c(\xi)) : \xi \in \mathbb{R}^{n_c},\ \|\xi\| \leq \epsilon\},$$

where $\epsilon > 0$ *is sufficiently small and* $h^c : \mathbb{R}^{n_c} \to \mathbb{R}^{n_s+n_u}$ *is a* C^k*-map satisfying*

$$h^c(0) = 0,\ Dh^c(0) = 0.$$

□

Actually, one can prove that h^c has the same smoothness as the RHS of system (6.43). Finally, we formulate without proof the Reduction Principle for ODEs.

Theorem 6.20 (Reduction Principle) *Consider a smooth system*

$$\begin{cases} \dot{\xi} = B\xi + f(\xi, \eta), \\ \dot{\eta} = C\eta + g(\xi, \eta), \end{cases} \tag{6.44}$$

where $\xi \in \mathbb{R}^{n_c}$, $\eta \in \mathbb{R}^{n_s+n_u}$, *the* $n_c \times n_c$ *matrix* B *has* n_c *critical eigenvalues with* $\operatorname{Re} \lambda = 0$, *while all eigenvalues of the* $(n_s + n_u) \times (n_s + n_u)$ *matrix* C *satisfy* $\operatorname{Re} \lambda \neq 0$, *and the functions* f *and* g *are smooth and have neither constant nor linear terms.*

The system (6.44) *is locally topologically conjugate near* $(0, 0)$ *to the system*

$$\begin{cases} \dot{\xi} = B\xi + f(\xi, h^c(\xi)), \\ \dot{\eta} = C\eta, \end{cases} \tag{6.45}$$

where h^c *is as in* Theorem 6.19. □

The systems for ξ and η are decoupled in (6.45). Therefore, the system (6.45) is locally topologically conjugate near $(0, 0)$ to a system

$$\begin{cases} \dot{\xi} = b(\xi), \\ \dot{\eta} = C\eta. \end{cases}$$

where $\dot{\xi} = b(\xi)$ is any system that is locally topologically conjugate near $\xi = 0$ to $\dot{\xi} = B\xi + f(\xi, h^c(\xi))$. Moreover, the second equation in the last system can be replaced by the *standard saddle* introduced in Sect. 2.3.1, i.e. the linear system

$$\begin{cases} \dot{\eta}_s = -\eta_s, & \eta_s \in \mathbb{R}^{n_s}, \\ \dot{\eta}_u = \eta_u, & \eta_u \in \mathbb{R}^{n_u}. \end{cases} \tag{6.46}$$

As for the discrete-time case, we have the following result.

Corollary 6.21 *If $\xi = 0$ is a stable equilibrium of the restriction of* (6.44) *to its centre manifold*

$$\dot{\xi} = B\xi + f(\xi, h^c(\xi)), \quad \xi \in \mathbb{R}^{n_c},$$

and all eigenvalues of C satisfy $\mathrm{Re}\,\lambda < 0$, *then* $(\xi, \eta) = (0, 0)$ *is a stable equilibrium of* (6.36). □

6.3 Critical Normal Forms on Centre Manifolds

We now address the problem of computing in practice the coefficients of normal forms of restrictions of multidimensional ODEs and maps to the corresponding centre manifolds at codim 1 bifurcations of equilibria and fixed points. The employed approach avoids the two-stage process, which first computes an approximate centre manifold and then performs the transformations to a normal form, but combines these stages. The resulting formulas are simple and allow one to perform all computations in the original coordinates, without any preliminary transformation. And they prepare for the analysis of bifurcations in parameter-dependent systems, as presented in the next sections.

Consider a smooth system of ODEs

$$\dot{u} = Au + F(u), \quad u \in \mathbb{R}^n, \tag{6.47}$$

where the matrix A has n_c eigenvalues with zero real parts, and write the Taylor expansion for F at $u_0 = 0$ as

$$F(u) = \frac{1}{2}B(u, u) + \frac{1}{6}C(u, u, u) + O(\|u\|^4),$$

where $B : \mathbb{R}^n \times \mathbb{R}^n \to \mathbb{R}^n$ and $C : \mathbb{R}^n \times \mathbb{R}^n \times \mathbb{R}^n \to \mathbb{R}^n$ are *multilinear functions* with the components:

$$B_i(p,q) = \sum_{j,k=1}^{n} \frac{\partial^2 F_i(0)}{\partial u_j \partial u_k} p_j q_k, \quad C_i(p,q,r) = \sum_{j,k,l=1}^{n} \frac{\partial^3 F_i(0)}{\partial u_j \partial u_k \partial u_l} p_j q_k r_l,$$

for $i = 1, 2, \ldots, n$. Note that in accordance with (6.6) we have $B(p,q) = D^2F(0)(p,q)$ and $C(p,q,r) = D^3F(0)(p,q,r)$.

In what follows, we use some results from Linear Algebra. Given an $n \times n$ complex matrix L, introduce its two linear invariant subspaces of $\mathbb{C}^n$: the *range*

$$R(L) = \{v \in \mathbb{C}^n : v = Lu \ \text{ for some } \ u \in \mathbb{C}^n\}$$

and the *null-space*

$$N(L) = \{w \in \mathbb{C}^n : Lw = 0\}.$$

We call two vectors $u, v \in \mathbb{C}^n$ *orthogonal* and write $u \perp v$, if their scalar product vanishes: $\langle u, v\rangle := \bar{u}^{\mathrm{T}} v = 0$. Denote by L^* the transposed matrix to the complex conjugate of L, i.e. $L^* = \overline{L}^{\mathrm{T}}$. If L is real, $L^* = L^{\mathrm{T}}$. The matrix L^* is called the *adjoint matrix* for L. We have

$$\langle u, Lv\rangle = \langle L^* u, v\rangle$$

for all $u, v \in \mathbb{C}^n$.

Lemma 6.22 (Fredholm Decomposition)

$$\mathbb{C}^n = R(L) \oplus N(L^*)$$

with $R(L) \perp N(L^*)$, *i.e., any vector* $x \in \mathbb{C}^n$ *can be uniquely represented as* $x = v + w$ *with* $v \in R(L), w \in N(L^*)$ *satisfying* $\langle w, v\rangle = 0$.

Proof Consider an orthogonal complement $W = R(L)^{\perp}$ in $\mathbb{C}^n$, so that $\mathbb{C}^n = R(L) \oplus W$ and $R(L) \perp W$. We are going to prove that $W = N(L^*)$.

(1) Suppose that $w \in N(L^*)$ meaning $L^* w = 0$. For any $v = Lu \in R(L)$, holds

$$\langle w, v\rangle = \langle w, Lu\rangle = \langle L^* w, u\rangle = \langle 0, u\rangle = 0.$$

Hence, $w \in W$.

(2) Suppose now that $w \in W$, i.e. $\langle w, v\rangle = 0$ for $v = Lu \in R(L)$ with any $u \in \mathbb{C}^n$. Take $u = L^* w$. Then we have

$$0 = \langle w, v\rangle = \langle w, Lu\rangle = \langle w, LL^* w\rangle = \langle L^* w, L^* w\rangle = \|L^* w\|^2.$$

Thus, $L^* w = 0$ and $w \in N(L^*)$. □

Lemma 6.22 implies that a linear system $Lu = v$ has a solution if and only if $\langle w, v\rangle = 0$ for all w satisfying $L^* w = 0$. This is known as the *Fredholm solvability condition.* If L is nonsingular, then L^* is also nonsingular, so that $N(L^*) = 0$ and $Lu = v$ has a unique solution for any $v \in \mathbb{R}^n$, $u = L^{-1} v$. If L

is singular, implying that both $N(L)$ and $N(L^*)$ are nontrivial, but v satisfies the Fredholm solvability condition, then a solution u to $Lu = v$ exists but is not unique. Indeed, $u + \xi$ is another solution for any $\xi \in N(L)$.

Theorem 6.23 (Critical fold coefficient) *Suppose $\lambda_1 = 0$ is a simple eigenvalue of A and assume that this matrix has no other critical eigenvalues. Introduce vectors $q, p \in \mathbb{R}^n$, such that*

$$Aq = 0, \ A^{\mathrm{T}} p = 0, \ \langle p, q \rangle = 1.$$

Then the restriction of (6.47) to a one-dimensional centre manifold $W^c(0)$ can be written in the form

$$\dot{\xi} = b\xi^2 + O(\xi^3), \quad \xi \in \mathbb{R}, \tag{6.48}$$

where

$$b = \frac{1}{2}\langle p, B(q,q)\rangle. \tag{6.49}$$

Proof Write (6.47) as

$$\dot{u} = Au + \frac{1}{2}B(u,u) + O(\|u\|^3), \quad u \in \mathbb{R}^n. \tag{6.50}$$

Any centre manifold is one-dimensional in this case and has the representation

$$u = \xi q + \frac{1}{2}\xi^2 h_2 + O(\xi^3), \quad \xi \in \mathbb{R}, \tag{6.51}$$

for some $h_2 \in \mathbb{R}^n$ that is independent of the choice of the manifold. Using (6.50), (6.51), and (6.48), we get

$$\dot{u} = \dot{\xi} q + \xi\dot{\xi} h_2 + \ldots = b\xi^2 q + b\xi^3 h_2 + \ldots$$

and

$$\dot{u} = Au + \frac{1}{2}B(u,u) + \ldots = \frac{1}{2}\xi^2 Ah_2 + \frac{1}{2}\xi^2 B(q,q) + \ldots$$

Comparing the ξ^2-terms we find

$$bq = \frac{1}{2}Ah_2 + \frac{1}{2}B(q,q)$$

or

$$Ah_2 = 2bq - B(q,q).$$

This linear system for h_2 is obviously singular but has a solution, since the centre manifold exists. The Fredholm solvability condition then implies that

$$\langle p, 2bq - B(q,q)\rangle = 0,$$

from which we obtain (6.49). □

Theorem 6.24 (Critical Andronov–Hopf coefficient) *Suppose*

$$\lambda_{1,2} = \pm i\omega_0, \quad \omega_0 > 0$$

is a simple pair of purely imaginary eigenvalues of A and assume that this matrix has no other critical eigenvalues. Introduce vectors $q, p \in \mathbb{C}^n$, such that

$$Aq = i\omega_0 q, \quad A^{\mathrm{T}} p = -i\omega_0 p, \quad \langle p, q\rangle = 1.$$

Then the restriction of (6.47) to a two-dimensional centre manifold $W^c(0)$ can be written in the form

$$\dot{\eta} = i\omega_0 \eta + c_1 \eta |\eta|^2 + O(|\eta|^4), \quad \eta \in \mathbb{C}, \tag{6.52}$$

where

$$c_1 = \frac{1}{2}\langle p, C(q,q,\bar{q}) - 2B(q, A^{-1}B(q,\bar{q})) + B(\bar{q}, (2i\omega_0 I - A)^{-1}B(q,q))\rangle. \tag{6.53}$$

Proof The normal form (6.52) was obtained in Lemma 5.15 with $\beta = 0$. It is only left to verify (6.53). Write (6.47) as

$$\dot{u} = Au + \frac{1}{2}B(u,u) + \frac{1}{6}C(u,u,u) + O(\|u\|^4), \quad u \in \mathbb{R}^n. \tag{6.54}$$

Any two-dimensional centre manifold can be parametrized by $\eta \in \mathbb{C}$ as

$$u = \eta q + \bar{\eta}\bar{q} + \frac{1}{2}\eta^2 h_{20} + \eta\bar{\eta}h_{11} + \frac{1}{2}\bar{\eta}^2 h_{02} + \frac{1}{2}\eta^2\bar{\eta}h_{21} + \dots, \quad \eta \in \mathbb{C}, \tag{6.55}$$

where the dots denote all inessential terms. The crucial assumption is that the system restricted to $W^c(0)$ has the normal form (6.52) in the η-variable. In other words, we employ the freedom in selecting $h_{ij} \in \mathbb{C}^n$ to put the restricted system into the normal form.

Using (6.54) and (6.52), we get by collecting the η^2-terms in (6.54):

$$(2i\omega_0 I - A)h_{20} = B(q,q).$$

The matrix of this system is nonsingular, since $2i\omega_0$ is not an eigenvalue of A. Thus,

$$h_{20} = (2i\omega_0 I - A)^{-1}B(q,q).$$

Collecting the $\eta\bar{\eta}$-terms gives another nonsingular system

$$Ah_{11} = -B(q, \bar{q})$$

or

$$h_{11} = -A^{-1}B(q, \bar{q}).$$

The $\bar{\eta}^2$-terms lead to $h_{02} = \bar{h}_{20}$, while collecting the coefficients in front of the $\eta|\eta|^2 = \eta^2\bar{\eta}$-term yields the linear system:

$$(i\omega_0 I - A)h_{21} = C(q, q, \bar{q}) + B(\bar{q}, h_{20}) + 2B(q, h_{11}) - 2c_1 q.$$

This system is singular but has a solution. Thus, the Fredholm solvability condition must be satisfied:

$$\langle p, C(q, q, \bar{q}) + B(\bar{q}, h_{20}) + 2B(q, h_{11})\rangle - 2c_1 = 0.$$

This gives (6.53). □

The first Lyapunov coefficient introduced by Definition 5.16 is, therefore,

$$l_1 = \frac{1}{\omega_0}\text{Re } c_1$$

with c_1 given by (6.53).

Consider now a smooth map

$$u \mapsto Au + F(u), \quad u \in \mathbb{R}^n, \tag{6.56}$$

where the matrix A has n_c critical eigenvalues satisfying $|\lambda| = 1$, and, as before,

$$F(u) = \frac{1}{2}B(u, u) + \frac{1}{6}C(u, u, u) + O(\|u\|^4).$$

Theorem 6.25 (Critical fold coefficient for maps) *Suppose $\lambda_1 = 1$ is a simple eigenvalue of A and assume that this matrix has no other critical eigenvalues. Introduce vectors $q, p \in \mathbb{R}^n$, such that*

$$Aq = q, \quad A^{\mathrm{T}}p = p, \quad \langle p, q\rangle = 1.$$

Then the restriction of (6.56) to a one-dimensional centre manifold $W^c(0)$ can be written in the form

$$\xi \mapsto \xi + b\xi^2 + O(\xi^3), \quad \xi \in \mathbb{R}, \tag{6.57}$$

where

$$b = \frac{1}{2}\langle p, B(q,q)\rangle. \tag{6.58}$$

Proof Write

$$f(H) = AH + \frac{1}{2}B(H,H) + O(\|H\|^3),$$

and locally represent a centre manifold W^c as the graph of a function $u = H(\xi)$, $H : \mathbb{R} \to \mathbb{R}^n$, where

$$H(\xi) = \xi q + \frac{1}{2}h_2\xi^2 + O(\xi^3), \quad \xi \in \mathbb{R}, \ h_2 \in \mathbb{R}^n.$$

The restriction of (6.56) to $W^c(0)$ is $\xi \mapsto G(\xi)$, where

$$G(\xi) = \xi + b\xi^2 + O(\xi^3).$$

The invariance equation for the centre manifold

$$f(H(\xi)) = H(G(\xi))$$

reads as

$$A\left(\xi q + \frac{1}{2}h_2\xi^2 + \cdots\right) + \frac{1}{2}B(\xi q + \cdots, \xi q + \cdots) + \cdots$$
$$= (\xi + b\xi^2 + \cdots)q + \frac{1}{2}h_2(\xi + \cdots)^2 + \cdots.$$

The ξ^2-terms give the equation for h_2:

$$(A - I)h_2 = -B(q,q) + 2bq.$$

It is singular but its solvability implies (6.58). □

Theorem 6.26 (Critical flip coefficient) *Suppose $\lambda_1 = -1$ is a simple eigenvalue of A and assume that it has no other critical eigenvalues. Introduce vectors $q, p \in \mathbb{R}^n$, such that*

$$Aq = -q, \ A^{\mathrm{T}}p = -p, \ \langle p, q\rangle = 1.$$

Then the restriction of (6.56) to a one-dimensional centre manifold $W^c(0)$ can be written in the normal form

$$\xi \mapsto -\xi + c\xi^3 + O(\xi^4), \quad \xi \in \mathbb{R}, \tag{6.59}$$

where

$$c = \frac{1}{6}\langle p, C(q,q,q)\rangle - \frac{1}{2}\langle p, B(q,(A-I)^{-1}B(q,q))\rangle. \tag{6.60}$$

Proof The normal form (6.59) was obtained in Theorem 5.25 with $\beta = 0$. It is only left to verify (6.60). Expand

$$f(H) = AH + \frac{1}{2}B(H,H) + \frac{1}{6}C(H,H,H) + O(\|H\|^4),$$

and parametrize a centre manifold $W^c(0)$ with $u = H(\xi)$, where

$$H(\xi) = \xi q + \frac{1}{2}h_2\xi^2 + \frac{1}{6}h_3\xi^3 + O(\xi^4), \quad \xi \in \mathbb{R},$$

and $h_{2,3} \in \mathbb{R}^n$. As in the Hopf case, we assume that the restricted map written in the ξ-coordinate has the normal form (6.59), i.e.

$$\tilde{\xi} = G(\xi) = -\xi + c\xi^3 + O(\xi^4).$$

The ξ^2-terms in the invariance equation

$$f(H(\xi)) = H(G(\xi))$$

give for h_2:

$$(A-I)h_2 = -B(q,q).$$

Since $\lambda = 1$ is not an eigenvalue of A, the matrix $A - I$ is nonsingular. Thus,

$$h_2 = -(A-I)^{-1}B(q,q).$$

The ξ^3-terms in the invariance equation give the linear system for h_3:

$$(A+I)h_3 = 6cq - C(q,q,q) - 3B(q,h_2).$$

This system is singular, since $(A+I)q = 0$, so it has a solution only if

$$\langle p, 6cq - C(q,q,q) - 3B(q,h_2)\rangle = 0,$$

which implies

$$c = \frac{1}{6}\langle p, C(q,q,q)\rangle + \frac{1}{2}\langle p, B(q,h_2)\rangle.$$

Taking into account $h_2 = -(A-I)^{-1}B(q,q)$, we obtain (6.60). □

Theorem 6.27 (Critical Neimark–Sacker coefficient) *Suppose that* $\lambda_{1,2} = e^{\pm i\theta_0}$, *where*

$$e^{ik\theta_0} \neq 1, \quad k = 1,2,3,4,$$

is a simple pair of purely imaginary eigenvalues of A and that A has no other critical eigenvalues. Introduce vectors $q, p \in \mathbb{C}^n$, such that

$$Aq = e^{i\theta_0} q, \quad A^{\mathrm{T}} p = e^{-i\theta_0} p, \quad \langle p, q \rangle = 1.$$

Then the restriction of (6.56) to a two-dimensional centre manifold $W^c(0)$ can be written in the form

$$\eta \mapsto e^{i\theta_0} \eta + c_1 \eta |\eta|^2 + O(|\eta|^4), \quad \eta \in \mathbb{C}, \tag{6.61}$$

where

$$c_1 = \frac{1}{2} \langle p, C(q, q, \bar{q}) + B(\bar{q}, (e^{2i\theta_0} I - A)^{-1} B(q, q)) + 2B(q, (I - A)^{-1} B(q, \bar{q})) \rangle. \tag{6.62}$$

Proof The invariance of a centre manifold $W^c(0)$ represented as the graph of $u = H(\eta, \bar{\eta})$ with $\eta \in \mathbb{C}$ can be written in the form

$$f(H(\eta, \bar{\eta})) = H(G(\eta, \bar{\eta})), \tag{6.63}$$

where

$$f(H) = AH + \frac{1}{2} B(H, H) + \frac{1}{6} C(H, H, H) + O(\|H\|^4)$$

and

$$H(\eta, \bar{\eta}) = \eta q + \bar{\eta}\, \bar{q} + \sum_{1 \leq j+k \leq 3} \frac{1}{j!k!} h_{jk} \eta^j \bar{\eta}^k + O(|\eta|^4).$$

As before, we use the freedom in selecting $h_{jk} \in \mathbb{C}^n$ to put the restricted map into the normal form

$$G(\eta, \bar{\eta}) = e^{i\theta_0} \eta + c_1 \eta |\eta|^2 + O(|\eta|^4).$$

This normal form is derived in Lemma 5.30, where one has to put $\beta = 0$.

Quadratic terms in (6.63) give

$$h_{20} = (e^{2i\theta_0} I - A)^{-1} B(q, q),$$
$$h_{11} = (I - A)^{-1} B(q, \bar{q}),$$

while the $\eta|\eta|^2 = \eta^2 \bar{\eta}$-terms lead to the singular system

$$(e^{i\theta_0} I - A) h_{21} = C(q, q, \bar{q}) + B(\bar{q}, h_{20}) + 2B(q, h_{11}) - 2c_1 q.$$

The solvability of this system is equivalent to

$$\langle p, C(q,q,\bar{q}) + B(\bar{q}, h_{20}) + 2B(q, h_{11}) - 2c_1 q\rangle = 0,$$

so the cubic normal form coefficient can indeed be expressed by (6.62). □

As shown in the proof of Theorem 5.31, the direction of the bifurcation of the closed invariant curve is determined by the sign of

$$a = \mathrm{Re}(e^{-i\theta_0} c_1).$$

6.4 Families of Centre Manifolds

Consider a smooth parameter-dependent system of ODEs

$$\begin{cases} \dot{\xi} = P(\xi, \eta, \alpha), \\ \dot{\eta} = Q(\xi, \eta, \alpha), \end{cases} \tag{6.64}$$

where $\xi \in \mathbb{R}^{n_c}, \eta \in \mathbb{R}^{n_s+n_u}, \alpha \in \mathbb{R}^m$, and suppose that (6.64) coincides with (6.44) at $\alpha = 0$:

$$P(\xi, \eta, 0) = B\xi + f(\xi, \eta),\ \ Q(\xi, \eta, 0) = C\eta + g(\xi, \eta).$$

Theorem 6.28 *The system* (6.64) *has a family of invariant manifolds, locally representable for small* $\|\alpha\|$ *as*

$$W^c_\alpha = \{(\xi, w(\xi, \alpha)) : \xi \in \mathbb{R}^{n_c}, \|\xi\| \leq \varepsilon\},$$

where $\varepsilon > 0$ *is sufficiently small and the map*

$$w : \mathbb{R}^{n_c} \times \mathbb{R}^m \to \mathbb{R}^{n_s+n_u}$$

is smooth. Moreover, $w(\xi, 0) = h^c(\xi)$, *i.e.* W^c_0 *coincides with a centre manifold* W^c *from* Theorem 6.19.

Proof Consider the following *extended system*:

$$\begin{cases} \dot{\xi} = P(\xi, \eta, \alpha), \\ \dot{\eta} = Q(\xi, \eta, \alpha), \\ \dot{\alpha} = 0, \end{cases} \tag{6.65}$$

where $\xi \in \mathbb{R}^{n_c}, \eta \in \mathbb{R}^{n_s+n_u}$, and $\alpha \in \mathbb{R}^m$. The equilibrium $(\xi, \eta, \alpha) = (0, 0, 0)$ of (6.65) is nonhyperbolic and has $n_c + m$ eigenvalues with $\mathrm{Re}\,\lambda = 0$ (m of them are equal to zero). Theorem 6.19 guarantees local existence of a $(n_c + m)$-dimensional invariant centre manifold in (6.65). This manifold is the union of n_c-dimensional manifolds W^c_α located in the invariant subspaces $\alpha = \text{const}$ of (6.65). □

Theorem 6.29 (Shoshitaishvilly, 1975) *The system* (6.64) *is locally topologically equivalent near* $(\xi, \eta, \alpha) = (0, 0, 0)$ *to the system*

$$\begin{cases} \dot{\xi} = P(\xi, w(\xi, \alpha), \alpha), \\ \dot{\eta} = C\eta. \end{cases} \tag{6.66}$$

□

This theorem—that we formulate without proof—means that all "essential" events near the bifurcation parameter value occur on the invariant manifold W^c_α and are captured by the n_c-dimensional *restricted system*:

$$\dot{\xi} = P(\xi, w(\xi, \alpha), \alpha), \quad \xi \in \mathbb{R}^{n_c}, \alpha \in \mathbb{R}^m. \tag{6.67}$$

Obviously, this system can be replaced in Theorem 6.29 by any smooth system

$$\dot{\xi} = r(\xi, \alpha), \quad \xi \in \mathbb{R}^{n_c}, \alpha \in \mathbb{R}^m,$$

that is locally topologically equivalent to (6.67), while the second equation in (6.66) can be replaced by the standard saddle (6.46).

A theorem similar to Theorem 6.29 can be formulated for discrete-time dynamical systems generated by smooth maps.

6.5 Implications for n-Dimensional ODEs

First we apply Theorem 6.29 to study the fold and Hopf bifurcations of equilibria in multidimensional systems. Then we describe implications of the discrete-time version of this theorem for local bifurcations of limit cycles in such systems. We mainly focus on cases with $n \leq 3$ for visualization reasons.

6.5.1 Codimension One Bifurcations of Equilibria

6.5.1.1 Generic Fold Bifurcation in Planar Systems

Consider a smooth planar system

$$\dot{x} = f(x, \alpha), \quad x \in \mathbb{R}^2, \ \alpha \in \mathbb{R}. \tag{6.68}$$

Assume that at $\alpha = 0$ it has the equilibrium $x_0 = 0$ with one eigenvalue $\lambda_1 = 0$ and one eigenvalue $\lambda_2 < 0$. Theorem 6.28 gives the existence of a smooth, locally defined, one-dimensional attracting invariant manifold W^c_α for (6.68) when $|\alpha|$ is small. Consider now the restriction of the system (6.68) to W^c_α, i.e. the Eq. (6.67) that is scalar in this case. At $\alpha = 0$, this equation takes the form

$$\dot{\xi} = b\xi^2 + O(\xi^3), \quad \xi \in \mathbb{R}.$$

If $b \neq 0$ and the restricted Eq. (6.67) depends generically on the parameter α (i.e. both genericity conditions (A.1) and (A.2) of Theorem 5.7 are satisfied), then (6.67) is locally topologically equivalent to the normal form

$$\dot{\xi} = \beta + \sigma\xi^2,$$

where $\sigma = \operatorname{sign} b = \pm 1$ and $\beta(0) = 0$ (see Theorems 5.7 and 5.6). Then Theorem 6.29 implies that (6.68) is locally topologically equivalent to the system

$$\begin{cases} \dot{\xi} = \beta + \sigma\xi^2, \\ \dot{\eta} = -\eta. \end{cases} \tag{6.69}$$

Equations (6.69) are decoupled. The resulting phase portraits are presented in Fig. 6.2 for the case $\sigma = 1$.

For $\beta < 0$, there are two hyperbolic equilibria in the ξ-axis: a stable node and a saddle. They collide at $\beta = 0$, forming a nonhyperbolic *saddle-node* point, and disappear. There are no equilibria for $\beta > 0$. The same events happen in (6.68) on some one-dimensional, parameter-dependent, invariant manifold W^c_α, that is locally attracting (see Fig. 6.3). All equilibria belong to this manifold. Incidentally, Figs. 6.2 and 6.3 explain why the fold bifurcation is often called the *saddle-node bifurcation*. It should be clear how to generalize these considerations to cover the case $\lambda_2 > 0$, as well as the n-dimensional case.

To apply the above theory to a particular n-dimensional ODE system,

$$\dot{u} = f(u, \alpha), \quad u \in \mathbb{R}^n, \ \alpha \in \mathbb{R},$$

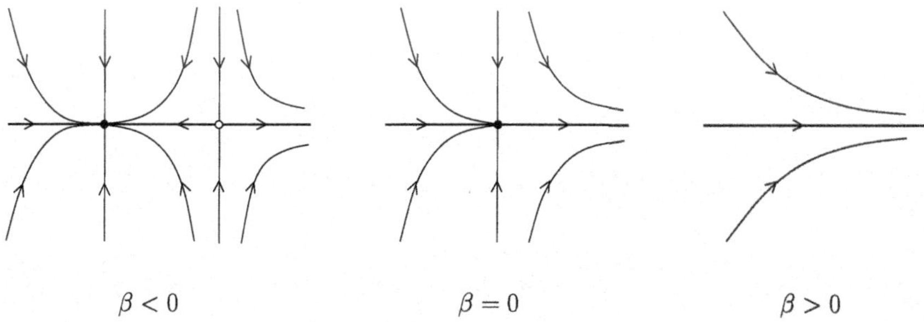

Fig. 6.2 Fold bifurcation in the standard system (6.69) for $\sigma = 1$

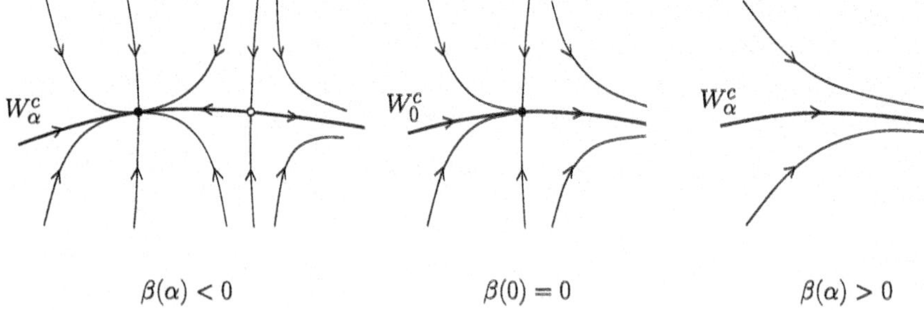

Fig. 6.3 Fold bifurcation in a generic planar system

we need to verify the genericity conditions (A.1) and (A.2) for the corresponding restricted system (6.67). This can be done in the original coordinates, without explicitly computing the restricted system. The nondegeneracy condition (A.1) is $b \neq 0$, where the critical normal form coefficient b is given by the formula (6.49). One can show that the transversality condition (A.2) amounts to

$$\langle p, D_\alpha f(0,0)q \rangle \neq 0,$$

where $q, p \in \mathbb{R}^n$ are as in Theorem 6.23. Then β can be considered as a smooth function of α, with $\beta(0) = 0$ but $\beta'(0) \neq 0$, which ensures that it changes sign at $\alpha = 0$.

6.5.1.2 Generic Andronov–Hopf Bifurcation in $\mathbb{R}^3$

Consider a smooth system

$$\dot{x} = f(x, \alpha), \quad x \in \mathbb{R}^3, \ \alpha \in \mathbb{R}. \tag{6.70}$$

Assume that at $\alpha = 0$ it has the equilibrium $x_0 = 0$ with eigenvalues $\lambda_{1,2} = \pm i\omega_0$, $\omega_0 > 0$ and one negative eigenvalue $\lambda_3 < 0$. Theorem 6.28 gives the existence of a parameter-dependent, smooth, local two-dimensional attracting invariant manifold W^c_α of (6.70) when $|\alpha|$ is small. Consider now the restriction of the system (6.70) to W^c_α, i.e. the planar system (6.67). At $\alpha = 0$, this equation can be written in the complex form as

$$\dot{z} = i\omega_0 z + g(z, \bar{z}), \quad z \in \mathbb{C},$$

where $g = O(|z|^2)$. If the first Lyapunov coefficient l_1 for this equation is nonzero and the restricted system (6.67) depends generically on the parameter α (i.e. both genericity conditions (B.1) and (B.2) of Theorem 5.13 are satisfied), then (6.67) is locally topologically equivalent to the complex normal form

$$\dot{z} = (\beta + i)z + sz^2\bar{z},$$

where $s = \text{sign } l_1 = \pm 1$ and $\beta(0) = 0$ (see Theorems 5.13 and 5.12). Then Theorem 6.29 implies that (6.70) is locally topologically equivalent to the system

$$\begin{cases} \dot{z} = (\beta + i)z + sz^2\bar{z}, \\ \dot{\eta} = -\eta. \end{cases} \tag{6.71}$$

The phase portrait of (6.71) is shown in Fig. 6.4 for $s = -1$. The supercritical Hopf bifurcation takes place in the invariant plane $\eta = 0$, which is attracting. The same events happen for (6.70) on some two-dimensional attracting manifold W_α^c (see Fig. 6.5). The construction can be generalized to an arbitrary dimension $n \geq 3$.

To apply the above results to a particular ODE system,

$$\dot{u} = f(u, \alpha), \quad u \in \mathbb{R}^n, \ \alpha \in \mathbb{R},$$

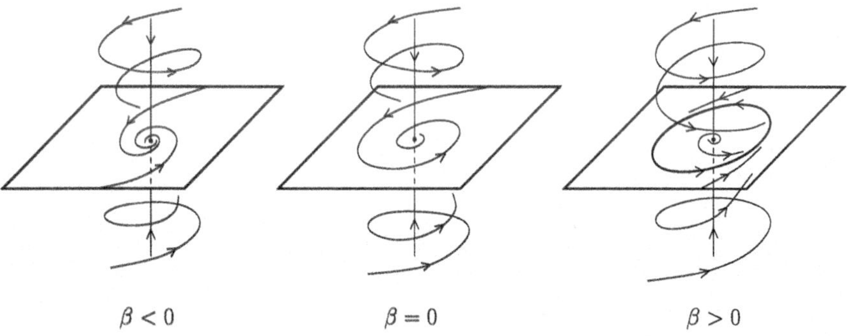

Fig. 6.4 Andronov–Hopf bifurcation in the standard system (6.71) for $s = -1$

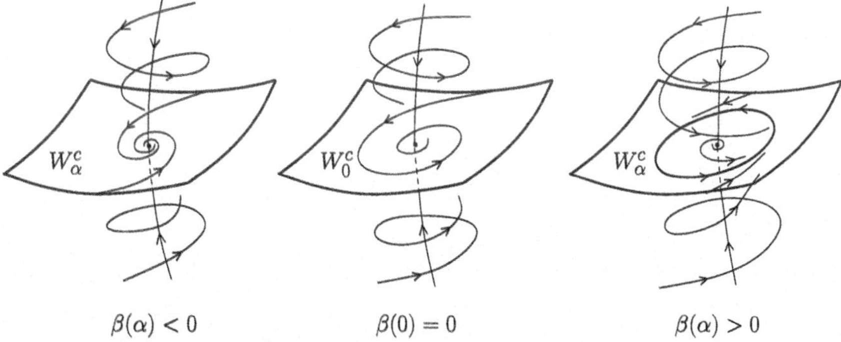

Fig. 6.5 Supercritical Andronov–Hopf bifurcation in a generic three-dimensional system

we need to verify the genericity conditions (B.1) and (B.2) for the corresponding restricted system (6.67). As in the fold case, this can be done in the original coordinates, without explicitly computing the restricted system and its complex form. The nondegeneracy condition (B.1) is

$$l_1 = \frac{1}{\omega_0} \mathrm{Re}\, c_1 \neq 0,$$

where the normal form coefficient c_1 is given by the formula (6.53). One can show that the transversality condition (B.2) is equivalent to

$$\mathrm{Re}\langle p, -B(A^{-1}[D_\alpha f](0,0), q) + [D_\alpha D_u f](0,0)q_0 \rangle \neq 0,$$

where $q, p \in \mathbb{C}^n$ are as in Theorem 6.24. Here $A = D_u f(0,0)$ and B are the usual linear and bilinear forms of $f(u,0)$ at $u = 0$. Then β can be considered as a smooth function of α, with $\beta(0) = 0$ but $\beta'(0) \neq 0$, which ensures that it changes sign at $\alpha = 0$. Recall that $\beta(\alpha)$ is proportional to the real part of the pair of the complex-conjugate eigenvalues crossing the imaginary axis at $\alpha = 0$.

6.5.2 Codimension One Bifurcations of Limit Cycles

A combination of the Poincaré map and the centre manifold reduction allows us to describe bifurcations of limit cycles in generic n-dimensional ODEs depending on one parameter. Contrary to the previous subsection, we will not explicitly write here the nondegeneracy and transversality conditions for the maps obtained by restriction to the centre manifolds. These are given in Chap. 5 and usually can only be verified numerically, see references in Sect. 6.6

Let L_0 be a limit cycle of a smooth system

$$\dot{x} = f(x, \alpha), \quad x \in \mathbb{R}^n, \ \alpha \in \mathbb{R}, \tag{6.72}$$

at $\alpha = 0$. Let P_α denote the associated Poincaré map for nearby α,

$$P_\alpha : \Sigma \to \Sigma,$$

where Σ is a local cross section to L_0. If some coordinates $\xi = (\xi_1, \ldots, \xi_{n-1})$ are introduced on Σ, then $\tilde{\xi} = P_\alpha(\xi)$ can be defined to be the point of the next intersection with Σ of the orbit of (6.72) having initial point with coordinates ξ on Σ. The intersection of Σ and L_0 gives a fixed point ξ_0 for P_0: $P_0(\xi_0) = \xi_0$. As we know, the map P_α is smooth and locally invertible.

Suppose that the cycle L_0 is nonhyperbolic, having n_0 *nontrivial* multipliers (i.e. eigenvalues of $DP_0(0)$) on the unit circle. The centre manifold theorems then give a parameter-dependent invariant n_0-dimensional manifold $W^c_\alpha \subset \Sigma$ of P_α on which the "essential" events take place. The Poincaré map P_α is locally topologically equivalent to the suspension of its restriction to this manifold by a linear hyperbolic map (i.e. a discrete-time analogue of the situation described in Theorem 6.29).

Fix $n = 3$, for simplicity, and consider the implications of this result for the limit cycle bifurcations in $\mathbb{R}^3$. The conclusions can be generalized to arbitrary dimension $n \geq 3$.

6.5.2.1 Generic Fold Bifurcation of Cycles in $\mathbb{R}^3$

Assume that at $\alpha = 0$ the cycle has a simple nontrivial multiplier $\mu_1 = 1$ and its other nontrivial multiplier satisfies $0 < \mu_2 < 1$. The restriction of P_α to the invariant manifold W^c_α is a one-dimensional map, having a fixed point with $\mu_1 = 1$ at $\alpha = 0$. As has been shown, this generically implies the collision and disappearance of two fixed points of P_α as α passes through zero. Under our assumption on μ_2, this happens on a one-dimensional attracting invariant manifold of P_α; thus, a stable and a saddle fixed point are involved in the bifurcation (see Fig. 6.6 for an illustration). Each fixed point of the Poincaré map corresponds to a limit cycle of the continuous-time system. Therefore, two limit cycles (stable and saddle) collide and disappear in system (6.72) at this *fold bifurcation of cycles*.

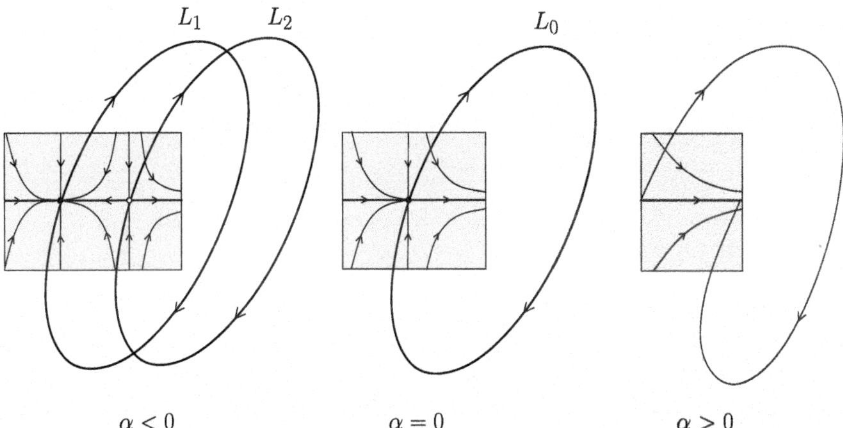

Fig. 6.6 Fold bifurcation of limit cycles

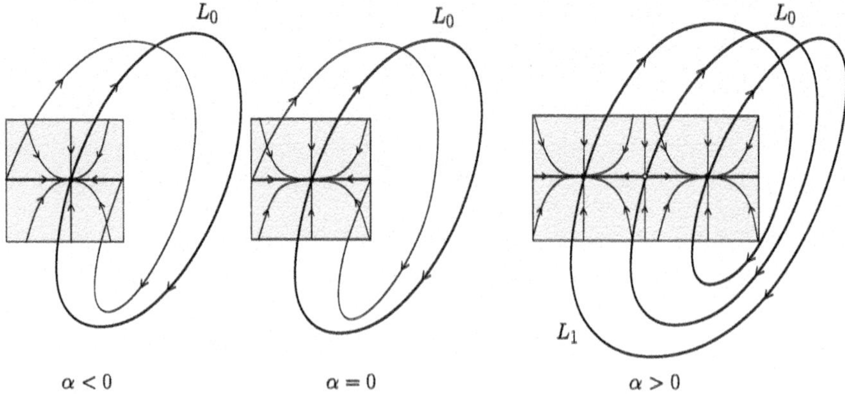

Fig. 6.7 Flip (period-doubling) bifurcation of limit cycles

6.5.2.2 Generic Flip (Period-Doubling) Bifurcation of Cycles in $\mathbb{R}^3$

Suppose that at $\alpha = 0$ the cycle has one multiplier $\mu_1 = -1$, while the other multiplier satisfies $-1 < \mu_2 < 0$. Then, the restriction of P_α to the invariant manifold will demonstrate generically the period-doubling (flip) bifurcation: A cycle of period-2 appears or disappears for the map, while the fixed point changes its stability (see Fig. 6.7, where the supercritical case is illustrated). Since the manifold is attracting, the stable fixed point, for example, loses stability and becomes a saddle point, while a stable cycle of period-2 appears. The fixed points correspond to limit cycles with the same stability type. The cycle of period-two points for the map corresponds to a unique stable limit cycle in (6.72) with *approximately* twice the period of the "basic" cycle L_0. The double-period cycle makes two big "excursions" near L_0 before the closure. The exact bifurcation scenario is determined by the normal form coefficient of the restricted Poincaré map evaluated at $\alpha = 0$.

6.5.2.3 Generic Neimark–Sacker (Torus) Bifurcation of Cycles in $\mathbb{R}^3$

The last codim 1 bifurcation corresponds to the case where both nontrivial multipliers are complex and simple and lie on the unit circle: $\mu_{1,2} = e^{\pm i\theta_0}$. The Poincaré map P_α then has a parameter-dependent, two-dimensional, invariant manifold on which a closed invariant curve generically bifurcates from the fixed point (see Fig. 6.8, where the supercritical bifurcation is shown). This closed curve corresponds to a two-dimensional *invariant torus* $\mathbb{T}^2$ in the system generated by (6.72). The bifurcation is determined by the normal form coefficient of the restricted Poincaré map at the critical parameter value. The

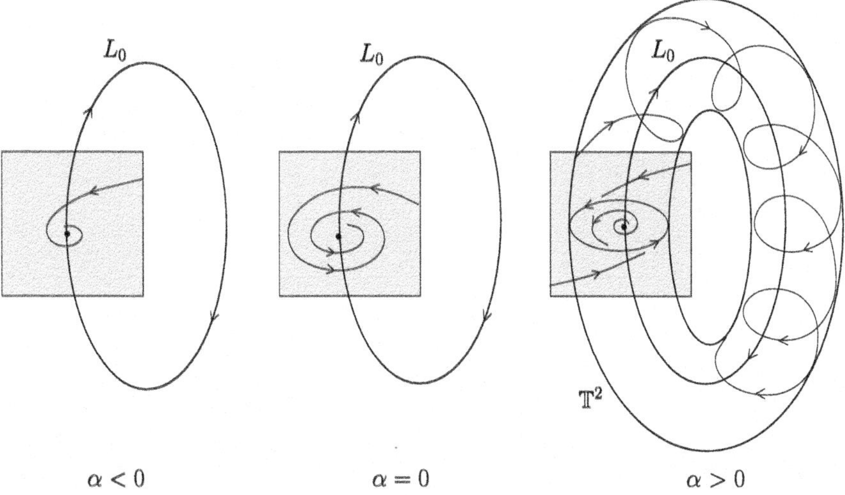

Fig. 6.8 Neimark–Sacker bifurcation of a limit cycle

orbit structure on the torus $\mathbb{T}^2$ depends on the restriction of the Poincaré map to this closed invariant curve. Thus, generically, there are long-period cycles of different stability types located on the torus, which appear and disappear pairwise via fold bifurcations.

6.6 References

Bifurcations of stationary points and periodic orbits in one- and two-parameter families of multidimensional ODEs and maps are treated in many textbooks, including Arnol'd (1983), Guckenheimer and Holmes (1983), Arrowsmith and Place (1990), Shilnikov et al. (2001), Wiggins (2003), Kuznetsov (2023). A useful summary is given in Arnol'd et al. (1994), while many technical issues are clarified in Iooss (1979), Vanderbauwhede (1989), Iooss and Adelmeyer (1992). For an alternative approach to the bifurcation theory based on the Lyapunov–Schmidt reduction, see Chow and Hale (1982), Iooss and Joseph (1990), Kielhöfer (2004).

The proofs of the Centre Manifold Theorems for maps in Sect. 6.1 are based on Sandstede and Theerakarn (2015). A recent account of Gagliardo–Nirenberg estimates is provided in Fiorenza et al. (2021). A direct proof of the existence of a local centre manifold near a nonhyperbolic equilibrium in ODEs, that does not depend on the corresponding result for maps, is given in Carr (1981); a proof of Theorem 6.20 (Reduction Principle for ODEs) can be found in Kirchgraber and Palmer (1990), while Theorem 6.29 (the parameter-dependent Reduction Principle for ODEs) was proven by Shoshitaishvili (1975). Various uniqueness and smoothness properties of the centre manifolds are discussed in Sijbrand (1985).

Numerical methods for bifurcations of stationary points and periodic orbits in multidimensional ODEs are summarized in Beyn et al. (2002), see also Chap. 10 in Kuznetsov (2023). Note that efficient numerical methods to study bifurcations of limit cycles do not use Poincaré maps but rely on *periodic normal forms*, see Kuznetsov et al. (2005).

6.7 Exercises

Exercise 6.1 (Andronov–Hopf bifurcation in 3D systems) Check that each of the following feedback control systems[1] has an equilibrium that exhibits an Andronov–Hopf bifurcation at $\mu = 0$, and compute the first Lyapunov coefficient of the restricted system on the centre manifold:

(a)

$$\begin{cases} \dot{x} = \mu x - y, \\ \dot{y} = \mu y + x + xz, \\ \dot{z} = -z + x^2. \end{cases}$$

(b)

$$\begin{cases} \dot{x} = \mu x - y - xz, \\ \dot{y} = \mu y + x, \\ \dot{z} = -z + y^2 + x^2 z. \end{cases}$$

Exercise 6.2 (Pitchfork bifurcation in Lorenz system) Compute the second-order approximation to the family of one-dimensional centre manifolds of the Lorenz system[2]

$$\begin{cases} \dot{x} = -\sigma x + \sigma y, \\ \dot{y} = -xz + rx - y, \\ \dot{z} = xy - bz, \end{cases} \tag{6.73}$$

near the origin $(x, y, z) = (0, 0, 0)$ for fixed (σ, b) and r close to $r_0 = 1$. Then, calculate the restricted system up to third-order terms in ξ and analyse its bifurcation.

Exercise 6.3 (Andronov–Hopf bifurcation in Lorenz system) (a) Show that for fixed $b > 0$, $\sigma > b + 1$, and

$$r_1 = \frac{\sigma(\sigma + b + 3)}{\sigma - b - 1}, \tag{6.74}$$

a nontrivial equilibrium of (6.73) exhibits an Andronov–Hopf bifurcation.

(b) Prove that this bifurcation is *subcritical* and, therefore, gives rise to a unique saddle limit cycle for $r < r_1$

---

[1] Moon, F. C., & Rand, R. H. (1985). Parametric stiffness control of flexible structures. In *Proceedings of the Workshop on Identification and Control of Flexible Space Structures* (Vol. II, pp. 329–342). Pasadena, CA: Jet Propulsion Laboratory Publication 85-29.

[2] Lorenz, E. (1963). Deterministic non-periodic flow. *Journal of the Atmospheric Sciences, 20*, 130–141.

(*Hints:* Shilnikov et al., 2001, pp. 877–880

(i) Write (6.73) as a single third-order equation

$$\dddot{x} + (\sigma + b + 1)\ddot{x} + b(1+\sigma)\dot{x} + b\sigma(1-r)x = \frac{(1+\sigma)\dot{x}^2}{x} + \frac{\dot{x}\ddot{x}}{x} - x^2\dot{x} - \sigma x^3.$$

(ii) Translate the origin to the equilibrium by introducing the new coordinate $\xi = x - x_0$, where $x_0 = \sqrt{b(r-1)}$, thus obtaining the equation

$$\dddot{\xi} + (\sigma + b + 1)\ddot{\xi} + [b(1+\sigma) + x_0^2]\dot{\xi} + [b\sigma(1-r) + 3\sigma x_0^2]\xi = f(\xi, \dot{\xi}, \ddot{\xi}), \tag{6.75}$$

where

$$f(\xi, \dot{\xi}, \ddot{\xi}) = -3\sigma x_0 \xi^2 - 2x_0 \xi\dot{\xi} + \frac{1+\sigma}{x_0}\dot{\xi}^2 + \frac{1}{x_0}\dot{\xi}\ddot{\xi} - \sigma\xi^3 - \xi^2\dot{\xi} - \frac{1+\sigma}{x_0^2}\xi\dot{\xi}^2 - \frac{1}{x_0^2}\xi\dot{\xi}\ddot{\xi} + \cdots$$

and the dots stand for all higher-order terms in $(\xi, \dot{\xi}, \ddot{\xi})$.

(iii) Rewrite (6.75) as a system

$$\dot{U} = AU + F(U), \quad U = (\xi, \dot{\xi}, \ddot{\xi}) \in \mathbb{R}^3. \tag{6.76}$$

Find the eigenvector and the adjoint eigenvector of A corresponding to its purely imaginary eigenvalues (when (6.74) is satisfied).

(iv) Compute the first Lyapunov coefficient l_1 for (6.76) using (6.53). Substitute $\sigma = \sigma^* + b + 1$ and show that l_1 is positive for all positive σ^* and b.)

Exercise 6.4 (No Neimark–Sacker bifurcation in Lorenz system) Prove that the Neimark–Sacker bifurcation of a limit cycle *never occurs* in (6.73), provided that (σ, r, b) are all positive.
(*Hint:* Use the formula for the multiplier product and the fact that div $f = -(\sigma + b + 1) < 0$, where f is the vector field given by the right-hand side of (6.73).)

Exercise 6.5 (Fold and flip bifurcations in Hénon map) Consider the Hénon map[3]:

$$\begin{pmatrix} x \\ y \end{pmatrix} \mapsto \begin{pmatrix} y \\ \alpha - \beta x - y^2 \end{pmatrix}. \tag{6.77}$$

(a) Find equations for the fold and flip bifurcation curves of fixed points in (6.77).

(b) Prove that the fold and flip bifurcations, occurring in (6.77) under variation of parameter α, are generic for fixed $\beta \neq \pm 1$.

Exercise 6.6 (Duopoly model of Kopel) Consider the following *Kopel map* from mathematical economics[4]:

$$\begin{pmatrix} x \\ y \end{pmatrix} \mapsto \begin{pmatrix} (1-\rho)x + \rho\mu y(1-y) \\ (1-\rho)y + \rho\mu x(1-x) \end{pmatrix}, \tag{6.78}$$

where (μ, ρ) are positive parameters.

[3] Hénon, M. (1976). A two-dimensional mapping with a strange attractor. *Journal Communications in Mathematical Physics, 50*, 69–77.

[4] Kopel, M. (1996). Simple and complex adjustment dynamics in Cournot duopoly models. *Chaos, Solitons & Fractals, 12*, 2031–2048.

(a) Find equations for period-doubling and Neimark–Sacker bifurcations of nonnegative fixed points in (6.78).

(b) Study the nondegeneracy of these bifurcations by computing the corresponding critical normal forms.

(c) Compute numerically bifurcation curves of fixed points, 2- and 4-cycles in the parameter domain

$$2.9 \le \mu \le 3.8, \quad 0.75 \le \rho \le 1.4.$$

Exercise 6.7 (Flip and Neimark–Sacker bifurcations in an adaptive control map) (a) Demonstrate that the fixed point $(x_0, y_0, z_0) = (1, 1, 1 - b - k)$ of the discrete-time dynamical system[5]

$$\begin{pmatrix} x \\ y \\ z \end{pmatrix} \mapsto \begin{pmatrix} y \\ bx + k + yz \\ z - \dfrac{ky}{c + y^2}(bx + k + zy - 1) \end{pmatrix}$$

exhibits a flip bifurcation at

$$b_F = 1 - \left[\frac{1}{2} + \frac{1}{4(c+1)}\right] k,$$

and a Neimark–Sacker bifurcation at

$$b_{NS} = -\frac{c+1}{c+2}.$$

(b) Determine the direction of the period-doubling bifurcation that occurs as b increases and passes through b_F.

(c) Show that the Neimark–Sacker bifurcation in the system under variation of the parameter b can be either sub- or supercritical depending on the values of the parameter (c, k).

[5] Golden, M. P., & Ydstie, B. E. (1988). Bifurcation in model reference adaptive control systems. *Systems and Control Letters, 11*, 413–430.

Chapter 7
Symbolic Dynamics and Global Bifurcations

Abstract So far in this book we mainly concentrated on the (changes in) *local behaviour* near steady states (including periodic orbits of maps and, in principle but not so easy in practice, periodic orbits of ODEs via the Poincaré map). The idea may arise that whenever orbits stay bounded the persistent behaviour is either constant in time or shows regular oscillations. The main aim of this chapter is to show that this idea is wrong: More complicated yet persistent behaviour of dynamical systems is possible and does occur. The main topics are

- continuous scalar maps and symbolic dynamics;
- the famous Lorenz attractor;
- Smale's horseshoe and how it is induced by homoclinic orbits of three-dimensional ODEs and planar maps.

In Appendices we will discuss general results about the bifurcation of periodic orbits from a homoclinic orbit in n-dimensional ODEs and the existence of complicated invariant sets of n-dimensional maps near their transverse homoclinic orbits.

7.1 One-Dimensional Maps

7.1.1 Periodic Orbits of Continuous Maps

Consider a (possibly noninvertible) continuous map

$$f : I \to I, \quad I = [0, 1], \tag{7.1}$$

and the associated discrete-time dynamical system $\{\mathbb{N}, I, f^k\}$, where $\mathbb{N}$ is the set of all nonnegative integers. Let $J, K \subset I$ be two closed intervals.

Definition 7.1 *We say that J* **covers** *K* **under** *f and write $J \rightharpoonup K$, if there exists a closed interval $L \subset J$ such that $f(L) = K$.*

Yu. Kuznetsov et al., *Dynamical Systems Essentials*, Texts in Applied Mathematics 83, https://doi.org/10.1007/978-3-032-04083-1_7

Fig. 7.1 Lemma 7.2

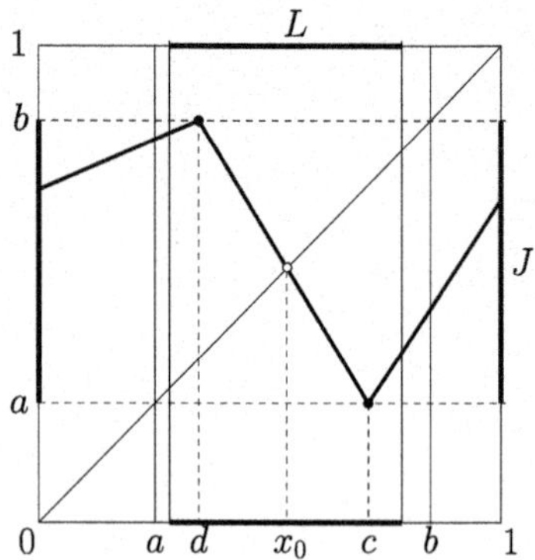

Lemma 7.2 *If $J \rightharpoonup J$ under f, then f has a fixed point $x \in J$.*

Proof Let $J = [a, b]$. By definition, there is an interval $L \subset J$ such that $f(L) = J$. This implies that there exist $c, d \in L$ satisfying $f(c) = a \leq c$ and $f(d) = b \geq d$ (see Fig. 7.1 for an illustration). Thus, by the Intermediate Value Theorem, the continuous function $g(x) = f(x) - x$ has a zero x_0 between c and d, so that $x_0 \in L \subset J$ is a fixed point: $f(x_0) = x_0$. □

By induction one can prove the following result.

Lemma 7.3 *If $I_0 \rightharpoonup I_1 \rightharpoonup I_2 \rightharpoonup \cdots \rightharpoonup I_n$, then there exists a closed interval $J \subset I_0$ such that $f^k(J) \subset I_k$ for $k = 1, 2, \ldots, n-1$, and $f^n(J) = I_n$.* □

Lemma 7.4 *If $I_0 \rightharpoonup I_1 \rightharpoonup I_2 \rightharpoonup \cdots \rightharpoonup I_{n-1} \rightharpoonup I_0$, then there exists $x \in I_0$ such that $x = f^n(x)$ and $f^k(x) \in I_k$ for $k = 0, 1, \ldots, n-1$.*

Proof From Lemma 7.3 it follows that there exists a closed interval $J \subset I_0$ such that $f^n(J) = I_0$ and $f^k(J) \subset I_k$ for $k = 0, 1, \ldots, n-1$. Applying Lemma 7.2 to the map f^n we get a fixed point $x \in J \subset I_0$ of f^n. Clearly, $f^k(x) \in f^k(J) \subset I_k$ for $k = 0, 1, 2, \ldots, n-1$. □

Note that n is not necessarily the minimal period of the cycle $\{x, \ldots, f^{n-1}(x)\}$ since we did not assume that $I_0 \cap I_k = \emptyset$ for $k = 1, 2, \ldots, n-1$.

Let us consider a collection $\{I_0, I_1, \ldots\}$ of closed intervals $I_i \subset I$ with pairwise disjoint interiors. The relation "$\rightharpoonup$" yields the edges of a directed graph, whose vertices are the intervals I_i (see Fig. 7.2). This graph is called a *Markov*

Fig. 7.2 A loop in the Markov graph corresponding to Lemma 7.4

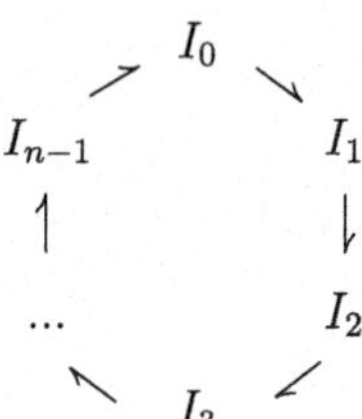

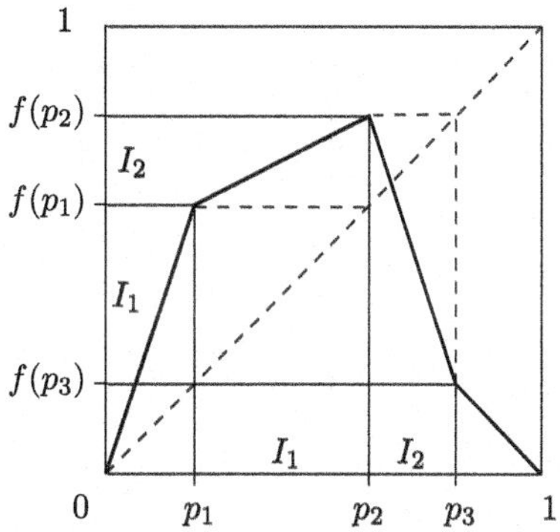

Fig. 7.3 Period three implies all periods

graph for f associated to $\{I_0, I_1, \ldots\}$. Lemma 7.4 implies that any loop in the Markov graph brings in a periodic orbit of f.

Theorem 7.5 (Li–Yorke part of Sharkovsky Theorem) *Suppose a continuous map $f : I \to I$ has a cycle of minimal period 3. Then f has a cycle of minimal period n for all $n \geq 1$.*

Proof Consider the period-3 orbit $\{p_1, p_2, p_3\}$,

$$p_2 = f(p_1),\ p_3 = f(p_2),\ p_1 = f(p_3),$$

and assume first that $p_1 < p_2 < p_3$ (see Fig. 7.3). Introduce two intervals with disjoint interiors:

$$I_1 = [p_1, p_2],\quad I_2 = [p_2, p_3].$$

Then, I_1 covers I_2 and I_2 covers both I_1 and I_2. Thus, the Markov graph associated to $\{I_1, I_2\}$ contains the graph

$$I_1 \rightleftharpoons I_2 \circlearrowleft$$

that has for any integer $n > 0$ a loop

$$I_1 \rightharpoonup I_2 \rightharpoonup I_2 \rightharpoonup \cdots \rightharpoonup I_2 \rightharpoonup I_1$$

with $(n-1)$ occurrences of I_2. By Lemma 7.4, the map f^n has a fixed point x. This fixed point cannot correspond to a cycle with a minimal period $k < n$ except $k = 3$, which existence is assumed.

Indeed, assume that the orbit of x is a cycle of minimal period $k < n$. Then necessarily $x \in I_1 \cap I_2$, since $x \in I_1$ implies $x = f^k(x) \in I_2$. Thus $x \in I_1 \cap I_2 = \{p_2\}$ and $k = 3$. This immediately excludes the possibility that the assumption holds when $n = 2$. Next note that $f^2(x) = f^2(p_2) = p_1 \notin I_2$ while for $n > 2$ our construction guarantees that $f^2(x) \in I_2$. So also for $n > 2$ the assumption leads to a contradiction.

The only other possible case, $p_1 < p_3 < p_2$, can be treated similarly by defining $I_1 = [p_3, p_2]$ and $I_2 = [p_1, p_3]$. □

Remark The proof reveals that one does not actually need a period-3 orbit. If for some $p \in I$ we have either $f^3(p) \geq p > f(p) > f^2(p)$ or $f^3(p) \leq p < f(p) < f^2(p)$ then the map f has cycles of minimal period n for all $n \geq 1$ (so including $n = 3$). ♢

Example 7.6 (Cycle of period 3 of the logistic map) Consider the scalar map

$$x \mapsto f_\alpha(x) = \alpha x(1 - x), \quad x \in [0, 1]. \tag{7.2}$$

This map is called the *logistic map*.

Proposition 7.7 *At* $\alpha_1 = 1 + 2\sqrt{2} < 4$ *the map* f_α^3 *exhibits a generic fold bifurcation, generating a stable period-3 cycle and an unstable period-3 cycle of* (7.2) *as* α *increases.*

Proof Introduce the function $G(x, \alpha) := f_\alpha^3(x)$. One gets

$$G(x, \alpha) = \alpha^3 x(1 - x)(1 - \alpha x + \alpha x^2)(1 - \alpha^2 x + \alpha^3 x^2 - 2\alpha^3 x^3 + \alpha^2 x^2 + \alpha^3 x^4).$$

The fold bifurcation condition for a period-3 cycle $\{x, f_\alpha(x), f_\alpha^2(x)\}$ translates into the following system of polynomial equations:

$$\begin{cases} G(x, \alpha) - x = 0, \\ \frac{\partial}{\partial x} G(x, \alpha) - 1 = 0. \end{cases}$$

Eliminating x from this system, we obtain a polynomial equation

$$(\alpha^2 - 2\alpha - 7)(\alpha - 1)^2(\alpha^2 + \alpha + 1)^2(\alpha^2 - 5\alpha + 7)^2 = 0,$$

whose real solutions coincide with those of the equation

$$(\alpha^2 - 2\alpha - 7)(\alpha - 1) = 0.$$

These solutions are

$$\alpha_1 = 1 + 2\sqrt{2}, \ \alpha_2 = 1 - 2\sqrt{2}, \ \alpha_3 = 1.$$

The second root is negative, while the third is related to the transcritical bifurcation of the fixed point $x = 0$. Thus, α_1 is the only possible critical value.

The fixed points of $f_{\alpha_1}^3$ satisfy the polynomial equation $G(x, \alpha_1) = x$, which has two trivial solutions, $x = 0$ and

$$x = 1 - \frac{1}{\alpha_1} = \frac{8}{7} - \frac{2\sqrt{2}}{7},$$

corresponding to the fixed points of f_{α_1}. Thus, the equation $G(x, \alpha_1) - x = 0$ is equivalent to

$$x(7x - 8 + 2\sqrt{2})(343x^3 - (490 + 49\sqrt{2})x^2 + (91 + 112\sqrt{2})x + 31 - 41\sqrt{2}) = 0.$$

Three real roots $x_1 < x_2 < x_3$ of the third factor

$$343x^3 - (490 + 49\sqrt{2})x^2 + (91 + 112\sqrt{2})x + 31 - 41\sqrt{2}$$

provide a period-3 cycle at $\alpha_1 = 1 + 2\sqrt{2}$. One can express x_1 in radicals and then verify that

$$\tfrac{\partial^2}{\partial x^2} G(x_1, \alpha_1) > 0, \ \tfrac{\partial}{\partial \alpha} G(x_1, \alpha_1) < 0.$$

Thus, two period-3 cycles of f_α, one stable and one unstable, bifurcate for small $|\alpha - \alpha_1|$ with $\alpha > \alpha_1$. □

Thus, at least for small positive $(\alpha - \alpha_1)$, the logistic map (7.2) has cycles of all periods. ◊

Definition 7.8 *The* ***Sharkovsky ordering*** *of the natural numbers is defined by*

$$\begin{aligned}
&3 \prec 5 \prec 7 \prec \cdots \prec 2k+1 \prec \cdots \\
&2 \cdot 3 \prec 2 \cdot 5 \prec 2 \cdot 7 \prec \cdots \prec 2 \cdot (2k+1) \prec \cdots \\
&2^2 \cdot 3 \prec 2^2 \cdot 5 \prec 2^2 \cdot 7 \prec \cdots \prec 2^2 \cdot (2k+1) \prec \cdots \\
&\cdots \\
&2^n \cdot 3 \prec 2^n \cdot 5 \prec 2^n \cdot 7 \prec \cdots \prec 2^n \cdot (2k+1) \prec \cdots \\
&\cdots \\
&\cdots \prec 2^n \prec 2^{n-1} \prec \cdots \prec 2 \prec 1.
\end{aligned}$$

Since any natural number can be written in the form $2^n \cdot (2k+1)$ for some integers $k, n \geq 0$, all natural numbers occur in this list. We state the following theorem without proof.

Theorem 7.9 (Sharkovsky, 1964) *If a continuous map $f : I \to I$ has a cycle of minimal period p, then it has a cycle of minimal period q, for any $q \succ p$.* □

This theorem implies Theorem 7.5 and can be proven using similar techniques. The Sharkovsky theorem is sharp, e.g. there are continuous maps from I into I which have period-5 cycles but no period-3 cycles.

7.1.2 One-Sided Symbolic Dynamics and Chaotic Maps

Let us introduce two scalar continuous maps on $I = [0, 1]$—one of which is quadratic while the other is piecewise-linear—that will play an important role in what follows.

Example 7.10 (The logistic map versus the tent map) Consider the logistic map (7.2) with $\alpha = 4$:

$$x \mapsto F(x) = 4x(1 - x), \quad x \in I, \tag{7.3}$$

and the *tent map*:

$$y \mapsto T(y) = 1 - |1 - 2y|, \quad y \in I. \tag{7.4}$$

(see Fig. 7.4). Clearly,

$$T(y) = \begin{cases} 2y & \text{for } 0 \le y \le \frac{1}{2}, \\ 2 - 2y & \text{for } \frac{1}{2} < y \le 1, \end{cases},$$

so that $F(I) = T(I) = I$ and both maps expand and fold I in such a way that the image covers I twice.

Proposition 7.11 *The maps F and T are topologically conjugate on $I = [0, 1]$.*

Proof Introduce

$$x = h(y) = \sin^2\left(\frac{\pi y}{2}\right).$$

This is a homeomorphism of the interval $[0, 1]$ onto itself. Moreover, for $y \in [0, \frac{1}{2}]$,

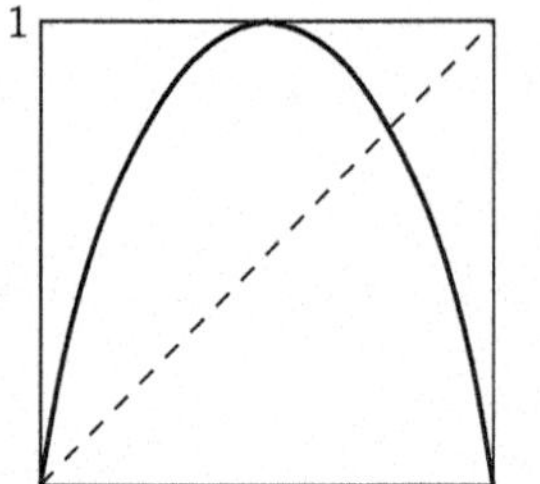

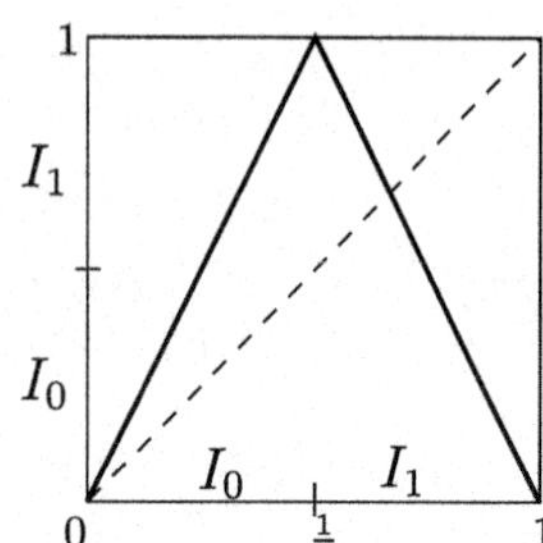

Fig. 7.4 The maps (7.3) and (7.4)

$$F(h(y)) = 4\sin^2\left(\frac{\pi y}{2}\right)\left(1-\sin^2\left(\frac{\pi y}{2}\right)\right) = 4\sin^2\left(\frac{\pi y}{2}\right)\cos^2\left(\frac{\pi y}{2}\right)$$
$$= \sin^2(\pi y) = h(T(y)).$$

By symmetry, $F(h(y)) = h(T(y))$ for $y \in [\frac{1}{2}, 1]$ as well. So $F \circ h = h \circ T$ on I. □

Therefore, all topological properties of F and T are the same and we can concentrate on the piecewise-linear tent map T. ◊

Consider the set Σ_2 of all one-sided infinite sequences of two symbols/digits $\{0, 1\}$. This set is a complete metric space with respect to the distance

$$d(\omega, \theta) = \sum_{k=0}^{\infty} \frac{|\omega_k - \theta_k|}{2^k}. \tag{7.5}$$

Introduce two closed intervals:

$$I_0 = \left[0, \tfrac{1}{2}\right], \quad I_1 = \left[\tfrac{1}{2}, 1\right].$$

Proposition 7.12 *For each sequence $\omega \in \Sigma_2$ there is a unique $x \in I$, such that*

$$T^k(x) \in I_{\omega_k}, \quad k = 0, 1, 2, \ldots.$$

Proof The Markov graph for T associated to $\{I_0, I_1\}$ is

$$\circlearrowleft I_0 \rightleftarrows I_1 \circlearrowright$$

Thus, by Lemma 7.3, for any $n \geq 0$, and any given finite sequence $\omega = \omega_0\omega_1\omega_2\ldots\omega_n$, we obtain a closed interval in I:

$$J_{\omega_0\omega_1\cdots\omega_n} = \{x \in I : T^k(x) \in I_{\omega_k} \text{ for } 0 \leq k \leq n\}.$$

Let $T^{-k}(J)$ denote the *full preimage* of $J \subset I$ under T^k, i.e.:

$$T^{-k}(J) = \{x \in I : T^k(x) \in J\}.$$

Then

$$J_{\omega_0\omega_1\cdots\omega_n} = I_{\omega_0} \cap T^{-1}(I_{\omega_1}) \cap T^{-2}(I_{\omega_2}) \cap \cdots \cap T^{-n}(I_{\omega_n}) = J_{\omega_0\omega_1\cdots\omega_{n-1}} \cap T^{-n}(I_{\omega_n}),$$

so that

$$J_{\omega_0\omega_1\cdots\omega_n} \subset J_{\omega_0\omega_1\cdots\omega_{n-1}}.$$

In words: The intervals $J_{\omega_0\omega_1\cdots\omega_n}$ are nested. Furthermore,

$$|J_{\omega_0\omega_1\cdots\omega_n}| = \frac{1}{2}|J_{\omega_0\omega_1\cdots\omega_{n-1}}|,$$

where $|J| = |a-b|$ is the length of $J = [a, b]$. Since $|I| = 1$, we have

$$|J_{\omega_0\omega_1\cdots\omega_n}| = \left(\frac{1}{2}\right)^n . \tag{7.6}$$

Thus, there exists a *unique* common point

$$\{x\} = \bigcap_{n\geq 0} J_{\omega_0\omega_1\cdots\omega_n}$$

and for this point $T^k(x) \in I_{\omega_k}$, $k = 0, 1, 2, \ldots$, by the construction. □

Therefore we have a well-defined map

$$\Phi : \Sigma_2 \to I, \quad \omega \mapsto \Phi(\omega) = x .$$

Moreover, from (7.6) it follows that Φ is continuous, i.e. two points $x = \Phi(\omega)$ and $x' = \Phi(\omega')$ are close if the sequences ω and ω' are close with respect to the distance (7.5). Note that the construction used in the above proof will be repeated several times in what follows.

The *shift map* $\sigma : \Sigma_2 \to \Sigma_2$ is defined by the formula $\sigma(\omega) = \theta$, where

$$\theta_k = \omega_{k+1}, \quad k = 0, 1, 2, \ldots . \tag{7.7}$$

The shift map $\sigma : \Sigma_2 \to \Sigma_2$ (cf. Example 1.4 in Chap. 1) is a continuous map with respect to the distance (7.5). It is noninvertible in Σ_2.

Theorem 7.13 *The discrete-time noninvertible dynamical system* $\{\mathbb{N}, \Sigma_2, \sigma^k\}$ *has*

(*i*) *a countable dense set of periodic orbits with arbitrarily long periods*;

(*ii*) *an uncountable set of nonperiodic orbits*;

(*iii*) *a dense orbit, i.e. an orbit that passes arbitrarily close to any given sequence.*

Proof (*i*) Any periodic sequence with a repeating block of length n defines an n-periodic orbit. The set of all periodic sequences is countable. To show that the periodic orbits are dense, take an arbitrary sequence and produce a periodic sequence by repeating a sufficiently long initial block of the given sequence.

(*ii*) Suppose that all nonperiodic sequences are listed. Then we can construct a sequence $\tilde{\omega}$ by choosing $\tilde{\omega}_0$ different from ω_0 in the first listed sequence, $\tilde{\omega}_1$ different from ω_1 in the second listed one, $\tilde{\omega}_2$ different from

ω_2 in the third listed one, etc. The sequence $\tilde{\omega}$ thus constructed is not in the list, since it differs from any listed one. A contradiction.

(*iii*) To produce a dense orbit, list all blocks of length k for all k and glue them together in the order of increasing k. The resulting sequence γ is called the *Morse sequence*. An appropriate shift of this sequence, say $\sigma^k(\gamma)$, agrees with any given sequence ω in the first l positions, where l is as large as we want. Therefore, the distance $d(\sigma^k(\gamma), \omega)$ defined by (7.5) can be made smaller than any given positive number. □

It happens that the shift map (7.7) acts on a one-sided sequence $\omega \in \Sigma_2$ as the tent map (7.4) on $x = \Phi(\omega) \in I$.

Proposition 7.14 $\Phi(\sigma(\omega)) = T(\Phi(\omega))$ *for all* $\omega \in \Sigma_2$.

Proof Let $x = \Phi(\omega)$ for some $\omega \in \Sigma_2$. Then, according to Proposition 7.12,

$$T^k(x) \in I_{\omega_k}, \quad k = 0, 1, 2, \ldots.$$

Thus

$$T^{l-1}(T(x)) \in I_{\omega_l}, \quad l = 1, 2, 3, \ldots,$$

or, equivalently,

$$T^k(T(x)) \in I_{\omega_{k+1}}, \quad k = 0, 1, 2, \ldots.$$

This implies that $T(x) = \Phi(\theta)$, where $\theta_k = \omega_{k+1}, \quad k = 0, 1, 2, \ldots$. Since $\theta = \sigma(\omega)$, we get

$$\Phi(\sigma(\omega)) = \Phi(\theta) = T(x) = T(\Phi(\omega))$$

for all $\omega \in \Sigma_2$. □

Proposition 7.14 immediately implies

$$\Phi(\sigma^k(\omega)) = T^k(\Phi(\omega)), \quad k = 1, 2, \ldots, \tag{7.8}$$

for all $\omega \in \Sigma_2$.

Our aim is now to transfer the properties listed in Theorem 7.13 for $\{\mathrm{N}, \Sigma_2, \sigma^k\}$ to $\{\mathrm{N}, I, T^k\}$ using Φ and Φ^{-1}. But unfortunately Φ is not invertible. Indeed, the sequences

$$01000\ldots \quad \text{and} \quad 11000\ldots$$

are both mapped to $x = \frac{1}{2}$. This reflects that $x = \frac{1}{2}$ belongs to both I_0 and I_1. Similarly, all preimages of $x = \frac{1}{2}$ under T have two symbolic representations.

In order to show that the multivaluedness of Φ^{-1} is in fact harmless, we begin by collecting the exceptional points in the set

$$E := \bigcup_{k=0}^{\infty} T^{-k}\left(\frac{1}{2}\right). \tag{7.9}$$

Note that E is countable and that every point of E is mapped to $0 \in I$ by a sufficiently high (depending on the point) iterate of T. For $x \in E$, the set $\Phi^{-1}(x)$ consists of two types of sequences,

$$\omega_0\omega_1 \ldots \omega_{k-1}01000\ldots$$

and

$$\omega_0\omega_1 \ldots \omega_{k-1}11000\ldots,$$

which differ only in the two symbols that precede the "tail" of zeroes. For $x \in I \setminus E$ the set $\Phi^{-1}(x)$ consists of a single sequence. Clearly, both E and—as a consequence—$I \setminus E$ are invariant sets for T.

Proposition 7.15 *If $\omega \in \Sigma_2$ is periodic, $x = \Phi(\omega)$ belongs to a periodic orbit of T and the minimal periods of this orbit and ω are the same.*

Proof The periodicity of $x = \Phi(\omega)$ with respect to T follows at once from (7.8). Since x is periodic, $x \notin E$ (all orbits that start in E end in 0). It follows that for every $k = 1, 2, \ldots$ $T^k(x) \notin E$. Hence $\Phi^{-1}(T^k(x))$ consists of a single sequence. So if $\sigma^k(\omega) \neq \omega$ for $k = 1, 2, \ldots, m-1$, while $T^k(x) = x$ for some $k \in \{1, 2, \ldots, m-1\}$, we get a contradiction by applying Φ^{-1} to the identity (7.8). We conclude that

$$x, T(x), T^2(x), \ldots$$

has the same minimal period as ω. □

Proposition 7.16 *The set of nonperiodic orbits of $\{\mathbb{N}, I, T^k\}$ is uncountable.*

Proof Assume that the set Y of nonperiodic orbits is countable. Since, for any $x \in I$, the set $\Phi^{-1}(x)$ contains one or two elements, $\Phi^{-1}(Y)$ is also countable. From Theorem 7.13(*ii*) we deduce by contradiction that $\{\mathbb{N}, I, T^k\}$ must have a nonperiodic orbit such that, when we apply Φ, we obtain a periodic orbit of T. A periodic orbit of T cannot belong to E. So we can apply Φ^{-1} as a single-valued map. Via (7.8) this leads to the contradiction that the orbit of σ, that is assumed to be nonperiodic, is in fact periodic. □

Proposition 7.17 *The dynamical system $\{\mathbb{N}, I, T^k\}$ has a dense orbit.*

Proof Let $\tilde{\omega} \in \Sigma_2$ be such that its orbit under σ is dense in Σ_2. Let $\tilde{x} = \Phi(\omega)$. Choose any $x \in I$. The claim is that, for every $\varepsilon > 0$, we can find k such that

$$|T^k(\tilde{x}) - x| < \varepsilon.$$

Let $\omega \in \Sigma_2$ be such that $\Phi(\omega) = x$. Choose $\delta = \delta(\varepsilon)$ such that $d(\tilde{\omega}, \omega) < \delta$ implies $|\Phi(\tilde{\omega}) - \Phi(\omega)| < \varepsilon$. Choose $k = k(\omega, \delta)$ such that $d(\sigma^k(\tilde{\omega}), \omega) < \delta$. Then

$$|T^k(\tilde{x}) - x| = |T^k(\Phi(\tilde{\omega})) - \Phi(\omega)| = |\Phi(\sigma^k(\tilde{\omega})) - \Phi(\omega)| < \varepsilon. \qquad \square$$

The above Propositions imply that all assertions in Theorem 7.13 are also valid for the dynamical system $\{\mathbb{N}, I, T^k\}$ generated by the tent map (7.4), and, taking into account Proposition 7.11, for the special logistic map (7.3). In particular, both maps have an infinite number of periodic orbits.

Next we introduce the notion of "sensitive dependence on initial conditions". Let X be a metric space with distance d.

Definition 7.18 *A dynamical system $\{\mathbb{T}, X, \varphi^t\}$ is said to exhibit* ***sensitive dependence on initial conditions*** *if there exists $r > 0$ such that for every point $x \in X$ and for each $\varepsilon > 0$ there is a point $y \in X$ with $d(x, y) < \varepsilon$ and a positive $t \in \mathbb{T}$ such that $d(\varphi^t(x), \varphi^t(y)) \geq r$.*

This phenomenon was observed by E.N. Lorenz in the early 1960s when he numerically integrated a system of three ODEs (see Sect. 7.2). He restarted some computations on the basis of printed intermediate coordinate values and found that the new results were in good agreement with the original outcome for small times, but did differ greatly for large times. He attributed this to the difference in decimal precision between the printed coordinate values and their machine representation.

It is easy to establish sensitive dependence on initial conditions for $\{\mathbb{N}, \Sigma_2, \sigma^k\}$. Indeed, if ω and θ agree up to symbol k, but differ at in the kth position, then $d(\omega, \theta) \leq 2^{1-k}$. For any $\varepsilon > 0$ we may choose k such that $2^{1-k} < \varepsilon$. Yet, since $\sigma^k(\omega)$ and $\sigma^k(\theta)$ differ in the first position, they are at least distance one apart. Since sensitive dependence on initial conditions is preserved under topological conjugacy, it follows that also the dynamical systems generated by F and T exhibit sensitive dependence on initial conditions. For the tent map T we see most clearly the underlying mechanism: At every time step small differences are increased by factor two, but the folding prevents that the orbits themself move to infinity. So there is a combination of local stretching and global bending that keeps orbits bounded, but expands the distance between nearby points.

If we make the tent bigger by raising the top, while preserving that both zero and one are mapped to zero, then part of the interval $I = [0, 1]$ is mapped outside that interval (see Fig. 7.5). A bigger part is mapped outside in two steps, etc. But something remains, i.e. there are points for which the entire forward half-orbit belongs to I. What is the structure of the set of such points ? Note that a similar behaviour—but in the plane—is demonstrated by the *Smale horseshoe map* to be treated in Sect. 7.3.1.

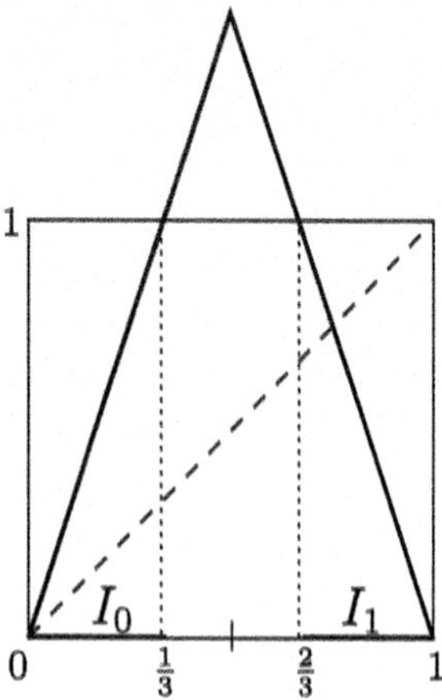

Fig. 7.5 'Big Tent Map' (7.10)

For definiteness let the map $\tilde{T}: \mathbb{R} \to \mathbb{R}$ shown in Fig. 7.5 be defined by

$$\tilde{T}(x) = \begin{cases} 3y & \text{for } x \leq \frac{1}{2}, \\ 3 - 3y & \text{for } \frac{1}{2} < x, \end{cases}, \tag{7.10}$$

so that once an orbit is outside $I = [0, 1]$ it tends to $-\infty$. Note that $\tilde{T}^{-1}(I) = I_0 \cup I_1$ with $I_0 = [0, \frac{1}{3}]$ and $I_1 = [\frac{2}{3}, 1]$. Next note that $\tilde{T}^{-1}(I_0) = I_{00} \cup I_{10}$ and $\tilde{T}^{-1}(I_1) = I_{01} \cup I_{11}$ with $I_{00} = [0, \frac{1}{9}], I_{10} = [\frac{8}{9}, 1], I_{01} = [\frac{2}{9}, \frac{1}{3}]$, and $I_{11} = [\frac{2}{3}, \frac{7}{9}]$. By induction we define for $\omega_i \in \{0, 1\}$ the intervals

$$J_{\omega_0\omega_1\ldots\omega_{n-1}} = \{x : \tilde{T}^k(x) \in I_{\omega_k} \quad \text{for} \quad 0 \leq k \leq n-1\}.$$

Let

$$S_n := \{x : \tilde{T}^k(x) \in I \quad \text{for} \quad 0 \leq k \leq n\},$$

then

$$S_n = \bigcap_{k=0}^{n} \tilde{T}^{-k}(I) = \bigcap_{k=0}^{n-1} \tilde{T}^{-k}(I_0 \cup I_1) = \bigcup_{\omega_i \in \{0,1\}} J_{\omega_0\omega_1\ldots\omega_{n-1}},$$

the union of 2^n closed intervals of length $(\frac{1}{3})^n$. Note that the S_n are nested: $S_{n+1} \subset S_n$. Finally, define

$$\Lambda = \bigcap_{n=0}^{\infty} S_n = \{x : \tilde{T}^k(x) \in I \quad \text{for all} \quad k \geq 0\}. \tag{7.11}$$

Our aim is to characterize the structure of Λ and the dynamics generated by $\tilde{T}$ on this set. In order to do so we need to recall some terminology.

Let X be a metric space and let $Y \subset X$. By definition, Y is *nowhere dense* if the interior of the closure of Y is empty. One calls Y *totally disconnected* if the connected components of Y are single points. A closed subset of $\mathbb{R}$ is nowhere dense if and only if it is totally disconnected (but a smooth curve in the plane is nowhere dense, yet not totally disconnected). Finally, Y is *perfect* if it is closed and every point $p \in Y$ is the limit of points $y_n \in Y$ with $y_n \neq p$ for all $n \in \mathbb{N}$. The set Y is called a *Cantor set* if it is compact, totally disconnected, and perfect. The classical example of a Cantor set is obtained by removing the middle open interval of length $\frac{1}{3}$ from the unit interval I, next removing the middle open interval of length $\frac{1}{9}$ from the remaining two closed intervals, and repeating this "surgery" indefinitely. So the following result should not come as a surprise.

Proposition 7.19 *Λ is a Cantor set.*

Proof Since S_n is closed, so is $\Lambda = \cap_{n=0}^{\infty} S_n$. Consider any $y \in Y$. For arbitrary $j \in \mathbb{N}$ choose $n = n(j)$ such that $3^{-n} < 2^{-j}$. Let $J_{\omega_0\omega_1\ldots\omega_{n-1}}$ be the component of S_n to which y belongs. Note that

$$J_{\omega_0\omega_1\ldots\omega_{n-1}} \cap S_{n+1} = J_{\omega_0\omega_1\ldots\omega_{n-1}0} \cup J_{\omega_0\omega_1\ldots\omega_{n-1}1}$$

and that $J_{\omega_0\omega_1\ldots\omega_{n-1}} \setminus S_{n+1}$ is not empty. For any choice of $p_j \in J_{\omega_0\omega_1\ldots\omega_{n-1}} \setminus S_{n+1}$ we have that $p_j \notin S_{n+1}$ hence $p_j \notin \Lambda$ and

$$|p_j - y| \leq 3^{-n} < 2^{-j}.$$

It follows that $p_j \to y$ for $j \to \infty$ and hence that Λ is nowhere dense.

Next choose q_j to be the endpoint of $J_{\omega_0\omega_1\ldots\omega_{n-1}\omega_n}$. Then $q_j \in \Lambda$ and

$$|q_j - y| \leq 3^{-n} < 2^{-j},$$

so also $q_j \to y$ for $j \to \infty$ but $q_j \neq y$. Thus, Λ is perfect. □

A Cantor set has "holes" everywhere and yet contains no isolated points. The set Λ is *self-similar*: If we restrict to $\Lambda \setminus [0, \frac{1}{3}]$ we recover Λ in the sense that any point in Λ is obtained from that in $\Lambda \cap [0, \frac{1}{3}]$ by multiplication by 3. Likewise we recover Λ if we zoom in on appropriate "smaller" parts of Λ: At every scale Λ contains a copy of itself. There are multiple inequivalent ways in which we can think about the "size" of Λ. As we will show below, Λ is not countable. Since the total length $(\frac{2}{3})^n$ of S_n converges to zero as $n \to \infty$, the Lebesgue measure of Λ is zero (but Cantor sets with positive Lebesgue measure do exist, so this is *not* a consequence of the Cantor structure).

Let us return to dynamics. Clearly, $\tilde{T}(\Lambda) \subset \Lambda$. If $\tilde{T}(x) \in \Lambda$, then $\tilde{T}^k(x) \in I$ for all k, so by the very definition of Λ the point x belongs to Λ. Hence $\tilde{T}(\Lambda) = \Lambda$, i.e. Λ is forward-invariant under $\tilde{T}$. We define $h : \Lambda \to \Sigma_2$ by

$$h(x) = \omega \quad \text{with} \quad \omega_k = \begin{cases} 0, \text{ if } \tilde{T}^k(x) \in I_0, \\ 1, \text{ if } \tilde{T}^k(x) \in I_1, \end{cases} \tag{7.12}$$

and call $h(x)$ the *itinerary* of x and h the *itinerary map.* From this definition it follows at once that

$$h(\tilde{T}(x)) = \sigma(h(x)) \tag{7.13}$$

for all $x \in \Lambda$. Hence $\{\mathbb{N}, \Lambda, \tilde{T}^k\}$ and $\{\mathbb{N}, \Sigma_2, \sigma^k\}$ are topologically conjugate if we can show that h is a homeomorphism. And this is exactly what we prove next.

Proposition 7.20 *The map $h : \Lambda \to \Sigma_2$ defined by (7.12) is a homeomorphism.*

Proof Let $\omega \in \Sigma_2$. The intervals $J_{\omega_0\omega_1\ldots\omega_n}$ introduced above are nested as n increases. If $x \in J_{\omega_0\omega_1\ldots\omega_n}$ the $h(x)$ and ω agree up to position n. So for $x_0 \in \cap_{n=0}^{\infty} J_{\omega_0\omega_1\ldots\omega_n}$ we have $h(x_0) = \omega$ and we conclude that h is onto.

Suppose that $h(x) = h(y) = \omega$. Then both x and y belong to $J_{\omega_0\omega_1\ldots\omega_n}$ and hence $|x - y| \leq 3^{-n}$. As this estimate holds for arbitrary n, we conclude that $x = y$ and hence that h is one-to-one.

Let $x \in \Lambda$ and $\omega = h(x)$. For every $\varepsilon > 0$ there exists $n = n(\varepsilon)$ such that $2^{-n} < \varepsilon$. Let $\delta = \delta(\varepsilon)$ be small enough to guarantee that from $y = \lambda$ and $|y - z| < \delta$ follows that $y \in J_{\omega_0\omega_1\ldots\omega_n}$. Let $\theta = h(y)$. Then $d(\omega, \theta) \leq \sum_{k=n+1}^{\infty} 2^{-k} = 2^{-n} < \varepsilon$. So h is continuous.

Likewise it follows that $d(\omega, \theta) < 2^{-n}$ implies that $h^{-1}(\theta) \in J_{\omega_0\omega_1\ldots\omega_n}$ and hence that $|h^{-1}(\omega) - h^{-1}(\theta)| \leq |J_{\omega_0\omega_1\ldots\omega_n}| = 3^{-(n+1)}$. So h^{-1} is also continuous. □

We can now apply Theorem 7.13 to conclude that the restriction of $\tilde{T}$ to the invariant set Λ has a very rich orbit structure. Also note that the diagonalization argument used to prove part (ii) of that theorem establishes that Σ_2, and hence Λ is not countable (in fact any perfect set is uncountable). Among the points in Λ are the boundary points of $J_{\omega_0\omega_1\ldots\omega_n}$. They form a countable subset of Λ and their itinerary ends with an infinite sequence of zeroes (since they are mapped to zero in finitely many iterates). So Λ is *not* just the limit of the ever-increasing set of boundary points.

The fact that $\tilde{T}$ restricted to Λ has a dense orbit shows that we cannot divide Λ into invariant pieces. To formulate this precisely, we introduce another important notions. Let X be a metric space and let $f : X \to X$ be continuous.

Definition 7.21 *The dynamical system $\{\mathbb{N}, X, f^k\}$ is called* ***topologically transitive*** *if for every pair of nonempty open subsets U and V of X there exists $n = n(U, V)$ such that $f^n(U) \cap V \neq \emptyset$.*

If f has a dense orbit and X has no isolated points then $\{\mathbb{N}, X, f^k\}$ is topologically transitive. The *Birkhoff Transitivity Theorem* asserts that, under some extra conditions on X, a topologically transitive dynamical system has a dense orbit.

Definition 7.22 *Let* Y *be an invariant subset of* X, *i.e.* $f(Y)=Y$. *The dynamical system* $\{\mathbb{N}, Y, f^k\}$ *is called* ***chaotic*** *if*
 (*i*) *it exhibits sensitive dependence on initial conditions*;
 (*ii*) *it is topologically transitive.*

It should be noted that some authors include the condition that periodic orbits are dense in Y in the definition of chaotic.

We can now summarize our earlier results in a simple statement: *The dynamical system* $\{\mathbb{N}, \Lambda, \tilde{T}^k\}$ *is chaotic.*

7.2 Lorenz Attractor

One-dimensional (piecewise-)continuous maps can be used to understand bifurcations leading to a chaotic attractor in the famous Lorenz system that appears in meteorology. We begin with the following—almost philosophical—question that attracted much attention.

Question *Can the climate and its fluctuations be predicted by using physical principles?*

To answer the question, one could proceed as follows:

Step 1: Take a fluid dynamics model of the atmosphere (coupled Navier–Stokes and heat equations). Consider Rayleigh–Bénard convection in a horizontal layer heated from below by absorbed sunlight.
Step 2: Galerkin approximation: Project solutions to the subspace spanned by the eigenfunctions (sometimes called "modes") of the linearization around a special solution. This gives a system of infinitely many ODEs. Crude simplification: Truncate, i.e. put coefficients corresponding to an infinite tail equal to zero.
Step 3: Analyse analytically equilibria and their stability in the truncated system.
Step 4: Perform numerical experiments: Define and study a Poincaré map numerically.
Step 5: Idealize the Poincaré map, reduce the dimension, use symbolic dynamics. This gives much qualitative insight.
Step 6: Prove correctness of the idealized description.
Step 7: Draw conclusions about the dynamics of the atmosphere.

Here we deal with **Step 3** and **Step 5**, starting from the *Lorenz system*[1]

[1] Lorenz, E. (1963). Deterministic non-periodic flow. *Journal of the Atmospheric Sciences, 20*, 130–141.

$$\begin{cases} \dot{x} = \sigma(y - x), \\ \dot{y} = rx - y - xz, \\ \dot{z} = -bz + xy, \end{cases} \tag{7.14}$$

where $\sigma = 10,\ b = \frac{8}{3}$, while $r > 0$ is a bifurcation parameter.

To begin with, we make some key observations:

(1) *Symmetry*: If $(x(t), y(t), z(t))$ is a solution to (7.14), so is $(-x(t), -y(t), z(t))$.
(2) *Bounds*: There exists a closed ball in the (x, y, z)-space such that every orbit of (7.14) enters it and then remains in this ball (Exercise 7.4).
(3) *Dissipativity*: The flow of (7.14) contracts volumes at a constant rate. Indeed, for the RHS vector field f,

$$\text{div } f = -(\sigma + 1 + b) < 0.$$

Application of Lemma 4.39 from Sect. 4.8 gives the result. We shall see that this does not preclude unstable equilibria and cycles, as well as complicated limit sets with Cantor structure.

There is always a trivial equilibrium: $O = (0, 0, 0)$. For $r > 1$, two nontrivial equilibria appear:

$$Q^{\pm} = \left(\pm\sqrt{b(r-1)}, \pm\sqrt{b(r-1)}, r - 1\right).$$

Thus, at $r = r_1 = 1$ a supercritical pitchfork bifurcation takes place. The eigenvalues for $r > 1$ are

$O : \lambda_2^0 > 0,\ \lambda_{1,3}^0 < 0$ satisfying for r somewhat larger than one

$$-\lambda_3^0 > \lambda_2^0 > -\lambda_1^0 > 0.$$

$Q^{\pm}$: For $1 < r \leq \tilde{r}$, three real eigenvalues with $\lambda_{1,2,3}^{\pm} < 0$. At $r = \tilde{r}$, there is a *double eigenvalue* $\lambda_2^{\pm} = \lambda_3^{\pm}$. For $\tilde{r} < r < r_H$, one real eigenvalue $\lambda_1^{\pm} < 0$ and two complex ones $\lambda_{2,3}^{\pm}$ with Re $\lambda_{2,3}^{\pm} < 0$. For $r > r_H$, one real eigenvalue $\lambda_1^{\pm} < 0$ and two complex eigenvalues $\lambda_{2,3}^{\pm}$ with Re $\lambda_{2,3}^{\pm} > 0$. At

$$r_H = \frac{\sigma(\sigma + b + 3)}{\sigma - b - 1}$$

we have a *subcritical Hopf bifurcation*, since the first Lyapunov coefficient $l_1 > 0$ (see Exercise 6.3 in Chap. 6). In what follows, we assume that $\tilde{r} < r < r_H$ and only consider those values of r for which $\lambda_1^{\pm} <$ Re $\lambda_{2,3}^{\pm} < 0$. This is certainly true near $r = r_H$.

Assume that we can choose a surface Σ containing the equilibria $Q^{\pm}$ and their one-dimensional *strong-stable manifolds* $W^{ss}(Q^{\pm})$, each composed of

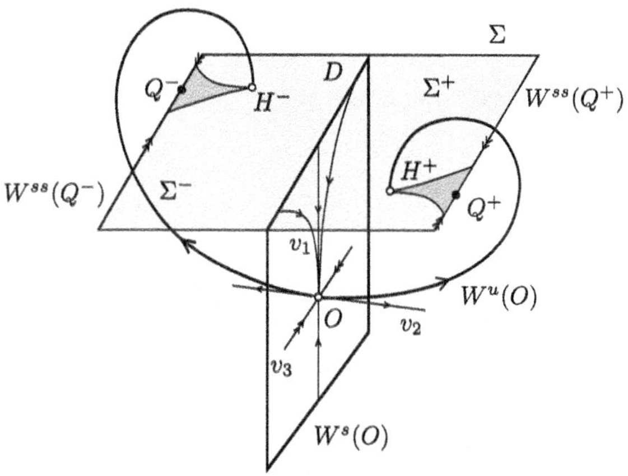

Fig. 7.6 Manifold configuration A: The line D is the intersection of the cross section Σ with the stable manifold $W^s(O)$; the strong-stable manifolds $W^{ss}(Q^{\pm})$ belong to the boundary of Σ

two exceptional orbits approaching $Q^{\pm}$ along the eigenvector corresponding to the real eigenvalue $\lambda_1^{\pm} < 0$. Further assume that the two-dimensional stable manifold $W^s(O)$ of the origin intersects Σ along a curve D, such that an orbit starting in a typical point in Σ returns to it from above at some time later (see Fig. 7.6). This defines a Poincaré map F on $\Sigma \setminus D$. The image $F(\Sigma \setminus D)$ intersects with Σ in two curvilinear triangles. Two lines $D^+ \subset \Sigma^+$ and $D^- \subset \Sigma^-$, which are "infinitesimally close" to D, are both mapped to points, but these points $H^{\pm}$ lie far from each other. This is not contradicting continuous dependence of the solutions on the initial conditions in finite time intervals, since it takes *infinite* time to be thus mapped. Note that $H^{\pm}$ are the points where the unstable manifold of O hits Σ from above for the first time.

From now on we focus on the so-called *geometric Lorenz attractor* using a reduction of F to a *scalar* map f that is called the *geometric Lorenz map*. This reduction cannot be done in a strict sense, so we move here to an idealized model system. Near O the flow contracts strongly in the leading eigendirection corresponding to λ_3. Assume that in suitable coordinates (u, v) on Σ the map takes the form:

$$F(u, v) = (f(u), g(u, v)),$$

i.e. the u-dynamics decouples, and

$$0 < D_v g(u, v) < c < 1$$

for $u \neq 0$, while

$$D_u g(u, v) \to 0 \text{ as } u \to 0.$$

The map F depends on the parameters (σ, b, r). In the figures below, u is the horizontal coordinate. The construction assumes the existence of a *global strong-stable foliation*. The equation $u = 0$ defines D in the new coordinates.

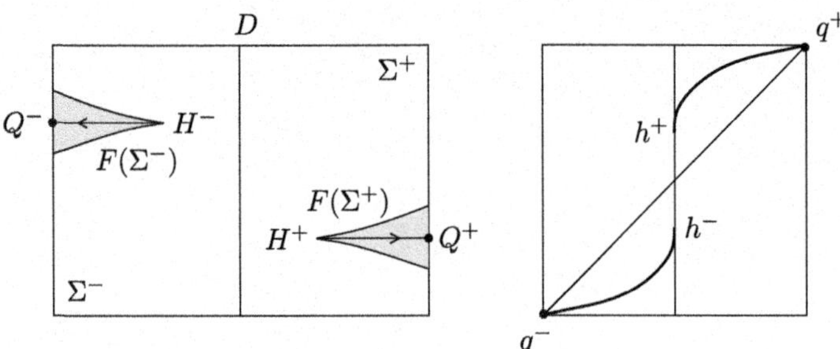

Fig. 7.7 Configuration A

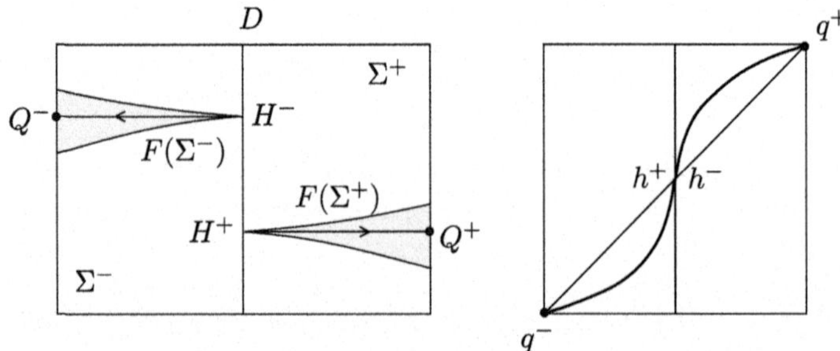

Fig. 7.8 Homoclinic transition A $\to$ B

Note that $f(0)$ is undefined, while the derivative of $f(u)$ is unbounded as $u \to 0^{\pm}$.

What happens to f if we increase r, keeping (σ, b) fixed as before? Figure 7.7 shows $F(\Sigma) \cap \Sigma$ for values $r > \tilde{r}$ for which $F(\Sigma^{\pm}) \subset \Sigma^{\pm}$, together with the graph of the corresponding map $f(u)$. The u-coordinates of $H^{\pm}$ are denoted by $h^{\pm}$, i.e. $h^{\pm} = f(0^{\pm})$. The fixed points $q^{\pm}$ of f are global one-side attractors.

For $r = r_2 = 13.296\ldots$, we have a pair of homoclinic orbits to O. Figure 7.8 shows schematically the manifold configuration at $r = r_2$: The unstable manifold $W^u(O)$ is contained in the stable manifold $W^s(O)$; $H^{\pm} \in D$ and $h^+ = h^- = 0$.

For r slightly bigger than r_2, the map f has two fixed points $p^{\pm}$, which are unstable (see Fig. 7.9). These points correspond to saddle limit cycles that "circle" $Q^{\pm}$ once and intersect Σ at points $P^{\pm}$. These cycles are born from the homoclinic orbits, so their period tends to infinity as $r \to r_2$ (see Appendix A). Notice that $h^- > p^+$ and $h^+ < p^-$ due to the fact that the derivative $f'(u)$ is infinite at both sides of $u = 0$. As in the tent map case, there exists an invariant Cantor set Λ of points that stay forever in $[p^-, p^+]$ (see Fig. 7.10), and the dynamics on Λ is conjugate to the shift on one-sided sequences of two symbols. This implies the existence of countably many periodic orbits (two of them, corresponding to $p^{\pm}$, have period one) and uncount-

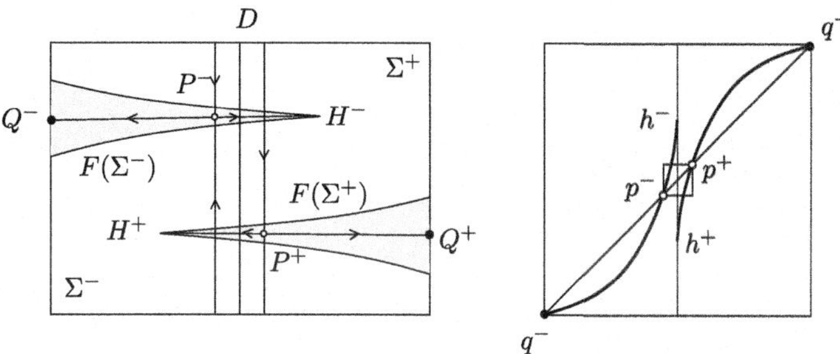

Fig. 7.9 Configuration B

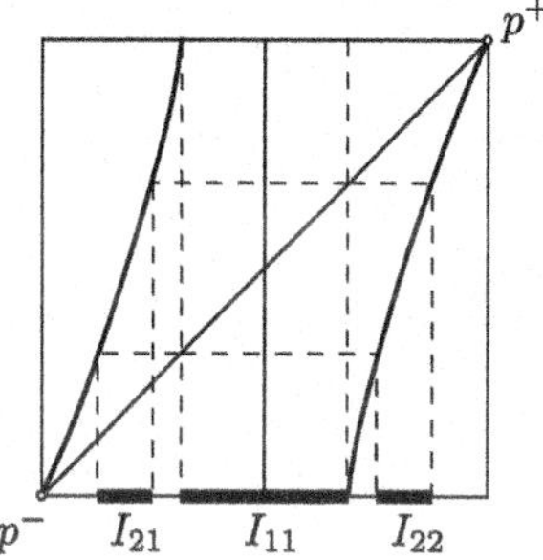

Fig. 7.10 Enlarged graph of f in configuration B. I_{11} consists of points leaving $[p^-, p^+]$ after one iterate; points in I_{21} or I_{22} leave $[p^-, p^+]$ in two iterates, etc.

ably many nonperiodic orbits. However, all these orbits are unstable: A tiny perturbation gives convergence towards either q^- or q^+. If the gap between h^+ and h^- is small, it takes a long time before such convergence to $q^\pm$ is finally visible. This phenomenon is called *preturbulence*, and is characterized by long chaotic transients before the solutions settle. Note that points in Λ correspond to orbits that circle forever, sometimes on the left, sometimes on the right, with arbitrary numbers of consecutive passages at a particular side. The A $\to$ B transition is sometimes called a *homoclinic explosion.*

When r increases, both $p^\pm$ and $h^\mp$ move towards $q^\pm$, but $h^\mp$ move *slower* than $p^\pm$ and are overtaken. This is an extra assumption on f, based on numerical evidence. When this happens at $r = r_3 = 24.06\ldots$ (see Fig. 7.11), the interval

$$[p^-, p^+] = [h^+, h^-]$$

is *invariant* and escape to $q^\pm$ becomes impossible. If we rectify the graph of f in this interval, we get the *saw-tooth map* (i.e. the multiplication by $2 (\mathrm{mod}\ 1)$, see Fig. 7.12). Note that the condition $p^\pm = h^\mp$ translates into: The unstable manifold $W^u(O)$ "hits" the stable manifold of the periodic orbits corresponding to $p^\pm$. We need the v-coordinate to see this; in fact, the unstable one-dimensional manifold of O is in the two-dimensional stable manifolds of the periodic orbits represented by $p^\pm$. Thus, there are heteroclinic orbits connecting the equilibrium O with the saddle cycles passing through $P^\pm$.

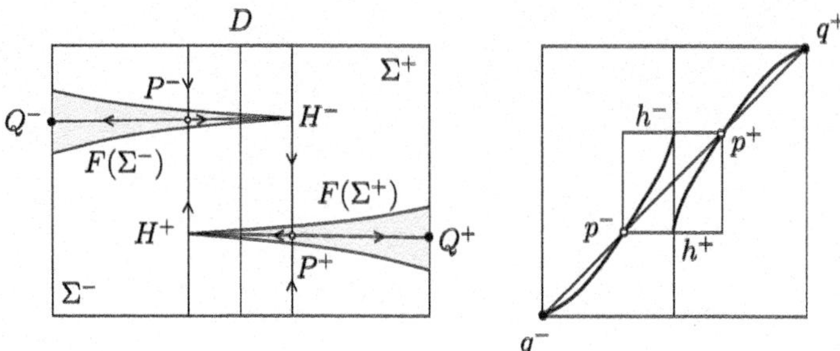

Fig. 7.11 Transition B → C

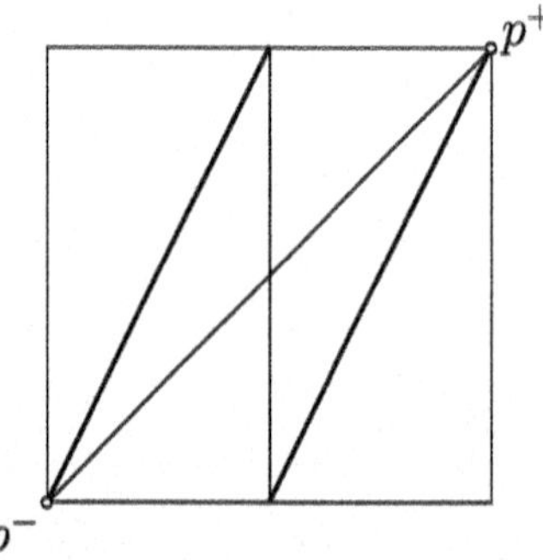

Fig. 7.12 Enlarged and rectified graph of f at the transition B → C

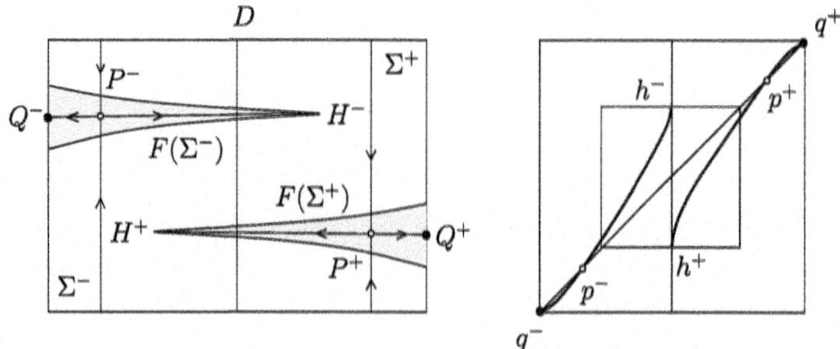

Fig. 7.13 Configuration C

For r slightly bigger than r_3, initial points in (p^-, p^+) are attracted to $[h^+, h^-]$ (see Figs. 7.13 and 7.14), while the points outside $[p^-, p^+]$ go to $q^\pm$. That is, the stable manifolds of the saddle periodic orbits $p^\pm$ separate the domain of attraction of the strange attractor corresponding to $[h^+, h^-]$ and the equilibria $Q^\pm$. When f is sufficiently expanding (e.g. $f' > \sqrt{2}$), the interval $[h^+, h^-]$ is indecomposable and we do have sensitive dependence on initial conditions within this interval.

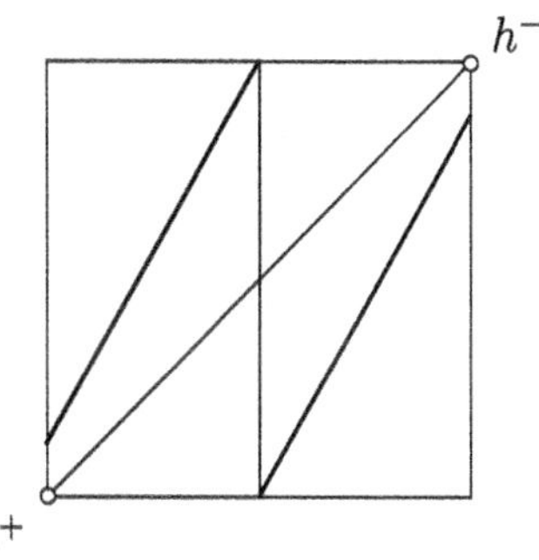

Fig. 7.14 Enlarged and rectified graph of f in the configuration C

Before discussing the dynamics in case C in more detail, we mention that eventually $p^{\pm}$ approach $q^{\pm}$ and disappear. This happens at $r = r_H$ corresponding to the subcritical Hopf bifurcation. This bifurcation ends the coexistence of the attractors. This, however, does not affect in any way the strange attractor itself.

Some observations concerning the Lorenz attractor in case C:

(I) For some integer $m > 0$, we have $f^{(m)}(h^+) > 0$. This means that the maximum length of a block of zeroes in the sequences coding points in Λ is $m - 1$. Consequently, not all sequences occur, there are restrictions.

(II) The preimage $f^{-n}(0)$ consists of 2^n points. Their union for all n is dense in $[h^+, h^-]$. In other words, the intersection of the unstable manifold of O with Σ is dense in the strange attractor.

(III) There is a relationship between (I) and (II): Families of orbits "disappear" in homoclinic explosions if $f(h^{\pm}) \in f^{-m}(0)$ for some m, i.e. if the unstable manifold of O is part of the stable manifold of O. The orbit on $W^u(O)$ departs from O, makes m turns, and then returns to O. When the position of $f^{(m)}(h^{\pm})$ relative to 0 changes, all periodic orbits with the corresponding block of zeroes and ones disappear.

(IV) The detailed structure of the attractor is very changeable: Systems with and without O-homoclinic orbits are both dense in the parameter space and neither of them is structurally stable! Details of the dynamics which do not persist under parameter variations may not correspond to verifiable physical properties of the system. There are many open questions here.

In the case of our geometric Lorenz attractor satisfying all assumptions above, we can fully describe the attractor by the *kneading sequence* of h^- or h^+. A kneading sequence is

$$k(x) = s_0 s_1 s_2 \cdots ,$$

where

$$s_i = \begin{cases} 0, \text{ if } f^{(i)}(x) < 0, \\ 1, \text{ if } f^{(i)}(x) > 0, \end{cases}$$

and the sequence ends if $f^{(m)}(x) = 0$ for some $m \in \mathbb{N}$. Assume that f is *locally eventually onto* $I = [h^+, h^-]$, i.e. for any $J \subset I$, there is n such that $f^{(n)}(J) = I$. It is sufficient to assume $f' > \sqrt{2}$ for this to hold. Direct consequences of this are that I is indecomposable and that the map $x \mapsto k(x)$ is injective. We have a natural (*lexicographic*) ordering on the sequences corresponding to the usual ordering of real numbers given by their binary expansion. The following lemma is due to the monotonicity (apart from one jump) of f:

Proposition 7.23 *If $x < y$, then $k(x) < k(y)$ in the lexicographic ordering.* □

From this it immediately follows that

$$k(h^+) < k(x) < k(h^-),$$

for all $x \in I$. Moreover,

$$k(h^+) \le \sigma^n k(x) \le k(h^-), \tag{7.15}$$

for all natural numbers $n \in \mathbb{N}$.

Theorem 7.24 *The condition* (7.15) *is necessary and sufficient for a sequence to correspond to some $x \in I$.* □

This means that $k(h^\pm)$ determine exactly which orbits can exist.

Theorem 7.25 *Two geometric Lorenz maps f_1 and f_2 are topologically conjugate if and only if*

$$k(h_1^\pm) = k(h_2^\pm).$$ □

Theorem 7.26 *The geometric Lorenz map f is topologically equivalent to the saw-tooth map.* □

Is there a geometric Lorenz attractor in the Lorenz system (7.14)? Does it evolve as described as r increases? The answers to these questions are certainly "yes", but available proofs involve numerical approximation of solutions. The existence of the primary homoclinic explosion is shown analytically.

7.3 Smale Horseshoe and Its Applications

7.3.1 Geometrical Smale Horseshoe

Let $S \subset \mathbb{R}^2$ be the unit square in the plane (see Fig. 7.15).

$$S = \{(\xi, \eta) \in \mathbb{R}^2 : 0 \le \xi \le 1,\ 0 \le \eta \le 1\}.$$

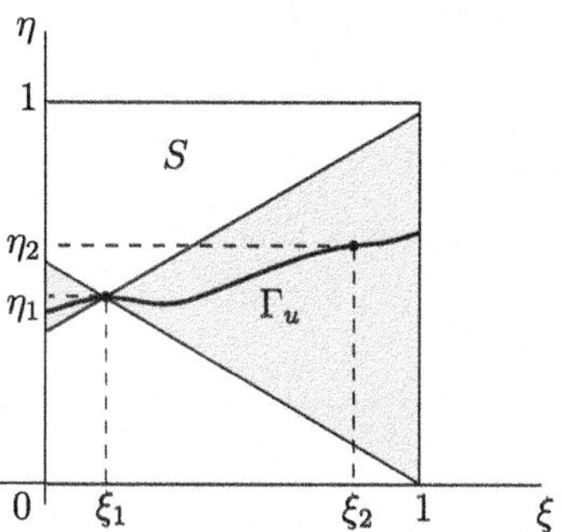

Fig. 7.15 A horizontal curve

Definition 7.27 *The graph $\Gamma_u \subset S$ of a scalar continuous function $\eta = u(\xi)$ satisfying*

(*i*) $0 \le u(\xi) \le 1$ *for* $0 \le \xi \le 1$,

(*ii*) $|u(\xi_1) - u(\xi_2)| \le \mu|\xi_1 - \xi_2|$ *for* $0 \le \xi_1 \le \xi_2 \le 1$, *where* $0 < \mu < 1$,

is called a μ-***horizontal curve*** *in* S.

Definition 7.28 *A graph $\Gamma_v \subset S$ of a scalar continuous function $\xi = v(\eta)$ satisfying:*

(*i*) $0 \le v(\eta) \le 1$ *for* $0 \le \eta \le 1$;

(*ii*) $|v(\eta_1) - v(\eta_2)| \le \mu|\eta_1 - \eta_2|$ *for* $0 \le \eta_1 \le \eta_2 \le 1$, *where* $0 < \mu < 1$,

is called a μ-***vertical curve*** *in* S.

Lemma 7.29 *A μ-horizontal curve $\Gamma_h \subset S$ and a μ-vertical curve $\Gamma_v \subset S$ intersect in precisely one point.*

Proof A point of intersection (ξ, η) corresponds to a zero ξ of $\xi - v(u(\xi))$ via $\eta = u(\xi)$. From the above definitions, one has for $0 \le \xi_1 < \xi_2 \le 1$:

$$|v(u(\xi_1) - v(u(\xi_2))| \le \mu|u(\xi_1) - u(\xi_2)| \le \mu^2|\xi_1 - \xi_2|$$

with $\mu^2 < 1$. Hence, the function $\psi(\xi) = \xi - v(u(\xi))$ is strictly monotonically increasing. Since $\psi(0) \le 0$ and $\psi(1) \ge 0$, it has precisely one zero. $\square$

For $x = (\xi, \eta)$, denote

$$\|x\|_1 = |\xi| + |\eta|.$$

Introduce

$$\|u\| = \max_{0 \le \xi \le 1} |u(\xi)|, \quad \|v\| = \max_{0 \le \eta \le 1} |v(\eta)|.$$

Lemma 7.30 *Let $\eta = u_{1,2}(\xi)$ define two μ-horizontal curves while $\xi = v_{1,2}$ (η) define two μ-vertical curves. Denote by $x_{1,2} = (\xi_{1,2}, \eta_{1,2})$ the intersection points of the corresponding horizontal and vertical curves. Then*

$$\|x_2 - x_1\|_1 \le \frac{1}{1-\mu}(\|u_2 - u_2\| + \|v_2 - v_1\|). \tag{7.16}$$

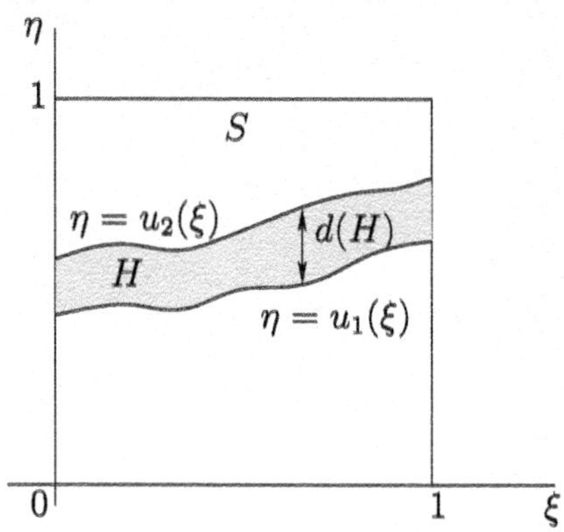

Fig. 7.16 A horizontal strip

Proof Since $\xi_i = v_i(\eta_i)$, one has

$$|\xi_2 - \xi_1| \leq |v_2(\eta_2) - v_1(\eta_2)| + |v_1(\eta_2) - v_1(\eta_1)| \leq \|v_2 - v_1\| + \mu|\eta_2 - \eta_1|.$$

Similarly,

$$|\eta_2 - \eta_1| \leq \|u_2 - u_1\| + \mu|\xi_2 - \xi_1|.$$

Addition of these inequalities gives (7.16), since $0 < \mu < 1$. □

Definition 7.31 *The closed set $H \subset S$ bounded by two nonintersecting μ-horizontal curves $\Gamma_{u_1,u_2} \subset S$ and two vertical line segments (see* Fig. 7.16) *is called a μ-**horizontal strip**. The number $d(H) = \|u_1 - u_2\|$ is called the **diameter** of H.*

The following lemma is obvious.

Lemma 7.32 *If*

$$H^{(1)} \supset H^{(2)} \supset H^{(3)} \supset \cdots$$

is a sequence of nested μ-horizontal strips $H^{(k)}$, $k = 1, 2, \ldots$, and $d(H^{(k)}) \to 0$ as $k \to \infty$ then

$$\bigcap_{k=1}^{\infty} H^{(k)}$$

is a μ-horizontal curve. □

One can analogously define a *μ-vertical strip V* and its *diameter $d(V)$*. A lemma similar to Lemma 7.32 applies to nested vertical strips.

Definition 7.33 *Suppose that $H_{0,1} \subset S$ are two disjoint μ-horizontal strips, while $V_{0,1} \subset S$ are two disjoint vertical strips. A homeomorphism $f : \mathbb{R}^2 \to \mathbb{R}^2$ is called a **horseshoe map** if it has the following properties:*

(H.1) $f(H_i) = V_i$, $i = 0, 1$, *and the horizontal/vertical boundaries of* H_i *are mapped onto the horizontal/vertical boundaries of* V_i, *respectively.*

(H.2) $H_{ij} = H_i \cap f^{-1}(H_j)$ *is a* μ*-horizontal strip and* $V_{ij} = f(V_i) \cap V_j$ *is a* μ*-vertical strip, and for some* ν *with* $0 < \nu < 1$

$$d(H_{ij}) \leq \nu d(H_k), \quad d(V_{ij}) \leq \nu d(V_k),$$

for $i, j, k \in \{0, 1\}$.

Example 7.34 (Geometric Smale horseshoe) Define a homeomorphism $f : \mathbb{R}^2 \to \mathbb{R}^2$ by the geometrical construction in Fig. 7.17. It acts on the unit square $S = ABCD$ as a strong contraction in the horizontal direction, followed by a strong expansion in the vertical direction, folding, and placing back over $ABCD$. Suppose that

$$f(S) \cap S = V_0 \cup V_1,$$

where $V_{0,1}$ are two disjoint vertical rectangles. Introduce also two disjoint horizontal rectangles $H_{0,1}$ such that

$$H_0 \cup H_1 = (f^{-1}(S)) \cap S,$$

and number them to have

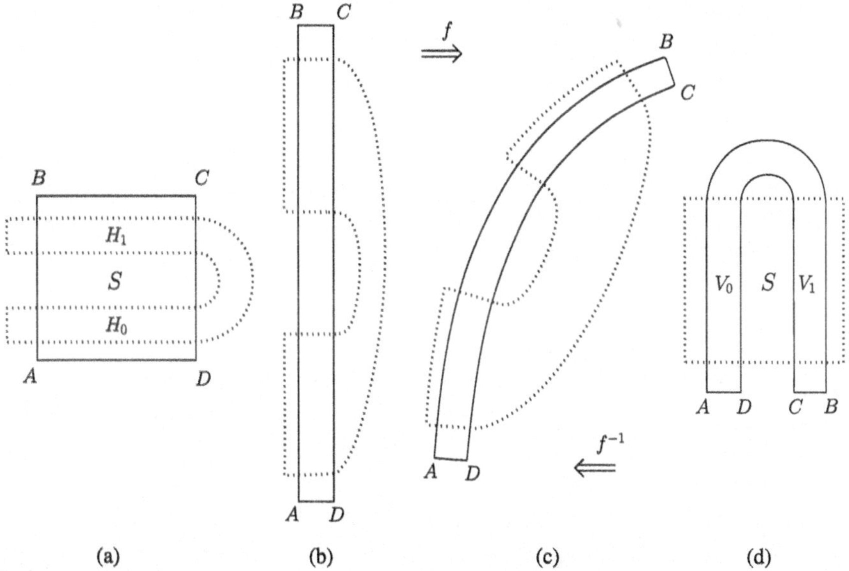

Fig. 7.17 Construction of the horseshoe map: $f(H_i) = V_i$, $i = 0, 1$

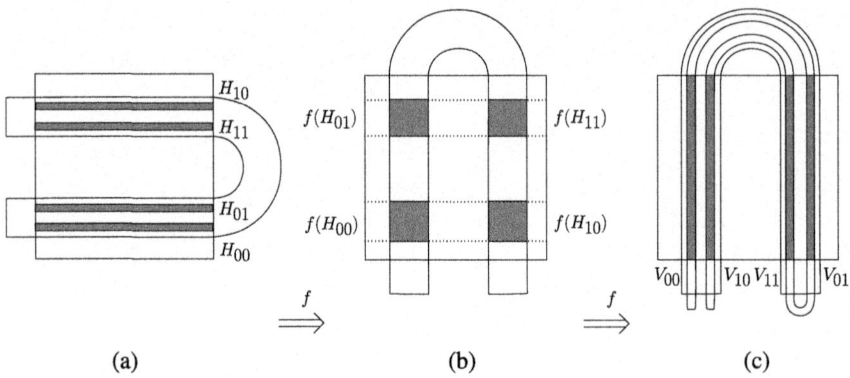

Fig. 7.18 $V_{ij} = f^2(H_{ij}),\ i, j = 0, 1$

$$f(H_i) = V_i, \quad i = 0, 1.$$

Assume that the map $f : H_i \to V_i$ is linear for $i = 0, 1$. Then, f possesses the properties $(H.1)$ and $(H.2)$ of a horseshoe map. (The second should become evident while carefully looking at Fig. 7.18.) ◊

The following considerations will resemble those in Sect. 7.1.2 concerning the "Big Tent Map" (7.10). For a general horseshoe map f, define an invariant set $\Lambda \subset S$:

$$\Lambda = \{x \in S : f^k(x) \in S, \text{ for all } k \in \mathbb{Z}\}$$

and recall from Example 1.4 in Chap. 1 that Ω_2 denotes the set of all bi-infinite sequences of two integers $\{0, 1\}$ with the distinguished zero position and the distance

$$d(\omega, \theta) = \sum_{k \in \mathbb{Z}} \frac{|\omega_k - \theta_k|}{2^{|k|}}. \tag{7.17}$$

Lemma 7.35 *The map $h : \Lambda \to \Omega_2$ defined by $h(x) = \omega$ with*

$$\omega_k = \begin{cases} 0 \text{ if } f^k(x) \in H_0, \\ 1 \text{ if } f^k(x) \in H_1, \end{cases}$$

for $k \in \mathbb{Z}$, is a homeomorphism.

Proof Clearly, the formula defines a map $h : \Lambda \to \Omega_2$, which assigns a sequence to each point of the invariant set.

To verify that this map is invertible, take a sequence $\omega \in \Omega_2$, fix $m > 0$, and consider the set $R_m(\omega)$ of all points $x \in S$, not necessarily belonging to Λ, such that

$$f^k(x) \in H_{\omega_k},$$

for $-m \leq k \leq m-1$. For example, if $m = 1$, the set R_1 is one of the four intersections $V_j \cap H_k$ shown in Fig. 7.18b. In general, R_m belongs to the intersection of a vertical and a horizontal strip. The property (H.2) implies that these strips are nested and that their diameters tend to zero as $m \to +\infty$. By Lemma 7.32, they approach in the limit a vertical and a horizontal curve, respectively. According to Lemma 7.29, such curves intersect at a single point x with $h(x) = \omega$. Thus, $h : \Lambda \to \Omega_2$ is a one-to-one map and hence h^{-1} exists.

To prove that h^{-1} is continuous, consider two sequences $\omega, \theta \in \Lambda$, which are close to each other (with respect to the distance (7.17)). This means that they have long coinciding central blocks:

$$\omega_k = \theta_k, \quad -m \leq k \leq m-1.$$

This implies that

$$x, y \in R_m(\omega) = R_m(\theta),$$

where $x = h^{-1}(\omega)$, $y = h^{-1}(\theta)$. Therefore, $\|y - x\|_1 \to 0$ as $m \to \infty$, i.e. $y \to x$ in $\mathbb{R}^2$ as $d(\omega, \theta) \to 0$.

Thus, $h^{-1} : \Omega_2 \to \Lambda$ is a continuous one-to-one map from the compact space Ω_2 to the compact set $\Lambda \subset \mathbb{R}^2$. Therefore, its inverse $h : \Lambda \to \Omega_2$ is also continuous.[2] □

Let $\sigma : \Omega_2 \to \Omega_2$ be the *shift map*:

$$\theta = \sigma(\omega), \ \theta_k = \omega_{k+1}.$$

The following is a direct implication of the definition of h.

Lemma 7.36 *$h(f(x)) = \sigma(h(x))$ for all $x \in \Lambda$.* □

From the basic properties of the discrete-time dynamical system $\{\mathbb{Z}, \Omega_2, \sigma^k\}$ established in Theorem 7.13 it now immediately follows:

Theorem 7.37 (Smale, 1963) *Any horseshoe map f has a closed invariant set Λ that contains a countable set of periodic orbits of arbitrarily long period, and an uncountable set of nonperiodic orbits, among which there is an orbit passing arbitrarily close to any point of Λ.* □

Our next aim is to replace (H.2) of Definition 7.33 by a more easily verifiable sufficient condition. For a constant $0 < \mu < 1$ define two *cones*:

$$K^+ = \{(\xi, \eta) \in \mathbb{R}^2 : |\xi| \leq \mu|\eta|\}, \quad K^- = \{(\xi, \eta) \in \mathbb{R}^2 : |\eta| \leq \mu|\xi|\}.$$

---

[2] This is a special case of the more general Open Map Lemma (Lee, 2011, Lemma 4.50) which states that a continuous bijection from a compact into a Hausdorff topological space has a continuous inverse. Note that metric spaces are Hausdorff.

Definition 7.38 *A diffeomorphism $f : \mathbb{R}^2 \to \mathbb{R}^2$ satisfies the* **μ-cone condition** *on $D \subset S$, if*

(i) *for all $p \in D$,*

$$[Df(p)]K^+ \subset K^+, \quad [Df(p)]^{-1}K^- \subset K^-;$$

(ii) *for all $p \in D$ and*

$$\begin{pmatrix} \xi_2 \\ \eta_2 \end{pmatrix} = [Df(p)] \begin{pmatrix} \xi_1 \\ \eta_1 \end{pmatrix},$$

the following inequalities are valid:

$$|\eta_1| \leq \mu|\eta_2|, \quad \begin{pmatrix} \xi_1 \\ \eta_1 \end{pmatrix} \in K^+,$$

and

$$|\xi_2| \geq \mu|\xi_1|, \quad \begin{pmatrix} \xi_1 \\ \eta_1 \end{pmatrix} \in K^-$$

(*see* Fig. 7.19).

Theorem 7.39 *Let $0 < \mu < \frac{1}{2}$, let $V_{1,2} \subset S$ be two disjoint μ-vertical strips with smooth vertical boundaries, and let $H_{0,1} \subset S$ be two disjoint μ-horizontal strips with smooth horizontal boundaries. Suppose that a map $f : \mathbb{R}^2 \to \mathbb{R}^2$ is a diffeomorphism of the domain*

$$D = H_0 \cup H_1$$

onto its image $f(D)$ with the following two properties:

(H.1) *$f(H_i) = V_i, \quad i = 0, 1,$ such that the horizontal/vertical boundaries of H_i are mapped onto the horizontal/vertical boundaries of V_i, respectively;*

(H.3) *f satisfies the μ-cone condition in D.*

Then the map f is a horseshoe map with $\nu = \mu(1 - \mu)^{-1}$.

Proof It suffices to show that the conditions (H.1) and (H.3) of the theorem imply Condition (H.2) from Definition 7.33.

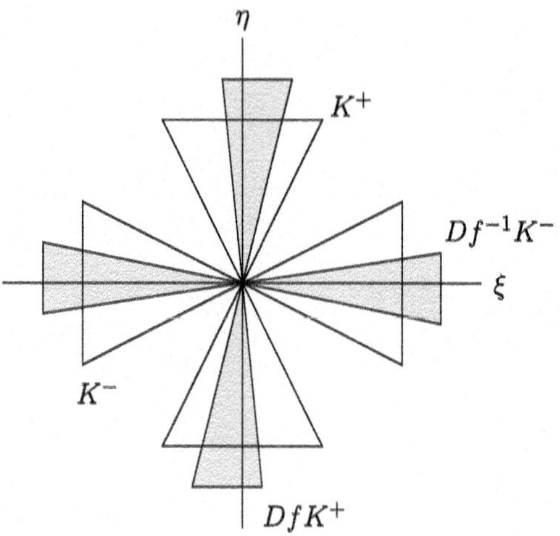

Fig. 7.19 The μ-cone condition

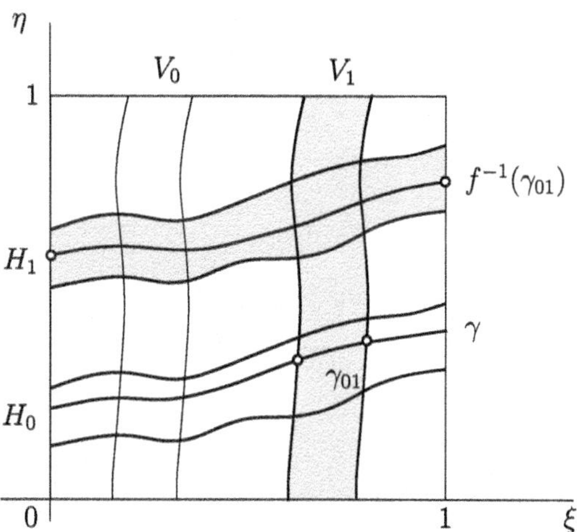

Fig. 7.20 $f^{-1}(\gamma_{01})$ is a horizontal curve

Let γ be a smooth μ-horizontal curve in a horizontal strip H_i with $i = 0$ or 1. (In Fig. 7.20, $i = 0$.) As such it intersects every vertical curve and, in particular, the boundaries of V_j, $j = 0, 1$ (in Fig. 7.20, we shade $j = 1$). Thus $\gamma_{ij} = \gamma \cap V_j$ connects the vertical boundaries of V_j, so necessarily $f^{-1}(\gamma_{ij})$ connects the vertical boundaries of $f^{-1}(V_j) = H_j$, i.e. the vertical sides of S.

We want to show that $f^{-1}(\gamma_{ij})$ is a horizontal curve. For this purpose we observe that γ is horizontal and is given, say, by a smooth function $\eta = u(\xi)$, satisfying

$$|u(\xi_1) - u(\xi_2)| \leq \mu|\xi_1 - \xi_2|.$$

Since $[Df]^{-1}$ maps K^- into K^-, it follows by application of the Intermediate Value Theorem that for any two points $(\xi_1, \eta_1), (\xi_2, \eta_2) \in f^{-1}(\gamma_{ij})$ one has

$$|\eta_1 - \eta_2| \leq \mu|\xi_1 - \xi_2|.$$

This shows that $f^{-1}(\gamma_{ij})$ is the graph of a function $\eta = w(\xi)$ defined for $\xi \in [0, 1]$ and satisfying

$$|w(\xi_1) - w(\xi_2)| \leq \mu|\xi_1 - \xi_2|.$$

We apply this observation to the horizontal boundaries of H_i and conclude that $f^{-1}(H_i) \cap H_j$ is a μ-horizontal strip. Similar arguments prove that $f(V_i) \cap V_j$, $i, j = 0, 1$, are μ-vertical strips.

To verify the statement about the diameters, take two points $p_{1,2} = (\xi, \eta_{1,2})$ with the same ξ-coordinate on the horizontal boundaries of $f^{-1}(H_i) \cap H_j$, such that

$$d(f^{-1}(H_i) \cap H_j) = |\eta_1 - \eta_2|.$$

The line segment

$$p(t) = (1 - t)p_1 + tp_2, \quad t \in [0, 1],$$

is parallel to the η-axis and hence $\dot{p} \in K^+$. Therefore, its image curve

$$z(t) = f(p(t))$$

has its tangent vector $\dot{z} = [Df(p(t))]\dot{p} \in K^+$ by Assumption (H.3) of the theorem. This shows that $z(0)$ and $z(1)$ lie on a vertical curve, for example, the one obtained by extending

$z(t)$ by two vertical segments. Thus, the points lie on a single vertical curve and two horizontal curves at a distance $d(H_i)$. From Lemma 7.30 follows that

$$\|z(0) - z(1)\|_1 \leq (1-\mu)^{-1} d(H_i).$$

Writing $z(t) = (\xi(t), \eta(t))$, we have by (H.3) that

$$|\dot{\eta}| \geq \frac{1}{\mu}\|\dot{p}\|_1 > 0$$

hence $\dot{\eta}$ does not change sign and

$$\begin{aligned}
|\eta_1 - \eta_2| = \|p(0) - p(1)\|_1 &= \int_0^1 \|\dot{p}\|_1 dt \\
&\leq \mu \int_0^1 |\dot{\eta}| dt = \mu|\eta(1) - \eta(0)| \\
&\leq \mu\|z(1) - z(0)\|_1 \leq \mu(1-\mu)^{-1} d(H_i),
\end{aligned}$$

which verifies $\nu = \mu(1-\mu)^{-1}$. □

The theorem implies that

$$\Lambda = \{x \in D : f^k(x) \in D \text{ for all } k \in \mathbb{Z}\}$$

is a nonempty invariant set for f. Moreover, the restriction of f to Λ is topologically conjugated to the shift σ on the set Ω_2.

7.3.2 Shilnikov Homoclinic Orbits to a Saddle-Focus

Smale Horseshoes appear frequently in Poincaré maps related to homoclinic bifurcations of ODEs.

Theorem 7.40 (Shilnikov, 1965) *Consider a smooth three-dimensional system*

$$\dot{x} = f(x, \alpha), \quad x \in \mathbb{R}^3, \alpha \in \mathbb{R}, \tag{7.18}$$

that has for all sufficiently small $|\alpha|$ an equilibrium $x_0 = 0$ with two complex eigenvalues $\lambda_{1,2}(\alpha) = \mu(\alpha) \pm i\omega(\alpha)$ such that $\mu(0) < 0, \omega(0) > 0$, and one real eigenvalue $\lambda_3(\alpha) = \gamma(\alpha)$ with $\gamma(0) > 0$. Suppose that at $\alpha = 0$ the system has an orbit Γ_0 homoclinic to x_0.

(i) *If $\mu(0) + \gamma(0) > 0$ then for all sufficiently small $|\alpha|$, system (7.18) has an infinite number of saddle limit cycles in a neighbourhood of $\Gamma_0 \cup x_0$.*

(ii) *If $\mu(0) + \gamma(0) < 0$ then, generically, a hyperbolic stable limit cycle appears in a neighbourhood of $\Gamma_0 \cup x_0$ for $\alpha > 0$ or $\alpha < 0$ with sufficiently small $|\alpha|$, while no other limit cycle exists in this neighbourhood for all sufficiently small $|\alpha|$.*

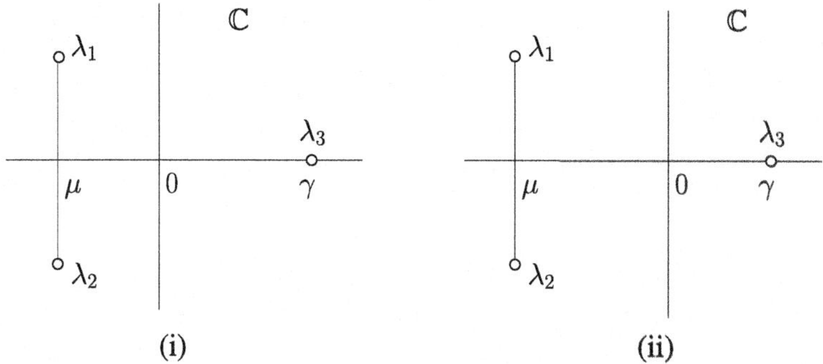

Fig. 7.21 Eigenvalues of a saddle-focus

Remarks

(1) An equilibrium x_0 of (7.18) with a pair of complex-conjugate eigenvalues and a real eigenvalue is called a *saddle-focus*. The real number

$$\sigma_0 = \mu(0) + \gamma(0) = \text{Re }\lambda_{1,2}(0) + \lambda_3(0)$$

is called the *saddle quantity* of x_0. In case (i), when $\sigma_0 > 0$, the two complex-conjugate eigenvalues $\lambda_{1,2}$ are closer to the imaginary axis than the real positive eigenvalue λ_3, see Fig. 7.21(i). In case (ii), when $\sigma_0 < 0$, the real positive eigenvalue λ_3 is closer to the imaginary axis than the two complex-conjugate eigenvalues $\lambda_{1,2}$, see Fig. 7.21(ii). An eigenvalue with the minimal $|\text{Re}\lambda|$ is called *determining*. Thus, each determining eigenvalue is complex in case (i) but real in case (ii).

(2) In Appendix A we prove that, generically, *at least one* periodic orbit (limit cycle) bifurcates from a homoclinic orbit to an hyperbolic equilibrium in smooth n-dimensional ODEs depending on a parameter α. In case (i), together with the "primary" cycle guaranteed by that result, many more cycles exist, as will be clear from the proof outlined below (that is specific for three-dimensional ODEs). ◇

Sketch of the proof of Theorem 7.40

Step 1: Local straightening of invariant manifolds.

Write system (7.18) in a basis, where its linear part assumes the real canonical form, i.e.:

$$\begin{cases} \dot{x}_1 = \mu(\alpha)x_1 - \omega(\alpha)x_2 + F_1(x_1, x_2, x_3, \alpha), \\ \dot{x}_2 = \omega(\alpha)x_1 + \mu(\alpha)x_2 + F_2(x_1, x_2, x_3, \alpha), \\ \dot{x}_3 = \gamma(\alpha)x_3 + F_3(x_1, x_2, x_3, \alpha), \end{cases} \tag{7.19}$$

where $F_k(x, \alpha) = O(\|x\|^2), k = 1, 2, 3,$ are smooth functions. According to Theorem 6.3, there exist stable and unstable parameter-dependent invariant manifolds of x_0 which can be represented locally as

$$W^u_{\text{loc}}(0) = \{x \in \mathbb{R}^3 : x_1 = U_1(x_3, \alpha), x_2 = U_2(x_3, \alpha),\ |x_3| \le d\}$$

and

$$W^s_{\text{loc}}(0) = \{x \in \mathbb{R}^3 : x_3 = V(x_1, x_2, \alpha),\ x_1^2 + x_2^2 \le d^2\},$$

where $d > 0$ is sufficiently small and the mappings $U_{1,2} : \mathbb{R} \times \mathbb{R} \to \mathbb{R}^2$ and $V : \mathbb{R}^2 \times \mathbb{R} \to \mathbb{R}$ are smooth, satisfy $U_{1,2}(0,0) = 0$ and $V(0,0) = 0$ and have vanishing first-order partial derivatives with respect to the x-variables at $(x, \alpha) = (0, 0)$. We shall assume that the homoclinic orbit corresponds to the part of $W^u_{\text{loc}}(0)$ with $x_3 > 0$.

Introduce new coordinates (u_1, u_2, u_3) in $\mathbb{R}^3$ by

$$\begin{cases} u_1 = x_1 - U_1(x_3, \alpha), \\ u_2 = x_2 - U_2(x_3, \alpha), \\ u_3 = x_3 - V(x_1, x_2, \alpha). \end{cases} \tag{7.20}$$

In the u-coordinate system the stable manifold $W^s(0)$ is (locally) the plane $u_3 = 0$, while $W^u(0)$ is (also locally) the line $u_1 = u_2 = 0$. In these coordinates, the system (7.19) takes the form:

$$\begin{cases} \dot{u}_1 = \mu(\alpha)u_1 - \omega(\alpha)u_2 + P_1(u_1, u_2, u_3, \alpha), \\ \dot{u}_2 = \omega(\alpha)u_1 + \mu(\alpha)u_2 + P_2(u_1, u_2, u_3, \alpha), \\ \dot{u}_3 = \gamma(\alpha)u_3 + P_3(u_1, u_2, u_3, \alpha), \end{cases} \tag{7.21}$$

with smooth[3] functions $P_k(u, \alpha) = O(\|u\|^2),\ k = 1, 2, 3$, satisfying for small $\|u\|$ and $|\alpha|$

$$P_1(0, 0, u_3, \alpha) = P_2(0, 0, u_3, \alpha) = 0, \quad P_3(u_1, u_2, 0, \alpha) = 0.$$

Step 2: Time reparametrization.

Since

$$\gamma(\alpha)u_3 + P_3(u_1, u_2, u_3, \alpha) = \gamma(\alpha)u_3(1 + R_3(u_1, u_2, u_3, \alpha)),$$

where $R_3(u, \alpha) = O(\|u\|)$, we can parametrize orbits of (7.21) near the origin by the new time τ with

$$d\tau = (1 + R_3(u, \alpha))\, dt$$

and consider a system that is locally orbitally equivalent to (7.21), namely:

[3] If the system (7.19) is C^r-smooth, the resulting system (7.21) is only C^{r-1}-smooth in general. Indeed, the mappings U and V are C^r-smooth, but we differentiate them once while transforming (7.19) into (7.21) using (7.20).

$$\begin{cases} \dot{u}_1 = \mu(\alpha)u_1 - \omega(\alpha)u_2 + Q_1(u_1, u_2, u_3, \alpha), \\ \dot{u}_2 = \omega(\alpha)u_1 + \mu(\alpha)u_2 + Q_2(u_1, u_2, u_3, \alpha), \\ \dot{u}_3 = \gamma(\alpha)u_3, \end{cases} \tag{7.22}$$

where the functions $Q_j(u, \alpha) = O(\|u\|^2)$ are smooth and satisfy

$$Q_1(0, 0, u_3, \alpha) = Q_2(0, 0, u_3, \alpha) = 0$$

for small $\|u\|$ and $|\alpha|$. The system (7.22) is smoothly orbitally equivalent to the system (7.19) near the origin.

Step 3: Local C^1-linearization of system (7.22).

According to the Grobman–Hartman Theorem 3.34, the flow generated by system (7.22) is locally C^0-conjugate near the origin to the flow

$$\begin{pmatrix} \xi_1 \\ \xi_2 \\ \xi_3 \end{pmatrix} \mapsto \begin{pmatrix} e^{\mu(\alpha)\tau}[\xi_1 \cos(\omega(\alpha)\tau) - \xi_2 \sin(\omega(\alpha)\tau)] \\ e^{\mu(\alpha)\tau}[\xi_1 \sin(\omega(\alpha)\tau) + \xi_2 \cos(\omega(\alpha)\tau)] \\ \xi_3 e^{\gamma(\alpha)\tau} \end{pmatrix} \tag{7.23}$$

of its linearization

$$\begin{cases} \dot{\xi}_1 = \mu(\alpha)\xi_1 - \omega(\alpha)\xi_2, \\ \dot{\xi}_2 = \omega(\alpha)\xi_1 + \mu(\alpha)\xi_2, \\ \dot{\xi}_3 = \gamma(\alpha)\xi_3. \end{cases} \tag{7.24}$$

To construct a C^1 local conjugating map $h: \mathbb{R}^3 \to \mathbb{R}^3$, $\xi = h(u)$, one can use the following geometric construction. Introduce new coordinates (r, φ, z) in the u-space by

$$\begin{cases} u_1 = r\cos\varphi, \\ u_2 = r\sin\varphi, \\ u_3 = z, \end{cases}$$

and consider a finite cylinder

$$\Omega = \{(r, \varphi, z) : 0 \le r \le d,\ |z| \le d\}$$

with sufficiently small $d > 0$ (see Fig. 7.22a).

Take a point $u \in \Omega$ that does not belong to the u_3-axis and consider the orbit of (7.22) arriving to this point at $\tau = 0$. Provided d is small enough, it can be shown that there is a smooth function $\tau^0 = \tau^0(u, \alpha) < 0$ such that for $\tau = \tau^0$ this orbit passes through the boundary $r = d$ of Ω at some point

$$u^0 = (d\cos\varphi^0(u, \alpha), d\sin\varphi^0(u, \alpha), z^0(u, \alpha)) \in \partial\Omega,$$

where

$$z^0(u, \alpha) = z e^{\gamma(\alpha)\tau^0(u, \alpha)}.$$

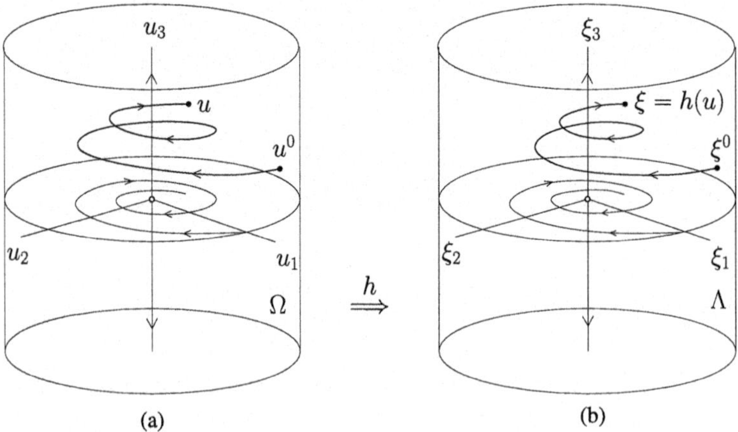

Fig. 7.22 Local linearization of the system near a saddle-focus

Also introduce the coordinates (ρ, ψ, ζ) in the ξ-space by

$$\begin{cases} \xi_1 = \rho \cos \psi, \\ \xi_2 = \rho \sin \psi, \\ \xi_3 = \zeta, \end{cases}$$

and consider the cylinder

$$\Lambda = \{(\rho, \psi, \zeta) : 0 \le \rho \le d, |\zeta| \le d\}$$

with the same d as before (see Fig. 7.22b). Map the point u^0 to the point

$$\xi^0 = (d \cos \varphi^0(u, \alpha), d \sin \varphi^0(u, \alpha), z^0(u, \alpha)) \in \partial\Lambda$$

and construct an orbit of the linear system (7.24) starting at ξ^0. The point $\xi = h(u)$ is by definition the image of ξ^0 under the flow (7.23) with $\tau = -\tau^0(u, \alpha) > 0$, i.e.:

$$\begin{pmatrix} \xi_1 \\ \xi_2 \\ \xi_3 \end{pmatrix} = \begin{pmatrix} d\, \mathrm{e}^{-\mu\tau^0(u,\alpha)} \cos\left[\varphi^0(u, \alpha) - \omega(\alpha)\tau^0(u, \alpha)\right] \\ d\, \mathrm{e}^{-\mu\tau^0(u,\alpha)} \sin\left[\varphi^0(u, \alpha) - \omega(\alpha)\tau^0(u, \alpha)\right] \\ u_3 \end{pmatrix}. \tag{7.25}$$

Let h map the z-axis into itself according to the identity, i.e. $\xi_3 = u_3$. The resulting map $u \mapsto \xi = h(u)$ is a local homeomorphism $h : \Omega \to \Lambda$. It obviously sends orbits of (7.22) into orbits of the linear system (7.24), preserving time parametrization. By construction, it is smooth away from the z-axis. It can be shown that h also has continuous first-order partial derivatives at points on the z-axis in Λ. In general, higher-order derivatives are not continuous but the C^1-smoothness of h is sufficient for our purposes. We can also extend the constructed map to a globally defined C^1-diffeomorphism $h : \mathbb{R}^3 \to \mathbb{R}^3$ that locally conjugates the flows of (7.22) and (7.24). We will use the (ρ, ψ, ζ)-coordinates in the following analysis.

Step 4: Construction of the Poincaré map at $\alpha = 0$.

To analyse the dynamics of the system near the bifurcation, we assume that for $\alpha = 0$ the homoclinic orbit Γ_0 returns into Λ at the point $(\rho, \psi, \zeta) = (d, 0, 0)$. Next represent the boundary without the bottom of the cylinder Λ as the union of two cross sections, namely:

$$S = \{(\rho, \psi, \zeta) : \rho = d, \ |\zeta| \le d\}$$

and

$$H = \{(\rho, \psi, \zeta) : 0 \le \rho \le d, \ \zeta = d\}$$

(see Fig. 7.23).

Define $\Sigma \subset S$ by

$$\Sigma = \{(\rho, \psi, \zeta) : \rho = d, \ |\psi| \le \delta, \ |\zeta| \le \varepsilon\},$$

where $\delta, \varepsilon > 0$ are sufficiently small, and denote its subset corresponding to $\zeta > 0$ by Σ^+. Now we can introduce a map along orbits of the system, $\mathcal{P} : \Sigma^+ \to S$, as a composition $\mathcal{P} = \mathcal{Q} \circ \Delta$ of two maps: a near-to-saddle map $\Delta : \Sigma^+ \to H$ and a global map $\mathcal{Q} : H \to S$.

The near-to-saddle map can be computed explicitly using the flow (7.23). Indeed, using (ψ, ζ) as coordinates on Σ^+ and (ξ_1, ξ_2) as the coordinates on

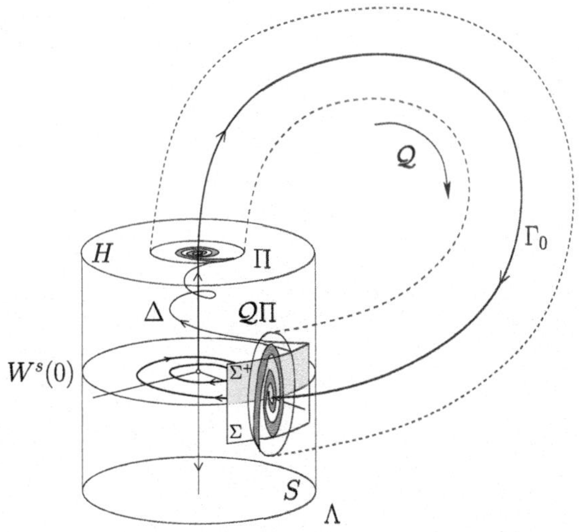

Fig. 7.23 Construction of the Poincaré map $\mathcal{P}$ in the saddle-focus case

H, we immediately see that the "flight time" from Σ^+ to H along the orbits of (7.24) is independent of ψ and is equal to

$$\tau = -\frac{1}{\gamma(\alpha)} \ln\left(\frac{\zeta}{d}\right).$$

Therefore, the map Δ can be explicitly computed for all sufficiently small $|\alpha|$ as $\Delta(\psi,\zeta) = \xi = (\xi_1,\xi_2)$ with

$$\begin{cases} \xi_1 = d\left(\frac{\zeta}{d}\right)^{q(\alpha)} \cos\left(\psi - \frac{\omega(\alpha)}{\gamma(\alpha)} \ln\frac{\zeta}{d}\right), \\ \xi_2 = d\left(\frac{\zeta}{d}\right)^{q(\alpha)} \sin\left(\psi - \frac{\omega(\alpha)}{\gamma(\alpha)} \ln\frac{\zeta}{d}\right), \end{cases} \tag{7.26}$$

where

$$q(\alpha) = -\frac{\mu(\alpha)}{\gamma(\alpha)} > 0. \tag{7.27}$$

It follows from (7.26) that the image $\Delta\Sigma^+$ of Σ^+ is a "solid spiral" in H as shown in Fig. 7.23. It is contained in the disc

$$\Pi = \left\{(\xi_1,\xi_2) : 0 \le \|\xi\| \le d\left(\frac{\varepsilon}{d}\right)^{q(\alpha)}\right\} \subset H.$$

The global map $\mathcal{Q}$ is a C^1-diffeomorphism that maps Π back to S. We can write this map as $\mathcal{Q}(\xi_1,\xi_2) = (\psi,\zeta)$ with

$$\begin{cases} \psi = \nu(\alpha) + a_{11}(\alpha)\xi_1 + a_{12}(\alpha)\xi_2 + o(\|\xi\|), \\ \zeta = \beta(\alpha) + a_{21}(\alpha)\xi_1 + a_{22}(\alpha)\xi_2 + o(\|\xi\|), \end{cases} \tag{7.28}$$

where $\nu(0) = \beta(0) = 0$ but $a_{11}(0)a_{22}(0) - a_{21}(0)a_{12}(0) \neq 0$. The function $\beta = \beta(\alpha)$ gives the relative ζ-displacement of the invariant manifolds $W^s(x_0)$ and $W^u(x_0)$ for small $\alpha \neq 0$. Generically $\beta'(0) \neq 0$, so its value can be considered as a new bifurcation parameter. The image $\mathcal{Q}\Pi$ is a deformed disc with the $O(\varepsilon^{q(\alpha)})$-height in the ζ-direction for $\varepsilon \to 0$.

If ε is sufficiently small, the composition $\mathcal{Q} \circ \Delta$ defines the Poincaré map $\mathcal{P} : \Sigma^+ \to S$ that can be written using (7.26) and (7.28) as $\mathcal{P}(\psi,\zeta) = (\tilde{\psi},\tilde{\zeta})$, where

$$\begin{cases} \tilde{\psi} = \nu(\alpha) + A(\alpha)d\left(\frac{\zeta}{d}\right)^{q(\alpha)} \cos\left(\psi - \frac{\omega(\alpha)}{\gamma(\alpha)} \ln\frac{\zeta}{d} + \psi_1(\alpha)\right) + o\left(\zeta^{q(\alpha)}\right), \\ \tilde{\zeta} = \beta(\alpha) + B(\alpha)d\left(\frac{\zeta}{d}\right)^{q(\alpha)} \cos\left(\psi - \frac{\omega(\alpha)}{\gamma(\alpha)} \ln\frac{\zeta}{d} + \psi_2(\alpha)\right) + o\left(\zeta^{q(\alpha)}\right), \end{cases} \tag{7.29}$$

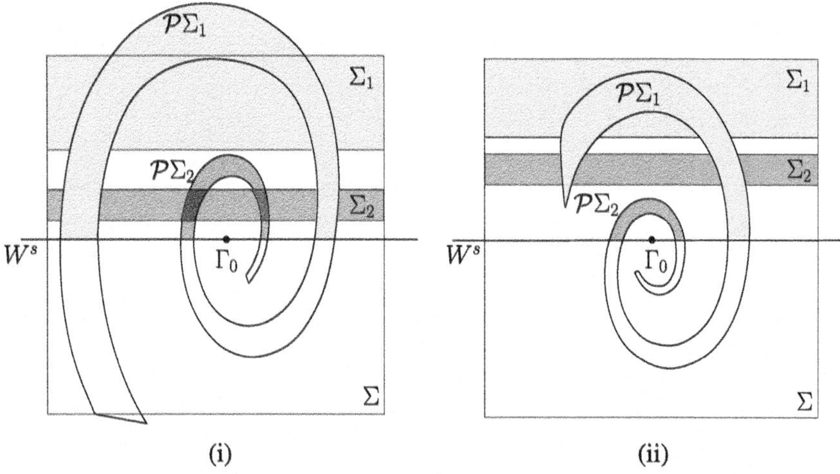

Fig. 7.24 Case $\alpha = 0$: (i) $q(0) < 1$, i.e. $\sigma_0 > 0$; (ii) $q(0) > 1$, i.e. $\sigma_0 < 0$

for $\zeta \to 0$, where $A(\alpha), B(\alpha), \psi_j(\alpha)$ are smooth functions determined by $a_{kl}(\alpha)$ and $q(\alpha)$.

Step 5 Analysis of the Poincaré map at $\alpha = 0$.

It is clear that no periodic orbit can start in Σ^-. Therefore, consider the intersection of the spiral image $\mathcal{P}\Sigma^+$ with Σ (see Fig. 7.24). The origin of the image is at the intersection of Γ_0 with Σ, i.e. at the point $(\rho, \psi, \zeta) = (d, 0, 0)$. The intersection of Σ with $W^s(x_0)$ corresponding to $\zeta = 0$ splits the spiral into an infinite number of *upper* and *lower* "half-turns". The preimages Σ_j of the upper "half-turns" $\mathcal{P}\Sigma_j$, $j = 1, 2, \ldots$, are almost horizontal strips in Σ^+, which are located between the lines

$$\zeta = \zeta_k = C(0)\mathrm{e}^{-\pi k \gamma(0)/\omega(0)}(1 + o(1)), \quad k \to \infty, \tag{7.30}$$

corresponding to $k = 2j$ and $k = 2j + 1$ (see (7.29)). Here $C(\alpha)$ is a smooth function. These strips accumulate on $\zeta = 0$ as $k \to \infty$. As follows from (7.30) and (7.29), the ζ-coordinate of the top of the jth half-spiral is estimated by

$$\zeta_{2j}^{q(0)} = O\left(\mathrm{e}^{-2\pi j \mu(0)/\omega(0)}\right), \quad j \to \infty.$$

Then we can distinguish two cases (see Fig. 7.24):

(i) If $\mu(0) + \gamma(0) > 0$ then $q(0) < 1$. Therefore, the intersection $\Sigma_j \cap \mathcal{P}\Sigma_j$ is nonempty and consists of two components for $j \geq j_0$, where j_0 is an integer number ($j_0 = 2$ in Fig. 7.24(i)). Hence, each intersection with sufficiently big index j forms a Smale horseshoe to which Theorem 7.37 applies. Each horseshoe implies an infinite number of saddle cycles. These cycles of $\mathcal{P}$ correspond to saddle limit cycles of (7.18).

(ii) If $\mu(0) + \gamma(0) < 0$ then $q(0) > 1$. Therefore, there is some index $j_0 > 0$ such that for $j \geq j_0$ the intersection $\Sigma_j \cap \mathcal{P}\Sigma_j$ is empty for $\alpha = 0$ ($j_0 = 2$ in Fig. 7.24(ii)). Thus, there are no fixed points (or cycles) of $\mathcal{P}$ in Σ^+ close to Γ_0.

Step 6: Analysis of the Poincaré map for small $|\alpha| \neq 0$.

If $\alpha \neq 0$, the point corresponding to the intersection of the unstable manifold of x_0 with S is generically displaced from the horizontal line $\zeta = 0$ in Σ and becomes $(\rho, \psi, \zeta) = (d, \nu(\alpha), \beta(\alpha))$ with $\beta \neq 0$. Also in this situation, no periodic orbit can start in Σ^-.

In case (i), there remains, in general, only a *finite* number of Smale horseshoes of $\mathcal{P}$ for small $|\beta(\alpha)|$. They still generate an infinite number of saddle limit cycles in (7.18) for all α with sufficiently small $|\alpha|$.

In case (ii), the map $\mathcal{P}$ is a contraction in Σ^+ for $\beta(\alpha) > 0$ and thus has a unique attracting fixed point corresponding to a stable limit cycle of (7.18). There are no other periodic orbits nearby. For small fixed $\beta(\alpha) < 0$, there is $\varepsilon > 0$ such that $\Sigma^+ \cap \mathcal{P}\Sigma^+ = \emptyset$, implying that no periodic orbits can exist in Σ^+ for small $\beta(\alpha) < 0$. □

Remark Fine bifurcation details in case (i) depend on another saddle quantity

$$\sigma_1 = 2\mu(0) + \gamma(0) = 2\text{Re } \lambda_{1,2}(0) + \lambda_3(0) = \text{div } f(0,0).$$

If $\sigma_1 < 0$ then for a countable number of parameter intervals the system (7.18) has a *stable cycle.* If $\sigma_1 > 0$ then for a countable number of parameter intervals the system (7.18) has a *repelling cycle.* ◇

Example 7.41 (Homoclinic orbits to saddle-foci in Arneodo's system) Both cases of Theorem 7.40 can be illustrated using the system

$$\begin{cases} \dot{x}_1 = x_2, \\ \dot{x}_2 = x_3, \\ \dot{x}_3 = -x_3 - bx_2 + cx_1 - x_1^2 \end{cases} \tag{7.31}$$

that appears in the analysis of bifurcations of an equilibrium with triple-zero eigenvalue.[4] This system exhibits many homoclinic bifurcations.

If $b = 0.5$ then at $c = 0.964149\ldots$ the system (7.31) has a homoclinic orbit Γ_0 to the saddle-focus equilibrium $x_0 = 0$. Since the eigenvalues of the corresponding Jacobian matrix

$$\begin{pmatrix} 0 & 1 & 0 \\ 0 & 0 & 1 \\ c & -b & -1 \end{pmatrix}$$

[4] Arneodo, A., Coullet, P. H., Spiegel, E. A., & Tresser, C. (1985). Asymptotic chaos. *Physica D, 14*, 327–347.

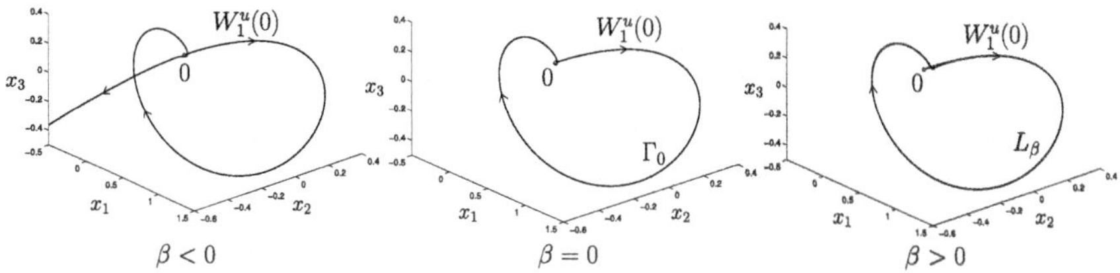

Fig. 7.25 The unstable invariant manifold $W_1^u(0)$ of the saddle-focus near a homoclinic bifurcation in (7.31) when $\mu(0) + \gamma(0) < 0$

are

$$\lambda_{1,2} = -0.81537\ldots \pm i0.92938\ldots, \quad \lambda_3 = 0.6307\ldots,$$

case (ii) applies. Thus, a unique and stable limit cycle bifurcates from Γ_0 for $\beta > 0$ (see Fig. 7.25). A branch $W_1^u(0)$ of the unstable invariant manifold of the saddle-focus tends to this cycle when it exists. This phenomenon can easily be seen by numerically integrating (7.31). The left picture in Fig. 7.25 corresponds to the parameter value $c = 0.960$, while for the right one the value $c = 0.965$ is used.

If $c = 5$ then at $b = 1.870532\ldots$ the system (7.31) also has a homoclinic orbit Γ_0 to the saddle-focus equilibrium x_0 at the origin. Here the eigenvalues of the corresponding Jacobian matrix are

$$\lambda_{1,2} = -1.0753\ldots \pm i1.78575\ldots, \quad \lambda_3 = 1.150677\ldots,$$

so we are in case (i). Thus, infinitely many saddle limit cycles ought to exist near Γ_0 for all α with sufficiently small $|\alpha|$. Since $\sigma_1 < 0$, infinitely many parameter intervals exist for $\beta > 0$, where (7.31) also has a stable cycle. This cycle, however, is hardly observable in numerical simulations, since its domain of attraction is very small.

As in the previous case, $W_1^u(0)$ leaves the neighbourhood of Γ_0 for $\beta < 0$. For small $\beta > 0$, $W_1^u(0)$ can either remain in this neighbourhood indefinitely (e.g. approaching one of the cycles) or leave it after a number of turns near Γ_0. The left and the right pictures in Fig. 7.26 correspond to $b = 1.870$ and $b = 1.875$, respectively. ◊

Example 7.42 (Homoclinic chaos in Rössler's system) A homoclinic orbit to a saddle-focus, with an infinite number of saddle cycles nearby, can be embedded in a larger attracting set, thus generating stable chaotic motions. A famous example of this phenomenon is provided by *Rössler's prototype chaotic system*[5]

---

[5] Rössler, O. E. (1979). Continuous chaos—four prototype equations. In *Bifurcation Theory and Applications in Scientific Disciplines* (vol. 316, pp. 376–392). New York: Annals of the New York Academy of Sciences.

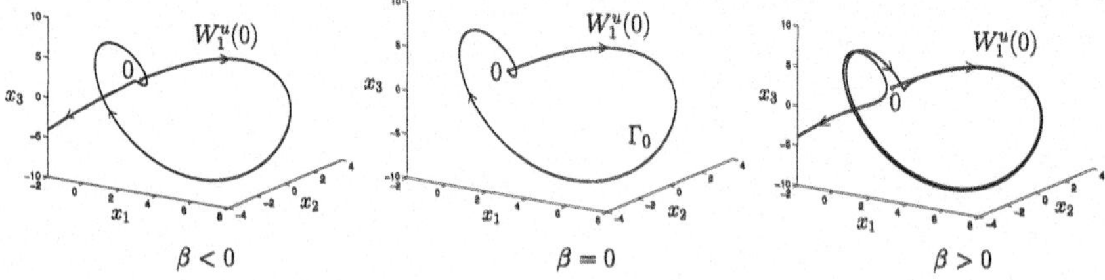

Fig. 7.26 The unstable invariant manifold $W_1^u(0)$ of the saddle-focus near a homoclinic bifurcation in (7.31) when $\mu(0) + \gamma(0) > 0$

$$\begin{cases} \dot{x}_1 = -x_2 - x_3, \\ \dot{x}_2 = x_1 + Ax_2, \\ \dot{x}_3 = Bx_1 - Cx_3 + x_1 x_3, \end{cases} \tag{7.32}$$

with fixed $A = 0.36$ and $C = 0.4$. For these parameter values, (7.32) has an orbit Γ_0 homoclinic to the saddle-focus equilibrium $x_0 = (0, 0, 0)$ at $B = 0.370322\ldots$. This equilibrium has eigenvalues

$$\lambda_{1,2} = 0.085618\ldots \pm i1.12034\ldots, \quad \lambda_3 = -0.211236\ldots,$$

so that one has to reverse time to apply Theorem 7.40. After that, we are evidently in case (i). A direct numerical integration of (7.32) with generic initial data reveals stable nonperiodic behaviour that persists for nearby values of the parameter B (see Fig. 7.27 corresponding to $B = 0.4$). Note that long-periodic stable cycles are also present here but hardly observable numerically. ◊

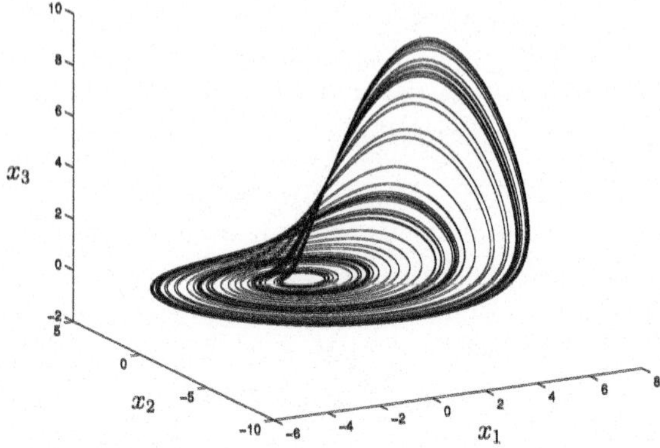

Fig. 7.27 Shilnikov chaos near a homoclinic orbit to a saddle-focus

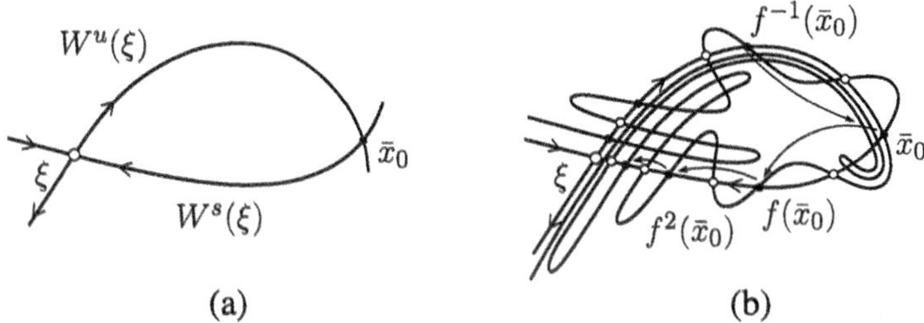

Fig. 7.28 Poincaré homoclinic structure

7.3.3 Transversal Poincaré Homoclinic Structure

Smale horseshoes naturally appear in maps with orbits homoclinic to hyperbolic fixed points. Consider a hyperbolic fixed point ξ of a planar diffeomorphism $f : \mathbb{R}^2 \to \mathbb{R}^2$ with eigenvalues $0 < \lambda_1 < 1 < \lambda_2$.[6] Its invariant manifolds $W^s(\xi)$ and $W^u(\xi)$ are one-dimensional and can *intersect* transversally (see Fig. 7.28a).

Moreover, if one transversal intersection occurs, it implies an *infinite number* of such intersections. Indeed, let $\bar{x}_0$ be a point of the intersection. By definition, it belongs to both invariant manifolds. Therefore, the orbit starting at this point converges to the fixed point ξ under repeated iteration of either f or f^{-1} : $f^k(\bar{x}_0) \to \xi$ as $k \to \pm\infty$. Each point of this orbit is a point of intersection of $W^s(\xi)$ and $W^u(\xi)$. This infinite number of intersections forces both manifolds to "oscillate" in a complex manner near ξ, as sketched in Fig. 7.28b. The resulting "web" is called the *Poincaré homoclinic structure*. The orbit starting at $\bar{x}_0$ is a *homoclinic orbit* to ξ. The presence of the homoclinic structure makes the intersection of $W^{s,u}(\xi)$ with any neighbourhood of ξ highly nontrivial.

The dynamical consequences of the existence of the homoclinic structure are also dramatic: We expect the appearance of an *infinite number* of periodic points with arbitrarily high periods near the homoclinic orbit. This follows from the presence of *Smale horseshoes*. Figure 7.29 illustrates how the horseshoes are formed. Take a (curvilinear) rectangle S near the stable manifold $W^s(\xi)$ and consider its iterations f^kS. If the homoclinic structure is present, for a sufficiently high number of iterations N, f^NS will look like the folded and expanded band Q shown in the figure. The intersection of S with Q forms several horseshoes satisfying Definition 7.33. Each of them implies (using Theorem 7.37) an infinite number of cycles with arbitrarily high periods.

[6] One can think that ξ is a fixed point of the (global) Poincaré map corresponding to a hyperbolic limit cycle of an ODE in $\mathbb{R}^3$.

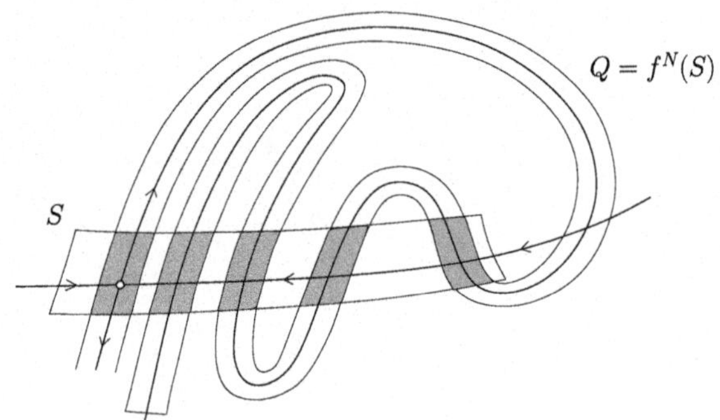

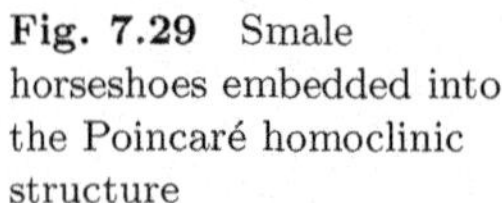
Fig. 7.29 Smale horseshoes embedded into the Poincaré homoclinic structure

In Appendix B, we prove that the dynamical system generated by the map f (in some neighbourhood of the homoclinic orbit) is topologically equivalent to a subshift dynamics (*topological Markov chain*) introduced in Example 1.4 of Chap. 1.

7.4 References

This chapter is a brief introduction to "chaotic dynamics" and global bifurcations, which is a huge and actively developing field. More extensive introductions are Katok and Hasselblatt (1995), Alligood et al. (1997), Hasselblatt and Katok (2003), Wiggins (2003).

The dynamics generated by one-dimensional maps is a classical mathematical subject, studied in detail in many introductory texts on dynamical systems, e.g. Guckenheimer and Holmes (1983), de Melo and van Strien (1993), Verhulst (1996), Brin and Stuck (2002), Hasselblatt and Katok (2003). Proofs of Sharkovsky's theorem can be found, for example, in Devaney (1989), Katok and Hasselblatt (1995), Brin and Stuck (2002).

For an account of analytical and numerical results on the Lorenz system, see Viana (2000), Shilnikov et al. (2001). Numerical methods for bifurcations of homoclinic orbits to equilibria in ODEs are summarized in Beyn et al. (2002), while various details can be found in Beyn (1990), Champneys et al. (1996), De Witte et al. (2012).

The Smale horseshoe is treated in the classical books (Nitecki, 1971; Moser, 1973), as well as in all textbooks mentioned above. A nice colour representation of the horseshoe map is given in Shub (2005).

The complicated structure formed by the intersecting invariant manifolds near a homoclinic orbit to a saddle fixed point has been discovered by Poincaré in 1890s. It is treated, together with Shilnikov's phenomena near a homoclinic orbit to a saddle-focus, in Guckenheimer and Holmes (1983), Wiggins (1988, 2003), Kuznetsov (2023). The theory of homoclinic bifurcations is systematically presented in Arnol'd et al. (1994), Ilyashenko and Li (1999) and, in particular, in Shilnikov et al. (1998, 2001).

Hyperbolics sets, the shadowing property, and exponential dichotomies are fundamental topics in dynamical systems theory that are known to be closely related to each other. A beautiful discussion of these interrelations from a modern perspective and an overview of the historical development is given in the foreword of Palmer's book (Palmer, 2000) on shadowing in dynamical systems. We decided to omit hyperbolic sets and shadowing as a general topic from this text. However, when discussing homoclinic phenomena we present some essentials from the theory of exponential dichotomies in continuous time in Appendix A (Sect. 7.6) and in discrete time in Appendix B (Sect. 7.7). Our concerns of these omissions are alleviated by the fact that there are by now quite a few books which treat these topics in depth. We mention, in particular, (Pilyugin, 1999; Palmer, 2000; Hasselblatt & Katok, 2003). For exponential dichotomies we refer to the classical book (Coppel, 1978) and the more recent treatment in Palmer (2000).

In Appendix A, we compile from Palmer (1984) and Beyn (1990) and follow an idea of Sandstede (1993), who directly analysed the perturbation of the associated Green's function. In Appendix B, we follow the work of Palmer (1988) on invertible systems as well as Steinlein and Walther (1990) and Beyn et al. (2016) on noninvertible systems. Their method of proof for symbolic dynamics employs the shadowing principle only for the specific instance when points leave or enter a neighbourhood of the fixed point.

7.5 Exercises

Exercise 7.1 (Smale horseshoe in Hénon map) Consider the following planar quadratic map depending on two parameters:

$$f_{(\alpha,\beta)}: \begin{pmatrix} x \\ y \end{pmatrix} \mapsto \begin{pmatrix} y \\ \alpha - \beta x - y^2 \end{pmatrix}. \tag{7.33}$$

An equivalent map was introduced by Hénon[7] as the simplest map with "chaotic behaviour".

1. Prove that $f_{(\alpha,\beta)}$ is invertible if $\beta \neq 0$.
2. Find a horseshoe for map (7.33) at $(\alpha, \beta) = (4.5, 0.2)$. (*Hint*: Consider the image $f_{(\alpha,\beta)}(R)$ of a rectangle R shown in Fig. 7.30.)

Exercise 7.2 (Schwarzian derivative)

Definition 7.43 *The **Schwarzian derivative** of a smooth function $f : \mathbb{R} \to \mathbb{R}$ at x is defined by*

$$(Sf)(x) := \frac{f'''(x)}{f'(x)} - \frac{3}{2}\left(\frac{f''(x)}{f'(x)}\right)^2.$$

[7] Hénon, M. (1976). A two-dimensional mapping with a strange attractor. *Journal of Communications in Mathematical Physics, 50*, 69–77.

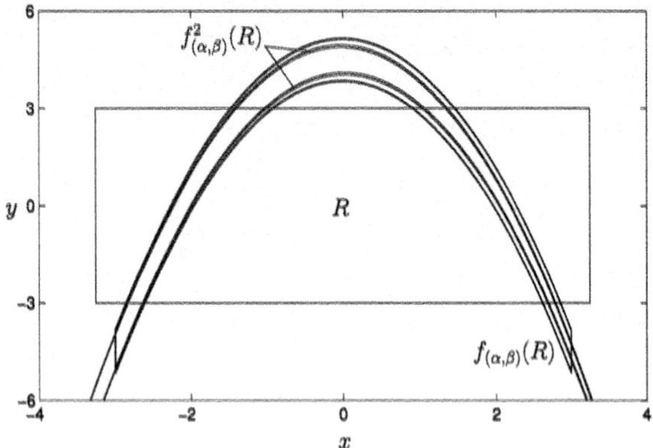

Fig. 7.30 Smale horseshoe in the Hénon map

We will write $Sf < 0$ if for all $x \in \mathbb{R}$ either $(Sf)(x) < 0$ or $\lim_{\xi \to x}(Sf)(\xi) = -\infty$. Prove the following statements:

1. Let $P(x)$ be a polynomial such that all roots of $P'(x)$ are real and distinct. Then $SP < 0$.
2. Suppose that $Sf < 0$ and f has n critical points. Then f has at most $n + 2$ attracting periodic orbits.
3. Let $Sf < 0$ and $Sg < 0$. Then holds $S(f \circ g) < 0$.
4. Only supercritical period-doubling bifurcations of fixed points and cycles can occur in a family of smooth functions $f_{(\alpha)} : \mathbb{R} \to \mathbb{R}$ with $Sf_{(\alpha)} < 0$. (*Hint*: Use the formula (5.72) for the critical normal form coefficient c at the period-doubling bifurcation.)

Exercise 7.3 (Travelling pulses in the FitzHugh–Nagumo model) The following system of partial differential equations is the FitzHugh–Nagumo simplification[8] of the Hodgkin–Huxley PDEs describing the nerve impulse propagation along an axon:

$$\begin{cases} \dfrac{\partial u}{\partial t} = \dfrac{\partial^2 u}{\partial x^2} - f_a(u) - v, \\ \dfrac{\partial v}{\partial t} = bu, \end{cases} \tag{7.34}$$

where $u = u(x,t)$ represents the membrane potential, $v = v(x,t)$ is a "recovery" variable, $f_a(u) = u(u-a)(u-1), 1 > a > 0, b > 0, -\infty < x < +\infty$, and $t > 0$.

Travelling waves are solutions to these equations of the form:

$$u(x,t) = U(\xi), \ v(x,t) = V(\xi), \ \xi = x + ct,$$

where $c > 0$ is an a priori unknown wave propagation speed.

1. Verify that $U(\xi)$ and $V(\xi)$ satisfy the following system of three autonomous ODEs:

$$\begin{cases} \dot{U} = W, \\ \dot{W} = cW + f_a(U) + V, \\ \dot{V} = \dfrac{b}{c}U, \end{cases} \tag{7.35}$$

[8] FitzHugh, R. (1961). Impulses and physiological states in theoretical models of nerve membrane. *Biophysical Journal, 1*, 445–446; Nagumo, J., Arimoto, S., & Yoshizawa, S. (1962). An active pulse transmission line simulating nerve axon. *Proceedings of the IRE, 50*, 2061–2070.

where the dot means differentiation with respect to "time" ξ. System (7.35) is called the *wave system* for (7.34); it depends on three positive parameters (a, b, c).

2. Check that for all $c > 0$ the wave system (7.35) has a unique equilibrium O with one positive eigenvalue λ_3 and two eigenvalues $\lambda_{1,2}$ with negative real parts. Conclude that this equilibrium has a one-dimensional unstable invariant manifold and a two-dimensional stable invariant manifold.
3. Show that for fixed $b > 0$ the eigenvalues $\lambda_{1,2}$ can either be real or form a complex-conjugate pair. Find a condition on the system parameters that defines a boundary between these two cases. Plot the boundaries in the (a, c)-plane corresponding to $b = 0.01, 0.005, 0.0025$, and specify in each case the region corresponding to saddle-foci. (*Hint:* At the curve

$$D_b = \{(a, c) : c^4(4b - a^2) + 2ac^2(9b - 2a^2) + 27b^2 = 0\}$$

the characteristic polynomial $h(\lambda)$ has a double root λ_0 satisfying $h(\lambda_0) = h'(\lambda_0) = 0$.)
4. A *travelling pulse* corresponds to a homoclinic orbit to O of (7.35), along which

$$(U(\xi), V(\xi), W(\xi)) \to (0, 0, 0)$$

as $\xi \to \pm\infty$. Sketch possible profiles of travelling pulses in both regions. (*Hint:* See Fig. 7.31.)
5. Show that the saddle quantity σ_0 of the equilibrium O in the wave system (7.35) is positive. (*Hint:* $\lambda_1 + \lambda_2 + \lambda_3 = c > 0$.)
6. Compute numerically curves $P_b^{(1)}$ in the (a, c)-plane for $b = 0.01, 0.005, 0.0025$, which correspond to the existence of the simplest homoclinic orbit in (7.35) and verify that these bifurcation curves look like in Fig. 7.32. (*Hint:* Use one of the standard software packages, e.g. AUTO with HomCont (Champneys et al., 1996) or MATCONT (De Witte et al., 2012).)

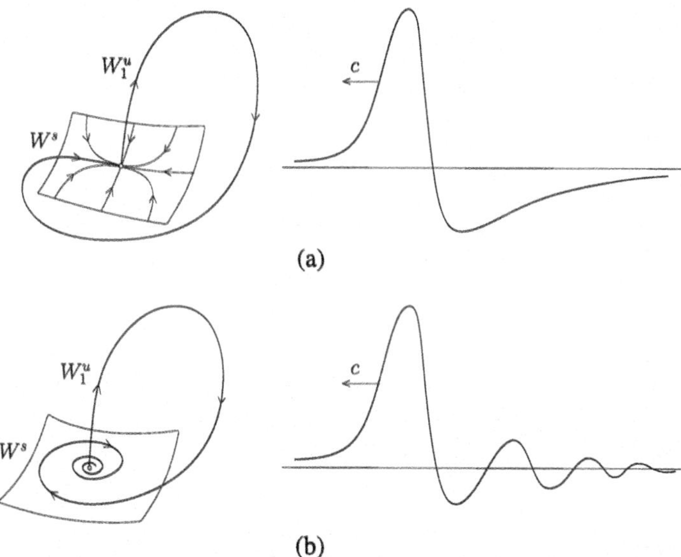

Fig. 7.31 Pulses with monotone (**a**) and oscillating (**b**) "tails"

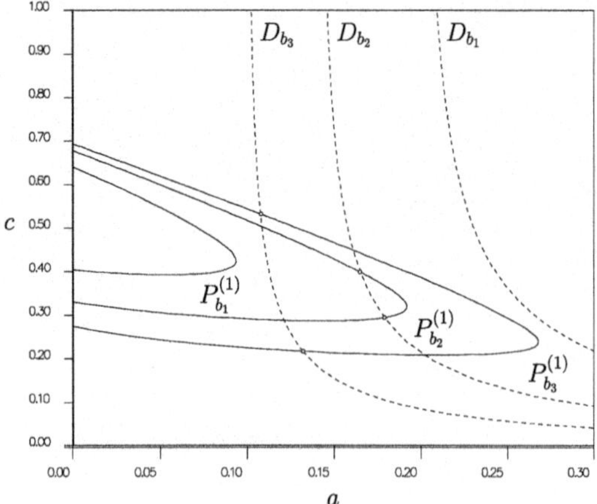

Fig. 7.32 Bifurcation curves of the wave system (7.35) for $b_1 = 0.01$, $b_2 = 0.005$, and $b_3 = 0.0025$

7. Conclude that the computed bifurcation curves $P_b^{(1)}$ pass through the saddle-focus region delimited by the corresponding curve D_b, and that for nearby parameter values the FitzHugh–Nagumo system (7.34) has infinitely many *periodic travelling waves*. (*Warning*: This analysis demonstrates the existence but says nothing about stability of these travelling waves as solutions to the PDE system (7.34).)

Exercise 7.4 (Bounds for Lorenz attractor) Let B_R be the closed ball of radius R centred at $(0, 0, \sigma + r)$. Prove that there exists $R > 0$ such that

(*i*) B_R is positively invariant set for Lorenz system (7.14), see Sect. 7.2.

(*ii*) Every orbit of (7.14) enters B_R.

(*Hint:* Consider $V(x, y, z) = x^2 + y^2 + z^2 - 2(\sigma + r)z$. Choose $\delta > 0$ and prove that for suitable $M > 0$

$$\frac{d}{dt}V(x(t), y(t), z(t)) < -\delta,$$

provided $V > M$.)

Exercise 7.5 (Homoclinic orbits in a slow–fast system in $\mathbb{R}^3$)

1. Check that the following slow–fast system[9]

$$\begin{cases} \dot{x} = (z+1) + (1-z)[(x-1) - y], \\ \dot{y} = (1-z)[(x-1) + y], \\ \varepsilon\dot{z} = (1-z^2)[z+1-m(x+1)] - \varepsilon z \end{cases} \tag{7.36}$$

has a homoclinic orbit to the equilibrium $(1, 0, -1)$ in the *singular limit* $\varepsilon = 0$, provided $m = 1$. (*Hint:* Formally set $\varepsilon = 0$ and analyse the equations on the *slow manifolds* defined by $z = \pm 1$.)

2. Determine the shape of the surface $\dot{z} = 0$ for small $\varepsilon > 0$. Convince yourself that there exists a continuous function $m = m(\varepsilon)$, $m(0) = 1$, defined for $\varepsilon > 0$, such that for the corresponding parameter value (7.36) has a saddle-focus with a homoclinic orbit.
3. Which case of Theorem 7.40 applies? How many periodic orbits can one expect near the bifurcation?

---

[9] Deng, B. (1994). Constructing homoclinic orbits and chaotic attractors. *International Journal of Bifurcation and Chaos, 4*, 823–841.

Exercise 7.6 (Local birth of Lorenz attractor)

1. Rewrite the Lorenz system (7.14) using new variables (ξ, η, ζ) such that

$$x = \sqrt{2}\xi, \ y = \sqrt{2}\left(\xi + \frac{\eta}{\sigma}\right), \ z = \frac{1}{\sigma}((2\sigma - b)\xi + \xi^2),$$

and assume that $2\sigma \neq b$. *Answer*: In the new variables the system will take the form:

$$\begin{cases} \dot{\xi} = \eta, \\ \dot{\eta} = \sigma(r-1)\xi - (1+\sigma)\eta - (2\sigma - b)\xi\zeta - \xi^3, \\ \dot{\zeta} = -b\zeta + \xi^2. \end{cases}$$

2. Introduce a small parameter $\varepsilon > 0$ and set back

$$x(t) = \varepsilon\xi(\varepsilon t), \ y(t) = \varepsilon^2\eta(\varepsilon t), \ z(t) = \varepsilon\zeta(\varepsilon t).$$

This yields

$$\begin{cases} \dot{x} = y, \\ \dot{y} = \varepsilon^2\sigma(r-1)x - \varepsilon(1+\sigma)y - \varepsilon(2\sigma - b)xz - x^3, \\ \dot{z} = -\varepsilon bz + x^2. \end{cases} \tag{7.37}$$

This allows us to consider (7.37) as a small perturbation of the singular system corresponding to $\varepsilon = 0$:

$$\begin{cases} \dot{x} = \ \ y\,, \\ \dot{y} = -x^3, \\ \dot{z} = \ \ x^2, \end{cases} \tag{7.38}$$

having an equilibrium $O = (0, 0, 0)$ with triple-zero eigenvalue.

3. Let (σ, r, b) belong to the chaotic region for the original Lorenz system (7.14). Argue that (7.37) has a small but diffeomorphic strange attractor for small $\varepsilon > 0$ which shrinks to O as $\varepsilon \to 0$.

Exercise 7.7 (Equivalent definition of Exponential Dichotomy in Appendix A)

1. Prove that conditions (7.57) and (7.58) in Definition 7.49 can be substituted by the following equivalent conditions:

$$\|Y(t)P\xi\| \leq Ke^{-\beta(t-s)}\|Y(s)P\xi\|, \quad s \leq t, \tag{7.39}$$

$$\|Y(t)(I-P)\xi\| \leq Ke^{-\beta(s-t)}\|Y(s)(I-P)\xi\|, \quad t \leq s, \tag{7.40}$$

where $\xi \in \mathbb{R}^n$, and

$$\|Y(t)PY^{-1}(t)\| \leq K. \tag{7.41}$$

2. Using this equivalent definition, show that operator $L_O x = \dot{x} - Ax$ with a constant hyperbolic matrix A has an exponential dichotomy on $J = \mathbb{R}$. What is P (and $(I-P)$) in this case? Give β explicitly in terms of the spectral bounds (7.45).

Exercise 7.8 (Lipschitz property of bifurcating periodic orbits in Appendix A)
Under the assumptions of the homoclinic bifurcation Theorem 7.45 derive the following estimate which compares periodic orbits with different periods $T_1 \geq T_0 \geq T$:

$$\|x_{T_0} - x_{T_1}|_{[-T_0,T_0]}\|_1 + |\alpha_{T_0} - \alpha_{T_1}| \leq C_\varepsilon e^{-\beta_\varepsilon T_0}|T_1 - T_0|, \tag{7.42}$$

where $\beta_\varepsilon = \min(\nu_s, \nu_u) - \varepsilon$ and C_ε is a suitable constant (*Hint*: Use the inequality (7.54) for $\xi_1 = (x_{T_0}, \alpha_{T_0})$ and $\xi_2 = (x_{T_1}|_{[-T_0,T_0]}, \alpha_{T_1})$).

Exercise 7.9 (Invariance of dichotomy projectors for one step in Appendix B) Show that the invariance property $P_{k+1}A_k = A_kP_k$, $k \in \mathbb{Z}$ (see (7.113)) holds for a linear dynamical system (7.109) if and only if the solution operator Φ satisfies the invariance condition (7.110) for all $j, k \in \mathbb{Z}$.

7.6 Appendix A: Birth of Cycles from Homoclinic Orbits

In this Appendix we prove that, generically, *at least one* periodic orbit (limit cycle) bifurcates from a homoclinic orbit to an equilibrium in smooth n-dimensional ODEs depending on a parameter α. The period T of the bifurcating cycle goes to infinity as $\alpha = \alpha_T$ approaches the bifurcation parameter value. The result will cover the Shilnikov saddle-focus homoclinic bifurcations in $\mathbb{R}^3$ studied in Sect. 7.3.2 of this chapter and will be applicable for general dimension $n \geq 2$. However, it only yields a primary branch of limit cycles which make one turn close to the homoclinic orbit. It will not specify how many *secondary* limit cycles with an arbitrary number of turns exist at a given parameter value near the bifurcation as provided by the geometrical construction in Sect. 7.3.2.

The dependence of α_T on the primary branch near the homoclinic bifurcation in $\mathbb{R}^n$ with $n \geq 3$ can either be oscillatory or monotone. These two cases actually occur near the saddle-focus homoclinic bifurcation in Sect. 7.3.2 and correspond to statements (i) and (ii) of Theorem 7.40, respectively. In the former case, the system can have many co-existing periodic orbits near the homoclinic orbit for each fixed α-value close to the bifurcation value. Moreover, there are periodic orbits on the secondary and higher-order branches which differ in the number of local turns near the equilibrium, as well as in the number of global excursions near the homoclinic orbit. Notably, the secondary limit cycles can appear and disappear via fold and period-doubling bifurcations, as well as from bifurcations of the multi-excursion homoclinic orbits.

7.6.1 *Nondegenerate Homoclinic Orbits and Main Results*

Consider a smooth system

$$\dot{x} = f(x, \alpha), \quad x \in \mathbb{R}^n, \alpha \in \mathbb{R}, \tag{7.43}$$

and assume without loss of generality that for $\alpha = \bar{\alpha}$ it has an equilibrium $x = 0$, i.e. $f(0, \bar{\alpha}) = 0$.

Suppose that at $\alpha = \bar{\alpha}$ the system (7.43) also has a solution $\bar{x}(t)$ satisfying

$$\lim_{t \to \pm\infty} \bar{x}(t) = 0. \tag{7.44}$$

The orbit Γ corresponding to $\bar{x}$ is homoclinic to the equilibrium at the origin. By differentiating (7.43) we find that the derivative $\dot{\bar{x}}(t)$ is a solution of the variational equation

$$\dot{y} = D_x f(\bar{x}(t), \bar{\alpha})y, \quad y \in \mathbb{R}^n.$$

Moreover, if we let $t \to \pm\infty$ in (7.43), then we obtain that $\dot{\bar{x}}(t) \to 0$ as $t \to \pm\infty$.

Definition 7.44 *The homoclinic solution $\bar{x}$ is called* **nondegenerate** *if the equilibrium $x = 0$ is hyperbolic and if the only bounded solutions $y(t)$, $t \in \mathbb{R}$ and $\mu \in \mathbb{R}$ of the system*

$$\dot{y} = D_x f(\bar{x}(t), \bar{\alpha})y + D_\alpha f(\bar{x}(t), \bar{\alpha})\mu$$

are given by

$$y(t) = c\dot{\bar{x}}(t),\ t \in \mathbb{R}, \quad \mu = 0$$

for some $c \in \mathbb{R}$.

For a hyperbolic equilibrium $x = 0$, introduce

$$A = D_x f(0, \bar{\alpha})$$

and define

$$\nu_s = -\max_{\lambda \in \sigma_s(A)} \mathrm{Re}(\lambda), \quad \nu_u = \min_{\lambda \in \sigma_u(A)} \mathrm{Re}(\lambda). \tag{7.45}$$

Consider the following *boundary value problem* on a finite interval:

$$\begin{cases} \dot{x}(t) - f(x(t), \alpha) = 0, & t \in [-T, T], \\ x(T) - x(-T) = 0, & \\ \langle \dot{\bar{x}}(0), x(0) - \bar{x}(0) \rangle = 0, & \end{cases} \tag{7.46}$$

where $T > 0$ is given. A solution (x, α) to this boundary value problem defines a *periodic orbit* (cycle) of (7.43) at the parameter value α with period $2T$ (see the first two equations). The last equation in (7.46) means that the point $x(0)$ in this cycle belongs to the plane orthogonal to the vector $\dot{\bar{x}}(0)$ at the point $\bar{x}(0) \in \Gamma$.

Denote by $C^1([-T, T], \mathbb{R}^n)$ the Banach space of bounded continuously differentiable functions $x : [-T, T] \to \mathbb{R}^n$ with the norm:

$$\|x\|_1 = \sup_{t \in [-T,T]} \|x(t)\| + \sup_{t \in [-T,T]} \|\dot{x}(t)\|. \tag{7.47}$$

Theorem 7.45 *Assume that system* (7.43) *has a nondegenerate homoclinic solution $\bar{x}$ at $\bar{\alpha}$. Then there exists $\rho > 0$ such that the boundary value problem* (7.46) *has a unique solution (x_T, α_T) in*

$$K_\rho = \{(x, \alpha) \in C^1([-T, T], \mathbb{R}^n) \times \mathbb{R} : \|x - \bar{x}|_{[-T,T]}\|_1 + |\alpha - \bar{\alpha}| \le \rho\}$$

for all sufficiently large T.

Moreover, for any $\varepsilon > 0$ there exists a constant C_ε such that the following estimate holds:

$$\|x_T - \bar{x}|_{[-T,T]}\|_1 + |\alpha_T - \bar{\alpha}| \le C_\varepsilon e^{-(\min\{\nu_s, \nu_u\} - \varepsilon)T}, \tag{7.48}$$

where ν_s and ν_u are defined by (7.45).

The theorem means that (7.43) has, for large T, a family of $2T$-periodic cycles that approaches the homoclinic orbit Γ as $\alpha \to \bar{\alpha}$. But uniqueness is only guaranteed in the function class K_ρ which does not rule out the existence of further limit cycles which follow several turns of the homoclinic orbit. The proof of Theorem 7.45 involves a number of preliminary results and will occupy the rest of this Appendix.

Before expanding on the proof we present an application of Theorem 7.45 to Hamiltonian systems which we recall from (4.61) with $x = (q, p) \in \mathbb{R}^{2m}$ as follows:

$$\dot{x} = \begin{pmatrix} \dot{q} \\ \dot{p} \end{pmatrix} = J_{2m}[DH(q, p)]^{\mathrm{T}} = \begin{pmatrix} [D_p H]^{\mathrm{T}} \\ -[D_q H]^{\mathrm{T}} \end{pmatrix} (x) = g(x). \tag{7.49}$$

Contrary to general systems, homoclinic orbits in Hamiltonian systems occur generically without involving a parameter. Accordingly, the definition of nondegeneracy is modified as follows.

Definition 7.46 *A homoclinic solution $\bar{x}$ of (7.49) satisfying (7.44) is called* **nondegenerate** *if the equilibrium $x = 0$ is hyperbolic and the only bounded solutions $y(t), t \in \mathbb{R}$ of the variational equation*

$$\dot{y} = D_x g(\bar{x})y$$

are $y(t) = c\dot{\bar{x}}(t)$ with $c \in \mathbb{R}$.

Theorem 7.47 *Assume that $\bar{x}$ is a nondegenerate homoclinic solution of the Hamiltonian system (7.49). Then there exist $T_0 > 0$ and constants $C, \nu > 0$ such that (7.49) has $2T$-periodic solutions $x_T \in C^1([-T,T], \mathbb{R}^n)$ for all $T \geq T_0$, and these solutions satisfy*

$$\|x_T - \bar{x}|_{[-T,T]}\|_1 \leq Ce^{-\nu T}. \tag{7.50}$$

The main idea of the proof is to apply Theorem 7.45 to the auxiliary parameter-dependent system

$$\dot{x} = g(x) + \alpha \begin{pmatrix} [D_q H]^{\mathrm{T}} \\ [D_p H]^{\mathrm{T}} \end{pmatrix} (x) = f(x, \alpha) \tag{7.51}$$

and to show that α vanishes for the resulting periodic orbits. The details will be given after we have collected the tools necessary to prove Theorem 7.45.

7.6.2 Preliminaries on Exponential Dichotomies

We begin with a local version of the Lipschitz Inverse Function Theorem 3.21.

Theorem 7.48 *Let X, Y be Banach spaces and let $L : X \to Y$ be a bounded linear operator with bounded inverse L^{-1}. Further, let $\xi_0 \in X$ and a function $F : B_\delta(\xi_0) \to Y$, defined on a δ-ball centred at ξ_0, be given with Lipschitz constant $\mathrm{Lip}(F)$ such that*

$$\mathrm{Lip}(F) < \|L^{-1}\|^{-1}, \tag{7.52}$$

$$\|(L+F)(\xi_0)\| \leq \left(\|L^{-1}\|^{-1} - \mathrm{Lip}(F)\right)\delta. \tag{7.53}$$

Then $(L + F)(\xi) = 0$ has a unique solution $\xi_* \in B_\delta(\xi_0)$. Moreover, the following estimate holds for all $\xi_1, \xi_2 \in B_\delta(\xi_0)$

$$\|\xi_1 - \xi_2\| \leq \left(\|L^{-1}\|^{-1} - \mathrm{Lip}(F)\right)^{-1} \|(L+F)(\xi_1) - (L+F)(\xi_2)\|, \tag{7.54}$$

so in particular

$$\|\xi_* - \xi_0\| \leq \left(\|L^{-1}\|^{-1} - \mathrm{Lip}(F)\right)^{-1} \|(L+F)(\xi_0)\|. \tag{7.55}$$

Proof The equation $(L + F)(\xi) = 0$ is equivalent to the fixed point problem

$$\xi = \Phi(\xi) = -L^{-1}F(\xi).$$

We will apply the Contraction Mapping Principle (Theorem 3.20 from Sect. 3.3) to prove the existence and uniqueness of the fixed point ξ_* of Φ in $B_\delta(\xi_0)$. First, Φ is Lipschitz with constant $\mathrm{Lip}(\Phi) = \|L^{-1}\|\mathrm{Lip}(F) < 1$ by (7.52). Then Φ maps $B_\delta(\xi_0)$ into itself, since (7.53) implies for $\xi \in B_\delta(\xi_0)$

$$\begin{aligned}\|\Phi(\xi)-\xi_0\| &\le \|\Phi(\xi)-\Phi(\xi_0)\|+\|\Phi(\xi_0)-\xi_0\|\\ &\le \mathrm{Lip}(\Phi)\|\xi-\xi_0\|+\|L^{-1}(F(\xi_0)+L\xi_0)\|\\ &\le \left[\mathrm{Lip}(F)\|L^{-1}\|+\|L^{-1}\|(\|L^{-1}\|^{-1}-\mathrm{Lip}(F))\right]\delta=\delta.\end{aligned}$$

The estimate (7.55) follows from (7.54) by setting $\xi_1=\xi_*, \xi_2=\xi_0$, and (7.54) is a consequence of the following estimate:

$$\begin{aligned}\|\xi_1-\xi_2\| &\le \|\xi_1-\Phi(\xi_1)-(\xi_2-\Phi(\xi_2))\|+\|\Phi(\xi_1)-\Phi(\xi_2)\|\\ &\le \|L^{-1}\|\|(L+F)(\xi_1)-(L+F)(\xi_2)\|+\mathrm{Lip}(\Phi)\|\|\xi_1-\xi_2\|.\end{aligned}$$

This finishes the proof. □

We also use several results on linear differential operators with *exponential dichotomy*. Exponential dichotomies separate solutions of linear differential equations into those that decay in forward and those that decay in backward time. In the notion of an exponential dichotomy one includes its data, i.e. the stable and unstable projectors onto initial vectors of forward and backward decaying solutions, the rates of decay and the bounding constants. Exponential dichotomies arise naturally when linearizing at a homoclinic orbit to a hyperbolic saddle with the dimensions of the ranges of the stable and unstable projectors determined by the linearization at the saddle. Therefore, they form the natural tool for investigating periodic orbits and chaotic behaviour near homoclinic orbits.

Let $J\subset\mathbb{R}$ be some interval, $J=\mathbb{R}$ and semi-infinite intervals allowed. By $X^k(J)$ we denote the space of bounded C^k-functions $x: J\to\mathbb{R}^n$. This is a Banach space with respect to the norm:

$$\|x\|_k := \sum_{j=0}^{k}\sup_{t\in J}\left\|\frac{d^j x(t)}{dt^j}\right\|.$$

For $k=1$ and $J=[-T,T]$, this norm coincides with (7.47). Consider the linear differential operator

$$L: X^1(\mathbb{R})\to X^0(\mathbb{R}),\quad (Lx)(t):=\dot{x}(t)-M(t)x(t),\ t\in\mathbb{R}, \tag{7.56}$$

where $t\mapsto M(t)$ is a continuous and bounded $n\times n$ matrix function. Let $Y(t), t\in\mathbb{R}$ be any *fundamental matrix* of L, i.e. Y satisfies $\dot{Y}=M(t)Y$ for $t\in\mathbb{R}$ and $Y(t)$ is invertible for some $t=t_0\in\mathbb{R}$ (and hence for all $t\in\mathbb{R}$).

Definition 7.49 *The operator L is said to have an* ***exponential dichotomy*** *on some interval $J\subset\mathbb{R}$ if there exist $K,\beta>0$ and a projector P in $\mathbb{R}^n$ such that for $s,t\in J$,*

$$\|Y(t)PY^{-1}(s)\|\le Ke^{-\beta(t-s)},\quad s\le t, \tag{7.57}$$

$$\|Y(t)(I-P)Y^{-1}(s)\|\le Ke^{-\beta(s-t)},\quad t\le s. \tag{7.58}$$

The triple (K,β,P) is called the data of the exponential dichotomy.

Please note that the two expressions at the left-hand side of the above inequalities reappear in Green's function introduced in the formula (7.70). Also note that this notion is independent of the matrix norm and of the fundamental matrix chosen. Another matrix norm will just change the constant K for the properties (7.57), (7.58) since all matrix norms are equivalent. Further, if $Z(t)$ is another fundamental matrix, then $Z(t)=Y(t)C$ holds for some invertible matrix C. Hence conditions (7.57) and (7.58) still hold with the same K and β but with Z instead of Y and with projector $C^{-1}PC$ instead of P.

An equivalent way of defining exponential dichotomy which makes this independence transparent uses the solution operator

$$S(t,s)=Y(t)Y^{-1}(s),\quad t,s\in J \tag{7.59}$$

and the t-dependent projectors

$$P(t) = Y(t)PY^{-1}(t), \quad t \in J. \tag{7.60}$$

With these settings and (7.57), (7.58) we obtain the following properties:

$$S(t,s)P(s) = P(t)S(t,s), \quad s,t \in J, \tag{7.61}$$

$$\|S(t,s)P(s)\| \le Ke^{-\beta(t-s)}, \quad s \le t, \tag{7.62}$$
$$\|S(t,s)(I-P(s))\| \le Ke^{-\beta(s-t)}, \quad t \le s. \tag{7.63}$$

The data of the exponential dichotomy are then written as $(K, \beta, P(t)_{t\in J})$. Conversely, let the properties (7.61)–(7.63) hold for some t-dependent projectors $P(t)$. Then we set $P = P(t_0)$ for some $t_0 \in J$ and let $Y(t) = S(t, t_0)$ be the fundamental matrix normalized by $Y(t_0) = I$. This implies the relation (7.60) as well as the inequalities (7.57) and (7.58).

Our first lemma shows that the range of P is uniquely determined but that we are still free to choose the null space of P, if the exponential dichotomy holds on a semi-infinite interval $J = [t_-, \infty)$. Similarly, the null space of P is unique and the range of P is free in case $J = (-\infty, t_+]$. As a consequence, the whole projector is unique if the exponential dichotomy holds on $J = \mathbb{R}$.

Lemma 7.50 *(i) If L has an exponential dichotomy on an interval $J = [t_-, \infty)$ with data (K, β, P) then the projectors $P(s)$, $s \in J$ from (7.60) satisfy*

$$R(P(s)) = \{x \in \mathbb{R}^n : \text{ there exists } C > 0 \text{ with } \|S(t,s)x\| \le C \text{ for all } t \ge s \in J\}. \tag{7.64}$$

If P_+ is a projector in $\mathbb{R}^n$ with $R(P_+) = R(P)$ then L also has an exponential dichotomy on J with data (K_+, β, P_+) for a suitable constant K_+.

(ii) If L has an exponential dichotomy on an interval $J = (-\infty, t_+]$ with data (K, β, P) then the projectors $P(s)$, $s \in J$ from (7.60) satisfy

$$N(P(s)) = \{x \in \mathbb{R}^n : \text{ there exists } C > 0 \text{ with } \|S(t,s)x\| \le C \text{ for all } t \le s \in J\}. \tag{7.65}$$

If P_- is a projector in $\mathbb{R}^n$ with $N(P) = N(P_-)$, then L also has an exponential dichotomy on $(-\infty, t_+]$ with data (K_-, β, P_-) for a suitable constant K_-.

Proof The relation $\subseteq$ in (7.64) follows directly from the estimate (7.62). For the converse consider $x \in \mathbb{R}^n$ with $\|S(t,s)x\| \le C$ for $t \ge s$ and estimate with (7.63), (7.61)

$$\|(I-P(s))x\| = \|S(s,t)(I-P(t))S(t,s)x\| \le Ke^{-\beta(t-s)}C.$$

Since the right-hand side converges to zero as $t \to \infty$ we obtain $(I - P(s))x = 0$, hence $x \in R(P(s))$.

Now assume $R(P) = R(P_+)$ and observe that $PP_+ = P_+$ and $P_+P = P$ follows. Then we obtain for $t \ge s$

$$\|Y(t)P_+Y^{-1}(s)\| = \|Y(t)PY^{-1}(s)Y(s)P_+Y^{-1}(s)\| \le Ke^{-\beta(t-s)}\|Y(s)P_+Y^{-1}(s)\|.$$

Further, using $PP_+(I-P) = P_+ - P$ we find

$$\begin{aligned}\|Y(s)P_+Y^{-1}(s)\| &\le \|Y(s)PY^{-1}(s)\| + \|Y(s)(P_+ - P)Y^{-1}(s)\| \\ &\le K + \|Y(s)PY^{-1}(t_-)Y(t_-)P_+Y^{-1}(t_-)Y(t_-)(I-P)Y^{-1}(s)\| \\ &\le K + K^2e^{-2\beta(s-t_-)}\|Y(t_-)P_+Y^{-1}(t_-)\|,\end{aligned}$$

which proves our assertion for $t \ge s$. For $t \le s$ one uses a similar argument.

The proof of part (ii) of the lemma is similar and left to the reader. □

Next we discuss one of the most important results in the theory: The *"roughness" theorem*. It shows that an exponential dichotomy persists under small perturbations and describes how its data are modified by the perturbation.

Theorem 7.51 (Roughness of Exponential Dichotomies) *Let the differential operator L defined by (7.56) have an exponential dichotomy on an unbounded interval $J \subset \mathbb{R}$ with data (K, β, P) and let $B : J \to \mathbb{R}^{n\times n}$ be a bounded continuous matrix function satisfying*

$$3b_0 < \beta, \quad \textit{where} \quad b_0 := K \sup_{t\in J} \|B(t)\|. \tag{7.66}$$

Then the perturbed operator $\tilde{L}x = Lx - B(\cdot)x$ has an exponential dichotomy on J with data $(\tilde{K}, \tilde{\beta}, \tilde{P})$ where

$$\tilde{\beta} = \beta - 2b_0, \quad \tilde{K} = K\left(2 + \frac{4b_0}{\beta - 3b_0}\right). \tag{7.67}$$

The solution operators S and $\tilde{S}$ of L and $\tilde{L}$ and the corresponding projectors $P(t)$, $\tilde{P}(t)$ satisfy

$$\left.\begin{array}{ll} t \geq s: & \|S(t,s)P(s) - \tilde{S}(t,s)\tilde{P}(s)\| \\ t \leq s: & \|S(t,s)(I - P(s)) - \tilde{S}(t,s)(I - \tilde{P}(s))\| \end{array}\right\} \leq K\tilde{K}\int_J e^{-\beta|t-\tau| - \tilde{\beta}|\tau - s|}\|B(\tau)\|d\tau \tag{7.68}$$

and

$$\|P(t) - \tilde{P}(t)\| \leq K\tilde{K}\int_J e^{-(\beta+\tilde{\beta})|t-\tau|}\|B(\tau)\|d\tau.$$

If J is unbounded from above then one has

$$\lim_{t\to\infty} P(t) - \tilde{P}(t) = 0 \quad \textit{if} \quad \lim_{t\to\infty} B(t) = 0, \tag{7.69}$$

and a corresponding result holds if J is unbounded from below.

Remark Before proving the theorem, let us note that our estimates show that the perturbed exponent $\tilde{\beta}$ is close to β if B is small. Moreover, the result (7.69) can be sharpened to $\tilde{P}(t) - P(t) = O(e^{-\varepsilon t})$ if $B(t) = O(e^{-\varepsilon t})$ for some $0 < \varepsilon < \beta + \tilde{\beta}$. ◇

Proof of Theorem 7.51 The proof proceeds through several steps.

Step 1: Let us first derive the fixed point equation which we solve by the Contraction Mapping Principle and which provides us with the perturbed data. We recall *Green's function* associated with L:

$$G(t,s) = \begin{cases} Y(t)PY^{-1}(s) = S(t,s)P(s), & t \geq s, \\ Y(t)(P - I)Y^{-1}(s) = S(t,s)(P(s) - I), & t < s, \end{cases} \quad t, s \in J. \tag{7.70}$$

Green's function acts as a solution kernel for solving inhomogeneous equations $Lx = y$ in the form $x(t) = \int_J G(t,s)y(s)ds$; see Proposition 7.52. We are looking for a Green's function $\tilde{G}$ of the perturbed operator $\tilde{L}$ in the form:

$$\tilde{G} = G + H,$$

where $H : J \times J \to \mathbb{R}^{n\times n}$ is a continuous kernel still to be determined. We set up an integral equation for H from the variation of constants formula. Suppose $z = \int_J (G + H)(\cdot, s)y(s)ds$ solves $\tilde{L}z = y$, then $Lz = y + B(\cdot)z$ and hence

$$
\begin{aligned}
z(t) &= \int_J [G+H](t,s)y(s)ds \\
&= \int_J G(t,s)y(s)ds + \int_J G(t,\tau)B(\tau)\left[\int_J (G+H)(\tau,s)y(s)ds\right]d\tau \\
&= \int_J \left[G(t,s) + \int_J G(t,\tau)B(\tau)(G+H)(\tau,s)d\tau\right] y(s)ds.
\end{aligned}
$$

Since we want this equality to hold for all y, we arrive at the following equation:

$$H(t,s) = \int_J G(t,\tau)B(\tau)(G+H)(\tau,s)d\tau, \quad t,s \in J. \tag{7.71}$$

The estimates (7.57) and (7.58) in the definition of the exponential dichotomy can be expressed in terms of Green's function as $\|G\|_\beta \le K < \infty$, where

$$\|G\|_\beta := \sup_{t,s\in J} e^{\beta|t-s|}\|G(t,s)\|$$

is a *weighted norm.* This suggests to try to solve Equation (7.71) by the Contraction Mapping Principle in the space of matrix-valued functions of two variables with a weighted norm:

$$\mathbb{X}^0_{\tilde\beta} = \{H \in C(J\times J, \mathbb{R}^{n\times n}) : \|H\|_{\tilde\beta} < \infty\},$$

where the constant $\tilde\beta$ is given by (7.67). Of course, this constant was chosen such that this approach is going to work! Note that $\mathbb{X}^0_{\tilde\beta}$ is a space of matrix-valued functions in two variables to be distinguished from the spaces X^0 and X^1 of vector-valued functions in one variable used in (7.56).

In the last steps, we will show that $\tilde G = G + H$, where $H \in \mathbb{X}^0_{\tilde\beta}$ is the solution of (7.71), has a representation like (7.70) and that the estimate of $\|H\|_{\tilde\beta}$ yields the exponential dichotomy for $\tilde L$.

Step 2: Let us introduce the space of kernels which have continuous first derivatives w.r.t. the first variable

$$\mathbb{X}^1_{\tilde\beta} = \{H \in \mathbb{X}^0_{\tilde\beta} : D_t H \in \mathbb{X}^0_{\tilde\beta}\}.$$

Proposition 7.52 *Let L from (7.56) have an exponential dichotomy on an unbounded interval J with data $(K,\beta,P(t)_{t\in J})$. Then for every $0 \le \tilde\beta < \beta$ and $R \in \mathbb{X}^0_{\tilde\beta}$ the inhomogeneous system*

$$
\begin{aligned}
LH(\cdot,s) &= R(\cdot,s) \text{ in } J,\ s\in J \\
0 &= \begin{cases} P(t_-)H(t_-,s),\ s\in J, & \text{if } J=[t_-,\infty), \\ (P(t_+)-I)H(t_+,s),\ s\in J & \text{if } J=(-\infty,t_+] \end{cases}
\end{aligned} \tag{7.72}
$$

has a unique solution $H \in \mathbb{X}^1_{\tilde\beta}$. This solution has the representation

$$H(t,s) = \int_J G(t,\tau)R(\tau,s)d\tau, \quad t,s\in J \tag{7.73}$$

and satisfies the estimate

$$\|H\|_{\tilde\beta} \le KC(\tilde\beta,\beta)\|R\|_{\tilde\beta} \text{ with } C(\tilde\beta,\beta) = \frac{2}{\tilde\beta+\beta} + \frac{1}{\beta-\tilde\beta}. \tag{7.74}$$

Proof Throughout the proof we assume $J = [t_-, \infty)$, the other cases then follow by analogous arguments. In particular, one finds that no condition other than $H \in \mathbb{X}^1_{\tilde{\beta}}$ is needed in case $J = \mathbb{R}$. First consider uniqueness. If $H \in \mathbb{X}^1_{\tilde{\beta}}$ solves the homogeneous system on $J = [t_-, \infty)$ then Lemma 7.50 (i) implies $H(t_-, s) \in R(P(t_-))$ for all $s \in J$. But the initial condition (7.72) yields $H(t_-, s) \in N(P(t_-))$ so that $H(t_-, s) = 0$ and $H(\cdot, s) \equiv 0$ follows.

Next we estimate the function H defined by (7.73). As a byproduct we find that the integral exists and is differentiable w.r.t. $t \in J$. Using the exponential dichotomy we find

$$\|H\|_{\tilde{\beta}} \leq \sup_{t,s\in J} e^{\tilde{\beta}|t-s|} \int_J \|G(t,\tau)\| \|R(\tau.s)\| ds \leq K\|R\|_{\tilde{\beta}} \sup_{t,s\in J} \gamma(t,s)$$

where

$$\gamma(t,s) = e^{\tilde{\beta}|t-s|} \int_J e^{-\beta|t-\tau|-\tilde{\beta}|\tau-s|} d\tau.$$

For $t \geq s$ we obtain

$$\begin{aligned}\gamma(t,s) &= e^{\tilde{\beta}(t-s)} \Big[\int_{\tau\leq s} e^{(\beta+\tilde{\beta})\tau-\beta t-\tilde{\beta}s} d\tau \\ &\quad + \int_s^t e^{(\beta-\tilde{\beta})\tau-\beta t+\tilde{\beta}s} d\tau + \int_{\tau\geq t} e^{-(\beta+\tilde{\beta})\tau+\beta t+\tilde{\beta}s} d\tau \Big] \\ &\leq e^{\tilde{\beta}(t-s)} \left[\frac{e^{-\beta(t-s)}}{\beta+\tilde{\beta}} + \frac{e^{-\tilde{\beta}(t-s)}}{\beta-\tilde{\beta}} + \frac{e^{-\tilde{\beta}(t-s)}}{\beta+\tilde{\beta}} \right] \leq C(\tilde{\beta},\beta).\end{aligned}$$

The same estimate is found for $t < s$ in a similar way. Next we differentiate the integrals to obtain the solution property

$$\begin{aligned}\frac{d}{dt}H(t,s) &= \frac{d}{dt}\Big[\int_{t\geq\tau\in J} S(t,\tau)P(\tau)R(\tau,s)d\tau + \int_{t<\tau\in J} S(t,\tau)(P(\tau)-I)R(\tau,s)d\tau \Big] \\ &= M(t)H(t,s) + S(t,t)P(t)R(t,s) - S(t,t)(P(t)-I)R(t,s) \\ &= M(t)H(t,s) + R(t,s).\end{aligned}$$

Since $P(t_-)$ is a projector, Eq. (7.61) finally implies the initial condition

$$P(t_-)H(t_-,s) = \int_{t_-}^{\infty} P(t_-)S(t_-,\tau)(P(\tau)-I)R(\tau,s)ds = 0. \qquad \square$$

Remark By inspection of the proof one can extend this proposition to right-hand sides $R: J \times J \to \mathbb{R}^{n\times n}$ for which $\|R\|_{\tilde{\beta}} < \infty$ and which are continuous in the closed triangle $\overline{\Delta}_+ = \{(t,s) \in J \times J : t \geq s\}$ and in the open triangle $\Delta_- = \{(t,s) \in J \times J : t < s\}$ and which have a continuous extension to the closure $\overline{\Delta}_-$. The differential equation then holds in $\Delta_+ \cup \Delta_-$ with continuous extension of the derivative $D_t H$ to $\overline{\Delta}_+$ (by the derivative from the right) and to $\overline{\Delta}_-$. We will apply this to the discontinuous kernel G in the integral of the fixed point Eq. (7.71). ◊

Step 3: Equation (7.71) is a linear operator equation of the form $H = \mathcal{L}(H) + \mathcal{L}(G), H \in \mathbb{X}^0_{\tilde{\beta}}$. Here $\mathcal{L}$ is the integral operator

$$\mathcal{L}(H)(t,s) = \int_J G(t,\tau)B(\tau)H(\tau,s)d\tau,$$

which can also be applied to the kernel G with a discontinuity on the diagonal (see the Remark above). We estimate the operator norm of $\mathcal{L}$ by using Proposition 7.52:

$$\|\mathcal{L}(H)\|_{\tilde{\beta}} \le KC(\tilde{\beta},\beta)\|B(\cdot)H\|_{\tilde{\beta}} \le KC(\tilde{\beta},\beta)\sup_{t\in J}\|B(t)\|\|H\|_{\tilde{\beta}}.$$

Using (7.74), (7.66), and (7.67), we conclude that $\mathcal{L}$ is a contraction with constant

$$q = b_0C(\tilde{\beta},\beta) = b_0\left(\frac{2}{2\beta-2b_0}+\frac{1}{2b_0}\right) = \frac{1}{2}+\frac{b_0}{\beta-b_0} < 1.$$

Therefore, (7.71) has a unique solution $H \in X^0_{\tilde{\beta}}$ which satisfies

$$\|H\|_{\tilde{\beta}} \le \frac{\|\mathcal{L}(G)\|_{\tilde{\beta}}}{1-q} \le \frac{qK}{1-q} = \left(1+\frac{4b_0}{\beta-3b_0}\right)K. \tag{7.75}$$

Step 4: With the solution H from above we define

$$\tilde{G} = G + H, \quad \tilde{P}(t) = \tilde{G}(t,t), \quad t \in J \tag{7.76}$$

and note that (7.75) implies

$$\|\tilde{G}\|_{\tilde{\beta}} \le \|G\|_{\tilde{\beta}} + \|H\|_{\tilde{\beta}} \le K\left(2+\frac{4b_0}{\beta-3b_0}\right) = \tilde{K}.$$

Next we show that $\tilde{G}$ is of the form (7.70) with $\tilde{P}$ from (7.76) and the solution operator $\tilde{S}$ of $\tilde{L}$. By Proposition 7.52 and (7.71) we obtain for $t \neq s$

$$\begin{aligned} D_t[G+H](t,s) &= D_tG(t,s) + M(t)H(t,s) + B(t)(G(t,s)+H(t,s)) \\ &= (M(t)+B(t))\tilde{G}(t,s). \end{aligned}$$

Hence we have $\tilde{L}\tilde{G}(t,s) = 0$ for $t \neq s$. Since $\tilde{G}$ and $D_t\tilde{G}$ extend continuously to the closed triangle $\overline{\Delta}_+$ we conclude

$$\tilde{G}(t,s) = \tilde{S}(t,s)\tilde{G}(s,s) = \tilde{S}(t,s)\tilde{P}(s), \quad t \ge s \in J. \tag{7.77}$$

In the following we consider only the case $J = [t_-,\infty)$. For $s > t_-$ we find with the continuity of H the lower limit

$$\lim_{t\uparrow s}\tilde{G}(t,s) = H(s,s) + \lim_{t\uparrow s}G(t,s) = H(s,s) + G(s,s) - I = \tilde{P}(s) - I,$$

and therefore

$$\tilde{G}(t,s) = \tilde{S}(t,s)(\tilde{P}(s)-I), \quad s > t \in J. \tag{7.78}$$

Using these representations and (7.70) we derive the estimate (7.68) directly from (7.71)

$$\begin{aligned} \|H(t,s)\| &\le \int_J e^{-\beta|t-\tau|}\|G\|_\beta\|B(\tau)\|e^{-\tilde{\beta}|\tau-s|}\|\tilde{G}\|_{\tilde{\beta}}d\tau \\ &\le K\tilde{K}\int_J e^{-\beta|t-\tau|-\tilde{\beta}||\tau-s|}\|B(\tau)\|d\tau. \end{aligned}$$

Let us further note that the initial condition $P(t_-)H(t_-,s) = 0$ and (7.70) lead to the relation

$$P(t_-)\tilde{G}(t_-,s) = P(t_-)G(t_-,s) = \delta_{t_-,s}P(t_-), \quad s \in J = [t_-,\infty). \tag{7.79}$$

Step 5: It remains to show that $\tilde{P}(t)$, $t \in J$ are projectors which satisfy the relation (7.60) with the solution operator $\tilde{S}$. For that purpose we define the modified kernel

$$\tilde{H}(t,s) = -G(t,s) + \begin{cases} \tilde{G}(t,s)\tilde{P}(s), & t \geq s, \\ \tilde{G}(t,s)(\tilde{P}(s)+I), & t < s, \end{cases} \tag{7.80}$$

and show that it belongs to $\mathbb{X}^0_{\tilde{\beta}}$ and satisfies (7.71) as well. The unique solvability then yields $H = \tilde{H}$ and thus

$$\tilde{P}(t) = \tilde{G}(t,t) = H(t,t) + G(t,t) = \tilde{H}(t,t) + G(t,t) = \tilde{G}(t,t)\tilde{P}(t) = \tilde{P}(t)^2.$$

The kernel $\tilde{H}$ is continuous at $t = s$ since

$$\begin{aligned} \tilde{H}(s,s) &= -G(s,s) + \tilde{G}(s,s)\tilde{P}(s) = -G(s,s) + \tilde{P}(s)^2, \\ \lim_{t \uparrow s} \tilde{H}(t,s) &= -\lim_{t \uparrow s} G(t,s) + \lim_{t \uparrow s} \tilde{G}(t,s)(\tilde{P}(s)+I) \\ &= I - G(s,s) + (\tilde{P}(s) - I)(\tilde{P}(s)+I) = -G(s,s) + \tilde{P}(s)^2. \end{aligned}$$

Further, from $\|\tilde{G}\|_{\tilde{\beta}} \leq \tilde{K}$ we obtain $\|\tilde{P}(t)\| \leq \tilde{K}$ for all $t \in J$ and thus

$$\|\tilde{H}\|_{\tilde{\beta}} \leq \|G\|_{\tilde{\beta}} + (\tilde{K}+1)\|\tilde{G}\|_{\tilde{\beta}} < \infty.$$

Using (7.77) and (7.78) we find that the kernel $\tilde{H}$ satisfies for $t \neq s$ the differential equation

$$\begin{aligned} D_t\tilde{H}(t,s) &= -D_tG(t,s) + (M(t)+B(t))\tilde{G}(t,s) \begin{cases} \tilde{P}(s), & t > s, \\ \tilde{P}(s)+I, & t < s \end{cases} \\ &= -M(t)G(t,s) + (M(t)+B(t))(\tilde{H}+G)(t,s) \\ &= M(t)\tilde{H}(t,s) + B(t)(\tilde{H}+G)(t,s). \end{aligned}$$

As above, the right-hand side is continuous on the triangles $\overline{\Delta}_+$, Δ_- and has a continuous extension to the closed triangle $\overline{\Delta}_-$. Finally, Eq. (7.79) leads to the initial condition $P(t_-)\tilde{H}(t_-,s) = 0$:

$$\begin{aligned} s > t_- : P(t_-)\tilde{H}(t_-,s) &= -P(t_-)G(t_-,s) + P(t_-)\tilde{G}(t_-,s)(\tilde{P}(s)+I) = 0, \\ s = t_- : P(t_-)\tilde{H}(t_-,t_-) &= -P(t_-)G(t_-,t_-) + P(t_-)\tilde{G}(t_-,t_-)^2 = 0. \end{aligned}$$

Therefore, Proposition 7.52 applies to the inhomogenous term $R(t,s) = B(t)(\tilde{H}+G)(t,s)$ and (7.73) shows that $\tilde{H}$ solves (7.71).

The proof of (7.61) proceeds in a similar manner. For $\delta \in (0,1]$ we consider the modified kernel

$$\begin{aligned} \tilde{H}(t,s) &= -G(t,s) + \tilde{S}(t,t+\delta)\tilde{G}(t+\delta,s+\delta)\tilde{S}(s+\delta,s) \\ &= -G(t,s) + \begin{cases} \tilde{S}(t,s+\delta)\tilde{P}(s+\delta)\tilde{S}(s+\delta,s), & t \geq s, \\ \tilde{S}(t,s+\delta)(\tilde{P}(s+\delta)-I)\tilde{S}(s+\delta,s), & t < s \end{cases} \end{aligned}$$

where (7.76) and (7.78) were used. Then we show that $\tilde{H}$ solves (7.71). By the uniqueness of the solution we obtain for $t \in J$

$$\begin{aligned} \tilde{P}(t)\tilde{S}(t,t+\delta) &= (H+G)(t,t)\tilde{S}(t,t+\delta) = (\tilde{H}+G)(t,t)\tilde{S}(t,t+\delta) \\ &= \tilde{S}(t,t+\delta)\tilde{P}(t+\delta)\tilde{S}(t+\delta,t)\tilde{S}(t,t+\delta)) = \tilde{S}(t,t+\delta)\tilde{P}(t+\delta). \end{aligned}$$

Therefore, the relation (7.61) holds for all $t \in J$ and $t \le s \le t+1$. By iterating this argument we obtain (7.61) for all $t, s \in J$. Continuity of $\tilde{H}$ at $t = s$ follows from

$$\begin{aligned} \tilde{H}(s,s) &= -G(s,s) + \tilde{S}(s,s+\delta)\tilde{P}(s+\delta)\tilde{S}(s+\delta,s), \\ \lim_{t \uparrow s} \tilde{H}(t,s) &= -(G(s,s) - I) + \tilde{S}(s,s+\delta)(\tilde{P}(s+\delta) - I)S(s+\delta,s) = \tilde{H}(s,s). \end{aligned}$$

Further, since $M(t)$ and $B(t)$ are uniformly bounded there exists a constant C such that $\|\tilde{S}(t,s)\| \le C$ holds for all $t, s \in J$ with $|t-s| \le 1$. Therefore, the definition of $\tilde{H}$ implies

$$\|\tilde{H}\|_{\tilde{\beta}} \le \|G\|_{\tilde{\beta}} + C^2\|\tilde{G}\|_{\tilde{\beta}} < \infty,$$

and hence $\tilde{H} \in \mathbb{X}^0_{\tilde{\beta}}$. Moreover, with the solution property of $\tilde{S}$ we find for $t \neq s$

$$\begin{aligned} D_t(\tilde{H}+G)(t,s) &= (M(t)+B(t)) \begin{cases} \tilde{S}(t,s+\delta)\tilde{P}(s+\delta)\tilde{S}(s+\delta,s), & t > s, \\ \tilde{S}(t,s+\delta)(\tilde{P}(s+\delta)-I)\tilde{S}(s+\delta,s), & t < s \end{cases} \\ &= M(t)(\tilde{H}+G)(t,s) + B(t)(\tilde{H}+G)(t,s). \end{aligned}$$

Note that the right-hand side is continuous in $\overline{\Delta}_+$ and Δ_- with a continuous extension to $\overline{\Delta}_-$.

Finally, we check the initial condition for $\tilde{H}$. For $s > t_-$ we have by (7.78) and (7.79)

$$\begin{aligned} P(t_-)\tilde{H}(t_-,s) = P(t_-)(\tilde{H}+G)(t_-,s) &= P(t_-)\tilde{S}(t_-,t_-+\delta)\tilde{G}(t_-+\delta,s+\delta)\tilde{S}(s+\delta,s) \\ &= P(t_-)\tilde{S}(t_-,t_-+\delta)\tilde{S}(t_-+\delta,s+\delta)(\tilde{P}(s+\delta)-I)\tilde{S}(s+\delta,s) \\ &= P(t_-)\tilde{G}(t_-,s+\delta)\tilde{S}(s+\delta,s) = 0. \end{aligned}$$

For $s = t_-$, Eq. (7.79) yields

$$\begin{aligned} P(t_-)\tilde{H}(t_-,t_-) &= P(t_-)(\tilde{H}+G)(t_-,t_-) - P(t_-) \\ &= P(t_-)\tilde{S}(t_-,t_-+\delta)\tilde{P}(t_-+\delta)\tilde{S}(t_-+\delta,t_-) - P(t_-) \\ &= P(t_-)\left(\tilde{G}(t_-,t_-+\delta) + \tilde{S}(t_-,t_-+\delta)\right)\tilde{S}(t_-+\delta,t_-) - P(t_-) = 0. \end{aligned}$$

Summarizing, we have shown that Proposition 7.52 applies with $R(t,s) = B(t)(\tilde{H}+G)(t,s)$ and yields $\tilde{H}$ as the unique solution. This finishes the proof. □

7.6.3 Exponential Dichotomies for Linearized Operators

In the following we assume that the matrices $M(t)$ in (7.56) converge as $t \to \pm\infty$ to the same limit, i.e.:

$$\lim_{t \to \pm\infty} M(t) = A. \tag{7.81}$$

The following theorem ensures that L inherits the exponential dichotomy from the constant coefficient operator $L_0 x = \dot{x} - Ax$ if the matrix A is hyperbolic. In the following we define Y as the fundamental matrix of L normalized by $Y(0) = I$.

Theorem 7.53 *Suppose that* (7.81) *holds with a hyperbolic matrix A (i.e. A has no eigenvalues λ with* $\mathrm{Re}(\lambda) = 0$*) and let $0 < \omega < \min(\nu_s, \nu_u)$ hold for the spectral bounds from* (7.45)*. Then the operator L defined by* (7.56) *has an exponential dichotomy on* $[0,\infty)$

with data (K, ω, P) *and on* $(-\infty, 0]$ *with data* (K, ω, Q). *The time-dependent projectors (see* (7.60)*) satisfy*

$$\lim_{t\to\infty} Y(t)PY^{-1}(t) = P_s \quad \text{and} \quad \lim_{t\to-\infty} Y(t)(I-Q)Y^{-1}(t) = P_u, \tag{7.82}$$

where P_s *and* $P_u = I - P_s$ *are the projectors onto the stable and unstable invariant eigenspaces* T^s *and* T^u *of* A, *respectively. All projectors have the same rank, i.e.* $\operatorname{rank} P = \operatorname{rank} Q = \operatorname{rank} P_s =: n_s$. *Moreover, there exist constants* $C, T \geq 0$ *such that the following estimates hold with the error terms* $b_\rho = \sup_{\tau\geq\rho} \|M(\tau) - A\|$ *and* $c_\rho = \sup_{\tau\leq\rho} \|M(\tau) - A\|$

$$\|Y(t)PY(s)^{-1} - e^{(t-s)A}P_s\| \leq C(e^{-\omega s}b_0 + e^{-\omega(t-s)}b_{\frac{t}{2}}), \quad T \leq \frac{t}{2} \leq s \leq t, \tag{7.83}$$

$$\|Y(t)(I-Q))Y(s)^{-1} - e^{(t-s)A}P_u\| \leq C(e^{\omega s}c_0 + e^{-\omega(s-t)}c_{\frac{t}{2}}), \quad t \leq s \leq \frac{t}{2} \leq -T.$$

Proof Consider first the case $[0, \infty)$ and choose ω_+ with $\omega < \omega_+ < \min(\nu_s, \nu_u)$. Since the matrix A is hyperbolic, Theorem 2.24 shows that the operator $L_0 x = x' - Ax$ has an exponential dichotomy on $\mathbb{R}_+$ with data (K_+, ω_+, P_s). Then take T so large that $\omega \leq \omega_+ - 3b_T$ holds for the error term b_T defined above. Thus we can apply the Roughness Theorem 7.51 to the operator L_0 and its perturbation $L = L_0 - (M(\cdot) - A)$ on the interval $[T, \infty)$. Therefore, L has an exponential dichotomy on $[T, \infty)$ with data $(\tilde{K}, \omega, P)$. Let us show the estimate (7.83) which then yields the first limit in (7.82) when taking $t = s \to \infty$. From the estimate (7.68) and $0 < \omega < \omega_+$ we obtain for $T \leq \frac{t}{2} \leq s \leq t$

$$\begin{aligned}
\|Y(t)PY(s)^{-1} - e^{(t-s)A}P_s\| &\leq K_+K \int_T^\infty e^{-\omega_+|t-\tau|-\omega|\tau-s|}\|M(\tau) - A\|d\tau \\
&\leq K_+K \left[b_0 \int_T^{\frac{t}{2}} e^{-\omega(t+s-2\tau)}d\tau + b_{\frac{t}{2}} \int_{\frac{t}{2}}^s e^{-\omega(t+s-2\tau)}d\tau \right. \\
&\quad \left. + b_s e^{-\omega_+ t + \omega s} \int_s^t e^{(\omega_+-\omega)\tau}d\tau + b_t \int_t^\infty e^{-\omega(2\tau-t-s)}d\tau \right] \\
&\leq \frac{K_+K}{2\omega} \left[b_0 e^{-\omega s} + b_{\frac{t}{2}} \left(e^{-\omega(t-s)} + \frac{2\omega}{\omega_+ - \omega} e^{-\omega(t-s)} + e^{-\omega(t-s)} \right) \right]
\end{aligned}$$

from which (7.83) follows with a suitable constant C.

Finally, the dichotomy can easily be extended to $[0, \infty)$ by keeping the data ω, P and adapting the constant: For $t, s \geq T$ there is nothing to prove, and for $0 \leq s, t \leq T$ all terms are bounded. Consider, for example, $t \geq T \geq s \geq 0$. Then we have

$$\begin{aligned}
\|Y(t)PY^{-1}(s)\| &= \|Y(t)PY^{-1}(T)Y(T)Y^{-1}(s)\| \\
&\leq \|Y(t)PY^{-1}(T)\|\|Y(T)Y^{-1}(s)\| \\
&\leq K_T K e^{-\omega(t-T)} \leq K_T K e^{\omega T} e^{-\omega(t-s)},
\end{aligned}$$

where $K_T = \sup_{t,s\in[0,T]} \|Y(t)Y^{-1}(s)\|$. A similar estimate can be done for $Y(t)(I-P)$ $Y^{-1}(s)$ with $0 \leq t \leq T \leq s$.

The case $(-\infty, 0]$ is treated in a similar manner. □

Introduce the *adjoint* differential operator for L:

$$L^* : X^1(\mathbb{R}) \to X^0(\mathbb{R}), \quad (L^*x)(t) := \dot{x}(t) + M^{\mathrm{T}}(t)x(t), \ t \in \mathbb{R}.$$

Lemma 7.54 *Under the assumptions of* Theorem 7.53, *the operator* L^* *has exponential dichotomies on* $\mathbb{R}_+ := [0, \infty)$ *(resp.* $\mathbb{R}_- := (-\infty, 0]$*) and suitable projectors are* $I - P^{\mathrm{T}}$ *(resp.* $I - Q^{\mathrm{T}}$*). Moreover*

$$N(L) = \{Y(t)x_0 : x_0 \in R(P) \cap N(Q)\}, \quad \dim N(L) = \dim N(L^*),$$

and

$$y \in R(L) \iff \int_{-\infty}^{\infty} \langle \varphi(t), y(t) \rangle dt = 0 \quad \text{for all} \quad \varphi \in N(L^*). \tag{7.84}$$

Proof By $Y(0) = I$ and (7.64) we can characterize the range of the projector as follows:

$$R(P) = \{x_0 \in \mathbb{R}^n : Y(t)x_0 \text{ is uniformly bounded for } t \geq 0\}. \tag{7.85}$$

Similarly, we obtain from (7.65)

$$N(Q) = \{x_0 \in \mathbb{R}^n : Y(t)x_0 \text{ is uniformly bounded for } t \leq 0\}. \tag{7.86}$$

From the definition of Y and the characterizations (7.85) and (7.86) it is clear that

$$N(L) = \{Y(\cdot)x_0 : x_0 \in R(P) \cap N(Q)\}. \tag{7.87}$$

Transposing and differentiating $Y^{-1}(t)Y(t) = I$ shows that $Y^*(t) = [Y^{\mathrm{T}}(t)]^{-1}$ is the fundamental matrix of the adjoint operator L^*, i.e.:

$$\dot{Y}^*(t) = -M^{\mathrm{T}}(t)Y^*(t), \quad Y^*(0) = I.$$

Using that $\|B\|_2 = \|B^{\mathrm{T}}\|_2$ holds for any matrix B and the spectral norm $\|\cdot\|_2$ (i.e. the matrix norm induced by the Euclidean norm in $\mathbb{R}^n$) we obtain for $t \geq s \geq 0$

$$\begin{aligned} \|Y^*(t)(I - P^{\mathrm{T}})(Y^*(s))^{-1}\|_2 &= \|\left(Y^*(t)(I - P^{\mathrm{T}})(Y^*(s))^{-1}\right)^{\mathrm{T}}\|_2 \\ &= \|Y(s)(I - P)Y^{-1}(t)\|_2 \leq K\mathrm{e}^{-\omega(t-s)}. \end{aligned}$$

Since all matrix norms are equivalent we can return to $\|\cdot\|$ and find the estimate (7.57) for the adjoint. In a similar manner one shows $\|Y^*(t)P^{\mathrm{T}}(Y^*(s))^{-1}\| \leq K\mathrm{e}^{-\omega(s-t)}$ for $s \geq t \geq 0$. Thus the adjoint L^* has exponential dichotomies on $\mathbb{R}_+$ with projector $P^* = I - P^{\mathrm{T}}$ and on $\mathbb{R}_-$ with projector $Q^* = I - Q^{\mathrm{T}}$.

Applying (7.87) to L^*, we find the representation

$$N(L^*) = \{Y^*(\cdot)\varphi_0 : \varphi_0 \in R(I - P^{\mathrm{T}}) \cap N(I - Q^{\mathrm{T}})\}. \tag{7.88}$$

By Fredholm's Decomposition (Lemma 6.22) we obtain

$$R(I - P^{\mathrm{T}}) \cap N(I - Q^{\mathrm{T}}) = N(I - P)^{\perp} \cap R(I - Q)^{\perp} = R(P)^{\perp} \cap N(Q)^{\perp} = (R(P) + N(Q))^{\perp} \tag{7.89}$$

where $^{\perp}$ denotes the orthogonal complement. Using Theorem 7.53 a count of dimensions yields

$$\begin{aligned} \dim N(L^*) &= n - \dim(R(P) + N(Q)) \\ &= n - [\dim R(P) + \dim N(Q) - \dim(R(P) \cap N(Q))] \\ &= n - [n_s + n - n_s - \dim(R(P) \cap N(Q))] = \dim N(L). \end{aligned}$$

It remains to prove (7.84). If we assume $y \in R(L)$ then there exists $z \in X^1(\mathbb{R})$ such that $Lz = y$. By the exponential dichotomy of L^*, the functions $\varphi \in N(L^*)$ decay exponentially in both directions. Moreover, since z is bounded we can integrate by parts,

$$\int_{-\infty}^{\infty} \langle \varphi, y \rangle dt = \int_{-\infty}^{\infty} \langle \varphi, \dot{z} - M(t)z \rangle dt = -\int_{-\infty}^{\infty} \langle \dot{\varphi} + M^{\mathrm{T}}(t)\varphi, z \rangle dt = 0.$$

Conversely, assume $\int_{-\infty}^{\infty} \langle \varphi, y \rangle dt = 0$ for all $\varphi \in N(L^*)$. Then we construct a solution $z \in X^1(\mathbb{R})$ of $Lz = y$ in the following form:

$$z(t) = \begin{cases} z_+(t) + Y(t)z_+^0, \; t \geq 0, \\ z_-(t) + Y(t)z_-^0, \; t < 0, \end{cases} \tag{7.90}$$

where $z_+^0 \in R(P), z_-^0 \in N(Q)$ are to be determined. The functions $z_\pm$ are defined by

$$z_+(t) = \int_0^{\infty} G_+(t,s)y(s)ds \quad (t \geq 0), \quad z_-(t) = \int_{-\infty}^{0} G_-(t,s)y(s)ds \quad (t \leq 0)$$

with Green's functions

$$G_+(t,s) = \begin{cases} Y(t)PY^{-1}(s), & 0 \leq s \leq t, \\ Y(t)(P-I)Y^{-1}(s), & 0 \leq t < s, \end{cases} \tag{7.91}$$

$$G_-(t,s) = \begin{cases} Y(t)QY^{-1}(s), & s \leq t \leq 0, \\ Y(t)(Q-I)Y^{-1}(s), & t < s \leq 0. \end{cases} \tag{7.92}$$

Note that $z_\pm \in X^1(\mathbb{R}_\pm)$ solve the equation $Lz_\pm = y$ on $\mathbb{R}_+$ and $\mathbb{R}_-$, respectively; cf. Proposition 7.52. If we arrange that z is continuous at $t = 0$, then it will be continuously different iable at 0 by using the differential equation on each half-line, and hence solve $Lz = y$ on $\mathbb{R}$. Continuity at $t = 0$ requires to choose $z_+^0 \in R(P), z_-^0 \in N(Q)$ such that $z_+(0) - z_-(0) = z_-^0 - z_+^0$. Therefore, we have to show $z_+(0) - z_-(0) \in N(Q) + R(P)$, or equivalently

$$0 = \langle \varphi_0, z_+(0) - z_-(0) \rangle \quad \forall\, \varphi_0 \in (N(Q) + R(P))^\perp.$$

For $\varphi(t) = Y^*(t)\varphi_0$ we have $\varphi \in N(L^*)$ by (7.88) and (7.89) so that our claim follows from the computation

$$\begin{aligned}
\langle \varphi_0, z_-(0) - z_+(0) \rangle \;&=\; \int_{-\infty}^{0} \frac{d}{dt}\langle \varphi, z_- \rangle dt + \int_0^{\infty} \frac{d}{dt}\langle \varphi, z_+ \rangle dt \\
&= \int_{-\infty}^{0} \langle \dot{\varphi}, z_- \rangle + \langle \varphi, \dot{z}_- \rangle dt + \int_0^{\infty} \langle \dot{\varphi}, z_+ \rangle + \langle \varphi, \dot{z}_+ \rangle dt \\
&= \int_{-\infty}^{0} \langle -M^{\mathrm{T}}(t)\varphi, z_- \rangle + \langle \varphi, M(t)z_- + y \rangle dt + \int_0^{\infty} \langle -M^{\mathrm{T}}(t)\varphi, z_+ \rangle + \langle \varphi, M(t)z_+ + y \rangle dt \\
&= \int_{-\infty}^{\infty} \langle \varphi, y \rangle dt \;=\; 0.
\end{aligned}$$

This finishes the proof. □

7.6.4 *Proof of General Homoclinic Bifurcation Theorem*

Now we are going to use the above results on the exponential dichotomies to study homoclinic orbits and their associated limit cycles.

Lemma 7.55 *Let $\bar{x}$ be a homoclinic solution of (7.43) at $\bar{\alpha}$. Then $\bar{x}$ is nondegenerate if and only if the linear operator*

$$L : X^1(\mathbb{R}) \to X^0(\mathbb{R}), \quad (Lx)(t) = \dot{x}(t) - D_x f(\bar{x}(t), \bar{\alpha})x(t), \quad t \in \mathbb{R},$$

satisfies the following conditions:

(i) $\dim N(L) = \dim N(L^*) = 1$ *and*

(ii) $\displaystyle\int_{-\infty}^{\infty} \langle \varphi(t), D_\alpha f(\bar{x}(t), \bar{\alpha})\rangle dt \neq 0$, *where φ spans $N(L^*)$.*

Proof Assuming that $\bar{x}$ is nondegenerate at $\bar{\alpha}$, we obtain from Definition 7.44

$$N(L) = \text{span}\{\dot{\bar{x}}\}, \quad \dim N(L) = 1.$$

Lemma 7.54 then shows $\dim(N(L^*)) = 1$, i.e. $N(L^*)) = \text{span}(\varphi)$ for $\varphi(t) = Y^*(t)\varphi_0$ and some $\varphi_0 \in (R(P) + N(Q))^\perp$. Moreover, the same definition shows that $D_\alpha f(\bar{x}, \bar{\alpha})\mu \in R(L)$ implies $\mu = 0$, hence

$$\left(\int_{-\infty}^{\infty} \langle \varphi(t), D_\alpha f(\bar{x}(t), \bar{\alpha})\rangle dt\right) \mu = 0$$

implies $\mu = 0$ by (7.84). This means that the integral in the above formula does not vanish.

Conversely, assume that conditions (i) and (ii) hold. Then using (7.84) we see that the homoclinic solution $\bar{x}$ is nondegenerate at $\bar{\alpha}$. □

Proof of Theorem 7.45 Let $J = [-T, T]$. We apply Theorem 7.48 for the spaces

$$X = X^1(J) \times \mathbb{R}, \; Y = X^0(J) \times \mathbb{R}^n \times \mathbb{R}$$

equipped with the norms

$$\begin{aligned} \|(x, \alpha)\| &:= \|x\|_1 + |\alpha|, \quad (x, \alpha) \in X, \\ \|(y, r, \tau)\| &:= \|y\|_0 + \|r\| + |\tau|, \quad (y, r, \tau) \in Y. \end{aligned}$$

Our operator is

$$(\hat{L} + F)(x, \alpha) = (\dot{x} - f(x, \alpha), x(-T) - x(T), \langle \dot{\bar{x}}(0), x(0) - \bar{x}(0)\rangle)$$

with $\xi = (x, \alpha)$ and the centre of the ball being $\xi_0 = (\bar{x}|_J, \bar{\alpha})$. The linear operator $\hat{L}$ is defined in such a way that solving the linear equation

$$\hat{L}(x, \alpha) = (y, r, \tau) \in Y$$

amounts to solving the following linear boundary value problem

$$\begin{cases} \dot{x}(t) - D_x f(\bar{x}(t), \bar{\alpha})x(t) - D_\alpha f(\bar{x}(t), \bar{\alpha})\alpha = y(t), \quad t \in J, \\ x(T) - x(-T) = r, \\ \langle \dot{\bar{x}}(0), x(0)\rangle = \tau. \end{cases} \tag{7.93}$$

The nonlinear operator is then given by

$$F(x, \alpha) = (D_x f(\bar{x}, \bar{\alpha})x + D_\alpha f(\bar{x}, \bar{\alpha})\alpha - f(x, \alpha), 0, -\langle \dot{\bar{x}}(0), \bar{x}(0)\rangle).$$

Below we show that there exists a constant $\sigma > 0$ such that for all $T > 0$ sufficiently large the boundary value problem (7.93) has a solution satisfying

$$\|x\|_1 + |\alpha| \leq \sigma^{-1}(\|y\|_0 + \|r\| + |\tau|). \tag{7.94}$$

Note that (7.93) is a linear three-point boundary value problem of dimension $n+1$ by adding the artificial equation $\dot{\alpha} = 0$. Thus we have the Fredholm alternative, i.e. the solution is unique, the operator $\hat{L}$ is invertible and $\|\hat{L}^{-1}\|$ has the bound σ^{-1}.

From the Mean Value Theorem we have for any two $(x_1, \alpha_1), (x_2, \alpha_2) \in X$ the equality

$$F(x_1, \alpha_1) - F(x_2, \alpha_2) = (d_{1,2}, 0, 0),$$

where the function $d_{1,2}$ is given by

$$d_{1,2} = \int_0^1 (D_x f(\bar{x}, \bar{\alpha}) - D_x f(\eta(s)))(x_1 - x_2) + (D_\alpha f(\bar{x}, \bar{\alpha}) - D_\alpha f(\eta(s)))(\alpha_1 - \alpha_2) ds,$$

$$\eta(s) = (x_2 + s(x_1 - x_2), \alpha_2 + s(\alpha_1 - \alpha_2)).$$

Using the smoothness of f we then find $\delta > 0$, independent of T, such that Lip $F < \sigma \leq \|\hat{L}^{-1}\|^{-1}$, i.e. condition (7.52) is satisfied on the ball $B_\delta(\xi_0)$. Finally, note that $\bar{x}$ satisfies the differential equation and the phase condition in (7.46), so that we end up with

$$\|(\hat{L} + F)(\xi_0)\| = \|\bar{x}(T) - \bar{x}(-T)\| \to 0 \quad \text{as } T \to \infty. \tag{7.95}$$

Here we use that $\bar{x}$ is the homoclinic solution to 0 at $\bar{\alpha}$. Hence, for T sufficiently large, we achieve $\|(\hat{L} + F)(\xi_0)\| \leq (\sigma - \text{Lip } F)\delta$, meaning that condition (7.53) is satisfied. Therefore, Theorem 7.48 implies that $\hat{L} + F$ has a unique zero (x_T, α_T) in $B_\delta(\xi_0)$. The estimate (7.48) follows from (7.55), (7.95) if we show the exponential decay rates for $\bar{x}(\pm T)$ as $T \to \infty$. For this purpose recall from Lemmas 7.55 and 7.54 that $\dot{\bar{x}}$ spans $N(L)$ and that

$$R(P) \cap N(Q) = \text{span}\{\dot{\bar{x}}(0)\}. \tag{7.96}$$

By the exponential dichotomies from Theorem 7.53 we obtain the estimate

$$\|\dot{\bar{x}}(t)\| \leq K e^{-\omega|t|} \|\dot{\bar{x}}(0)\|, \quad t \in \mathbb{R},$$

for any $\omega < \min(\nu_s, \nu_u)$ where K depends on ω. This leads to

$$\|\bar{x}(T)\| \leq \int_T^\infty \|\dot{\bar{x}}(t)\| dt \leq K \|\dot{\bar{x}}(0)\| \int_T^\infty e^{-\omega t} dt = \frac{K}{\omega} \|\dot{\bar{x}}(0)\| e^{-\omega T}.$$

A similar estimate holds for $\bar{x}(-T)$. In what follows we prove (7.94). Let us first apply Lemma 7.50 to modify the projectors P and Q from Theorem 7.53 without changing $R(P)$ and $N(Q)$. In view of (7.96) we can choose subspaces $V_k, k = 1, 2, 3, 4$ of $\mathbb{R}^n$, such that

$$\begin{aligned}
V_1 &= R(P) \cap N(Q), \quad \dim(V_1) = 1,, \\
V_1 \oplus V_2 &= R(P), \quad \dim(V_2) = n_s - 1, \\
V_1 \oplus V_3 &= N(Q), \quad \dim(V_3) = n - n_s - 1, \\
(R(P) + N(Q)) \oplus V_4 &= \mathbb{R}^n, \quad \dim(V_4) = 1.
\end{aligned}$$

Thus we have the direct sum $\mathbb{R}^n = V_1 \oplus V_2 \oplus V_3 \oplus V_4$ and we define modified projectors $\tilde{P}$ and $\tilde{Q}$ by

$$\tilde{P}v = v_1 + v_2, \quad \tilde{Q}v = v_2 + v_4 \quad \text{for} \quad v = v_1 + v_2 + v_3 + v_4 \in V_1 \oplus V_2 \oplus V_3 \oplus V_4.$$

Then $R(P) = R(\tilde{P})$ and $N(\tilde{Q}) = N(Q)$ hold and by Lemma 7.50 the operator L has exponential dichotomies with respect to these projectors with the same exponent. For simplicity we drop the tilde in the following and assume $\tilde{P} = P$ and $\tilde{Q} = Q$.

Next we reduce the linear boundary value problem (7.93) to a linear system of dimension $2n+1$ for the unknowns

$$z_+^0 \in V_1 \oplus V_2, \quad z_+^\infty \in R(P_u), \quad z_-^0 \in V_1 \oplus V_3, \quad z_-^\infty \in R(P_s), \quad \alpha \in \mathbb{R}. \tag{7.97}$$

Our Ansatz for a solution (x, α) of (7.93) is a generalization of (7.90) and given by

$$x(t) = \begin{cases} z_+(t) + \psi_+(t,T)\alpha + Y(t)z_+^0 + Y(t)(I-P)Y^{-1}(T)z_+^\infty, & 0 \le t \le T, \\ z_-(t) + \psi_-(t,T)\alpha + Y(t)z_-^0 + Y(t)QY^{-1}(-T)z_-^\infty, & -T \le t < 0, \end{cases} \tag{7.98}$$

where

$$z_+(t) = \int_0^T G_+(t,s)y(s)ds, \ \psi_+(t,T) = \int_0^T G_+(t,s)D_\alpha f(\bar{x}(s),\bar{\alpha})ds, 0 \le t \le T,$$

$$z_-(t) = \int_{-T}^0 G_-(t,s)y(s)ds, \ \psi_-(t,T) = \int_{-T}^0 G_-(t,s)D_\alpha f(\bar{x}(s),\bar{\alpha})ds, -T \le t < 0,$$

with Green's functions $G_\pm$ from (7.91) and (7.92). By construction, the pair (x, α) solves the differential equation in (7.93) on both intervals $[-T, 0)$ and $[0, T]$. It remains to choose the $2n+1$ unknowns (7.97) such that continuity at $t=0$ holds and the periodic boundary condition as well as the phase condition is satisfied, i.e.:

$$\begin{aligned} z_-(0) - z_+(0) &= (\psi_+(0,T) - \psi_-(0,T))\alpha + z_+^0 - z_-^0 \\ &+ \ (I-P)Y^{-1}(T)z_+^\infty - QY^{-1}(-T)z_-^\infty, \\ r + z_-(-T) - z_+(T) &= (\psi_+(T,T) - \psi_-(-T,T))\alpha + Y(T)z_+^0 - Y(-T)z_-^0 \\ &+ \ Y(T)(I-P)Y^{-1}(T)z_+^\infty - Y(-T)QY^{-1}(-T)z_-^\infty, \\ \tau - \langle \dot{\bar{x}}(0), z_+(0) \rangle &= \langle \dot{\bar{x}}(0), \psi_+(0,T)\alpha + z_+^0 + (I-P)Y^{-1}(T)z_+^\infty \rangle. \end{aligned} \tag{7.99}$$

We consider the limit $T \to \infty$ of the linear map between $(2n+1)$-dimensional spaces defined by the right-hand side and prove below that it is nonsingular. Then we conclude that (7.99) has a unique solution $(z_+^0, z_+^\infty, z_-^0, z_-^\infty, \alpha)$ for T sufficiently large, which can be estimated uniformly in T in terms of the left-hand sides $r, \tau, z_\pm(\pm T), z_\pm(0)$. By the exponential estimates of G_+ and G_- we can bound $\|z_\pm\|_0$ by $\|y\|_0$ uniformly in T. Using this in the definition (7.98) of x we find a uniform bound for $\|x\|_0$ in terms of $\|y\|_0 + \|r\| + |\tau|$. Note that the exponential dichotomies as well as the choice of unknowns in (7.97) show that $\psi_\pm$ are uniformly bounded and the following estimates hold:

$$\begin{aligned} \|Y(t)z_+^0\| + \|Y(t)(I-P)Y(T)^{-1}z_+^\infty\| &\le Ke^{-\omega t}(\|z_+^0\| + \|z_+^\infty\|), \\ \|Y(t)z_-^0\| + \|Y(t)QY(-T)^{-1}z_-^\infty\| &\le Ke^{-\omega t}(\|z_-^0\| + \|z_-^\infty\|). \end{aligned} \tag{7.100}$$

Finally, the differential equation in (7.93) has bounded coefficients, so that we can also bound $\|\dot{x}\|_0$ in terms of $\|x\|_0, |\alpha|, \|y\|_0, \|r\|, |\tau|$.

We introduce the quantities

$$\begin{aligned} \psi_+^0 &= \textstyle\int_0^\infty G_+(0,s)D_\alpha f(\bar{x}(s),\bar{\alpha})ds, & \psi_-^0 &= \textstyle\int_{-\infty}^0 G_-(0,s)D_\alpha f(\bar{x}(s),\bar{\alpha})ds, \\ \psi_+^\infty &= \lim_{T\to\infty} \psi_+(T,T), & \psi_-^\infty &= \lim_{T\to\infty} \psi_-(-T,T). \end{aligned} \tag{7.101}$$

The existence of the limits $\psi_+^\infty, \psi_-^\infty$ will be established below. In the limit $T \to \infty$, the homogeneous equation corresponding to (7.99) is the following system for the unknowns from (7.97)

$$\begin{cases} 0 = (\psi_+^0 - \psi_-^0)\alpha + z_+^0 - z_-^0, \\ 0 = (\psi_+^\infty - \psi_-^\infty)\alpha + z_+^\infty - z_-^\infty, \\ 0 = \langle \dot{\bar{x}}(0), \psi_+^0\alpha + z_+^0 \rangle. \end{cases} \tag{7.102}$$

Here we used the exponential decay from (7.100) and we employed the fact that the projectors $Y(T)(I-P)Y^{-1}(T)$ and $Y(-T)QY^{-1}(-T)$ converge to the identity on $R(P_u)$ and on $R(P_s)$, respectively, according to (7.82). The limits $\lim_{T\to\infty}\psi_+(0,T)=\psi_+^0$ and $\lim_{T\to\infty}\psi_-(0,T)=\psi_-^0$ are readily established, for example,

$$\begin{aligned}\|\psi_+(0,T)-\psi_+^0\| &= \left\|\int_T^\infty G_+(0,s)D_\alpha f(\bar{x}(s),\bar{\alpha})ds\right\| \\ &\le K\int_T^\infty e^{-\omega s}ds\|D_\alpha f(\bar{x},\bar{\alpha})\|_0 = \frac{K}{\omega}e^{-\omega T}\|D_\alpha f(\bar{x},\bar{\alpha})\|_0.\end{aligned}$$

Now let us assume that $(z_+^0, z_+^\infty, z_-^0, z_-^\infty, \alpha)$ solves the system (7.102). From the first equation we find that the function z defined by

$$z(t)=\begin{cases}\int_0^\infty G_+(t,s)D_\alpha f(\bar{x}(s),\bar{\alpha})ds\,\alpha+Y(t)z_+^0, & t\ge 0,\\ \int_{-\infty}^0 G_-(t,s)D_\alpha f(\bar{x}(s),\bar{\alpha})ds\,\alpha+Y(t)z_-^0, & t<0\end{cases}$$

is continuous at $t=0$ and a bounded solution of

$$\dot{z}-D_x f(\bar{x},\bar{\alpha})z-D_\alpha f(\bar{x},\bar{\alpha})\alpha=0.$$

By our assumption that the homoclinic orbit $(\bar{x}(t))_{t\in\mathbb{R}}$ is nondegenerate (cf. Definition 7.44) this implies $\alpha=0$ and $z_+^0=z_-^0=c\dot{\bar{x}}(0)$ for some $c\in\mathbb{R}$. From the second equation in (7.102) we then infer $z_+^\infty=0$, $z_-^\infty=0$ since the spaces $R(P_u)$ and $R(P_s)$ are complementary. Finally, the last equation in (7.102) yields

$$0=\langle\dot{\bar{x}}(0),z_+^0\rangle=|c|\|\dot{\bar{x}}(0)\|^2.$$

Thus we have $c=0$ and $z_+^0=z_-^0=0$. Summing up, we have shown that the homogeneous system (7.102) has only the trivial solution and that the associated linear map is invertible.

It remains to identify the limits ψ_+^∞ and ψ_-^∞ in (7.101). We define $g(t)=D_\alpha f(\bar{x}(t),\bar{\alpha})$ and $g(\infty)=\lim_{t\to\pm\infty}g(t)=D_\alpha f(0,\alpha)$ and we claim

$$\psi_+^\infty=-A^{-1}P_s g(\infty),\quad \psi_-^\infty=-A^{-1}P_u g(\infty).$$

We consider only the first limit since the second one follows by essentially the same arguments. Recall from Theorem 7.53 that there exist constants $C, T_0\ge 0$ such that for $T_0\le\frac{T}{2}\le s\le T$,

$$\|G_+(T,s)-e^{(T-s)A}P_s\|\le C\left[e^{-\omega s}b_0+e^{-\omega(T-s)}b_{\frac{T}{2}}\right],$$

where $b_\rho=\sup_{\tau\ge\rho}\|M(\tau)-A\|$. With this we can estimate as follows:

$$\begin{aligned}&\|\psi_+(T,T)+A^{-1}P_s g(\infty)\|\\ &=\left\|\int_0^T G_+(T,s)g(s)ds-\int_0^T e^{(T-s)A}P_s g(\infty)ds+A^{-1}e^{TA}P_s g(\infty)\right\|\\ &\le\left\|\int_0^{T/2}G_+(T,s)g(s)ds\right\|+\left\|\int_0^{T/2}e^{(T-s)A}P_s g(\infty)ds\right\|\\ &\quad+\|A^{-1}e^{TA}P_s g(\infty)\|+\left\|\int_{T/2}^T G_+(T,s)(g(s)-g(\infty))ds\right\|\\ &\quad+\left\|\int_{T/2}^T(G_+(T,s)-e^{(T-s)A}P_s)ds\,g(\infty)\right\|\end{aligned}$$

$$
\begin{aligned}
&\leq \int_0^{T/2} K\mathrm{e}^{-\omega(T-s)}ds\|g\|_0 + \int_0^{T/2} K_+\mathrm{e}^{-\omega_+(T-s)}ds\|g\|_0 \\
&+ \ K_+\mathrm{e}^{-\omega T}\|A^{-1}\|\|g(\infty)\| + \int_{T/2}^{T} K\mathrm{e}^{-\omega(T-s)}ds \sup_{\tau\geq T/2} \|g(\tau)-g(\infty)\| \\
&+ \ C\int_{T/2}^{T} \mathrm{e}^{-\omega s}b_0 + \mathrm{e}^{-\omega(T-s)}b_{\frac{T}{2}}\,ds.
\end{aligned}
$$

It is now obvious that all five terms on the right-hand side converge to zero as $T \to \infty$.
This completes the proof of (7.94) and that of Theorem 7.45. □

7.6.5 *Proof of Hamiltonian Homoclinic Bifurcation Theorem*

Let us show that $(\bar{x}, \bar{\alpha} = 0)$ is a nondegenerate solution of (7.51) in the sense of Definition 7.44. For this purpose we verify conditions (i) and (ii) of Lemma 7.55 for the operator

$$
Lx = \dot{x} - D_x f(\bar{x}, 0)x = \dot{x} - D_x g(\bar{x})x.
$$

By the nondegeneracy of $\bar{x}$ we have $N(L) = \mathrm{span}\{\dot{\bar{x}}\}$. From Lemma 7.54 we then infer $\dim(N(L^*)) = \dim(N(L)) = 1$. Next, we claim that $N(L^*)$ is spanned by $\varphi = J_{2m}\dot{\bar{x}}$. This follows from the equality

$$
\begin{aligned}
\dot{\varphi} = J_{2m}\ddot{\bar{x}} = J_{2m}D_x g(\bar{x})\dot{\bar{x}} &= J_{2m}\begin{pmatrix} D_q(D_pH)^{\mathrm{T}} & D_p^2H \\ -D_q^2H & -D_p(D_qH)^{\mathrm{T}} \end{pmatrix}(\bar{x})\,\dot{\bar{x}} \\
&= -\begin{pmatrix} D_p(D_qH)^{\mathrm{T}} & -D_q^2H \\ D_p^2H & -D_p(D_qH)^{\mathrm{T}} \end{pmatrix}(\bar{x})J_{2m}\dot{\bar{x}} = -D_x g(\bar{x})^{\mathrm{T}} J_{2m}\dot{\bar{x}} \\
&= -\,D_x g(\bar{x})^{\mathrm{T}}\varphi.
\end{aligned}
$$

Note that the last equation uses the symmetry of the Hessians D_q^2H, D_p^2H and the fact that $D_p(D_qH)^{\mathrm{T}}$ is the transpose of $D_q(D_pH)^{\mathrm{T}}$. Since $D_\alpha f(\bar{x}, 0) = -J_{2m}\dot{\bar{x}}$ and $\dot{\bar{x}}$ does not vanish we obtain

$$
\int_{-\infty}^{\infty} \langle \varphi, D_\alpha f(\bar{x}, 0)\rangle dt = -\int_{-\infty}^{\infty} \|\varphi(t)\|^2 dt < 0.
$$

Hence condition (ii) of Lemma 7.55 is satisfied. For T sufficiently large, Theorem 7.45 yields a branch of $2T$-periodic solutions $x_T = (q_T, p_T) \in C^1([-T, T], \mathbb{R}^{2m})$ of (7.51) with parameter values α_T. To show that α_T vanishes, we consider the Hamiltonian along the orbit:

$$
\begin{aligned}
\tfrac{d}{dt}H(x_T) &= D_qH(x_T)\dot{q}_T + D_pH(x_T)\dot{p}_T \\
&= D_qH(x_T)((D_pH)^{\mathrm{T}} + \alpha_T(D_qH)^{\mathrm{T}})(x_T) \\
&\quad + D_pH(x_T)(-(D_qH)^{\mathrm{T}} + \alpha_T(D_pH)^{\mathrm{T}})(x_T) \\
&= \alpha_T(\|D_qH(x_T)\|^2 + \|D_pH(x_T)\|^2).
\end{aligned}
$$

If α_T is nonzero, the value $H(x_T(t))$ will be either strictly increasing or strictly decreasing, which contradicts the periodicity in both cases. Hence we have $\alpha_T = 0$ for all T sufficiently large. Together with the estimate (7.48) this completes the proof of Theorem 7.47. □

Remark In case $m = 1$, $n = 2$, the nondegeneracy of a homoclinic orbit $\bar{x}$ in the sense of Definition 7.46 follows from the saddle property of the origin. According to Theorem 7.53 both projectors P_s and P_u then have rank 1. Correspondingly, the one-sided operators

$L|_{[0,\infty)}$ and $L|_{(-\infty,0]}$ have one-dimensional spaces of bounded solutions in forward, resp., backward time, see Lemma 7.50. Since elements of $N(L)$ lie in both spaces when restricted to $[0,\infty)$, resp., $(-\infty,0]$, we conclude that there is no further solution in $N(L)$ which is independent of $\dot{\bar{x}}$. ◊

7.7 Appendix B: Symbolic Dynamics Near Homoclinic Orbit

In this Appendix we consider smooth dynamical systems $\{\mathbb{Z}, \mathbb{R}^n, f^k\}$ for which a transversal homoclinic orbit exists, i.e. an orbit which converges in forward and in backward time to a fixed point. Transversality means that the fixed point is a hyperbolic saddle and that its stable and unstable manifold intersect transversally (at one point and hence) at all points of the homoclinic orbit. Figure 7.28b displays the resulting Poincaré homoclinic structure in the two-dimensional case. Here our goal is to show that for general $n \geq 2$ there exists a compact and invariant set, containing the homoclinic orbit, on which f is conjugate to the shift map σ restricted to a certain subset of the set Ω_m of all symbolic sequences of m integers $\{0,\dots,m-1\}$, see Chap. 1. The specific subset used here is a *subshift* associated with the topological Markov chain from Example 1.4. It retains all the chaotic features of the full shift as discussed in Sects. 7.1 and 7.3. We will first specify the assumptions and the statement of this famous result. Then, by contrast to the geometric arguments in Sects. 7.1 and 7.3, we will use analytical tools based on the theory of exponential dichotomies for maps to provide the proof. This theory is quite analogous to the continuous-time theory in Sect. 7.6.2. Therefore, we occasionally skip details.

7.7.1 Subshifts and the Main Result

Recall from Example 1.4 and Sects. 7.1 and 7.3 the following facts and definitions. The space Ω_m is compact with respect to the metric

$$d(\omega,\theta) = \sum_{k\in\mathbb{Z}} \frac{|\omega_k - \theta_k|}{2^{|k|}}, \quad \omega,\theta \in \Omega_m. \tag{7.103}$$

Definition 7.56 *Any closed subset $\Omega \subset \Omega_m$ which is invariant under the* ***shift***

$$\sigma(\omega)_i = \omega_{i+1}, \quad i \in \mathbb{Z}, \quad \omega \in \Omega_m$$

is called a ***subshift****, and the triple $(\mathbb{Z}, \Omega, \sigma^k)$ is called the* ***subshift dynamical system****. For any binary $m \times m$-matrix $A = (a_{ij})_{i,j=0,\dots,m-1}$, $a_{ij} \in \{0,1\}$, the subshift*

$$\Omega_A = \{\omega \in \Omega_m : a_{\omega_i,\omega_{i+1}} = 1 \;\text{ for all }\; i \in \mathbb{Z}\} \tag{7.104}$$

defines a subshift dynamical system $\{\mathbb{Z}, \Omega_A, \sigma^k\}$, called the ***topological Markov chain*** *associated with A.*

It is not difficult to see that Ω_A defined in (7.104) is closed with respect to the metric (7.103) and is invariant under the shift. For our purpose the following binary $m \times m$ matrix is relevant; cf. (1.7):

$$A(m) = \begin{pmatrix} 1 & 1 & 0 & \dots & 0 \\ 0 & 0 & 1 & \ddots & \vdots \\ \vdots & & \ddots & \ddots & 0 \\ 0 & & & \ddots & 1 \\ 1 & 0 & \dots & \dots & 0 \end{pmatrix}. \tag{7.105}$$

The symbol 0 in a sequence from $\Omega_{A(m)}$ is either followed by 0 or 1 while a symbol $j \in \{1, \dots, m-1\}$ is successively followed by $j+1, \dots, m-1$ until it finally jumps back to 0, see the associated Markov graph in Fig. 1.2. More formally, we have

$$\Omega_{A(m)} = \{\omega \in \Omega_m : \omega_i = 0 \Rightarrow \omega_{i+1} \in \{0,1\}, \omega_i \geq 1 \Rightarrow \omega_{i+1} = \omega_i + 1 (\text{mod } m)\}.$$

With these preparations we can formulate the main result.

Theorem 7.57 *Let $f \in C^1(\mathcal{U}, \mathbb{R}^n)$ be a diffeomorphism defined on some open set $\mathcal{U} \subset \mathbb{R}^n$ and let $\{\bar{x}_k = f^k(\bar{x}_0) : k \in \mathbb{Z}\}$ be a nontrivial transversal homoclinic orbit in $\mathcal{U}$ converging to a fixed point $\xi \in \mathcal{U}$. Then there exist $m \in \mathbb{N}$ and a neighbourhood $\mathcal{O}$ of $H = \{\xi\} \cup \{\bar{x}_k\}_{k \in \mathbb{Z}}$ such that its maximal invariant set*

$$\Lambda = \{x \in \mathcal{O} : f^k(x) \in \mathcal{O} \quad \text{for all } k \in \mathbb{Z}\} \tag{7.106}$$

is compact, and there exists a homeomorphism

$$h : \Omega_{A(m)} \to \Lambda \tag{7.107}$$

such that

$$h \circ \sigma(\omega) = f \circ h(\omega) \tag{7.108}$$

for all $\omega \in \Omega_{A(m)}$.

By a nontrivial homoclinic orbit we mean a non-constant orbit satisfying

$$\lim_{k \to \pm\infty} \bar{x}_k = \xi.$$

The property of transversality will be made precise in Definition 7.63. Also note that the map h from $\Omega_{A(m)}$ to Λ is such that h^{-1} will satisfy $\sigma \circ h^{-1} = h^{-1} \circ f$; cf. Lemma 7.36. The reason for starting conceptually with h rather than with h^{-1} is that we construct the map h and then show that it is invertible.

The intuition behind the theorem is that the map h associates to any element $\omega \in \Omega_{A(m)}$ an orbit in $\mathcal{O}$ such that a sequence of zeroes specifies when the orbit stays close to the fixed point while a sequence $1, \dots, m-1, 0$ stands for an excursion close to the homoclinic orbit, see Fig. 7.33 for an illustration. In particular, every sequence $\omega \in \Omega_{A(m)}$ which has leading and trailing zeroes (i.e. $\omega_j = 0$ for $|j| \geq k$ and some $k \in \mathbb{N}$) is sent by h to a homoclinic orbit of the original system $\{\mathbb{Z}, \Lambda, f^k\}$. These infinitely many homoclinic orbits coexist with the infinitely many periodic orbits of arbitrarily large period which one can construct as in Theorem 7.13.

Remark The assumption of a diffeomorphism is not very restrictive. For example, consider $f \in C^1(V, \mathbb{R}^n)$ for some open neighbourhood V of $H = \{\xi\} \cup \{\bar{x}_k : k \in \mathbb{Z}\}$ and assume $Df(\xi)$ to be hyperbolic and $Df(\bar{x}_k)$, $k \in \mathbb{Z}$ to be invertible. Then f will be diffeomorphic on a set $\mathcal{U}$ which is the union of a small neighbourhood of ξ and of sufficiently small balls around the points outside of this neighbourhood. ◇

Corollary 7.58 *Under the assumptions of* Theorem 7.57, *the dynamical system $(\mathbb{Z}, \Lambda, f^k)$ is chaotic in the sense of* Definition 7.22.

Proof As in Sect. 7.1 it is enough to show sensitive dependence on initial conditions and topological transitivity for the symbolic dynamical system $\{\mathbb{Z}, \Omega_{A(m)}, \sigma^k\}$. By the conjugacy (7.108) the system $\{\mathbb{Z}, \Lambda, f^k\}$ inherits these properties. Given any $\varepsilon > 0$ and $\omega \in \Omega_{A(m)}$ we take $k \in \mathbb{N}$ such that $m2^{1-k} \le \varepsilon$. Since symbolic sequences in $\Omega_{A(m)}$ always return to 0 after m steps there exists $k_0 = \min\{j > k : \omega_j = 0\} \le k + m$. Then we define $\theta \in \Omega_m$ by $\theta_j = \omega_j$ for $j \le k_0$ and $\theta_{k_0+1} = 1 - \omega_{k_0+1}$. Since $\omega_{k_0+1} \in \{0, 1\}$ and $\omega_{k_0} = 0$ we can continue θ_j for $j > k_0 + 1$ such that $\theta \in \Omega_{A(m)}$ holds. With this setting we obtain $d(\omega, \theta) \le m2^{1-k} \le \varepsilon$ as well as $d(\sigma^{k_0+1}\omega, \sigma^{k_0+1}\theta) \ge 1$.

A dense orbit for $(\mathbb{Z}, \Omega_{A(m)}, \sigma^k)$ is constructed as in the proof of Theorem 7.13 by patching together all finite segments from sequences in $\Omega_{A(m)}$ which begin and end with the symbol 0. The last condition ensures that the concatenated sequence belongs to $\Omega_{A(m)}$ again. Finally, the construction of θ above shows that there are no isolated points in $\Omega_{A(m)}$. These properties guarantee that $\{\mathbb{Z}, \Omega_{A(m)}, \sigma^k\}$ is topologically transitive. □

7.7.2 *Exponential Dichotomies and Transversality*

Similar to Sect. 7.6.2, we collect some basic results on exponential dichotomies, now for maps, which form the main tool for proving Theorem 7.57. Proofs are omitted when they are rather analogous to the continuous-time case treated in Sect. 7.6. Consider a linear discrete-time dynamical system

$$x_{j+1} = A_j x_j, \quad j \in \mathbb{Z}, \tag{7.109}$$

where we assume the matrices $A_j \in \mathbb{R}^{n \times n}$ to be uniformly bounded and to have uniformly bounded inverses

$$\|A_j\|, \|A_j^{-1}\| \le C \quad \text{for all } j \in \mathbb{Z}.$$

The solution of the initial value problem

$$x_{j+1} = A_j x_j, \quad x_k = y \in \mathbb{R}^n$$

is then given by $x_j = \Phi(j, k)y$, $j \in \mathbb{Z}$, where Φ is the solution operator defined by

$$\Phi(j,k) = \begin{cases} A_{j-1} \cdot \ldots \cdot A_k, & j > k, \\ I_n, & j = k, \\ A_j^{-1} \cdot \ldots \cdot A_{k-1}^{-1}, & j < k. \end{cases}$$

Note the semigroup property $\Phi(j, k) \circ \Phi(k, \ell) = \Phi(j, \ell)$ for all $j, k, \ell \in \mathbb{Z}$. In the following we will call any set of the type $J = [n_-, \infty) \cap \mathbb{Z}$ for some $n_- > -\infty$ or $J = (-\infty, n_+] \cap \mathbb{Z}$ for some $n_+ < \infty$ or $J = \mathbb{Z}$ an *unbounded discrete interval.*

Definition 7.59 *The system* (7.109) *has an* ***exponential dichotomy*** *on an unbounded discrete interval J if there exist constants $K, \alpha > 0$ and projectors P_j, $j \in J$, with the following properties for $j, k \in J$:*

$$P_j \Phi(j,k) = \Phi(j,k) P_k, \tag{7.110}$$

$$\|\Phi(j,k)P_k\| \le K e^{-\alpha(j-k)}, \quad \text{if } j \ge k, \tag{7.111}$$

$$\|\Phi(j,k)(I - P_k)\| \le K e^{-\alpha(k-j)}, \quad \text{if } j \le k. \tag{7.112}$$

The triple $(K, \alpha, (P_j)_{j \in J})$ is called the data of the exponential dichotomy.

As in the continuous-time case, the existence of an exponential dichotomy does not depend on the choice of matrix norm. Just the constant K must be adapted when taking a different norm. In any case, we have $K \geq 1$ since (7.111) and (7.112) with $j = k$ imply $K \geq \max(\|P_k\|, \|I - P_k\|) \geq 1$, where the last inequality holds for every projector.

In case $j = k + 1$ the invariance condition (7.110) reads

$$P_{k+1}A_k = A_k P_k, \quad k \in \mathbb{Z}. \tag{7.113}$$

By an induction one shows that this relation is also sufficient for the condition (7.110) to hold; see Exercise 7.9. Moreover, by the invertibility of A_k, we have $P_{k+1} = A_k P_k A_k^{-1}$, hence the projectors P_k and P_{k+1} are similar and have the same rank.

In the autonomous case we have $A_j = A$ for all $j \in \mathbb{Z}$. Then we obtain from Theorem 2.13 that an exponential dichotomy holds with exponent $\alpha = -\log(\rho)$ if A is hyperbolic. In fact, the hyperbolicity of A is also a necessary condition.

Further, as in Lemma 7.50, the range of the projectors is unique if J is unbounded above and the null space is unique if J is unbounded below. The following holds:

$$R(P_j) = \{y \in \mathbb{R}^n : \text{there exists } C > 0 \text{ with } \|\Phi(k, j)y\| \leq C \text{ for all } k \geq j\}, \tag{7.114}$$
$$N(P_j) = \{y \in \mathbb{R}^n : \text{there exists } C > 0 \text{ with } \|\Phi(k, j)y\| \leq C \text{ for all } k \leq j\}. \tag{7.115}$$

In the following we use the Banach space of bounded sequences in $\mathbb{R}^n$:

$$X_{\mathbb{Z}} = \{x = (x_j)_{j\in\mathbb{Z}} : \|x\|_\infty = \sup_{j\in\mathbb{Z}} \|x_j\| < \infty\}. \tag{7.116}$$

An exponential dichotomy on $\mathbb{Z}$ allows to construct a bounded solution of an inhomogeneous equation as shown in the following analogue of Proposition 7.52.

Proposition 7.60 *Assume that* (7.109) *has an exponential dichotomy on* $\mathbb{Z}$ *with data* $(K, \alpha, (P_j)_{j\in\mathbb{Z}})$. *Then for each* $r \in X_{\mathbb{Z}}$ *the system*

$$x_{j+1} = A_j x_j + r_j, \quad j \in \mathbb{Z}$$

has a unique solution $x \in X_{\mathbb{Z}}$ *given by*

$$x_j = \sum_{k\in\mathbb{Z}} G(j, k + 1)r_k, \quad j \in \mathbb{Z} \tag{7.117}$$

with Green's function defined by

$$G(j, k) = \begin{cases} \Phi(j, k)P_k, & j \geq k, \\ \Phi(j, k)(P_k - I), & j < k. \end{cases}$$

The solution satisfies the estimate

$$\|x\|_\infty \leq K\frac{1 + e^{-\alpha}}{1 - e^{-\alpha}}\|r\|_\infty. \tag{7.118}$$

Proof To prove uniqueness of the solution, let $x \in X_{\mathbb{Z}}$ solve the homogenous system (7.109). Since x is bounded, we infer from (7.114) that $x_0 \in R(P_0)$ holds and similarly, we find $x_0 \in N(P_0)$ from (7.115). Thus we have $x_0 = 0$ which then implies $x_j = 0$ for all $j \in \mathbb{Z}$.

Defining x_j by (7.117) for $j \in \mathbb{Z}$ we obtain

$$\begin{aligned}
x_{j+1} &= \sum_{k \le j} \Phi(j+1, k+1) P_{k+1} r_k + \sum_{k > j} \Phi(j+1, k+1)(P_{k+1} - I) r_k \\
&= \sum_{k<j} A_j \Phi(j, k+1) P_{k+1} r_k + P_{j+1} r_j + (I - P_{j+1}) r_j \\
&\quad + \sum_{k \ge j} A_j \Phi(j, k+1)(P_{k+1} - I) r_k \\
&= A_j x_j + r_j.
\end{aligned}$$

Finally, the estimate (7.118) follows from the exponential dichotomy

$$\|x_j\| \le \sum_{k \in \mathbb{Z}} K e^{-\alpha|j-k|} \|r_k\| \le K \frac{1 + e^{-\alpha}}{1 - e^{-\alpha}} \|r\|_\infty$$

by taking the maximum over $j \in \mathbb{Z}$. □

The most important result is a *"roughness" theorem*, similar to Theorem 7.51, which specifies the persistence of an exponential dichotomy under small perturbations.

Theorem 7.61 *Let the system (7.109) have an exponential dichotomy with data $(K, \alpha, (P_j)_{j \in J})$ on an unbounded discrete interval J and let $0 < \beta < \alpha$. Then there exist positive constants $\gamma = \gamma(K, \alpha, \beta)$ and $\mu = \mu(K, \alpha, \beta)$ such that for any sequence $B_j \in \mathbb{R}^{n \times n}$, $j \in J$, satisfying $\sup_{j \in J} \|B_j\| \le \gamma$, the perturbed system*

$$x_{j+1} = (A_j + B_j) x_j, \quad j, j+1 \in J$$

has an exponential dichotomy on J with data $(2K, \beta, (\tilde{P}_j)_{j \in J})$, where

$$\sup_{j \in J} \|P_j - \tilde{P}_j\| \le 2K\mu \sup_{j \in J} \|B_j\|. \tag{7.119}$$

If J is unbounded from above then one has

$$\lim_{j \to \infty} P_j - \tilde{P}_j = 0 \quad \textit{if} \quad \lim_{j \to \infty} B_j = 0,$$

and a corresponding result holds if J is unbounded from below. □

The proof of the theorem is rather similar to that of Theorem 7.51 and will be omitted. Instead, we proceed with an application to the variational equation along a homoclinic orbit.

Proposition 7.62 *Let $(\bar{x}_k)_{k \in \mathbb{Z}}$ be a homoclinic orbit of $f \in C^1(\mathcal{U}, \mathbb{R}^n)$ converging to a saddle $\xi \in \mathcal{U}$ and let P^s be the stable projector associated with $Df(\xi)$. Then the variational equation*

$$x_{j+1} = Df(\bar{x}_j) x_j \tag{7.120}$$

has an exponential dichotomy both on $\mathbb{Z}_-$ with suitable data $(K, \bar{\alpha}, (P_j^-)_{j \in \mathbb{Z}_-})$ and on $\mathbb{Z}_+$ with data $(K, \bar{\alpha}, (P_j^+)_{j \in \mathbb{Z}_+})$. Moreover, the projectors P_j^+ and P_j^- have the same rank $n_s = \operatorname{rank}(P^s)$ and satisfy

$$\lim_{j \to -\infty} P_j^- = P^s = \lim_{j \to \infty} P_j^+. \tag{7.121}$$

Proof Let us define

$$A_j = Df(\xi) \quad \text{and} \quad B_j = Df(\bar{x}_j) - Df(\xi).$$

Then $Df(\bar{x}_j) = A_j + B_j$ holds and $B_j \to 0$ as $j \to \pm\infty$. So Theorem 7.61 applies to (7.120) on the interval $J_+ = [n_-, \infty) \cap \mathbb{Z}$ for $n_- \in \mathbb{Z}_+$ sufficiently large and to $J_- = (-\infty, n_+] \cap \mathbb{Z}$ for $n_+ \in \mathbb{Z}_-$ sufficiently negative. This establishes the exponential dichotomies on $J_\pm$ with suitable data $(\tilde{K}, \bar{\alpha}, (P_j^\pm)_{j\in J_\pm})$ as well as the limit behaviour (7.121). The exponent satisfies $0 < \bar{\alpha} < \alpha = -\log(\rho)$ where ρ is determined by the hyperbolic matrix $Df(\xi)$ as above. Then we continue the projectors to $[0, n_-) \cap \mathbb{Z}$ and to $(n_+, 0] \cap \mathbb{Z}$ by using the relation (7.110):

$$P_j^+ = \Phi(j, n_-) P_{n_-}^+ \Phi(n_-, j), \quad 0 \le j < n_-,$$
$$P_j^- = \Phi(j, n_+) P_{n_+}^- \Phi(n_+, j), \quad n_+ < j \le 0.$$

With these settings, the exponential dichotomies still hold on the semiaxes $\mathbb{Z}_\pm$ with the same exponent $\bar{\alpha}$ but a new constant K (cf. the proof of Theorem 7.53). Finally, the equality of ranks follows from the limits in (7.121). □

It is important to note that the projectors need not fit together at $j = 0$, i.e. we have $P_0^+ \neq P_0^-$, in general. However, the range $R(P_0^+)$ and the null space $N(P_0^-)$ are uniquely determined by (7.114) and (7.115). In particular, if Φ has an exponential dichotomy on $\mathbb{Z}$ then the projectors are uniquely determined.

The characterizations (7.114), (7.115) also show that the following definition is independent of the choice of projectors for the dichotomy.

Definition 7.63 *A homoclinic orbit $(\bar{x}_k)_{k\in\mathbb{Z}}$ of $f \in C^1(\mathcal{U}, \mathbb{R}^n)$ converging to a fixed point $\xi \in \mathcal{U}$ is called **transversal** if ξ is a saddle and the projectors from* Proposition 7.62 *satisfy*

$$R(P_0^+) \cap N(P_0^-) = \{0\}. \tag{7.122}$$

Condition (7.122) has an immediate geometric interpretation. One can show that the stable manifold of ξ has tangent space $R(P_0^+)$ at $\bar{x}_0$ and the unstable manifold has tangent space $N(P_0^-)$ at $\bar{x}_0$. Therefore, condition (7.122) specifies the transversal intersection of the two manifolds at $\bar{x}_0$. Further, note that the projectors P_j^+, $j \in \mathbb{Z}_+$, and P_j^-, $j \in \mathbb{Z}_-$, have rank $n_s = \mathrm{rank}(P^s)$ due to (7.121), so that (7.122) implies $\mathbb{R}^n = R(P_0^+) \oplus N(P_0^-)$. This situation is different from the linearization about a continuous homoclinic orbit in Lemma 7.55 where the tangent spaces have a nontrivial intersection given by the tangent of the orbit.

Similar to Lemma 7.50, if (7.122) holds we can modify the null space $N(P_0^+)$ to $N(P_0^-)$ and the range $R(P_0^-)$ to $R(P_0^+)$ while keeping the exponential dichotomies on $\mathbb{Z}_+$ and on $\mathbb{Z}_-$. Thus we obtain an exponential dichotomy on the whole axis $\mathbb{Z}$ with the same exponent $\bar{\alpha}$ and a new constant $\bar{K}$. This allows us to drop the upper indices $\pm$ of the projectors in the following.

Lemma 7.64 *Let $(\bar{x}_k)_{k\in\mathbb{Z}}$ be a homoclinic orbit of $f \in C^1(\mathcal{U}, \mathbb{R}^n)$ converging to a fixed point $\xi \in \mathcal{U}$. Then the orbit is transversal if and only if ξ is a saddle and the variational equation* (7.120) *has an exponential dichotomy on $\mathbb{Z}$.* □

Let $\bar{\Phi}$ denote the solution operator of the variational Eq. (7.120) and let $\Phi_\infty(j, k) = Df(\xi)^{j-k}$ be the solution operator for the linearization at ξ. For later use, we summarize and quantify their dichotomy properties for $j, k \in \mathbb{Z}$, given a transversal homoclinic orbit. Due to Propositions 7.64 and 7.62 the following holds:

$$\bar{P}_j \bar{\Phi}(j, k) = \bar{\Phi}(j, k) \bar{P}_k, \tag{7.123}$$
$$P^s \Phi_\infty(j, k) = \Phi_\infty(j, k) P^s, \tag{7.124}$$
$$\|\bar{\Phi}(j, k)\bar{P}_k\|, \|\Phi_\infty(j, k) P^s\| \le \bar{K} e^{-\bar{\alpha}(j-k)}, \quad \text{if } j \ge k, \tag{7.125}$$
$$\|\bar{\Phi}(j, k)(I - \bar{P}_k)\|, \|\Phi_\infty(j, k)(I - P^s)\| \le \bar{K} e^{-\bar{\alpha}(k-j)}, \quad \text{if } j \le k. \tag{7.126}$$

As can be seen in the proof of Proposition 7.62, the operator Φ_∞ satisfies estimates with a smaller constant and a larger exponent, but (7.125) and (7.126) are good enough for our purposes.

7.7.3 Pseudo-Orbits and Exponential Dichotomies

In this section, we consider the setting of Theorem 7.57 and begin with the construction of h in (7.107). We shall associate to any symbolic sequence $\omega \in \Omega_{A(m)}$ suitable points $(y_j)_{j\in\mathbb{Z}}$ in the neighbourhood of $H = \{\xi\} \cup \{\bar{x}_k\}_{k\in\mathbb{Z}}$ which form a so-called pseudo-orbit, i.e. $y_{j+1} - f(y_j)$ is small uniformly in $j \in \mathbb{Z}$. Finding a true orbit of f near such a pseudo-orbit is the topic of a general theory called *shadowing*. Rather than expanding on this general theory, we will find these orbits in our specific situation by using the theory of exponential dichotomies.

Since the homoclinic orbit is nontrivial and f is a diffeomorphism, we have $\bar{x}_k \neq \xi$ for all $k \in \mathbb{Z}$. For every $m \in \mathbb{N}$ we determine a subset H_m of H which consists of $m-1$ points and forms a "principal part" of the orbit, i.e.:

$$H_m = \{\bar{x}_{a(m)+1}, \ldots \bar{x}_{a(m)+m-1}\}, a(m) = -\lfloor \tfrac{m}{2} \rfloor, a(m) + m - 1 = \lfloor \tfrac{m-1}{2} \rfloor. \tag{7.127}$$

So the bounds $a(m), a(m)+m$ for the orbit indices are given by $-\frac{m}{2}, \frac{m}{2}$ for m even and by $-\frac{m-1}{2}, \frac{m+1}{2}$ for m odd. It is instructive, though not necessary, to number the points of the orbit such that $\|\bar{x}_0 - \xi\| = \sup_{j\in\mathbb{Z}} \|\bar{x}_j - \xi\|$, i.e. $\bar{x}_0$ is the point furthest from the fixed point. In any case, our construction immediately leads to the following lemma.

Lemma 7.65 *For each $\varepsilon > 0$ there exists some $m \in \mathbb{N}$ such that $\|\bar{x}_j - \xi\| \le \varepsilon$ holds for all $j \le a(m)+1$ and $j \ge a(m)+m$. Moreover, for $m_1 \le m_2$ we have*

$$[a(m_1), a(m_1)+m_1] \subseteq [a(m_2), a(m_2)+m_2]. \tag{7.128}$$

□

For any $\omega \in \Omega_{A(m)}$ and $m \in \mathbb{N}$ we define the *pseudo-orbit* $y = (y_j)_{j\in\mathbb{Z}}$ as follows:

$$y_j = y_j(\omega, m) = \begin{cases} \xi, & \text{if } \omega_j = 0, \\ \bar{x}_{a(m)+\omega_j}, & \text{if } \omega_j \in \{1, \ldots, m-1\}. \end{cases} \tag{7.129}$$

For this pseudo-orbit the values $y_{j+1} - f(y_j)$ are nonzero whenever $\omega_j = 0$ is followed by $\omega_{j+1} = 1$ or whenever $\omega_j = m-1$, $\omega_{j+1} = 0$. In the following we show how to obtain an exponential dichotomy for the linearization about such a pseudo-orbit. This is not obvious, since we need projectors that satisfy the invariance condition (7.110) despite the presence of jumps. To construct these, we use the following lemma.

Lemma 7.66 *Let $B \in \mathbb{R}^{n\times n}$ be invertible and let three projectors $P_{-1}, P_0, Q_0 \in \mathbb{R}^{n\times n}$ be given with $BP_{-1} = P_0 B$. Then the perturbed matrix*

$$B_0 = (I + \Delta)B, \quad \textit{where } \Delta = 2Q_0P_0 - P_0 - Q_0$$

satisfies $B_0 P_{-1} = Q_0 B_0$ and the estimate

$$\|\Delta\| \le (\|Q_0\| + \|P_0\|)\|P_0 - Q_0\|. \tag{7.130}$$

If $\|\Delta\| < 1$ then B_0 is invertible, it maps $N(P_{-1})$ into $N(P_0)$ and $R(P_{-1})$ into $R(P_0)$ and satisfies the estimates

$$\|B_0\| \le (1+\|\Delta\|)\|B\|, \quad \|B_0^{-1}\| \le \frac{\|B^{-1}\|}{1-\|\Delta\|}. \tag{7.131}$$

Proof A direct computation shows

$$I+\Delta = Q_0P_0 + (I-Q_0)(I-P_0) = I + Q_0(P_0-Q_0) + (Q_0-P_0)P_0.$$

From the first equality and our assumption we obtain

$$B_0P_{-1} = (I+\Delta)P_0B = Q_0P_0B = Q_0(I+\Delta)B = Q_0B_0,$$

while the second equality leads to the estimate (7.130). Moreover, we conclude that $I+\Delta$ maps $N(P_0)$ and $R(P_0)$ into $N(Q_0)$ and $R(Q_0)$, respectively. If, in addition, we have $\|\Delta\| < 1$ then $I+\Delta$ is invertible and $\|(I+\Delta)^{-1}\| \le \frac{1}{1-\|\Delta\|}$. Thus the projectors P_0 and Q_0 have the same rank and we obtain that $B_0 : N(P_{-1}) \to N(Q_0)$ is bijective. The estimate (7.131) follows from $B_0^{-1} = B^{-1}(I+\Delta)^{-1}$. □

Theorem 7.67 *Let $(\bar{K}, \bar{\alpha}, (\bar{P}_j)_{j\in\mathbb{Z}})$ and $(\bar{K}, \bar{\alpha}, P^s)$ be the data of the dichotomies associated with, respectively, $\bar{\Phi}$ and Φ_∞ and recall (7.123)–(7.126). Further, let α, β be given with $0 < \beta < \alpha < \bar{\alpha}$. Then there exists $m_{\mathrm{ED}} \in \mathbb{N}$ such that for all $m \ge m_{\mathrm{ED}}$ and for all $\omega \in \Omega_{A(m)}$ the variational equation*

$$x_{j+1} = Df(y_j(\omega, m))x_j, \quad j \in \mathbb{Z} \tag{7.132}$$

(with y_j defined by (7.129)) has an exponential dichotomy with data $(8\bar{K}^4, \alpha, (Q_j)_{j\in\mathbb{Z}})$. For any $\bar{\varepsilon} > 0$ there exists $\overline{m} \ge m_{\mathrm{ED}}$ such that for all $m \ge \overline{m}$ the dichotomy projectors satisfy for $j \in \mathbb{Z}$

$$\begin{aligned} \|Q_j - P^s\| &\le \bar{\varepsilon}, \ \text{if} \ \ \omega_j = 0, \\ \|Q_j - \bar{P}_{a(m)+k}\| &\le \bar{\varepsilon}, \ \text{if} \ \ \omega_j = k \in \{1, \ldots, m-1\}. \end{aligned} \tag{7.133}$$

There exists an $\varepsilon_{\mathrm{ED}} > 0$ such that the system

$$z_{j+1} = \int_0^1 Df(x_j + \tau d_j)d\tau z_j, \quad j \in \mathbb{Z} \tag{7.134}$$

has an exponential dichotomy on $\mathbb{Z}$ with data $(32\bar{K}^4, \beta, (\tilde{Q}_j)_{j\in\mathbb{Z}})$ whenever

$$\|x_j - y_j(\omega, m)\|, \|d_j\| \le \varepsilon_{\mathrm{ED}} \quad \text{for all } j \in \mathbb{Z}. \tag{7.135}$$

Remark We have suppressed the dependence of the projectors Q_j on ω and m in the notation. Note that the constants $\bar{K}, \alpha, \beta, m_{\mathrm{ED}}, \varepsilon_{\mathrm{ED}}$ are independent of ω and m. ◇

Proof Our main tool is the roughness Theorem 7.61. We proceed in several steps.

Step1: Choice of constants.

Choose α_0 with $0 < \alpha < \alpha_0 < \bar{\alpha}$ and then $m_0 \in \mathbb{N}$ with

$$4\bar{K}^2 \exp(-(\bar{\alpha} - \alpha_0)m_0) < 1. \tag{7.136}$$

Define for $m \in \mathbb{N}$ the matrices

$$\begin{aligned} \Delta_-(m) &= 2\bar{P}_{a(m)+1}P^s - P^s - \bar{P}_{a(m)+1}, \Delta_+(m) \\ &= 2P^s\bar{P}_{a(m)+m-1} - \bar{P}_{a(m)+m-1} - P^s, \end{aligned}$$

which by (7.130) satisfy the estimates

$$\|\Delta_-(m)\| \le 2\bar{K}\|\bar{P}_{a(m)+1} - P^s\|, \quad \|\Delta_+(m)\| \le 2\bar{K}\|\bar{P}_{a(m)+m-1} - P^s\|. \tag{7.137}$$

Since the right-hand sides converge to zero for $m \to \infty$ by (7.121) and Lemma 7.65, we find $m_1 \in \mathbb{N}$ such that for all $m \geq m_1$

$$\|\Delta_{\pm}(m)\| \leq \frac{1}{2}, \quad \|I + \Delta_{\pm}(m)\| \leq \frac{3}{2} < 2, \tag{7.138}$$

estimates we shall use below. Similarly, we choose $m_2 \in \mathbb{N}$ such that

$$\max(\|\Delta_{-}(m)Df(\xi)\|, \|\Delta_{+}(m)Df(\bar{x}_{a(m)+m-1})\|) \leq \gamma(4\bar{K}^4, \alpha_0, \alpha) \quad \forall m \geq m_2, \tag{7.139}$$

where γ is the function from the roughness Theorem 7.61. Finally, we set $m_{\mathrm{ED}} = \max(m_0, m_1, m_2)$.

Step 2: Definition of modified system.
For $m \geq m_{\mathrm{ED}}$, $\omega \in \Omega_{A(m)}$ and $j \in \mathbb{Z}$ we define

$$A_j = \begin{cases} (I + \Delta_{-}(m))Df(\xi), & \text{if } \omega_{j+1} = 1, \\ (I + \Delta_{+}(m))Df(\bar{x}_{a(m)+m-1}), & \text{if } \omega_j = m-1, \\ Df(y_j(\omega, m)), & \text{otherwise} \end{cases} \tag{7.140}$$

and $B_j = Df(y_j(\omega, m)) - A_j$, $j \in \mathbb{Z}$. Using (7.139) and the construction (7.129) of the pseudo-orbit, we obtain $\|B_j\| \leq \gamma(4\bar{K}^4, \alpha_0, \alpha)$ for all $j \in \mathbb{Z}$. Then Theorem 7.61 yields the exponential dichotomy for the system (7.132) with data $(8\bar{K}^4, \alpha, (Q_j)_{j\in\mathbb{Z}})$, provided we have shown that the modified system

$$z_{j+1} = A_j z_j, \quad j \in \mathbb{Z} \tag{7.141}$$

has an exponential dichotomy with constants $(4\bar{K}^4, \alpha_0)$.

Step 3: Exponential dichotomy of the modified system.
Let Φ be the solution operator of the system (7.141). Define the modified projectors

$$\tilde{P}_j = \begin{cases} P^s, & \text{if } \omega_j = 0, \\ \bar{P}_{a(m)+k}, & \text{if } \omega_j = k \in \{1, \ldots, m-1\}. \end{cases} \tag{7.142}$$

We claim that these projectors and A_j satisfy the condition (7.113), which then implies the invariance condition (7.110) as noted after (7.113). If $\omega_j = 1, \ldots, m-2$ then (7.113) follows with $k = j$ directly from (7.142), (7.140), and (7.129). We now check the cases $\omega_j = 0$, $\omega_{j+1} = 1$ and $\omega_j = m-1$, $\omega_{j+1} = 0$. In the first case, we apply Lemma 7.66 to the setting $(B, P_{-1}, P_0, Q_0) = (Df(\xi), P^s, P^s, P_{a(m)+1})$ and find

$$A_j\tilde{P}_j = A_j P^s = (I + \Delta_{-}(m))Df(\xi)P^s = \bar{P}_{a(m)+1}(I + \Delta_{-}(m))Df(\xi) = \tilde{P}_{j+1}A_j.$$

The same relation is obtained in the second case by applying Lemma 7.66 to the setting

$$(B, P_{-1}, P_0, Q_0) = (Df(\bar{x}_{a(m)+m-1}), \bar{P}_{a(m)+m-1}, \bar{P}_{a(m)+m-1}, P^s).$$

Having verified (7.110) for Φ corresponding to (7.141) and projectors given by (7.142), our aim is to derive, with α_0 as in **Step 1**, the inequality

$$\|\Phi(\ell, j)\tilde{P}_j\| \leq 4\bar{K}^4 e^{-\alpha_0(\ell-j)}, \quad \ell \geq j, \tag{7.143}$$

corresponding to (7.111) with constants $(4\bar{K}^4, \alpha_0)$. The derivation is based on viewing $\omega \in \Omega_{A(m)}$ as a concatenation of segments and a corresponding multiplicative decomposition of solution operators.

When $\omega_j, \ldots, \omega_\ell$ all belong to $(1, \ldots, m-1)$, the estimate

$$\|\Phi(\ell, j)\tilde{P}_j\| \leq \bar{K}e^{-\bar{\alpha}(\ell-j)} \tag{7.144}$$

follows from (7.125), since Φ agrees with $\bar{\Phi}$ for these indices. Exactly the same estimate holds when $\omega_j, \ldots, \omega_\ell$ are all equal to zero.

When $\omega_j, \ldots, \omega_{\ell-1}$ belong to $(1, \ldots, m-1)$ and $\omega_{\ell-1} = m-1$, we necessarily have $\omega_\ell = 0$ and hence the estimate (7.144) involves an additional factor $\|I + \Delta_+(m)\|$ bounded by 2; cf. (7.138). When $\omega_j, \ldots, \omega_{\ell-1}$ are all zero and $\omega_\ell = 1$, the estimate involves an additional factor $\|I + \Delta_-(m)\|$ which likewise is bounded by 2.

For any k with $j \leq k \leq \ell$ we have $\Phi(\ell, j) = \Phi(\ell, k)\Phi(k, j)$. Moreover,

$$\Phi(\ell, j)\tilde{P}_j = \Phi(\ell, k)\tilde{P}_k\Phi(k, j)\tilde{P}_j, \tag{7.145}$$

since $\tilde{P}_j$, $j \in \mathbb{Z}$ are projectors for which we have verified the invariance condition (7.110). This decomposition enables us to extend estimates to larger time stretches on the basis of estimates for smaller parts. The price we pay is that constants get multiplied. So in order to obtain suitable bounds, we need to lower the exponent from $\bar{\alpha}$ to α_0 and use (7.136). We now elaborate on the details.

Assume that $\omega_j \geq 1$ and $\omega_i \neq 1$ for $j < i < \ell$. If $\omega_\ell = 0$ then there exists an index k with $j < k \leq \ell$ such that $\omega_{k-1} = m-1$. Using this k in (7.145), we obtain from our estimates above

$$\|\Phi(\ell, j)\tilde{P}_j\| \leq \bar{K}e^{-\bar{\alpha}(\ell-k)}2\bar{K}e^{-\bar{\alpha}(k-j)} = 2\bar{K}^2e^{-\bar{\alpha}(\ell-j)}. \tag{7.146}$$

When we actually know that $\omega_j = 1$ and further assume $\omega_\ell = 1$ (hence $\omega_{\ell-1} = 0$) then we have an extra 0 to 1 transition and an additional factor 2. Moreover, the length $\ell - j$ exceeds m and hence (7.136) allows us to estimate

$$\|\Phi(\ell, j)\tilde{P}_j\| \leq 4\bar{K}^2e^{-\bar{\alpha}(\ell-j)} = 4\bar{K}^2e^{-(\bar{\alpha}-\alpha_0)(\ell-j)}e^{-\alpha_0(\ell-j)} < e^{-\alpha_0(\ell-j)}. \tag{7.147}$$

In general, we decompose $\omega_j, \ldots, \omega_\ell$ into a head, a certain number of excursions and a tail.

An excursion begins at an index k in

$$\mathbb{Z}_1 = \mathbb{Z}_1(\omega) = \{k \in \mathbb{Z} : \omega_k = 1\} \tag{7.148}$$

and ends at its successor $\nu(k) = \min\{r \in \mathbb{Z}_1 : k < r\}$ (where $\nu(k) = \infty$ if the set is empty). If $\omega_j, \ldots, \omega_\ell$ contains no index in $\mathbb{Z}_1$, one of the estimates (7.144) and (7.146) applies. Otherwise, we let

$$k_- = \min\{k \in \mathbb{Z}_1 : j \leq k \leq \ell\}, \quad k_+ = \max\{k \in \mathbb{Z}_1 : j \leq k \leq \ell\}.$$

Then we decompose $\omega_j, \ldots, \omega_\ell$ into the tail $\omega_j, \ldots, \omega_{k_-}$, the excursions $\omega_k, \ldots, \omega_{\nu(k)}$ for $k \in \mathbb{Z}_1$, $k_- \leq k < k_+$, and the head $\omega_{k_+}, \ldots, \omega_\ell$. Note that there is no tail in case $\omega_j = 1$ and no head in case $\omega_\ell = 1$. We split the segment $\omega_j, \ldots, \omega_\ell$ at the interfaces $k_- \leq k < k_+$, $k \in \mathbb{Z}_1$ and use the decomposition (7.145) repeatedly. Then our previous estimates (7.144), (7.146), (7.147) yield

$$\begin{aligned}
\|\Phi(\ell, j)\tilde{P}_j\| &= \|\Phi(\ell, k_+)\tilde{P}_{k_+}\Phi(k_+, k_-)\tilde{P}_{k_-}\Phi(k_-, j)\tilde{P}_j\| \\
&\leq 2\bar{K}^2e^{-\bar{\alpha}(\ell-k_+)}\left[\Pi_{k_-\leq k<k_+, k\in\mathbb{Z}_1}e^{-\alpha_0(\nu(k)-k)}\right]2\bar{K}^2e^{-\bar{\alpha}(k_--j)} \\
&\leq 4\bar{K}^4e^{-\bar{\alpha}(\ell-k_++k_--j)}e^{-\alpha_0(k_+-k_-)} \leq 4\bar{K}^4e^{-\alpha_0(\ell-j)}.
\end{aligned}$$

Thus we have verified (7.143) and the dichotomy properties for the systems (7.141) and (7.132).

Step 4: To prove (7.133), let $\bar{\varepsilon} > 0$ and let $\mu = \mu(8\bar{K}^4, \alpha_0, \alpha)$ be given by Theorem 7.61. Then by the inequalities (7.137) from **Step 1** we find $\bar{m} \geq m_{\mathrm{ED}}$ such that for all $m \geq \bar{m}$

$$\max\{\|\Delta_-(m)Df(\xi)\|, \|\Delta_+(m)Df(\bar{x}_{a(m)+m-1})\|\} \leq \frac{\bar{\varepsilon}}{16\bar{K}^4\mu}.$$

The roughness Theorem 7.61 then yields an exponential dichotomy for the system (7.132) with data $(16\bar{K}^4, \alpha, (\hat{P}_k)_{k\in\mathbb{Z}})$ as well as the estimate (7.119)

$$\sup_{j\in\mathbb{Z}} \|\tilde{P}_j - \hat{P}_j\|) \leq 16\bar{K}^4\mu\frac{\bar{\varepsilon}}{16\bar{K}^4\mu} = \bar{\varepsilon} \quad \text{for all} \quad m \geq \bar{m}.$$

By the uniqueness of the projectors on $\mathbb{Z}$ we have $\hat{P}_j = Q_j$ for all $j \in \mathbb{Z}$ and (7.133) follows from (7.142).

Step 5: We choose a compact domain $\mathcal{H}$ containing H in its interior and define the modulus of continuity of Df by

$$M(Df, \mathcal{H}, \varepsilon) = \max\{\|Df(x) - Df(y)\| : x, y \in \mathcal{H}, \|x - y\| \leq \varepsilon\}. \tag{7.149}$$

By the uniform continuity of Df on $\mathcal{H}$ there exists $\varepsilon_{\mathrm{ED}} > 0$ such that

$$\left\|\int_0^1 Df(x_k + \tau d_k)d\tau - Df(y_k(\omega, m))\right\| \leq M(Df, \mathcal{H}, \varepsilon_{\mathrm{ED}}) \leq \gamma(16\bar{K}^4, \alpha, \beta)$$

for all $k \in \mathbb{Z}$, $m \geq m_{\mathrm{ED}}$, $\omega \in \Omega_{A(m)}$ and x_j, d_j with (7.135). Here γ is the function from Theorem 7.61, which then implies an exponential dichotomy for the system (7.134) with constants $(32\bar{K}^4, \beta)$. □

7.7.4 Construction of Shadowing Orbits

In this section, we construct f-orbits which lie in a small neighbourhood

$$\mathcal{B}_\varepsilon(y(\omega, m)) = \{x = (x_j)_{j\in\mathbb{Z}} \in X_{\mathbb{Z}} : \|x - y(\omega, m)\|_\infty \leq \varepsilon\}$$

of the pseudo-orbits $y = y(\omega, m)$ defined by (7.129).

Theorem 7.68 *Let the assumptions of* Theorem 7.67 *hold. Then there exists an* $\varepsilon_{\mathrm{sol}} > 0$ *and for any* $0 < \varepsilon \leq \varepsilon_{\mathrm{sol}}$ *a number* $m_0 = m_0(\varepsilon) \in \mathbb{N}$ *such that the system*

$$x_{j+1} = f(x_j), \quad j \in \mathbb{Z} \tag{7.150}$$

has a unique solution $x = x(\omega, m, \varepsilon) \in \mathcal{B}_\varepsilon(y(\omega, m))$ *for all* $m \geq m_0$ *and* $\omega \in \Omega_{A(m)}$. *Moreover, we have*

$$\mathcal{B}_\varepsilon(y(\omega, m)) \cap \mathcal{B}_\varepsilon(y(\theta, m)) = \emptyset \text{ if } \theta \neq \omega. \tag{7.151}$$

Proof Our goal is to apply the local inverse Lipschitz Theorem 7.48 to the settings

$$L : X_{\mathbb{Z}} \to X_{\mathbb{Z}},\ L(x)_j = x_{j+1} - Df(y_j(\omega, m))x_j,\ j \in \mathbb{Z},$$
$$F : X_{\mathbb{Z}} \to X_{\mathbb{Z}},\ F(x)_j = Df(y_j(\omega, m))x_j - f(x_j),\ j \in \mathbb{Z}$$

on the ball $\mathcal{B}_\varepsilon(y(\omega, m))$. Note the dependence of L and F on m and $\omega \in \Omega_{A(m)}$ which we do not express in the notation. Let m_{ED} be given by Theorem 7.67. Then Proposition 7.60 implies that L is invertible for all $m \geq m_{\mathrm{ED}}$ and satisfies

$$\|L^{-1}\| \le C_1 := 8\bar{K}^4 \frac{1+e^{-\alpha}}{1-e^{-\alpha}}.$$

Recall the modulus of continuity from (7.149) and choose $\varepsilon_1 > 0$ such that

$$M(Df, \mathcal{H}, \varepsilon_1) \le \frac{1}{2C_1}.$$

Further, $\bar{x}_j$ has a finite distance from the compact set $H \setminus \{\bar{x}_j\}$, so we find an $\varepsilon_2 > 0$ with

$$\max_{j=1,\dots,m_{\mathrm{ED}}-1} \operatorname{dist}\left(\bar{x}_{a(m_{\mathrm{ED}})+j}, H \setminus \{\bar{x}_{a(m_{\mathrm{ED}})+j}\}\right) > 2\varepsilon_2. \tag{7.152}$$

With these choices we define $\varepsilon_{\mathrm{sol}} = \min(\varepsilon_1, \varepsilon_2, \frac{1}{2}\varepsilon_{\mathrm{ED}})$.

For two elements $x, z \in \mathcal{B}_\varepsilon(y(\omega, m))$ with $0 < \varepsilon \le \varepsilon_{\mathrm{sol}}$ and $m \ge m_{\mathrm{ED}}$ we obtain

$$\begin{aligned} \|F(x) - F(z)\|_\infty &\le \sup_{j\in\mathbb{Z}} \left\| \int_0^1 [Df(y_j) - Df(z_j + \tau(x_j - z_j))]\, d\tau (x_j - z_j) \right\| \\ &\le M(Df, \mathcal{H}, \varepsilon)\|x - z\|_\infty \le \frac{1}{2C_1}\|x - z\|_\infty, \end{aligned}$$

and hence $\mathrm{Lip}(F) \le \frac{1}{2C_1} \le \frac{1}{2}\|L^{-1}\|^{-1} < \|L^{-1}\|^{-1}$. By Lemma 7.65 we find $m_0 \ge m_{\mathrm{ED}}$ such that for all $m \ge m_0$

$$\|\bar{x}_{a(m)+1} - \xi\|, \|\xi - \bar{x}_{a(m)+m}\| \le \frac{\varepsilon}{2C_1}.$$

The final condition (7.53) then follows from

$$\begin{aligned} \|(L+F)(y(\omega,m))\| &= \sup_{j\in\mathbb{Z}}\{\|y_{j+1}(\omega,m) - f(y_j(\omega,m))\|\} \\ &= \max(\|\bar{x}_{a(m)+1} - \xi\|, \|\xi - \bar{x}_{a(m)+m}\|) \\ &\le \frac{\varepsilon}{2C_1} \le (\|L^{-1}\|^{-1} - \mathrm{Lip}(F))\varepsilon, \end{aligned}$$

since $\frac{1}{2C_1} = \frac{1}{C_1} - \frac{1}{2C_1} \le \|L^{-1}\|^{-1} - \mathrm{Lip}(F)$. Thus Theorem 7.48 yields a unique solution $x(\omega, m) \in \mathcal{B}_\varepsilon(y(\omega, m))$ of (7.150).

Next we consider $\omega, \theta \in \Omega_{A(m)}$ with $\omega \ne \theta$ and without loss of generality assume $\theta \ne 0$. Then we claim that there exists some $\ell \in \mathbb{Z}$ such that

$$\omega_{\ell+j} \ne \theta_{\ell+j} = j \text{ for } j = 1, \dots, m-1. \tag{7.153}$$

Since $\mathbb{Z}_1(\omega)$ from (7.148) determines ω uniquely we conclude $\mathbb{Z}_1(\omega) \ne \mathbb{Z}_1(\theta)$ and there exists an index $\ell + 1 \in \mathbb{Z}_1(\theta)$ with $\theta_{\ell+j} = j$ for $j = 1, \dots, m-1$ and $\omega_{\ell+1} \ne 1$. Both when $\omega_{\ell+1} = 0$ and when $\omega_{\ell+1} \in \{2, \dots, m-1\}$, the symbolic subsequence $\omega_{\ell+j}, j = 1, \dots, m-1$ differs from $\theta_{\ell+j}$ at every position. By the definition (7.129) of the pseudo-orbits, we obtain for the index ℓ from (7.153)

$$y_{\ell+j}(\theta, m) = \bar{x}_{a(m)+j} \ne y_{\ell+j}(\omega, m) \in H \setminus \{\bar{x}_{a(m)+j}\}, \quad j = 1, \dots, m-1.$$

Using (7.128) with $m_1 = m_{\mathrm{ED}}$, $m_2 = m$, and (7.152) we find

$$\begin{aligned}\max_{j=1,\dots,m-1} &\|y_{\ell+j}(\theta,m)-y_{\ell+j}(\omega,m)\| \\ &\geq \max_{j=1,\dots,m-1} \operatorname{dist}(\bar{x}_{a(m)+j}, H\setminus\{\bar{x}_{a(m)+j}\}) \\ &\geq \max_{j=1,\dots,m_{\mathrm{ED}}-1} \operatorname{dist}(\bar{x}_{a(m_{\mathrm{ED}})+j}, H\setminus\{\bar{x}_{a(m_{\mathrm{ED}})+j}\}) > 2\varepsilon.\end{aligned}$$

This contradicts $\mathcal{B}_\varepsilon(y(\theta,m))\cap\mathcal{B}_\varepsilon(y(\omega,m))\neq\emptyset$. □

In the following, we endow $X_{\mathbb{Z}}$ with a metric analogous to (7.103)

$$d_{\mathbb{Z}}(x,z)=\sum_{j\in\mathbb{Z}}2^{-|j|}\|x_j-z_j\|,\quad x,z\in X_{\mathbb{Z}}. \tag{7.154}$$

This generates the product topology on $X_{\mathbb{Z}}$ which is weaker than the topology induced by the strong norm $\|\cdot\|_\infty$ in (7.116). The following subset of orbits inherits this metric from $X_{\mathbb{Z}}$:

$$X_{\mathbb{Z}}(m,\varepsilon)=\Big\{x\in\bigcup_{\omega\in\Omega_{A(m)}}\mathcal{B}_\varepsilon(y(\omega,m)) : x_{j+1}=f(x_j)\ \forall j\in\mathbb{Z}\Big\}. \tag{7.155}$$

For $0<\varepsilon\leq\varepsilon_{\mathrm{sol}}$ and $m\geq m_0$ we find that this set consists of the solutions from Theorem 7.68, i.e.:

$$X_{\mathbb{Z}}(m,\varepsilon)=\{x(\omega,m,\varepsilon):\omega\in\Omega_{A(m)}\}.$$

On $X_{\mathbb{Z}}(m,\varepsilon)$ we define the orbit shift by

$$\mathcal{F}:X_{\mathbb{Z}}(m,\varepsilon)\to X_{\mathbb{Z}}(m,\varepsilon),\quad \mathcal{F}(x)_j=x_{j+1},\quad j\in\mathbb{Z}.$$

Indeed, $\mathcal{F}(x)$ is an f-orbit if $x=(x_j)_{j\in\mathbb{Z}}$ is. By the construction of the pseudo-orbits we have

$$\|\mathcal{F}(x(\omega,m,\varepsilon))_j-y_j(\sigma(\omega),m)\|=\|x_{j+1}(\omega,m,\varepsilon)-y_{j+1}(\omega,m)\|\leq\varepsilon\quad\forall\, j\in\mathbb{Z}.$$

Hence the uniqueness in Theorem 7.68 guarantees the relation

$$\mathcal{F}(x(\omega,m,\varepsilon))=x(\sigma(\omega),m,\varepsilon)\quad\text{for all }\omega\in\Omega_{A(m)},m\geq m_0. \tag{7.156}$$

This leads us to the first conjugacy result.

Theorem 7.69 *Let the assumptions of* Theorem 7.68 *hold. Then the mapping*

$$h_1:\Omega_{A(m)}\to X_{\mathbb{Z}}(m,\varepsilon),\quad h_1(\omega)=x(\omega,m,\varepsilon) \tag{7.157}$$

is a homeomorphism of the metric spaces $(\Omega_{A(m)},d)$ and $(X(m,\varepsilon),d_{\mathbb{Z}})$ which satisfies

$$h_1\circ\sigma=\mathcal{F}\circ h_1 \tag{7.158}$$

for all $0<\varepsilon\leq\varepsilon_{\mathrm{sol}}$, $m\geq m_0$.

Proof First note that (7.158) is just a restatement of (7.156). By Theorem 7.68 the mapping h_1 is onto. Further, the mapping is one-to-one, since any f-orbit in $X_{\mathbb{Z}}(m,\varepsilon)$ lies in exactly one ball $\mathcal{B}_\varepsilon(y(\omega,m))$ due to (7.151) and hence coincides with $h_1(\omega)=x(\omega,m,\varepsilon)$. Next we show that h_1 is continuous. Then the inverse h_1^{-1} is also continuous since a continuous bijection from a compact metric space into a metric space has a continuous inverse (this follows from the Open Map Lemma).

Given $\delta>0$ we determine $k=k(\delta)\in\mathbb{N}$ such that $d(\omega,\theta)\leq 2^{-k-1}$ for $\omega,\theta\in\Omega_{A(m)}$ implies

$$d_{\mathbb{Z}}(x(\omega,m,\varepsilon),x(\theta,m,\varepsilon))=\sum_{j\in\mathbb{Z}}2^{-|j|}\|x_j(\omega,m,\varepsilon)-x_j(\theta,m,\varepsilon)\|\le\delta.$$

From $d(\omega,\theta)\le 2^{-k-1}$ we infer $\omega_j=\theta_j$ and $y_j(\omega,m)=y_j(\theta,m)$ for $|j|\le k$. The modified sequence

$$\tilde{x}_j=\begin{cases}x_j(\theta,m,\varepsilon), & |j|\le k,\\ x_j(\omega,m,\varepsilon), & |j|>k\end{cases}$$

then satisfies $\tilde{x}\in\mathcal{B}_\varepsilon(y(\omega,m))$. In the following we consider $d_j=\tilde{x}_j-x_j(\omega,m,\varepsilon)$ and the system

$$z_{j+1}=A_jz_j,\ \text{where}\ A_j=\int_0^1 Df(x_j(\omega,m,\varepsilon)+\tau d_j)d\tau.$$

By Theorem 7.67 the solution operator $\tilde{\Phi}$ of this system has an exponential dichotomy on $\mathbb{Z}$ with data $(32\bar{K}^4,\beta,(\tilde{Q}_j)_{j\in\mathbb{Z}})$. Note that (7.135) is satisfied since $\|x_j(\omega,m,\varepsilon)-y_j(\omega,m)\|\le\varepsilon\le\varepsilon_{\mathrm{ED}}$ holds for all $j\in\mathbb{Z}$, and for $|j|\le k$ we obtain from $2\varepsilon\le 2\varepsilon_{\mathrm{sol}}\le\varepsilon_{\mathrm{ED}}$

$$\|d_j\|=\|x_j(\omega,m,\varepsilon)-x_j(\theta,m,\varepsilon)\|\le 2\varepsilon\le\varepsilon_{\mathrm{ED}}.$$

Moreover, we have for $-k\le j<k$

$$d_{j+1}=x_{j+1}(\theta,m,\varepsilon)-x_{j+1}(\omega,m,\varepsilon)=f(\tilde{x}_j)-f(x_j(\omega,m,\varepsilon))=A_jd_j,$$

and hence for $|j|\le k$

$$\begin{aligned}d_j&=\tilde{\Phi}(j,0)(\tilde{Q}_0d_0+(I-\tilde{Q}_0)d_0)\\&=\tilde{\Phi}(j,-k)\tilde{\Phi}(-k,0)\tilde{Q}_0d_0+\tilde{\Phi}(j,k)\tilde{\Phi}(k,0)(I-\tilde{Q}_0)d_0\\&=\tilde{\Phi}(j,-k)\tilde{Q}_{-k}d_{-k}+\tilde{\Phi}(j,k)(I-\tilde{Q}_k)d_k.\end{aligned}$$

Therefore, the exponential dichotomy and $\|d_{-k}\|,\|d_k\|\le 2\varepsilon$ yield the estimate

$$\|d_j\|\le C_2\left(e^{-\beta(j+k)}+e^{-\beta(k-j)}\right),\quad C_2=64\bar{K}^4\varepsilon,\quad |j|\le k.$$

Since all solutions lie in a compact domain we have a global bound $\|d\|_\infty\le C_3$. Then we estimate for any index $0<\ell<k$

$$\begin{aligned}d_{\mathbb{Z}}(x(\theta,m,\varepsilon),x(\omega,m,\varepsilon))&\le\sum_{|j|\le\ell}2^{-|j|}\|d_j\|+C_3\sum_{|j|>\ell}2^{-|j|}\\&\le C_2\sum_{|j|\le\ell}2^{-|j|}\left(e^{-\beta(j+k)}+e^{-\beta(k-j)}\right)+2C_32^{-\ell}\\&\le C_2e^{-\beta(k-\ell)}\sum_{|j|\le\ell}2^{-|j|}\left(e^{-\beta(j+\ell)}+e^{-\beta(\ell-j)}\right)+2C_32^{-\ell}\\&\le 2C_2e^{-\beta(k-\ell)}\sum_{|j|\le\ell}2^{-|j|}+2C_32^{-\ell}\\&\le 6C_2e^{-\beta(k-\ell)}+2C_32^{-\ell}\le\delta,\end{aligned}$$

if we choose $\ell>0$ such that $2C_32^{-\ell}\le\frac{\delta}{2}$ and next $k>\ell$ such that $6C_2e^{-\beta(k-\ell)}\le\frac{\delta}{2}$. Thus we verified that, with this choice of k, we have $d_{\mathbb{Z}}(h_1(\theta),h_1(\omega))\le\delta$ when $d(\omega,\theta)\le 2^{-k-1}$. □

7.7.5 Proof of Theorem 7.57

Step 1: Construction of $\mathcal{O}$.

Let $\bar{\varepsilon} = \varepsilon_{\text{sol}}$ and $m_0 = m_0(\bar{\varepsilon})$ be given by Theorem 7.68. Recall how we defined in (7.127) the principal part $H_m = \{\bar{x}_{a(m)+1}, \ldots, \bar{x}_{a(m)+m-1}\}$ of $H = \{\bar{x}_j : j \in \mathbb{Z}\} \cup \{\xi\}$ and recall Lemma 7.65. We choose $\overline{m} \geq m_0$ such that

$$\|\bar{x}_{a(\overline{m})+j} - \xi\| \leq \frac{\bar{\varepsilon}}{2} \quad \text{for all } j \in \mathbb{Z} \setminus \{1, \ldots, \overline{m} - 1\}. \tag{7.159}$$

Then choose $\varepsilon \leq \frac{1}{2}\bar{\varepsilon}$ with the following properties:

$$\|\bar{x}_{a(\overline{m})+j} - x_{a(\overline{m})+k}\| \geq 3\varepsilon \quad \text{for } j \in \{1, \ldots, \overline{m} - 1\}, k \in \mathbb{Z} \setminus \{j\}, \tag{7.160}$$

$$\|f(z) - \bar{x}_{a(\overline{m})+j}\| > \varepsilon \text{ for } j \in \{2, \ldots, \overline{m} - 1\}, \ \text{dist}(z, H \setminus H_{\overline{m}}) \leq \varepsilon. \tag{7.161}$$

Note that (7.161) can be achieved since $H \setminus H_{\overline{m}}$ is compact and $\bar{x}_{a(\overline{m})+j} \notin f(H \setminus H_{\overline{m}})$ for $j = 2, \ldots, \overline{m} - 1$. Let $m_0(\varepsilon)$ be given by Theorem 7.68 and set $m = \max(m_0(\varepsilon), \overline{m} + 1)$. Further, we introduce abbreviations for ε-balls as follows:

$$\begin{aligned} B_i &= B_\varepsilon(\bar{x}_{a(\overline{m})+i}) = \{z \in \mathbb{R}^n : \|z - \bar{x}_{a(\overline{m})+i}\| \leq \varepsilon\}, \ i \in \mathbb{Z}, \\ B_\infty &= B_\varepsilon(\xi) = \{z \in \mathbb{R}^n : \|z - \xi\| \leq \varepsilon\}. \end{aligned}$$

With these preparations we define the following sets (see Fig. 7.33)

$$V_0 = B_\infty \cup \bigcup_{i \in \mathbb{Z} \setminus \{1, \ldots, \overline{m}-1\}} B_i, \tag{7.162}$$

$$V_{\overline{m}-1} = B_{\overline{m}-1} \cap \bigcap_{j=1}^{m-\overline{m}} f^{-j}(V_0), \tag{7.163}$$

$$V_i = B_i \cap f^{-1}(V_{i+1}), \quad i = \overline{m} - 2, \overline{m} - 3, \ldots, 1, \tag{7.164}$$

$$\mathcal{O} = \bigcup_{i=0}^{\overline{m}-1} V_i. \tag{7.165}$$

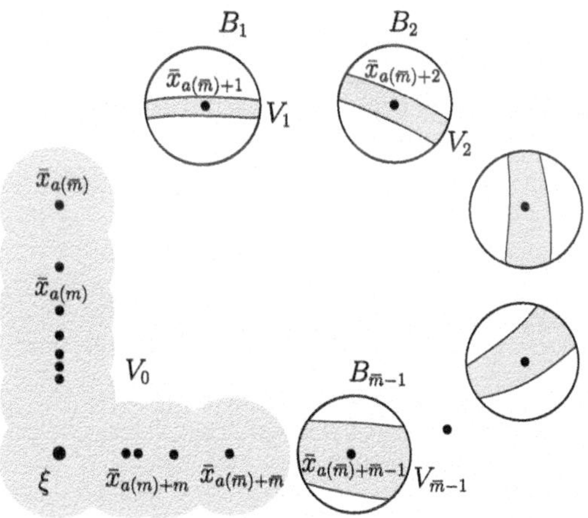

Fig. 7.33 Construction of the neighbourhood $\mathcal{O} = \bigcup_{i=0}^{\overline{m}-1} V_i$ of the homoclinic orbit

We claim that the assertion of Theorem 7.57 holds with this choice of $\mathcal{O}$. The set $\mathcal{O}$ is composed of $\overline{m}$ components: the base component V_0; the exit component V_1; the intermediate components V_i, $i = 2, \ldots, \overline{m}-2$; and the re-entry component $V_{\overline{m}-1}$. First we take $\bar{\varepsilon}$ small and $\overline{m}$ large such that existence and uniqueness of shadowing orbits is guaranteed by Theorem 7.68. Then we make sure that components are disjoint and that orbits leave the base via the exit component V_1. This is achieved by choosing ε sufficiently small to guarantee that (7.160) and (7.161) hold. However, we also need existence and uniqueness of shadowing orbits from Theorem 7.57 for this ε-value. Therefore, we increase m to a value which is generally larger than the value $\overline{m}$ we started with. To make the difference between m and $\overline{m}$ harmless, we force the orbits to stay in the base for a sojourn time of at least $m - \overline{m}$, once they have entered it, by the choice of the re-entry component in (7.163). The definition (7.164) then forces points in the re-entry component to directly evolve from the exit component through the intermediate components. We now provide the details of these arguments.

Step 2: Properties of $\mathcal{O}$.

Note that $f^j(\bar{x}_{a(\overline{m})+\overline{m}-1}) \in V_0$ holds for all $j \geq 1$, so that $V_{\overline{m}-1}$ as defined in (7.163) is a neighbourhood of $\bar{x}_{a(\overline{m})+\overline{m}-1}$, and, consequently, V_i is a neighbourhood of $\bar{x}_{a(\overline{m})+i}$ for $i = \overline{m}-2, \ldots, 1$.

The sets V_0 and $B_i, i = 1, \ldots, \overline{m}-1$ are pairwise disjoint due to (7.160), and hence this also holds for the components $V_0, \ldots, V_{\overline{m}-1}$. We show that an orbit in $\mathcal{O}$ can leave V_0 only via V_1:

$$x \in V_0, f(x) \in \mathcal{O} \setminus V_0 \text{ implies } f(x) \in V_1. \tag{7.166}$$

Indeed, for $x \in V_0$ we have $x \in B_i$ for some $i \notin \{1, \ldots, \overline{m}-1\}$ or $x \in B_\infty$, hence $\operatorname{dist}(x, H \setminus H_{\overline{m}}) \leq \varepsilon$. Then (7.161) guarantees $f(x) \notin B_j$ for $j = 2, \ldots, \overline{m}-1$ so that, when $f(x) \in \mathcal{O}$, necessarily $f(x) \in V_1$. The next two properties follow immediately from (7.163) and (7.164)

$$x \in V_i, i \in \{1, \ldots, \overline{m}-2\} \text{ implies } \quad f(x) \in V_{i+1}, \tag{7.167}$$

$$x \in V_{\overline{m}-1} \text{ implies } \quad f^j(x) \in V_0, j = 1, \ldots, m - \overline{m}. \tag{7.168}$$

Step 3: The orbits in $\mathcal{O}$.

We show that the solution set $X_{\mathbb{Z}}(m, \varepsilon)$ from (7.155) coincides with the orbits in $\mathcal{O}$, i.e.:

$$X_{\mathbb{Z}}(m, \varepsilon) = \{(f^j(z))_{j\in\mathbb{Z}} : f^j(z) \in \mathcal{O}\ \forall\, j \in \mathbb{Z}\} =: \operatorname{Orb}(\mathcal{O}). \tag{7.169}$$

Consider first an orbit $x(\omega, m, \varepsilon) \in X_{\mathbb{Z}}(m, \varepsilon)$ from Theorem 7.69 and recall $x(\omega, m, \varepsilon) \in \mathcal{B}_\varepsilon(y(\omega, m))$ where $\omega \in \Omega_{A(m)}$. For $j \in \mathbb{Z}$ we consider three cases:

(i) $x_j(\omega, m, \varepsilon) \notin \bigcup_{i=1}^{\overline{m}-1} B_i$: From $\|x_j(\omega, m, \varepsilon) - y_j(\omega, m)\| \leq \varepsilon$ we infer $x_j(\omega, m, \varepsilon) \in B_\infty \cup \bigcup_{i\notin\{1,\ldots,\overline{m}-1\}} B_i$, hence $x_j(\omega, m, \varepsilon) \in V_0 \subset \mathcal{O}$.

(ii) $x_j(\omega, m, \varepsilon) \in B_{\overline{m}-1}$: By the disjointness as observed in Step 2, we have that $x_j(\omega, m, \varepsilon)$ lies neither in V_0 nor in some $B_i, i = 1, \ldots, \overline{m}-2$. Combining this with $\|x_j(\omega, m, \varepsilon) - y_j(\omega, m)\| \leq \varepsilon$ and $y_j(\omega, m) = \bar{x}_{a(m)+\omega_j}$ we obtain

$$a(m) + \omega_j = a(\overline{m}) + \overline{m} - 1.$$

By iteration with f the shadowing property yields

$$x_{j+\nu}(\omega, m, \varepsilon) = f^\nu(x_j(\omega, m, \varepsilon)) \in B_\varepsilon(\bar{x}_{a(\overline{m})+\overline{m}-1+\nu})$$

for $\nu = 1, \ldots, m-1-\omega_j =: \nu_0$. In particular, $f^{\nu_0}(x_j(\omega, m, \varepsilon)) \in B_\varepsilon(\bar{x}_{a(m)+m-1})$ and $\omega_{j+\nu_0} = \omega_j + \nu_0 = m-1$. Then the symbolic sequence ω continues with at least one zero before starting a new excursion $1, \ldots, m-1$ with $1, \ldots, a(\overline{m}) - a(m)$ (note that $a(\overline{m}) - a(m) \leq -a(m) \leq \lfloor \frac{m}{2} \rfloor \leq m-1$ follows from (7.127)). The corresponding orbit points lie in ε-balls of the pseudo-orbit with centres ξ and $\bar{x}_{a(m)+1}, \ldots, \bar{x}_{a(\overline{m})}$.

Hence they belong to V_0. Counting the iterates we obtain $f^\nu(x_j(\omega, m, \varepsilon)) \in V_0$ for $\nu = 1, \ldots, \nu_1$, where

$$\begin{aligned}\nu_1 \geq \nu_0 + 1 + a(\overline{m}) - a(m) &= m - 1 - (a(\overline{m}) + \overline{m} - 1 - a(m)) + 1 + a(\overline{m}) - a(m) \\ &= m - \overline{m} + 1.\end{aligned}$$

It follows that $x_j(\omega, m, \varepsilon) \in V_{\overline{m}-1} \subset \mathcal{O}$.

(iii) $x_j(\omega, m, \varepsilon) \in B_i$ for some $i \in \{1, \ldots, \overline{m} - 2\}$: Similar to case (ii), since the neighbourhoods V_0 and B_k, $k \neq i$ are disjoint we find $\omega_j = a(\overline{m}) - a(m) + i$. By the definition of the pseudo-orbit $y(\omega, m)$ this implies $x_{j+\nu}(\omega, m, \varepsilon) = f^\nu(x_j(\omega, m, \varepsilon)) \in B_{i+\nu}$ for $\nu = 1, \ldots, \nu_0 = \overline{m} - 1 - i$. Thus, we obtain from case (ii) the relation $x_{j+\nu_0}(\omega, m, \varepsilon) \in V_{\overline{m}-1}$. Using the definition (7.164) of $V_{i+\nu}$, $\nu = \nu_0 - 1, \ldots, 0$ an induction shows $x_j(\omega, m, \varepsilon) \in B_i \cap f^{-1}(V_{i+1}) = V_i \subset \mathcal{O}$.

We conclude that $X_{\mathbb{Z}}(m, \varepsilon) \subseteq \mathrm{Orb}(\mathcal{O})$.

Next we show $\mathrm{Orb}(\mathcal{O}) \subseteq X_{\mathbb{Z}}(m, \varepsilon)$. Let $z_j = f^j(z_0), j \in \mathbb{Z}$ be an orbit in $\mathrm{Orb}(\mathcal{O})$. Define a symbolic sequence $\bar{\omega} \in \Omega_{\overline{m}}$ by

$$\bar{\omega}_j = \begin{cases} k, & \text{if } z_j \in V_k, k \in \{1, \ldots, \overline{m} - 1\}, \\ 0, & \text{if } z_j \in V_0. \end{cases} \tag{7.170}$$

The properties (7.166), (7.167), (7.168) show that $\bar{\omega}$ belongs to the subshift $\Omega_{A(\overline{m})}$. Moreover, by (7.170) and the definitions (7.163), (7.164) we obtain $\|z_j - \bar{x}_{a(\overline{m})+k}\| \leq \varepsilon \leq \bar{\varepsilon}$ if $\bar{\omega}_j = k \in \{1, \ldots, \overline{m} - 1\}$. If $\bar{\omega}_j = 0$ then $z_j \in V_0$ implies $\|z_j - \xi\| \leq \varepsilon \leq \bar{\varepsilon}$ in case $z_j \in B_\infty$, and

$$\|z_j - \xi\| \leq \|z_j - \bar{x}_{a(\overline{m})+i}\| + \|\bar{x}_{a(\overline{m})+i} - \xi\| \leq \varepsilon + \frac{\bar{\varepsilon}}{2} \leq \bar{\varepsilon}$$

in case $z_j \in B_i$ for some $i \notin \{1, \ldots, \overline{m} - 1\}$. Altogether, we find $\|(z_j)_{j\in\mathbb{Z}} - y(\bar{\omega}, \overline{m})\|_\infty \leq \bar{\varepsilon}$ for the pseudo-orbit $y(\bar{\omega}, \overline{m})$ from (7.129). According to Theorem 7.68 solutions in $\mathcal{B}_{\bar{\varepsilon}}(y(\bar{\omega}, \overline{m}))$ are unique and we arrive at

$$z = (z_j)_{j\in\mathbb{Z}} = x(\bar{\omega}, \overline{m}, \bar{\varepsilon}). \tag{7.171}$$

Next we extend $\bar{\omega}$ to a sequence $\omega \in \Omega_{A(m)}$. The table below illustrates how this is done. We abbreviate

$$\ell = a(\overline{m}) - a(m)$$

and show both symbol sequences $\bar{\omega}, \omega$ and the corresponding pseudo-orbits $y(\bar{\omega}, \overline{m}), y(\omega, m)$. The second row displays an excursion $1, \ldots, \overline{m} - 1$ of $\bar{\omega}$ with a (finite or infinite) number of zeroes to the left and to the right (indicated by dots). The corresponding segment of the pseudo-orbit is shown in the first row. Our constructive definition of $\mathcal{O}$ guarantees that any segment of zeroes (i.e. of the point ξ on the pseudo-orbit) has at least length $m - \overline{m} + 1$. Indeed, a finite segment is in between two excursions and the definition (7.163) guarantees that after entering V_0 the orbit stays in V_0 for at least $m - \overline{m}$ time units. To transform an excursion in $\bar{\omega}$ into an excursion in ω, we want to start at index $a(m) + 1$ with symbol 1 and finish at $a(m) + m - 1$ with symbol $m - 1$. So we sacrifice ℓ of the zeroes in $\bar{\omega}$ preceding the excursion and $m - \overline{m} - \ell$ zeroes in $\bar{\omega}$ following the excursion. We next adjust the numbering to reflect that now we consider a subshift on m symbols. Consequently, the pseudo-orbit changes too. The sequence thus extended has m symbols and is shown in the third row, accompanied by its pseudo-orbit in row 4. Note that the third row has 0 as a leftmost and as a rightmost entry to indicate that the part of the extended symbol sequence ω on which we focus our attention begins and finishes with at least one zero.

$y(\bar{\omega}, \bar{m})$	ξ	ξ	$\cdot$	ξ	$\bar{x}_{a(\bar{m})+1}$	$\cdot$	$\bar{x}_{a(\bar{m})+\bar{m}-1}$	ξ	$\cdot$	ξ	ξ
$\bar{\omega}$	0	0	$\cdot$	0	1	$\cdot$	$\bar{m}-1$	0	$\cdot$	0	0
ω	0	1	$\cdot$	ℓ	$\ell+1$ $\cdot$		$\ell+\bar{m}-1$	$\ell+\bar{m}$	$\cdot$	$m-1$	0
$y(\omega, m)$	ξ	$\bar{x}_{a(m)+1}$	$\cdot$	$\bar{x}_{a(\bar{m})}$	$\bar{x}_{a(\bar{m})+1}$	$\cdot$	$\bar{x}_{a(\bar{m})+\bar{m}-1}$	$\bar{x}_{a(\bar{m})+\bar{m}}$	$\cdot$	$\bar{x}_{a(m)+m-1}$	ξ

By Theorem 7.68 there exists for this extended ω a unique f-orbit $x(\omega, m, \varepsilon) \in \mathcal{B}_\varepsilon$ $(y(\omega, m))$. Below we show

$$\mathcal{B}_\varepsilon(y(\omega, m)) \subseteq \mathcal{B}_{\bar{\varepsilon}}(y(\bar{\omega}, \bar{m})). \tag{7.172}$$

Then the unique solvability in Theorem 7.68 implies $x(\bar{\omega}, \bar{m}, \varepsilon) = x(\omega, m, \varepsilon)$. In view of (7.171), we arrive at $z = x(\omega, m, \varepsilon)$ which proves that $\mathrm{Orb}(\mathcal{O}) \subseteq X_{\mathbb{Z}}(m, \varepsilon)$. Combining the two inclusions we obtain the equality (7.169).

It remains to prove (7.172). The table shows that the pseudo-orbits $y(\bar{\omega}, \bar{m})$ and $y(\omega, m)$ differ at two types of two positions:

(i) $y_{a(m)+i}(\bar{\omega}, \bar{m}) = \xi$, $y_{a(m)+i}(\omega, m) = \bar{x}_{a(m)+i}$ for some $i \in \{1, \ldots, a(\bar{m}) - a(m)\}$.
By (7.159) we then have for $\|z - y_{a(m)+i}(\omega, m))\| \le \varepsilon$

$$\|z - \xi\| \le \|z - y_{a(m)+i}(\omega, m)\| + \|\bar{x}_{a(m)+i} - \xi\| \le \varepsilon + \frac{\bar{\varepsilon}}{2} \le \bar{\varepsilon}.$$

(ii) $y_{a(\bar{m})+\bar{m}+i}(\bar{\omega}, \bar{m}) = \xi$, $y_{a(\bar{m})+\bar{m}+i}(\omega, m) = \bar{x}_{a(\bar{m})+\bar{m}+i}$ for some $i \in \{0, \ldots, a(m) - a(\bar{m}) - 1\}$.
By (7.159) we conclude for $\|z - y_{a(\bar{m})+\bar{m}+i}(\omega, m))\| \le \varepsilon$

$$\|z - \xi\| \le \|z - y_{a(\bar{m})+\bar{m}+i}(\omega, m)\| + \|\bar{x}_{a(\bar{m})+\bar{m}+i} - \xi\| \le \varepsilon + \frac{\bar{\varepsilon}}{2} \le \bar{\varepsilon}.$$

Step 4: The conjugacy.

Consider the operator which evaluates a sequence at zero

$$\mathrm{ev} : X_{\mathbb{Z}} \to \mathbb{R}^n, \quad x = (x_j)_{j\in\mathbb{Z}} \mapsto \mathrm{ev}(x) = x_0. \tag{7.173}$$

Since $\|\mathrm{ev}(x) - \mathrm{ev}(z)\| \le d_{\mathbb{Z}}(x, z)$ this map is continuous w.r.t. the metric from (7.154). Defining $\Lambda = \{x : (f^j(x))_{j\in\mathbb{Z}} \in \mathcal{O}\}$ we find that the restriction $\mathrm{ev} : \mathrm{Orb}(\mathcal{O}) = X_{\mathbb{Z}}(m, \varepsilon) \to \Lambda$ is continuous and onto. Actually, it is one-to-one because every $x \in \Lambda$ uniquely determines its orbit $\mathrm{ev}^{-1}(x) = (f^j(x))_{j\in\mathbb{Z}}$. Finally, we compose ev with the homeomorphism h_1 from (7.157)

$$h = \mathrm{ev} \circ h_1 : \Omega_{A(m)} \to \Lambda, \quad h(\omega) = x_0(\omega, m, \varepsilon).$$

As above, a continuous bijection h between compact metric spaces is a homeomorphism. By the conjugacy (7.158) in Theorem 7.69 we obtain for $\omega \in \Omega_{A(m)}$

$$\begin{aligned} h(\sigma(\omega)) = {} & \mathrm{ev}(h_1(\sigma(\omega)) = \mathrm{ev}(\mathcal{F}(h_1(\omega))) = \mathrm{ev}(\mathcal{F}(x(\omega, m, \varepsilon))) \\ = {} & \mathrm{ev}((x_{j+1}(\omega, m, \varepsilon)_{j\in\mathbb{Z}}) = x_1(\omega, m, \varepsilon) = f(x_0(\omega, m, \varepsilon)) = f(h(\omega)). \end{aligned}$$

This finishes the proof. □

References

Alligood, K. T., Sauer, T. D., & Yorke, J. A. (1997). *Chaos: An introduction to dynamical systems.* New York: Springer.

Amann, H. (1990). *Ordinary differential equations: An introduction to nonlinear analysis.* de Gruyter Studies in Mathematics (Vol. 13). Berlin: Walter de Gruyter & Co.

Andronov, A. A., Leontovich, E. A., Gordon, I. I., & Maier, A. G. (1971). *Qualitative theory of second-order dynamic systems.* Jerusalem: Israel Program of Scientific Translations.

Andronov, A. A., Leontovich, E. A., Gordon, I. I., & Maier, A. G. (1973). *Theory of bifurcations of dynamic systems on a plane.* Jerusalem: Israel Program for Scientific Translations.

Anosov, D. V., Bronshtein, I. U., Aranson, S. K., & Grines, V. Z. (1988). Smooth Dynamical Systems. In D. V. Anosov & V. I. Arnol'd (Eds.), *Dynamical systems I. Encyclopaedia of mathematical sciences.* New York: Springer.

Arnol'd, V. I. (1973). *Ordinary differential equations.* Cambridge, MA: MIT Press.

Arnol'd, V. I. (1983). *Geometrical methods in the theory of ordinary differential equations.* New York: Springer.

Arnol'd, V. I. (1989). *Mathematical methods of classical mechanics.* Graduate texts in mathematics (Vol. 60, 2nd edn) New York: Springer.

Arnol'd, V. I., & Il'yashenko, Y. S. (1988). Ordinary differential equations. In D. V. Anosov & V. I. Arnol'd (Eds.), *Dynamical systems I.* Encyclopaedia of mathematical sciences New York: Springer.

Arnol'd, V. I., Afraimovich, V. S., Il'yashenko, Yu. S., & Shil'nikov, L. P. (1994). Bifurcation theory. In V. I. Arnol'd (Eds.), *Dynamical systems V. Encyclopaedia of mathematical sciences* New York: Springer.

Arrowsmith, D. K., & Place, C. M. (1990). *An introduction to dynamical systems.* Cambridge: Cambridge University Press.

Babenko, K. I., & Petrovich, V. Y. (1983). *Demonstrative computations on a computer, Preprint 83–133.* USSR Academy of Sciences, Moscow: Institute of Applied Mathematics. In Russian.

Bazykin, A. D. (1998). *Nonlinear dynamics of interacting populations.* River Edge, NJ: World Scientific.

Beyn, W.-J., Hüls, T., & Schenke, A. (2016). Symbolic coding for noninvertible systems: Uniform approximation and numerical computation. *Nonlinearity, 29*(11), 3346–3384.

Beyn, W.-J. (1990). The numerical computation of connecting orbits in dynamical systems. *IMA Journal on Numerical Analysis, 9*, 379–405.

Yu. Kuznetsov et al., *Dynamical Systems Essentials*, Texts in Applied Mathematics 83, https://doi.org/10.1007/978-3-032-04083-1

Beyn, W.-J., Champneys, A., Doedel, E., Govaerts, W., Kuznetsov, Y. A., & Sandstede, B. (2002). Numerical continuation, and computation of normal forms. In B. Fiedler (Ed.), *Handbook of dynamical systems* (Vol. 2, pp. 149–219). Amsterdam: Elsevier Science.

Brin, M., & Stuck, G. (2002). *Introduction to dynamical systems.* Cambridge: Cambridge University Press.

Carr, J. (1981). *Applications of center manifold theory.* New York: Springer.

Champneys, A. R., Kuznetsov, Y. A., & Sandstede, B. (1996). A numerical toolbox for homoclinic bifurcation analysis. *International Journal of Bifurcation and Chaos in Applied Sciences and Engineering, 6*(5), 867–887.

Chow, S. N., & Hale, J. K. (1982). Methods of bifurcation theory. *Grundlehren der Mathematischen Wissenschaften [Fundamental Principles of Mathematical Science]* (Vol. 251). New York: Springer.

Coppel, W. (1978). *Dichotomies in Stability Theory.* Lecture Notes in Mathematics (Vol. 629). Berlin-New York: Springer.

Coullet, P., & Eckmann, J.-P. (1980). *Iterated Maps on the interval as a dynamical system.* Boston, MA: Birkhauser.

de Melo, W., & van Strien, S. (1993). One-dimensional Dynamics. *Ergebnisse der Mathematik und ihrer Grenzgebiete (3) [Results in Mathematics and Related Areas (3)]* (Vol. 25). Berlin: Springer.

De Witte, V., Govaerts, W., Kuznetsov, Yu. A., & Friedman, M. (2012). Interactive initialization and continuation of homoclinic and heteroclinic orbits in MATLAB. *ACM Transactions on Mathematical Software,38*(3), Art. 18, 34.

Devaney, R. L. (1989). *An introduction to chaotic dynamical systems. Addison-Wesley studies in nonlinearity* (2nd ed.). Redwood City, CA: Addison-Wesley Publishing Company Advanced Book Program.

Dumortier, F., Llibre, J., & Artés, J. C. (2006). *Qualitative theory of planar differential systems.* Springer, Berlin: Universitext.

Feigenbaum, M. (1978). Quantitative universality for a class of nonlinear transformations. *Journal of Statistical Physics, 19*, 25–52.

Fiorenza, A., Formica, M. R., Roskovec, T. G., & Soudský, F. (2021). Detailed proof of classical Gagliardo-Nirenberg interpolation inequality with historical remarks. *Zeitschrift für Analysis, 40*(2), 217–236.

Guckenheimer, J., & Holmes, P. (1983). *Nonlinear oscillations. Dynamical systems and bifurcations of vector fields.* New York: Springer.

Hale, J., & Koçak, H. (1991). *Dynamics and Bifurcations.* New York: Springer.

Hartman, P. (2002). *Ordinary differential equations.* Classics in Applied Mathematics (Vol. 38). Philadelphia, PA: Society for Industrial and Applied Mathematics (SIAM).

Hasselblatt, B., & Katok, A. (2003). *A first course in dynamics: With a panorama of recent developments.* New York: Cambridge University Press.

Hirsch, M., & Smale, S. (1974). *Differential equations. Dynamical systems and linear algebra.* New York: Academic.

Ilyashenko, Yu., & Li, W. (1999). *Nonlocal bifurcations.* Providence, RI: American Mathematical Society.

Iooss, G. (1979). *Bifurcations of maps and applications.* Amsterdam: North-Holland.

Iooss, G., & Adelmeyer, M. (1992). *Topics in bifurcation theory and applications.* Singapore: World Scientific.

Iooss, G., & Joseph, D. D. (1990). *Elementary stability and bifurcation theory. Undergraduate texts in mathematics* (2nd ed.). New York: Springer.

Irwin, M. (1980). *Smooth dynamical systems.* New York: Academic.

Kato, T. (1980). *Perturbation theory for linear operators.* New York: Springer.

Katok, A. & Hasselblatt, B. (1995). *Introduction to the modern theory of dynamical systems.* Encyclopedia of Mathematics and Its Applications (Vol. 54). Cambridge: Cambridge University Press.

Kielhöfer, H. (2004). *Bifurcation theory: An introduction with applications to PDEs.* New York: Springer.

Kirchgraber, U. & Palmer, K. J. (1990). *Geometry in the neighborhood of invariant manifolds of maps and flows and linearization.* Pitman research notes in mathematics series (Vol. 233). Harlow: Longman Scientific & Technical.

Kuznetsov, Yu. A. (2023). *Elements of applied bifurcation theory. Applied mathematical sciences (Vol. 112* (4th ed.). New York: Springer.

Kuznetsov, Yu. A., Govaerts, W., Doedel, E. J., & Dhooge, A. (2005). 'Numerical periodic normalization for codim 1 bifurcations of limit cycles. *SIAM Journal of Numerical Analysis,43,* 1407–1435.

Lanford, O. (1980). A computer-assisted proof of the Feigenbaum conjectures. *Bulletin of the American Mathematical Society, 6,* 427–434.

Lee, J. L. (2011). *Introduction to topological manifolds.* Springer mathematics and statistics: Springer.

Lyubich, M. (2000). The quadratic family as a qualitatively solvable model of chaos. *Notices of the American Mathematical Society, 47,* 1042–1052.

Marsden, J. E., & Ratiu, T. S. (1999). *Introduction to mechanics and symmetry. Texts in applied mathematics (Vol. 17* (2nd ed.). New York: Springer.

Moser, J. (1973). *Stable and random motions in dynamical systems.* Princeton, NJ: Princeton University Press.

Nitecki, Z. (1971). *Differentiable dynamics.* Cambridge, MA: MIT Press.

Palmer, K. (1984). Exponential dichotomies and transversal homoclinic points. *Journal of Differential Equations, 55,* 225–256.

Palmer, K. (2000). *Shadowing in dynamical systems.* Mathematics and its applications, theory and applications (Vol. 501). Dordrecht: Kluwer Academic Publishers.

Palmer, K. J. (1988). Exponential dichotomies, the shadowing lemma and transversal homoclinic points. In Dynamics reported. Dynamics Reported. New Series Dynamic Systems and Applications (Vol. 1, pp. 265–306) Chichester: Wiley.

Perko, L. (2001). *Differential equations and dynamical systems. Texts in applied mathematics (Vol. 7* (3rd ed.). New York: Springer.

Petrovich, V. Yu. (1990). Numerical spectral analysis of the differential of the doubling operator by K.I. Babenko's method, Preprint 90-81. Institute of Applied Mathematics, USSR Academy of Sciences, Moscow. In Russian.

Pilyugin, S. Y. (1999). *Shadowing in dynamical systems.* Lecture notes in mathematics (Vol. 1706). Berlin: Springer.

Sandstede, B. (1993). Verzweigungstheorie homokliner Verdopplungen, PhD thesis, Institut für Angewandte Analysis und Stochastik, Berlin.

Sandstede, B., & Theerakarn, T. (2015). Regularity of center manifolds via the graph transform. *Journal of Dynamics and Differential Equations, 27*(3–4), 989–1006.

Shilnikov, L. P., Shilnikov, A. L., Turaev, D. V., & Chua, L. (1998). *Methods of qualitative theory in nonlinear dynamics: Part I.* Singapore: World Scientific.

Shilnikov, L. P., Shilnikov, A. L., Turaev, D. V., & Chua, L. (2001). *Methods of qualitative theory in nonlinear dynamics: Part II.* Singapore: World Scientific.

Shoshitaishvili, A. N. (1975). Bifurcations of topological type of singular points of vector fields that depend on parameters. *Proceedings of Petrovskii Seminar, 1,* 279–309. In Russian.

Shub, M. (2005). What is ...a horseshoe? *Notices of the American Mathematical Society,52*(5), 516–517.

Sijbrand, J. (1985). Properties of center manifolds. *Transactions of the American Mathematical Society, 289*(2), 431–469.

Steinlein, H., & Walther, H.-O. (1990). Hyperbolic sets, transversal homoclinic trajectories, and symbolic dynamics for C^1-maps in Banach spaces. *Journal of Dynamics and Differential Equations, 2*(3), 325–365.

Vanderbauwhede, A. (1989). Centre manifolds, normal forms and elementary bifurcations. *Dynamics Reported, 2*, 89–169.

Verhulst, F. (1996). *Nonlinear differential equations and dynamical systems, Universitext* (2nd ed.). Berlin: Springer.

Viana, M. (2000). What's new on Lorenz strange attractors? *The Mathematical Intelligencer, 22*(3), 6–19.

Wiggins, S. (1988). *Global bifurcations and chaos.* New York: Springer.

Wiggins, S. (2003). *Introduction to applied nonlinear dynamical systems and chaos. Texts in applied mathematics. (Vol. 2* (2nd ed.). New York: Springer.

Index

Yu. Kuznetsov et al., *Dynamical Systems Essentials*, Texts in Applied Mathematics 83, https://doi.org/10.1007/978-3-032-04083-1

C

D

E

The manufacturer's authorised representative in the EU is Springer
Nature Customer Service Centre GmbH, Europaplatz 3, 69115 Heidelberg,
Germany. If you have any concerns regarding our products, please
contact ProductSafety@springernature.com

Printed and bound by CPI Group (UK) Ltd, Croydon, CR0 4YY
07/07/2026
02160915-0001